CONTINUUM MECHANICS

This is a modern textbook for courses in continuum mechanics. It provides both the theoretical framework and the numerical methods required to model the behavior of continuous materials. This self-contained textbook is tailored for advanced undergraduate or first-year graduate students with numerous step-by-step derivations and worked-out examples. The author presents both the general continuum theory and the mathematics needed to apply it in practice. The derivation of constitutive models for ideal gases, fluids, solids, and biological materials and the numerical methods required to solve the resulting differential equations are also detailed. Specifically, the text presents the theory and numerical implementation for the finite difference and the finite element methods in the Matlab® programming language. It includes thirteen detailed Matlab® programs illustrating how constitutive models are used in practice.

Dr. Franco M. Capaldi received his PhD in Mechanical Engineering from the Massachusetts Institute of Technology. He taught Mechanical Engineering at Drexel University from 2006 to 2011. He is currently an Associate Professor of Civil and Mechanical Engineering at Merrimack College. His research focuses on the modeling of biological and polymeric materials at various length scales.

Continuum Mechanics

CONSTITUTIVE MODELING OF STRUCTURAL AND BIOLOGICAL MATERIALS

Franco M. Capaldi
Merrimack College

CAMBRIDGE
UNIVERSITY PRESS

CAMBRIDGE
UNIVERSITY PRESS

32 Avenue of the Americas, New York NY 10013-2473, USA

Cambridge University Press is part of the University of Cambridge.

It furthers the University's mission by disseminating knowledge in the pursuit of education, learning and research at the highest international levels of excellence.

www.cambridge.org
Information on this title: www.cambridge.org/9781107480995

First published 2012
First paperback edition 2015

A catalogue record for this publication is available from the British Library

Library of Congress Cataloguing in Publication data
Capaldi, Franco M., 1977–
Continuum mechanics : constitutive modeling of structural and biological materials / Franco M. Capaldi.
p. cm.
Includes index.
ISBN 978-1-107-01181-6
1. Continuum mechanics. I. Title.
QA808.2.C327 2012
531–dc23 2012015690

ISBN 978-1-107-01181-6 Hardback
ISBN 978-1-107-48099-5 Paperback

To Irene, Emma, and Nina with love.

Contents

Preface *page* xiii

1 Mathematics 1

1.1 Vectors 1
1.2 Second-Order Tensors 8
1.3 Eigenvalues and Eigenvectors 12
1.4 Spectral Decomposition of a Symmetric Tensor 15
1.5 Coordinate Transformation 18
1.6 Invariants 24
1.7 Cayley-Hamilton Theorem 24
1.8 Scalar, Vector, and Tensor Functions and Fields 25
1.9 Integral Theorems 30
Exercises 32
Matlab® Exercises 33

2 Kinematics 35

2.1 Configurations 35
2.2 Velocity and Acceleration 38
2.3 Displacement 43
2.4 Deformation Gradient 45
2.5 Jacobian 51
2.6 Nanson's Formula 52
2.7 Homogenous Deformation, Isochoric Deformation, and Rigid Body Rotation 54
2.8 Material and Spatial Derivatives 55
2.9 Polar Decomposition of the Deformation Gradient 58
2.10 Stretch Ratios 61
2.11 Left and Right Cauchy Deformation Tensor 63
2.12 Green Strain Tensor 64
2.13 Almansi Strain Tensor 64
2.14 Infinitesimal Strain Tensor 65
2.15 Velocity Gradient, Rate of Deformation, Vorticity 65

2.16 Reynolds' Transport Theorem 68
Exercises 71
Matlab® Exercises 73

3 The Stress Tensor 74

3.1 Mass, Density, and Forces 74
3.2 Traction Vector 75
3.3 Cauchy Stress Tensor 77
3.4 First Piola-Kirchhoff Stress Tensor 80
3.5 Second Piola-Kirchhoff Stress Tensor 80
3.6 Maximum Normal and Shear Stress 81
3.7 Decomposition of the Stress Tensor 82
Exercises 83

4 Introduction to Material Modeling 85

4.1 Forces and Fields 86
4.2 Balance Laws 86
4.2.1 Conservation of Mass 87
4.2.2 Conservation of Linear Momentum 88
4.2.3 Conservation of Angular Momentum 90
4.2.4 Conservation of Energy 93
4.2.5 The Second Law of Thermodynamics 96
4.2.6 Summary of the Field Equations 98
4.3 Stress Power 99
4.4 Jump Conditions 100
4.5 Constitutive Modeling 104
4.5.1 Constitutive Modeling Principles 106
4.5.2 Principle of Dissipation 107
4.5.3 Principle of Material Frame Indifference 109
4.6 Material Symmetry 114
4.6.1 Isotropic Scalar-Valued Functions 115
4.6.2 Isotropic Tensor-Valued Functions 116
4.7 Internal Variables 119
4.8 Thermodynamics of Materials 120
4.9 Heat Transfer 120
Exercises 121

5 Ideal Gas 122

5.1 Historical Perspective 122
5.2 Forces and Fields 124
5.3 Balance Laws 124
5.4 Constitutive Model 124
5.4.1 Constraints 126
5.4.2 Constitutive Relations 127
5.4.3 Molecular Model of an Ideal Gas 129
5.5 Governing Equations 131

5.6 Acoustic Waves 132
5.6.1 Finite Difference Method 135
5.6.2 Explicit Algorithm 137
5.6.3 Implicit Algorithm 138
5.6.4 Example Problem 140
5.6.5 Matlab® File – Explicit Algorithm 141
5.6.6 Matlab® File – Implicit Algorithm 143

6 Fluids 146

6.1 Historical Perspective 147
6.2 Forces and Fields 148
6.3 Balance Laws 148
6.4 Constitutive Model 149
6.4.1 Constraints 150
6.4.2 Constitutive Relations for the Newtonian Fluid 155
6.4.3 Stokes Condition 157
6.5 Governing Equations 157
6.5.1 Compressible Newtonian Fluid 158
6.5.2 Incompressible Newtonian Fluid 159
6.5.3 Irrotational Steady Flow of an Incompressible Newtonian Fluid 160
6.6 Non-Newtonian Fluid Models 160
6.6.1 Power Law Model 161
6.6.2 Cross Model 162
6.6.3 Bingham Model 162
6.7 Couette Viscometer 162
6.7.1 Newtonian Fluid 162
6.7.2 Power Law Fluid Model 170
6.7.3 General Non-Newtonian Fluid 175

7 Elastic Material Models 183

7.1 Historical Perspective 183
7.2 Finite Thermoelastic Material Model 184
7.2.1 Forces and Fields 184
7.2.2 Balance Laws 185
7.2.3 Constitutive Model 186
7.2.4 Constraints Due to Material Frame Indifference 186
7.2.5 Constraints Due to the Second Law of Thermodynamics 186
7.3 Hyperelastic Material Model 188
7.3.1 Balance Laws 188
7.3.2 Constitutive Model 188
7.3.3 Constraints Due to Material Frame Indifference 189
7.3.4 Clausius-Duhem Inequality 189
7.3.5 Material Symmetry 190
7.3.6 Isotropic Materials 190
7.3.7 Transversely Isotropic Materials 193

7.3.8 Incompressible Materials 195
7.3.9 Common Hyperelastic Constitutive Models 196
7.3.10 Freely Jointed Chain 196
7.4 Linear Thermoelastic Material Model 201
7.4.1 Balance Laws 201
7.4.2 Constitutive Model 202
7.4.3 Clausius-Duhem Inequality 202
7.4.4 Linear Thermoelastic Constitutive Relation 204
7.4.5 Material Symmetry 206
7.4.6 Governing Equations for the Isotropic Linear Elastic Material 207
7.5 Uniaxial Tension Test 208
7.5.1 Kinematics 209
7.5.2 Isotropic Linear Thermoelastic Material 210
7.5.3 Incompressible Isotropic Neo-Hookean Model 211

8 Continuum Mixture Theory 214
8.1 Forces and Fields 214
8.2 Balance Laws 216
8.2.1 Conservation of Mass 217
8.2.2 Conservation of Momentum 218
8.2.3 Conservation of Angular Momentum 219
8.2.4 Conservation of Energy 220
8.2.5 Second Law of Thermodynamics 221
8.3 Biphasic Model 222
8.4 Isothermal Biphasic Model 222
8.5 Application to Soft Tissue 224
8.5.1 Confined Compression Experiment 225
8.5.2 Unconfined Compression 235

9 Growth Models 244
9.1 Forces and Fields 244
9.2 Balance Laws 244
9.2.1 Conservation of Mass 245
9.2.2 Reynolds' Transport Theorem 246
9.2.3 Conservation of Momentum 247
9.2.4 Conservation of Angular Momentum 247
9.2.5 Conservation of Energy 248
9.3 Decomposition of the Deformation Gradient 248
9.4 Summary of the Field Equations 249
9.5 Constitutive Model 250
9.6 Uniaxial Loading 251
9.6.1 Kinematics 251
9.6.2 Governing Equation 253

9.6.3 Finite Difference Algorithm 254
9.6.4 Example Problem 255
9.6.5 Matlab® File 256

10 Parameter Estimation and Curve Fitting 258

10.1 Propagation of Error 258
10.2 Least Squares Fit 260

11 Finite Element Method 269

11.1 Introduction 269
11.1.1 Element Types 270
11.1.2 Natural Versus Global Coordinates for a Quadrilateral Element 271
11.1.3 Field Variable Representation Within an Element 272
11.1.4 Matrix Representation 273
11.1.5 Integration of a Field Variable 274
11.1.6 Gaussian Quadrature 276
11.1.7 Differentiation of a Field Variable 277
11.2 Formulation of the Governing Equations 278
11.3 Plane Strain Deformation 279
11.3.1 Statement of Virtual Work 280
11.3.2 Discretization of Space 280
11.3.3 Approximation of the Field Variables 280
11.3.4 FEM Formulation 282
11.3.5 Element Stiffness Tensor 282
11.3.6 Body Force Vector 283
11.3.7 Traction Force Vector 283
11.3.8 Single Element Implementation 284
11.4 Axisymmetric Deformation 290
11.4.1 Statement of Virtual Work 291
11.4.2 Discretization of Space 291
11.4.3 Approximation of the Field Variables 291
11.4.4 FEM Formulation 293
11.4.5 Element Stiffness Tensor 293
11.4.6 Body Force Vector 294
11.4.7 Single Element Implementation 297
11.4.8 Multiple Element Implementation 305
11.5 Infinitesimal Plane Strain FEM with Material Nonlinearity 314
11.5.1 Statement of Virtual Work 314
11.5.2 Discretization of Space 314
11.5.3 Approximation of the Field Variables 315
11.5.4 FEM Formulation 316
11.6 Plane Strain Finite Deformation 319
11.6.1 Total Lagrangian Method 320
11.6.2 Updated Lagrangian Method 323

11.6.3 Updated Lagrangian Method Single Element Implementation 324

12 Appendix 333

12.1 Introduction to Matlab® 333

12.2 Reference Tables 334

Index 341

Preface

This textbook is designed to give students an understanding and appreciation of continuum-level material modeling. The mathematics and continuum framework are presented as a tool for characterizing and then predicting the response of materials. The textbook attempts to make the connection between experimental observation and model development in order to put continuum-level modeling into a practical context. This comprehensive treatment of continuum mechanics gives students an appreciation for the manner in which the continuum theory is applied in practice and for the limitations and nuances of constitutive modeling.

This book is intended as a text for both an introductory continuum mechanics course and a second course in constitutive modeling of materials. The objective of this text is to demonstrate the application of continuum mechanics to the modeling of material behavior. Specifically, the text focuses on developing, parameterizing, and numerically solving constitutive equations for various types of materials. The text is designed to aid students who lack exposure to tensor algebra, tensor calculus, and/or numerical methods. This text provides step-by-step derivations as well as solutions to example problems, allowing a student to follow the logic without being lost in the mathematics.

The first half of the textbook covers notation, mathematics, the general principles of continuum mechanics, and constitutive modeling. The second half applies these theoretical concepts to different material classes. Specifically, each application covers experimental characterization, constitutive model development, derivation of governing equations, and numerical solution of the governing equations. For each material application, the text begins with the experimental observations, which outline the behavior of the material and must be captured by the material model. Next, we formulate the continuum model for the material and present general constitutive equations. These equations often contain parameters that must be determined experimentally. Therefore, the textbook has a chapter covering the theory and application of experimental error analysis and simple curve fitting. For each material class, the continuum model is then applied to a specific application and the resulting differential equations are solved numerically. Complete descriptions of the finite difference and finite element methods are included. Numerical solutions are implemented in Matlab® and provided in the text along with flow charts illustrating the logic in the Matlab® scripts.

1 Mathematics

As scientists and engineers, we make sense of the world around us through observation and experimentation. Using mathematics, we attempt to describe our observations and make useful predictions based on these observations. For example, a simple experimental observation that the distance traversed by an object traveling at a constant velocity is linearly related to both the velocity and the time can be formalized using the relation, $\boldsymbol{d} = \boldsymbol{v}t$, where $\boldsymbol{d}$ is the distance vector, $\boldsymbol{v}$ is the velocity vector, and t is the time. The distance, velocity, and time are physical quantities that can be measured or controlled. Physical quantities such as distance, velocity, and time are represented mathematically as tensors. A scalar, for example, is a zeroth-order tensor. Only a magnitude is required to specify the value of a zeroth-order tensor. In our previous example, time is such a quantity. If you are told that the duration of an event was 3 seconds, you need no other information to fully characterize this physical quantity. Velocity, on the other hand, requires both a magnitude and a direction to specify its meaning. The velocity would be represented using a first-order tensor, also known as a vector. The internal stress in a material is a second-order tensor, which requires a magnitude and two directions to specify its value. You may recognize that the two required directions are the normal of the surface on which the stress acts and the direction of the traction vector on this surface. Tensors of higher order require additional information to specify their physical meaning. In this chapter, we will review the basic tensor algebra and tensor calculus that will be used in the formulation of continuum representations.

1.1 Vectors

A first-order tensor, also known as a vector, is used to represent a physical quantity whose representation requires both direction and magnitude. However, additional requirements must be satisfied. First, two vectors must add according to the parallelogram rule. Second, if a vector is defined within a given reference frame, and a second rotated reference frame is defined, it must be possible to express the components of a vector in one reference frame in terms of the components within another reference frame.

Whereas the physical meaning of a vector, such as the velocity of a car, is independent of coordinate system, the components of a vector are not. If we define a set of ***orthonormal basis vectors***, $\{\boldsymbol{e}_1, \boldsymbol{e}_2, \boldsymbol{e}_3\}$, we can express a vector as a linear combination of the basis vectors such that

$$\boldsymbol{a} = a_1\boldsymbol{e}_1 + a_2\boldsymbol{e}_2 + a_3\boldsymbol{e}_3,$$

where a_1, a_2, a_3 are scalars representing the components of the vector in the $\boldsymbol{e}_1$, $\boldsymbol{e}_2$, and $\boldsymbol{e}_3$ directions respectively. The ***magnitude*** of a vector, $|\boldsymbol{a}|$, is a measure of the length of a vector and is defined as

$$|\boldsymbol{a}| = \sqrt{(a_1^2 + a_2^2 + a_3^2)}.$$

It is often necessary to compare the relative size of two physical quantities whether they be scalars, vectors, or a general nth-order tensor. In each case, we may compare the ***norm*** of the two tensors. The norm of a scalar is equal to the absolute value of the scalar, whereas the norm of a vector, denoted as $\|\boldsymbol{a}\|$, is equal to its magnitude. Both the magnitude and the norm of a vector are zero if and only if each of the components of the vector is zero.

Whereas magnitude specifies the size of the vector, the direction of the vector may be represented by a ***unit vector***, $\hat{\boldsymbol{a}}$, parallel to the original vector, $\boldsymbol{a}$, such that

$$\hat{\boldsymbol{a}} = \frac{\boldsymbol{a}}{|\boldsymbol{a}|}.$$

This unit vector captures the directional information contained within the vector but discards the magnitude. The magnitude of any unit vector is equal to one. If two vectors, $\boldsymbol{a}$ and $\boldsymbol{b}$, are parallel, one vector can be written as a scalar, α, times the other vector,

$$\boldsymbol{a} = \alpha\boldsymbol{b}.$$

Vector and tensor equations can become quite complicated. It is often possible to use ***index notation*** to simplify and manipulate the representation of vector or tensor equations. Let us begin with the assumption that we are modeling physical quantities in a three-dimensional space that is spanned by the orthonormal basis vectors, $\{\boldsymbol{e}_1, \boldsymbol{e}_2, \boldsymbol{e}_3\}$. In order to write a vector equation in index notation, we introduce an index, i, which in this case is a variable that can assume the value of 1, 2, or 3. The representation of a vector as a linear combination of the basis vectors can be written in the compact form,

$$\boldsymbol{a} = a_1\boldsymbol{e}_1 + a_2\boldsymbol{e}_2 + a_3\boldsymbol{e}_3 = \sum_{i=1}^{3} a_i\boldsymbol{e}_i.$$

The summation from 1 to 3 over a repeated index is quite common and may be represented in a more compact form using the ***abbreviated summation convention*** which is also termed ***Einstein notation*** as

$$\boldsymbol{a} = a_i\boldsymbol{e}_i. \tag{1.1}$$

The abbreviated summation convention is implied if and only if an index appears exactly twice within the same term of an equation.

The sum of two vectors, $\boldsymbol{b}$ and $\boldsymbol{c}$, is equal to a vector such that

$$\boldsymbol{a} = \boldsymbol{b} + \boldsymbol{c}.$$

The addition of two vectors is both communitive, $\boldsymbol{b}+\boldsymbol{c}=\boldsymbol{c}+\boldsymbol{b}$, and consistent with the parallogram rule. The components of the vectors $\boldsymbol{b}$ and $\boldsymbol{c}$ parallel to the same basis vector can be added. Vector addition can be written in terms of components such that

$$a_1\boldsymbol{e}_1+a_2\boldsymbol{e}_2+a_3\boldsymbol{e}_3=(b_1+c_1)\boldsymbol{e}_1+(b_2+c_2)\boldsymbol{e}_2+(b_3+c_3)\boldsymbol{e}_3.$$

This gives three separate equations for the components of the vector $\boldsymbol{a}$,

$$a_1=b_1+c_1,$$
$$a_2=b_2+c_2,$$
$$a_3=b_3+c_3.$$

In index notation, this set of three equations is represented as

$$a_i=b_i+c_i,$$

where i can take on a value of 1, 2, or 3. The subscript i, termed a ***free index***, appears exactly once in each of the terms in the equation. In contrast, the subscript i, appears twice in the right term in Equation (1.1). In that equation, the subscript is termed a ***dummy index*** which signifies a summation from 1 to 3 over the repeated indices.

The scalar valued ***dot product***, also known as a ***scalar product*** or ***inner product***, of two vectors is defined as

$$\boldsymbol{a}\cdot\boldsymbol{b}=|\boldsymbol{a}||\boldsymbol{b}|\cos\theta_{\boldsymbol{ab}}=a_1b_1+a_2b_2+a_3b_3=a_ib_i,$$

where $\theta_{\boldsymbol{ab}}$ is the angle between the two vectors. There are no free indices in this equation, but there is a single dummy index, i. When written in index notation, a scalar-valued function will have no free indices, and a vector valued function will have a single free index. In the general case, an nth-order tensor-valued function will have n free indices. From the definition of the dot product, we can see that the dot product of two perpendicular vectors ($\theta_{\boldsymbol{ab}}=90°$) is equal to zero. In addition, the dot product of a vector with itself gives the magnitude of the vector squared, $|\boldsymbol{a}|^2=\boldsymbol{a}\cdot\boldsymbol{a}$. The dot product of a unit vector with itself will then be equal to one, $\boldsymbol{e}_1\cdot\boldsymbol{e}_1=1$. An ***orthonormal basis set*** has the property that each basis vector is perpendicular to the others. Therefore, the dot product of each basis vector with all other basis vectors is zero and the dot product of each basis vector with itself is equal to one giving

$$\boldsymbol{e}_i\cdot\boldsymbol{e}_j=\delta_{ij},$$

where we have introduced the ***Kronecker delta***, δ_{ij}, which has the property

$$\delta_{ij}=\begin{cases}0 & if \quad i\neq j\\ 1 & if \quad i=j\end{cases}. \tag{1.2}$$

The components of a vector, $\boldsymbol{a}$, along the direction of a unit vector $\boldsymbol{e}_1$, is given by

$$\boldsymbol{a}\cdot\boldsymbol{e_1}=|\boldsymbol{a}|\cos\theta_{a e_1}=a_1,$$

where $\theta_{a\mathbf{e}_1}$ is the angle between vector $\mathbf{a}$ and the basis vector $\boldsymbol{e}_1$.

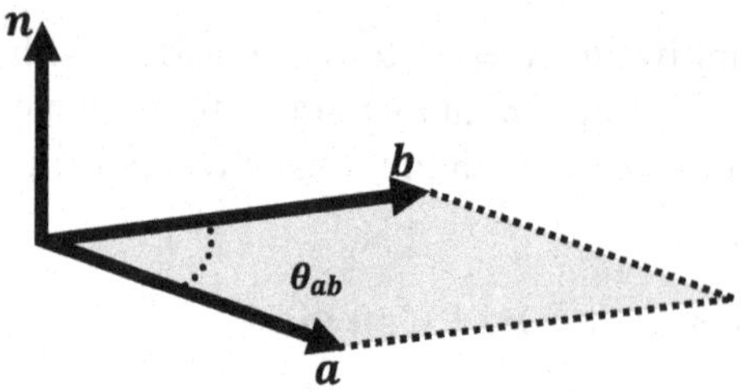

Figure 1.1. Illustration of a parallelogram bounded by vectors $\boldsymbol{a}$ and $\boldsymbol{b}$.

The result of the ***vector product*** or ***cross product***, $\boldsymbol{c}$, of two vectors, $\boldsymbol{a}$ and $\boldsymbol{b}$, is a vector that is perpendicular to each of the original vectors. The cross product is written as

$$\boldsymbol{c} = \boldsymbol{a} \times \boldsymbol{b} = (a_2 b_3 - a_3 b_2)\,\boldsymbol{e}_1 + (a_3 b_1 - a_1 b_3)\,\boldsymbol{e}_2 + (a_1 b_2 - a_2 b_1)\,\boldsymbol{e}_3.$$

The magnitude of the cross product is equal to

$$|\boldsymbol{c}| = |\boldsymbol{a}||\boldsymbol{b}| \sin\theta_{ab},$$

where θ_{ab} is the angle between the two vectors.

The magnitude of the cross product is a measure of the area within a parallelogram defined by the two vectors $\boldsymbol{a}$ and $\boldsymbol{b}$, Figure 1.1. The unit normal perpendicular to the parallelogram is defined by the direction of the cross product, $\boldsymbol{n} = \frac{\boldsymbol{a}\times\boldsymbol{b}}{|\boldsymbol{a}\times\boldsymbol{b}|}$. Two parallel vectors will have a cross product equal to zero.

In this textbook, we will always employ a ***right-handed orthonormal basis set***, which has the properties that each basis vector is perpendicular to the other two, the magnitude of each basis vector is equal to one, and the basis vectors are related according to $\boldsymbol{e}_1 \times \boldsymbol{e}_2 = \boldsymbol{e}_3$. If these conditions are satisfied, the cross product between any two unit vectors can be written as

$$\boldsymbol{e}_i \times \boldsymbol{e}_j = \varepsilon_{ijk} \boldsymbol{e}_k,$$

where ε_{ijk} is the ***Levi-Civita symbol***, also known as the ***permutation symbol*** or the ***alternating symbol.*** The Levi-Civita symbol has the values

$$\varepsilon_{ijk} = \begin{cases} 1 & \text{if} \quad ijk = 123, 231, \text{or } 312 \\ -1 & \text{if} \quad ijk = 132, 213, \text{or } 321 \\ 0 & \text{for repeated indices} \end{cases}.$$

A commonly used identity relating the permutation symbol and the Kronecker delta is

$$\varepsilon_{ijk}\varepsilon_{ipq} = \delta_{jp}\delta_{kq} - \delta_{jq}\delta_{kp}. \tag{1.3}$$

EXAMPLE 1.1. *Determine whether each term in the following equation is a scalar, vector, or tensor and identify the free and dummy indices.*

$$B_{ij} = a_m a_m I_{ij} + \beta C_{ij}.$$

Solution:
The indices i and j both appear exactly once within each term of this equation. They are each free indices. The index m appears exactly twice within the second term. This is a dummy index and signifies a summation over the index m. The summation may be expanded to obtain

$$B_{ij} = (a_1a_1 + a_2a_2 + a_3a_3)I_{ij} + \beta C_{ij}.$$

The variable, a_m, has a single index which signifies that a_m is the scalar component of the vector **a**. The variable, B_{ij}, has two indices, which means B_{ij} is a scalar component of the second-order tensor, **B**. The variable β has no index and is therefore a scalar.

EXAMPLE 1.2. *Find the value of* δ_{ii}.

Solution:
Expanding this equation using the summation convention, we find that

$$\begin{aligned}\delta_{ii} &= \sum_{i=1}^{3} \delta_{ii}\\ &= \delta_{11} + \delta_{22} + \delta_{33}\\ &= 3.\end{aligned}$$

EXAMPLE 1.3. *Show that* $\delta_{ij}a_i = a_j$.

Solution:
In this equation, there is both a dummy index, i and a free index, j. Therefore, this is a compact representation of the following three equations:

$$\begin{aligned}\delta_{i1}a_i &= \delta_{11}a_1 + \delta_{21}a_2 + \delta_{31}a_3\\ &= 1 \times a_1 + 0 \times a_2 + 0 \times a_3\\ &= a_1,\\ \delta_{i2}a_i &= \delta_{12}a_1 + \delta_{22}a_2 + \delta_{32}a_3\\ &= 0 \times a_1 + 1 \times a_2 + 0 \times a_3\\ &= a_2,\\ \delta_{i3}a_i &= \delta_{13}a_1 + \delta_{23}a_2 + \delta_{33}a_3\\ &= 0 \times a_1 + 0 \times a_2 + 1 \times a_3\\ &= a_3.\end{aligned}$$

This result can be compactly written as

$$\delta_{ij}a_i = a_j.$$

EXAMPLE 1.4. *Express the square of the magnitude of a vector in terms of its components.*

Solution:
We begin with the definition of the magnitude of a vector $|\boldsymbol{a}| = \sqrt{\boldsymbol{a} \cdot \boldsymbol{a}}$. Squaring this gives

$$|\boldsymbol{a}|^2 = \boldsymbol{a} \cdot \boldsymbol{a}.$$

The vector, $\boldsymbol{a}$, may be written in terms of its components as $\boldsymbol{a} = a_i \boldsymbol{e}_i$. Note that we cannot use the same index for the vector on the left and right of the dot product. Each is independently equal to the sum of the projections along each basis vector such that

$$|\boldsymbol{a}|^2 = a_i e_i \cdot a_j e_j.$$

The components, a_i and a_j, are scalar terms and commute freely. The dot product of the basis vectors e_i and e_j gives the Kronecker delta such that

$$\begin{aligned} |\boldsymbol{a}|^2 &= a_i a_j \delta_{ij} \\ &= a_i a_i \\ &= a_1^2 + a_2^2 + a_3^2. \end{aligned}$$

EXAMPLE 1.5. *Obtain an equation for the cross product of the two vectors* ***a*** *and* ***b*** *in terms of the components of the vectors.*

Solution:
The cross product, $\boldsymbol{c}$, of vectors $\boldsymbol{a}$ and $\boldsymbol{b}$ can be written as

$$\boldsymbol{c} = \boldsymbol{a} \times \boldsymbol{b}.$$

Expanding the vectors in terms of their components gives

$$\boldsymbol{c} = a_i \boldsymbol{e}_i \times b_j \boldsymbol{e}_j.$$

The components a_i and b_j are scalars and can be shifted in the equation to give

$$\boldsymbol{c} = a_i b_j \left(\boldsymbol{e}_i \times \boldsymbol{e}_j\right).$$

The cross product of the two basis vectors can be written in terms of the permutation symbol as

$$\varepsilon_{ijk} \boldsymbol{e}_k = \boldsymbol{e}_i \times \boldsymbol{e}_j,$$

which gives the result

$$\begin{aligned} \boldsymbol{c} &= a_i b_j \varepsilon_{ijk} \boldsymbol{e}_k \\ &= (a_2 b_3 - a_3 b_2)\boldsymbol{e}_1 + (a_3 b_1 - a_1 b_3)\boldsymbol{e}_2 + (a_1 b_2 - a_2 b_1)\boldsymbol{e}_3. \end{aligned}$$

EXAMPLE 1.6. *Show that* $\boldsymbol{a} \times (\boldsymbol{b} \times \boldsymbol{c}) = (\boldsymbol{a} \cdot \boldsymbol{c})\boldsymbol{b} - (\boldsymbol{a} \cdot \boldsymbol{b})\boldsymbol{c}$.

Solution:
Drawing from the previous example, we can write

$$\boldsymbol{d} = \boldsymbol{b} \times \boldsymbol{c} = \varepsilon_{jmn} b_m c_n \boldsymbol{e}_j.$$

The components of the vector $\boldsymbol{d}$ are given by

$$d_j = \varepsilon_{jmn} b_m c_n.$$

We can also write

$$\boldsymbol{a} \times (\boldsymbol{d}) = \varepsilon_{ijk} a_i d_j \boldsymbol{e}_k.$$

Substituting the components of $\boldsymbol{d}$ into this equation gives the result

$$\boldsymbol{a} \times (\boldsymbol{b} \times \boldsymbol{c}) = \varepsilon_{ijk} a_i \left(\varepsilon_{mnj} b_m c_n\right) \boldsymbol{e}_k.$$

The scalar terms can be rearranged to give

$$\boldsymbol{a} \times (\boldsymbol{b} \times \boldsymbol{c}) = \varepsilon_{kij} \varepsilon_{mnj} a_i b_m c_n \boldsymbol{e}_k.$$

Using the identity in Equation (1.3), we obtain

$$\begin{aligned} \boldsymbol{a} \times (\boldsymbol{b} \times \boldsymbol{c}) &= (\delta_{km}\delta_{in} - \delta_{kn}\delta_{im}) a_i b_m c_n \boldsymbol{e}_k \\ &= a_n b_k c_n \boldsymbol{e}_k - a_m b_m c_k \boldsymbol{e}_k \\ &= (a_n c_n) b_k \boldsymbol{e}_k - (a_m b_m) c_k \boldsymbol{e}_k \\ &= (\boldsymbol{a} \cdot \boldsymbol{c})\boldsymbol{b} - (\boldsymbol{a} \cdot \boldsymbol{b})\boldsymbol{c}. \end{aligned}$$

The final step makes use of the fact that $a_n c_n$ and $a_m b_m$ represent the dot product of two vectors, whereas $b_k \boldsymbol{e}_k$ and $c_k \boldsymbol{e}_k$ are component expansions of the vectors $\boldsymbol{b}$ and $\boldsymbol{c}$.

EXAMPLE 1.7. *Find the volume enclosed by the rhombus defined by vectors* $\boldsymbol{a}, \boldsymbol{b}$, *and* $\boldsymbol{c}$.

Solution:
The volume of a rhombus is simply the base area multiplied by the height perpendicular to the base:

$$Volume = height \times base\, area.$$

The area of the base can be found by taking the magnitude of the cross product of the bounding vectors, $base\, area = |\boldsymbol{a} \times \boldsymbol{b}|$. The height perpendicular to the base

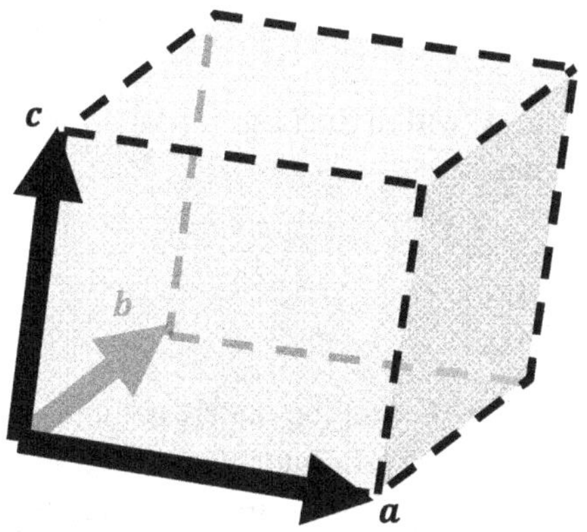

Figure 1.2. Illustration of a rhombus bounded by vectors $\boldsymbol{a}$, $\boldsymbol{b}$, and $\boldsymbol{c}$.

is found by projecting the vector, $\boldsymbol{c}$, onto the vector perpendicular to the base, $\boldsymbol{n}$. Substitution gives

$$V = (\boldsymbol{c} \cdot \boldsymbol{n})(|\boldsymbol{a} \times \boldsymbol{b}|).$$

Since $|\boldsymbol{a} \times \boldsymbol{b}|$ is a scalar, we can slide it into the dot product

$$V = (\boldsymbol{c} \cdot |\boldsymbol{a} \times \boldsymbol{b}|\boldsymbol{n}).$$

Finally, we note that the normal to the base, $\boldsymbol{n}$, is equal to the magnitude of the cross product of the vectors $\boldsymbol{a}$ and $\boldsymbol{b}$ such that $\boldsymbol{n} = \frac{\boldsymbol{a} \times \boldsymbol{b}}{|\boldsymbol{a} \times \boldsymbol{b}|}$. Therefore, we can write

$$V = \boldsymbol{c} \cdot (\boldsymbol{a} \times \boldsymbol{b}).$$

1.2 Second-Order Tensors

A second-order tensor is used to represent a physical quantity whose representation requires a magnitude and two directions. For example, each component of the stress tensor is the resolution of a traction vector onto a surface. One needs the magnitude of the traction vector, as well as the direction of both the traction vector and the normal of the surface in order to define each component of the stress tensor. A more general and mathematically rigorous definition would be that a tensor, $\boldsymbol{A}$, is a linear operator that transforms a vector, $\boldsymbol{b}$, into another vector, $\boldsymbol{a}$,

$$\boldsymbol{a} = \boldsymbol{A} \cdot \boldsymbol{b}.$$

In the remainder of the textbook, we will use bolded capital roman letters to signify tensors while using lowercase bold letters to signify vectors. The only exception will be $\boldsymbol{X}$, which will be reserved for a vector. The ***tensor product*** or ***dyadic product*** of two vectors, $\boldsymbol{a}$ and $\boldsymbol{b}$, operating on any vector $\boldsymbol{c}$, produces a vector along the direction of $\boldsymbol{a}$ such that

$$(\boldsymbol{a} \otimes \boldsymbol{b}) \cdot \boldsymbol{c} = (\boldsymbol{b} \cdot \boldsymbol{c})\,\boldsymbol{a}. \tag{1.4}$$

The components of the dyadic product of two vectors are given by

$$(\boldsymbol{a} \otimes \boldsymbol{b})_{ij} = a_i b_j.$$

Any second-order tensor may be expressed as a linear combination of the dyadic product of the basis vectors as

$$\boldsymbol{A} = A_{ij} \boldsymbol{e}_i \otimes \boldsymbol{e}_j\,, \tag{1.5}$$

where A_{ij} are the scalar components of the tensor, $\boldsymbol{A}$, within the basis set $\{\boldsymbol{e}_1, \boldsymbol{e}_2, \boldsymbol{e}_3\}$. These components may be expressed as a 3×3 matrix

$$[\boldsymbol{A}] = \begin{bmatrix} A_{11} & A_{12} & A_{13} \\ A_{21} & A_{22} & A_{23} \\ A_{31} & A_{32} & A_{33} \end{bmatrix}.$$

The component indices, A_{ij}, represent the matrix row, i, and the matrix column, j. Whereas we have expressed the components of a tensor as a matrix, not all 3×3 matrices are tensors. The physical quantity, such as stress or velocity, measured by either a vector or a tensor remains invariant when the coordinate system is changed.

The vector representing velocity does not change, only the components expressed within the coordinate system change. The same must be true for any higher order tensor. In fact, the transformation relating components of any tensor within two reference frames is quite exact and will be explored in more detail in a later section.

The majority of the equations within this textbook will require the manipulation of tensor and vector equations. A number of useful tensor and vector identities can be found in Table 12.3. In this section, we will introduce the commonly used tensor notation. The ***transpose***, $\boldsymbol{A}^T$, of a tensor, $\boldsymbol{A}$, is denoted by the superscript T and is defined in terms of the components of the original tensor $\boldsymbol{A}$ by

$$\boldsymbol{A}^T = A_{ji}\boldsymbol{e}_i \otimes \boldsymbol{e}_j = \begin{bmatrix} A_{11} & A_{21} & A_{31} \\ A_{12} & A_{22} & A_{32} \\ A_{13} & A_{23} & A_{33} \end{bmatrix}.$$

If a tensor is defined as a dyadic product of two vectors, $\boldsymbol{A} = \boldsymbol{a} \otimes \boldsymbol{b}$, then the transpose is given by $\boldsymbol{A}^T = \boldsymbol{b} \otimes \boldsymbol{a}$. Generally, a tensor is not equal to its transpose and the dyadic product does not commute

$$\boldsymbol{a} \otimes \boldsymbol{b} \neq \boldsymbol{b} \otimes \boldsymbol{a}.$$

There are special names reserved for tensors, which possess additional symmetry properties. A ***symmetric tensor*** is a tensor that is equal to its transpose giving $\boldsymbol{A} = \boldsymbol{A}^T$. A ***skew-symmetric tensor*** is a tensor that is equal to the negative of its transpose giving $\boldsymbol{A} = -\boldsymbol{A}^T$. Notice that the condition for a skew-symmetric tensor requires that each diagonal term of the tensor is equal to its negative. This requires that each diagonal term must be equal to zero for the skew-symmetric tensor. In later sections, we will introduce a number of physical quantities that may be represented by symmetric tensors such as the Cauchy stress and infinitesimal strain.

It is sometimes convenient to take a general tensor, $\boldsymbol{B}$, which has no symmetry properties, and break it up into a symmetric, $\boldsymbol{B}^s$, and a skew-symmetric tensor, $\boldsymbol{B}^a$. The superscript a stands for antisymmetric. This can be done through a simple decomposition of the tensor, $\boldsymbol{B}$,

$$\boldsymbol{B} = \frac{1}{2}(\boldsymbol{B} + \boldsymbol{B}^T) + \frac{1}{2}(\boldsymbol{B} - \boldsymbol{B}^T).$$

The first bracketed term gives the symmetric part, $\boldsymbol{B}^s = \frac{1}{2}(\boldsymbol{B} + \boldsymbol{B}^T)$ of the tensor, whereas the second bracketed term gives the skew-symmetric part, $\boldsymbol{B}^a = \frac{1}{2}(\boldsymbol{B} - \boldsymbol{B}^T)$.

The ***trace*** of a tensor is the sum of the diagonal components of the tensor

$$tr(\mathbf{A}) = A_{11} + A_{22} + A_{33} = A_{ii}.$$

We also introduce a new operator, :, which represents the ***contraction*** of two tensors

$$\boldsymbol{A} : \boldsymbol{B} = tr(\mathbf{A}^T \cdot \mathbf{B}) = A_{ij}B_{ij}.$$

The contraction of two tensors gives a scalar. The contraction of any symmetric tensor with any skew-symmetric tensor is always equal to zero. In the case of the decomposition described earlier, we have $\boldsymbol{B}^s : \boldsymbol{B}^a = 0$. We will use the contraction operator to define the ***tensor norm***, $||\boldsymbol{A}||$, as

$$||\mathbf{A}|| = \sqrt{(\mathbf{A} : \mathbf{A})}.$$

This norm has the property that it is a positive scalar and has a value of zero if and only if each of the components of the tensor $\boldsymbol{A}$ is zero.

The ***determinant*** of a tensor is given by

$$\det(\boldsymbol{A}) = \epsilon_{ijk} A_{i1} A_{j2} A_{k3}.$$

The determinate of a tensor is equal to the determinate of its transpose

$$\det(\boldsymbol{A}) = \det(\boldsymbol{A}^T)$$

A tensor is ***singular*** if and only if the determinant of that tensor is zero. Any ***nonsingular*** tensor will possess a unique ***inverse***, $\mathbf{A}^{-1}$, which satisfies

$$\mathbf{A} \cdot \mathbf{A}^{-1} = \mathbf{I}.$$

As mentioned earlier, a tensor is a linear operator that gives a vector when acting on a vector. Assuming we have a nonsingular tensor, it uniquely transforms vectors from one vector space to another vector space. An ***orthogonal tensor***, $\boldsymbol{Q}$, acting on a vector will change the direction but not the magnitude of the vector

$$|\boldsymbol{x}| = |\boldsymbol{Q} \cdot \boldsymbol{x}|.$$

This is a tensor that rotates vectors from one vector space into another. Orthogonal tensors have the property that the transpose is equal to the inverse

$$\boldsymbol{Q}^T \cdot \boldsymbol{Q} = \mathbf{I}. \tag{1.6}$$

This has the interesting implication that the determinant of the orthogonal tensor must be equal to plus or minus one. This is obtained by taking the determinant of Equation (1.6). This gives

$$\begin{aligned} \det(\boldsymbol{Q}^T \cdot \boldsymbol{Q}) &= \det(\mathbf{I}) \\ \det(\boldsymbol{Q}^T)\det(\boldsymbol{Q}) &= 1 \\ (\det(\boldsymbol{Q}))^2 &= 1. \end{aligned}$$

If the determinant of the tensor is equal to one, the tensor is a ***proper orthogonal tensor***. A proper orthogonal tensor preserves the right-handedness (or left-handedness) of a coordinate system. An ***improper orthogonal tensor*** acting on a system will invert the spatial relationship between vectors.

EXAMPLE 1.8. *Given the tensor,* $\mathbf{A}$, *where*

$$\mathbf{A} = \begin{bmatrix} 4 & 3 & 9 \\ -5 & 2 & 3 \\ 7 & 6 & -4 \end{bmatrix},$$

find a) the transpose, $\mathbf{A}^T$, *b) the trace,* $tr\mathbf{A}$, *c) the tensor norm,* $\|\mathbf{A}\|$, *and d) the symmetric and skew-symmetric tensors* $\mathbf{A}^s$ *and* $\mathbf{A}^a$ *such that* $\mathbf{A} = \mathbf{A}^s + \mathbf{A}^a$.

Solution:

a) $\boldsymbol{A}^T = \begin{bmatrix} 4 & -5 & 7 \\ 3 & 2 & 6 \\ 9 & 3 & -4 \end{bmatrix}$

b) $tr\mathbf{A} = A_{ii} = 4+2-4 = 2$

c) $\|\mathbf{A}\| = \sqrt{A:A} = \sqrt{4^2+(-5)^2+7^2+3^2+2^2+6^2+9^2+3^2+(-4)^2} = \sqrt{245}$

d) $$\mathbf{A}^s = \frac{1}{2}(\mathbf{A}+\mathbf{A}^T) = \frac{1}{2}\left(\begin{bmatrix} 4 & 3 & 9 \\ -5 & 2 & 3 \\ 7 & 6 & -4 \end{bmatrix} + \begin{bmatrix} 4 & -5 & 7 \\ 3 & 2 & 6 \\ 9 & 3 & -4 \end{bmatrix}\right) = \begin{bmatrix} 4 & -1 & 8 \\ -1 & 2 & \frac{9}{2} \\ 8 & \frac{9}{2} & -2 \end{bmatrix}$$

$$\mathbf{A}^a = \frac{1}{2}(\mathbf{A}-\mathbf{A}^T) = \frac{1}{2}\left(\begin{bmatrix} 4 & 3 & 9 \\ -5 & 2 & 3 \\ 7 & 6 & -4 \end{bmatrix} - \begin{bmatrix} 4 & -5 & 7 \\ 3 & 2 & 6 \\ 9 & 3 & -4 \end{bmatrix}\right) = \begin{bmatrix} 0 & 8 & 2 \\ -8 & 0 & -3 \\ -2 & 3 & 0 \end{bmatrix}$$

EXAMPLE 1.9. *Prove that* $det\mathbf{A} = \frac{(\mathbf{A}\cdot\mathbf{c})\cdot((\mathbf{A}\cdot\mathbf{a})\times(\mathbf{A}\cdot\mathbf{b}))}{\mathbf{c}\cdot(\mathbf{a}\times\mathbf{b})}$ *where* $\mathbf{a}, \mathbf{b}$, *and* $\mathbf{c}$ *can be any nonzero vector.*

Solution:
We begin by expanding the numerator in index notation. Noting that $\mathbf{A}\cdot\mathbf{c} = A_{ir}c_r$, $\mathbf{A}\cdot\mathbf{a} = A_{js}a_s$, and $\mathbf{A}\cdot\mathbf{b} = A_{kt}b_t$, we can write

$$\begin{aligned}(\mathbf{A}\cdot\mathbf{c})\cdot((\mathbf{A}\cdot\mathbf{a})\times(\mathbf{A}\cdot\mathbf{b})) &= A_{ir}c_r\varepsilon_{ijk}A_{js}a_sA_{kt}b_t \\ &= \varepsilon_{ijk}A_{ir}A_{js}A_{kt}c_ra_sb_t. \qquad (1.7)\end{aligned}$$

The result is a scalar and has no free indices. Equation (12.20) gives the determinant of a tensor in terms of the components of the tensor such that

$$\varepsilon_{ijk}A_{i1}A_{j2}A_{k3} = \det\mathbf{A}. \qquad (1.8)$$

Note that we can also rewrite this as

$$\varepsilon_{ijk}A_{i3}A_{j2}A_{k1} = -\det\mathbf{A}.$$

Substitution of Equation (1.8) into Equation (1.7) gives

$$\varepsilon_{ijk}A_{ir}A_{js}A_{kt}c_ra_sb_t = \det\mathbf{A}\varepsilon_{rst}c_ra_sb_t.$$

Finally, we recognize that

$$\varepsilon_{rst}c_ra_sb_t = \mathbf{c}\cdot(\mathbf{a}\times\mathbf{b}).$$

Combining these equations, we find that

$$\det\mathbf{A} = \frac{(\mathbf{A}\cdot\mathbf{c})\cdot\big((\mathbf{A}\cdot\mathbf{a})\times(\mathbf{A}\cdot\mathbf{b})\big)}{\mathbf{c}\cdot(\mathbf{a}\times\mathbf{b})}.$$

EXAMPLE 1.10. *Show that the components of a second-order tensor along the orthonormal basis vectors* $\{\mathbf{e}_1, \mathbf{e}_2, \mathbf{e}_3\}$ *are given by* $A_{ij} = \mathbf{e}_i\cdot(\mathbf{A}\cdot\mathbf{e}_j)$.

Solution:
Using Equation (1.5), the tensor, $\mathbf{A}$, can be written as $\mathbf{A} = A_{lk}(\mathbf{e}_l\otimes\mathbf{e}_k)$. Then we can write

$$\mathbf{A}\cdot\mathbf{e}_j = A_{lk}(\mathbf{e}_l\otimes\mathbf{e}_k)\mathbf{e}_j.$$

Using Equation (1.4) to eliminate the dyadic product gives

$$\begin{aligned}\mathbf{A}\cdot\boldsymbol{e}_j &= A_{lk}(\boldsymbol{e}_k\cdot\boldsymbol{e}_j)\boldsymbol{e}_l\\ &= A_{lk}\delta_{kj}\boldsymbol{e}_l\\ &= A_{lj}\boldsymbol{e}_l.\end{aligned}$$

Therefore, we can write

$$\begin{aligned}\boldsymbol{e}_i\cdot(\mathbf{A}\cdot\boldsymbol{e}_j) &= \boldsymbol{e}_i\cdot A_{lj}\boldsymbol{e}_l\\ &= A_{lj}\,(\boldsymbol{e}_i\cdot\boldsymbol{e}_l)\\ &= A_{lj}\delta_{il}\\ &= A_{ij}.\end{aligned}$$

1.3 Eigenvalues and Eigenvectors

In physics and engineering, we commonly encounter equations that have the form

$$\mathbf{A}\cdot\boldsymbol{n} = \lambda\boldsymbol{n}, \tag{1.9}$$

where $\mathbf{A}$ is a tensor, $\boldsymbol{n}$ is a vector, and λ is a scalar. Recalling that a tensor is a linear operator that transforms a vector into another vector, Equation (1.9) implies that the tensor operator $\mathbf{A}$ has not altered the direction of the vector $\boldsymbol{n}$ since $\mathbf{A}\cdot\boldsymbol{n}$ is parallel to the resultant vector $\boldsymbol{n}$. This equation will only be satisfied for particular ***eigenpairs***, which are the combinations of λ, ***eigenvalues***, and $\boldsymbol{n}$, ***eigenvectors***. Rearranging Equation (1.9) gives

$$\begin{aligned}\mathbf{A}\cdot\boldsymbol{n} - \lambda\mathbf{I}\cdot\boldsymbol{n} &= 0\\ (\mathbf{A} - \lambda\mathbf{I})\cdot\boldsymbol{n} &= \mathbf{0}.\end{aligned} \tag{1.10}$$

The equation will always have the trivial solution, $\boldsymbol{n} = \mathbf{0}$, but a nontrivial solution will exist if and only if $(\mathbf{A} - \lambda\mathbf{I})$ is singular. This requires

$$\det(\mathbf{A} - \lambda\mathbf{I}) = \begin{vmatrix} A_{11}-\lambda & A_{12} & A_{13}\\ A_{21} & A_{22}-\lambda & A_{23}\\ A_{31} & A_{32} & A_{33}-\lambda\end{vmatrix} = 0. \tag{1.11}$$

Since the tensor components of $\mathbf{A}$ are known, Equation (1.11) becomes a single cubic equation in terms of the unknown eigenvalues. This equation is called the ***characteristic equation*** with the three roots λ_1, λ_2, and λ_3 known as the eigenvalues or ***principal values***. For each of these eigenvalues, λ_i, the corresponding eigenvector, $\boldsymbol{n}^{(i)}$, can be found. This is accomplished by substituting the eigenvalues back into Equation (1.10) to give a set of equations

$$(\mathbf{A} - \lambda_i\mathbf{I})\cdot\boldsymbol{n}^{(i)} = \mathbf{0}.$$

Written in matrix form, this becomes

$$\begin{bmatrix} A_{11}-\lambda_i & A_{12} & A_{13}\\ A_{21} & A_{22}-\lambda_i & A_{23}\\ A_{31} & A_{32} & A_{33}-\lambda_i\end{bmatrix}\begin{bmatrix} n_1^{(i)}\\ n_2^{(i)}\\ n_3^{(i)}\end{bmatrix} = 0.$$

For each eigenvalue, this gives three equations for the three unknown components of the eigenvector, $n_j^{(i)}$. Finally, the normalized eigenvector, $\hat{n}^{(i)}$, can be computed by dividing the eigenvalue by its magnitude such that

$$\hat{\boldsymbol{n}}^{(i)} = \frac{\boldsymbol{n}^{(i)}}{|\boldsymbol{n}^{(i)}|}.$$

In addition to Equation (1.9), we may also encounter the similar but distinct equation

$$\boldsymbol{m} \cdot \boldsymbol{A} = \lambda \boldsymbol{m}, \tag{1.12}$$

where $\boldsymbol{m}$ is the ***left eigenvector*** and $\boldsymbol{n}$ from Equation (1.9) is the ***right eigenvector***. While the eigenvalues, λ, corresponding to Equations (1.9) and (1.12) are the same, the left and right eigenvectors are not generally equal. However, for the case of a symmetric tensor, the right and left eigenvectors are equivalent since taking the transpose of both sides of Equation (1.12) gives $\boldsymbol{A}^T \boldsymbol{m} = \lambda \boldsymbol{m}$ and for a symmetric tensor $\boldsymbol{A} = \boldsymbol{A}^T$. When left or right is unspecified, the term *eigenvector* is used to refer to the right eigenvector.

EXAMPLE 1.11. *Find the eigenvalues and eigenvectors of the tensor, $\boldsymbol{B}$, where*

$$\boldsymbol{B} = \begin{bmatrix} 1 & 2 & 4 \\ 1 & 1 & 5 \\ 1 & 5 & 1 \end{bmatrix}.$$

Solution:
First, we find the eigenvalues given by the solution to Equation (1.11). For the tensor, $\boldsymbol{B}$, we obtain

$$\begin{vmatrix} 1-\lambda & 2 & 4 \\ 1 & 1-\lambda & 5 \\ 1 & 5 & 1-\lambda \end{vmatrix} = 0.$$

Expansion of the determinate gives

$$(1-\lambda)(1-\lambda)(1-\lambda) + 2(5)(1) + 4(1)(5) - (1-\lambda)(5)(5) - 2(1)(1-\lambda)$$
$$-4(1-\lambda)(1) = 0.$$

Rearranging terms gives

$$1 - 3\lambda - 3\lambda^2 - \lambda^3 + 30 - 31 + 31\lambda = 0,$$
$$28\lambda - 3\lambda^2 - \lambda^3 = 0,$$
$$(28 - 3\lambda - \lambda^2)\lambda = 0.$$

The eigenvalues are the roots of the characteristic equation giving $\lambda_1 = 0$, $\lambda_2 = 7$, and $\lambda_3 = -4$. The corresponding eigenvectors are found by substituting each of the eigenvalues into Equation (1.11). The first eigenvalue gives

$$\begin{bmatrix} 1-0 & 2 & 4 \\ 1 & 1-0 & 5 \\ 1 & 5 & 1-0 \end{bmatrix} \begin{bmatrix} n_1^{(1)} \\ n_2^{(1)} \\ n_3^{(1)} \end{bmatrix} = 0.$$

The corresponding eigenvector, $\boldsymbol{n}^{(1)}$, is given by the solution of these three simultaneous equations

$$n_1^{(1)} + 2n_2^{(1)} + 4n_3^{(1)} = 0,$$
$$n_1^{(1)} + 1n_2^{(1)} + 5n_3^{(1)} = 0,$$
$$n_1^{(1)} + 5n_2^{(1)} + 1n_3^{(1)} = 0.$$

This gives $\hat{\boldsymbol{n}}^{(1)} = \frac{1}{\sqrt{38}}\langle 6 \quad 1 \quad 1\rangle$.

The second eigenvector, $\hat{\boldsymbol{n}}^{(2)}$, is found by substituting $\lambda_2 = 7$ into Equation (1.11) giving

$$\begin{bmatrix} 1-7 & 2 & 4 \\ 1 & 1-7 & 5 \\ 1 & 5 & 1-7 \end{bmatrix} \begin{bmatrix} n_1^{(2)} \\ n_2^{(2)} \\ n_3^{(2)} \end{bmatrix} = 0.$$

The solution is $\hat{n}^{(2)} = \frac{1}{\sqrt{3}}\langle 1 \quad 1 \quad 1\rangle$.

Finally, the third eigenvector, $\hat{\boldsymbol{n}}^{(3)}$, is found by substituting $\lambda_3 = -4$ into Equation (1.11) giving

$$\begin{bmatrix} 1-4 & 2 & 4 \\ 1 & 1-4 & 5 \\ 1 & 5 & 1-4 \end{bmatrix} \begin{bmatrix} n_1^{(3)} \\ n_2^{(3)} \\ n_3^{(3)} \end{bmatrix} = 0.$$

The solution is $\hat{\boldsymbol{n}}^{(3)} = \frac{1}{\sqrt{1070}}\langle -10 \quad -21 \quad 23\rangle$.

EXAMPLE 1.12. *Using Matlab®, find the eigenvalues and eigenvectors of the tensor*

$$\boldsymbol{B} = \begin{bmatrix} 1 & 2 & 4 \\ 1 & 1 & 5 \\ 1 & 5 & 1 \end{bmatrix}.$$

Solution:

At the command prompt, we define the tensor $\boldsymbol{B}$.

```
>> B = [1 2 4; 1 1 5; 1 5 1];
```

If you remove the semicolon, Matlab® will display the matrix you have just defined allowing you to check for data entry errors. Solving for the eigenvalues and eigenvectors is accomplished with the "eig" function.

```
>> [eigenVectors, eigenValues] = eig(B)
```

The function "eig" stores the eigenvalues as the diagonals of the tensor eigenValues with the corresponding eigenvectors stored as the columns of the tensor eigenVectors. These can be recovered in scalar and tensor form using the following variable assignments.

```
>> lambda_1 = eigenValues(1,1)
>> lambda_2 = eigenValues(2,2)
>> lambda_3 = eigenValues(3,3)
>> n_1 = eigenVectors(:,1)
```

```
>> n_2 = eigenVectors(:,2)
>> n_3 = eigenVectors(:,3)
```

The results are:

lambda_1 = 0.0

lambda_2 = 7.0

lambda_3 = −4.0

n_1 = [−0.9733; 0.1622; 0.1622]

n_2 = [0.5774; 0.5774; 0.5774]

n_3 = [−0.3057; −0.6420; 0.7031]

These results are identical to those found in Example 1.11. Also, note that these are the normalized eigenvectors.

1.4 Spectral Decomposition of a Symmetric Tensor

Many common stress and strain measures can be expressed as real symmetric tensors. The eigenvalues of real symmetric tensors are real numbers, and three linearly independent, orthogonal, eigenvectors can always be found. Therefore, the normalized eigenvectors, $\hat{\boldsymbol{n}}^{(i)}$, form a complete basis set. Therefore, the sum of the tensorial products of the eigenvectors gives the identity matrix

$$\mathbf{I} = \sum_{i=1}^{3} \hat{\boldsymbol{n}}^{(i)} \otimes \hat{\boldsymbol{n}}^{(i)}. \tag{1.13}$$

Any real symmetric tensor, $\boldsymbol{A}$, can be represented as a function of its eigenvalues, λ_i, and the corresponding normalized eigenvectors, $\hat{\boldsymbol{n}}^{(i)}$. This is known as ***spectral decomposition*** or as the ***spectral representation*** of a tensor. The spectral representation of tensor, $\boldsymbol{A}$, can be found by making use of Equation (1.13) such that

$$\boldsymbol{A} = \boldsymbol{A} \cdot \mathbf{I} = (\boldsymbol{A} \cdot \hat{\boldsymbol{n}}^{(i)}) \otimes \hat{\boldsymbol{n}}^{(i)} = \sum_{i=1}^{3} \lambda_i \hat{\boldsymbol{n}}^{(i)} \otimes \hat{\boldsymbol{n}}^{(i)}. \tag{1.14}$$

The square of a symmetric tensor can then be written as

$$\begin{aligned}
\boldsymbol{A}^2 = \boldsymbol{A} \cdot \boldsymbol{A} &= \left(\sum_{i=1}^{3} \lambda_i \hat{\boldsymbol{n}}^{(i)} \otimes \hat{\boldsymbol{n}}^{(i)} \right) \cdot \left(\sum_{j=1}^{3} \lambda_j \hat{\boldsymbol{n}}^{(j)} \otimes \hat{\boldsymbol{n}}^{(j)} \right) \\
&= \sum_{i=1}^{3} \sum_{j=1}^{3} \lambda_i \lambda_j (\hat{\boldsymbol{n}}^{(i)} \otimes \hat{\boldsymbol{n}}^{(i)}) \cdot (\hat{\boldsymbol{n}}^{(j)} \otimes \hat{\boldsymbol{n}}^{(j)}) \\
&= \sum_{i=1}^{3} \sum_{j=1}^{3} \lambda_i \lambda_j (\hat{\boldsymbol{n}}^{(i)} \cdot \hat{\boldsymbol{n}}^{(j)}) (\hat{\boldsymbol{n}}^{(i)} \otimes \hat{\boldsymbol{n}}^{(j)}).
\end{aligned}$$

Since the eigenvectors are orthonormal, $\hat{\boldsymbol{n}}^{(i)} \cdot \hat{\boldsymbol{n}}^{(j)} = \delta_{ij}$, and the previous equation may be written as

$$\boldsymbol{A} \cdot \boldsymbol{A} = \sum_{i=1}^{3} \lambda_i^2 \hat{\boldsymbol{n}}^{(i)} \otimes \hat{\boldsymbol{n}}^{(i)}.$$

Similarly, any power of $\boldsymbol{A}$ can be expressed as

$$\boldsymbol{A}^{\mathrm{n}} = \sum_{i=1}^{3} \lambda_i^{\mathrm{n}} \hat{\boldsymbol{n}}^{(i)} \otimes \hat{\boldsymbol{n}}^{(i)}$$

for all nonnegative integers $\mathrm{n} = 0, 1, 2, \ldots$.

The square root of a symmetric tensor, $\boldsymbol{A}$, can be written as

$$\sqrt{\boldsymbol{A}} = \sum_{i=1}^{3} \sqrt{\lambda_i} \hat{\boldsymbol{n}}^{(i)} \otimes \hat{\boldsymbol{n}}^{(i)}. \tag{1.15}$$

The square root of a symmetric tensor will be a real symmetric tensor if the eigenvalues are all nonnegative. Given a negative eigenvalue, the square root will be complex.

The inverse of a symmetric tensor with nonzero eigenvalues can be written as

$$\boldsymbol{A}^{-1} = \sum_{i=1}^{3} \lambda_i^{-1} \hat{\boldsymbol{n}}^{(i)} \otimes \hat{\boldsymbol{n}}^{(i)}.$$

The real symmetric tensor will have three eigenvalues, and they may include repeated eigevenalues. If there is a single repeated eigenvalue, $\lambda_1 = \lambda_2$, then the identity tensor can be written as $\mathbf{I} = 2\hat{\boldsymbol{n}}^{(1)} \otimes \hat{\boldsymbol{n}}^{(1)} + \hat{\boldsymbol{n}}^{(3)} \otimes \hat{\boldsymbol{n}}^{(3)}$, and the spectral representation of tensor $\boldsymbol{A}$ becomes

$$\boldsymbol{A} = \lambda_1 (\mathbf{I} - \hat{\boldsymbol{n}}^{(3)} \otimes \hat{\boldsymbol{n}}^{(3)}) + \lambda_3 \hat{\boldsymbol{n}}^{(3)} \otimes \hat{\boldsymbol{n}}^{(3)}.$$

If all three eigenvalues are equal, $\lambda_1 = \lambda_2 = \lambda_3$, then the spectral representation becomes

$$\boldsymbol{A} = \lambda_1 \mathbf{I}.$$

EXAMPLE 1.13. *Prove that a real symmetric tensor, **A**, has real eigenvalues.*

Solution:

Substitution of a particular eigenvalue, λ_1, and its corresponding eigenvector, $\mathbf{m}^{(1)}$, into Equation (1.9) gives

$$\boldsymbol{A} \cdot \boldsymbol{n}^{(1)} = \lambda_1 \boldsymbol{n}^{(1)}. \tag{1.16}$$

Taking the complex conjugate of this equation gives

$$\overline{\boldsymbol{A} \cdot \boldsymbol{n}^{(1)}} = \overline{\lambda_1 \boldsymbol{n}^{(1)}}. \tag{1.17}$$

The tensor $\boldsymbol{A}$ is real, which means the tensor is equal to its complex conjugate, $\boldsymbol{A} = \overline{A}$, and it is symmetric, $\boldsymbol{A} = \boldsymbol{A}^T$. Taking the transpose of Equation (1.17) and making use of these two properties gives

$$(\overline{\boldsymbol{A} \cdot \boldsymbol{n}^{(1)}})^T = (\overline{\lambda_1 \boldsymbol{n}^{(1)}})^T,$$

$$\overline{n}^{(1)^T} \cdot \boldsymbol{A} = \overline{\lambda_1} \overline{n}^{(1)^T}. \tag{1.18}$$

Multiplying Equation (1.16) by $\overline{\boldsymbol{n}}^{(1)^T}$ on the left gives

$$\overline{\boldsymbol{n}}^{(1)^T} \cdot \boldsymbol{A} \cdot \boldsymbol{n}^{(1)} = \lambda_1 \overline{\boldsymbol{n}}^{(1)^T} \boldsymbol{n}^{(1)}. \tag{1.19}$$

Multiplying Equation (1.18) by $\boldsymbol{n}^{(1)}$ on the right gives

$$\overline{\boldsymbol{n}}^{(1)^T} \cdot \boldsymbol{A} \cdot \boldsymbol{n}^{(1)} = \overline{\lambda_1} \overline{\boldsymbol{n}}^{(1)^T} \boldsymbol{n}^{(1)}. \tag{1.20}$$

Subtracting Equation (1.19) from Equation (1.20) gives

$$\overline{\boldsymbol{n}}^{(1)^T} \cdot \boldsymbol{A} \cdot \boldsymbol{n}^{(1)} - \overline{\boldsymbol{n}}^{(1)^T} \cdot \boldsymbol{A} \cdot \boldsymbol{n}^{(1)} = \overline{\lambda_1} \overline{\boldsymbol{n}}^{(1)^T} \boldsymbol{n}^{(1)} - \lambda_1 \overline{\boldsymbol{n}}^{(1)^T} \boldsymbol{n}^{(1)}, \tag{1.21}$$

$$0 = \left(\overline{\lambda}_1 - \lambda_1\right) \overline{\boldsymbol{n}}^{(1)^T} \boldsymbol{n}^{(1)}.$$

Since the eigenvector $\boldsymbol{n}^{(1)}$ is a nontrivial solution for Equation (1.9), its magnitude is greater than zero. Therefore, $\overline{\boldsymbol{n}}^{(1)^T} \boldsymbol{n}^{(1)} \neq \boldsymbol{0}$. Then Equation (1.21) requires that $0 = \left(\overline{\lambda}_1 - \lambda_1\right)$, which says that the eigenvalue is equal to its complex conjugate.

EXAMPLE 1.14. *Prove that real symmetric tensor, **A**, has orthogonal eigenvectors if the eigenvalues are distinct.*

Solution:
Substitution of a particular eigenvalue, λ_1, and its corresponding eigenvector, $\boldsymbol{n}^{(1)}$, into Equation (1.9) gives

$$\boldsymbol{A} \cdot \boldsymbol{n}^{(1)} = \lambda_1 \boldsymbol{n}^{(1)}. \tag{1.22}$$

Substitution of a second eigenvalue, λ_2, and its corresponding eigenvector, $\boldsymbol{n}^{(2)}$, into Equation (1.9) gives

$$\boldsymbol{A} \cdot \boldsymbol{n}^{(2)} = \lambda_2 \boldsymbol{n}^{(2)}. \tag{1.23}$$

Multiplying Equation (1.22) by $\boldsymbol{n}^{(2)^T}$ on the left gives

$$\boldsymbol{n}^{(2)^T} \boldsymbol{A} \cdot \boldsymbol{n}^{(1)} = \lambda_1 \boldsymbol{n}^{(2)^T} \boldsymbol{n}^{(1)}. \tag{1.24}$$

Multiplying Equation (1.23) by $\boldsymbol{n}^{(1)^T}$ on the left gives

$$\boldsymbol{n}^{(1)^T} \boldsymbol{A} \cdot \boldsymbol{n}^{(2)} = \lambda_2 \boldsymbol{n}^{(1)^T} \boldsymbol{n}^{(2)}. \tag{1.25}$$

Taking the transpose of Equation (1.25) gives

$$(\boldsymbol{n}^{(1)^T} \boldsymbol{A} \cdot \boldsymbol{n}^{(2)})^T = (\lambda_2 \boldsymbol{n}^{(1)^T} \boldsymbol{n}^{(2)})^T,$$

$$\boldsymbol{n}^{(2)^T} \boldsymbol{A} \cdot \boldsymbol{n}^{(1)} = \lambda_2 \boldsymbol{n}^{(2)^T} \boldsymbol{n}^{(1)}. \tag{1.26}$$

Subtracting Equation (1.25) from Equation (1.24) gives

$$\boldsymbol{n}^{(2)^T} \boldsymbol{A} \cdot \boldsymbol{n}^{(1)} - \boldsymbol{n}^{(2)^T} \boldsymbol{A} \cdot \boldsymbol{n}^{(1)} = \lambda_1 \boldsymbol{n}^{(2)^T} \boldsymbol{n}^{(1)} - \lambda_2 \boldsymbol{n}^{(2)^T} \boldsymbol{n}^{(1)},$$

$$0 = (\lambda_2 - \lambda_1) \boldsymbol{n}^{(2)^T} \boldsymbol{n}^{(1)}.$$

Since the eigenvalues of a symmetric tensor are distinct, $(\lambda_2 - \lambda_1) \neq 0$. This requires $\boldsymbol{n}^{(2)^T} \boldsymbol{n}^{(1)} = 0$, which means that the two eigenvectors are orthogonal.

EXAMPLE 1.15. *Find $\boldsymbol{B}^2$ using spectral decomposition where*

$$\boldsymbol{B} = \begin{bmatrix} 1 & 2 & 4 \\ 2 & 1 & 3 \\ 4 & 3 & 1 \end{bmatrix}.$$

Solution:
First, we must find the eigenvectors and eigenvalues using Matlab®

$$\begin{aligned} \lambda_1 &= \mathbf{-3.1879}, & \hat{\boldsymbol{n}}^{(1)} &= \langle \mathbf{0.6013} \quad \mathbf{0.2552} \quad \mathbf{-0.7572} \rangle, \\ \lambda_2 &= \mathbf{-0.8868}, & \hat{\boldsymbol{n}}^{(2)} &= \langle \mathbf{0.5449} \quad \mathbf{-0.8240} \quad \mathbf{0.1550} \rangle, \\ \lambda_3 &= \mathbf{7.0747}, & \hat{\boldsymbol{n}}^{(3)} &= \langle \mathbf{0.5844} \quad \mathbf{0.5058} \quad \mathbf{0.6346} \rangle. \end{aligned}$$

Next, we write the square of the tensor in terms of its eigenvalues and eigenvectors

$$\begin{aligned} \boldsymbol{B}^2 &= \sum_{i=1}^{3} \lambda_i^2 \hat{\boldsymbol{n}}^{(i)} \otimes \hat{\boldsymbol{n}}^{(i)} \\ &= \lambda_1^2 \hat{\boldsymbol{n}}^{(1)} \otimes \hat{\boldsymbol{n}}^{(1)} + \lambda_2^2 \hat{\boldsymbol{n}}^{(2)} \otimes \hat{\boldsymbol{n}}^{(2)} + \lambda_3^2 \hat{\boldsymbol{n}}^{(3)} \otimes \hat{\boldsymbol{n}}^{(3)} \\ &= \lambda_1^2 \hat{\boldsymbol{n}}^{(1)} (\hat{\boldsymbol{n}}^{(1)})^T + \lambda_2^2 \hat{\boldsymbol{n}}^{(2)} (\hat{\boldsymbol{n}}^{(2)})^T + \lambda_3^2 \hat{\boldsymbol{n}}^{(3)} (\hat{\boldsymbol{n}}^{(3)})^T. \end{aligned}$$

Plugging in the eigenvalues and eigenvectors gives

$$\begin{aligned} \boldsymbol{B}^2 = (\mathbf{-3.1879})^2 &\begin{bmatrix} \mathbf{0.6013} \\ \mathbf{0.2552} \\ \mathbf{-0.7572} \end{bmatrix} \begin{bmatrix} \mathbf{0.6013} & \mathbf{0.2552} & \mathbf{-0.7572} \end{bmatrix} \\ + (\mathbf{-0.8868})^2 &\begin{bmatrix} \mathbf{0.5449} \\ \mathbf{-0.8240} \\ \mathbf{0.1550} \end{bmatrix} \begin{bmatrix} \mathbf{0.5449} & \mathbf{-0.8240} & \mathbf{0.1550} \end{bmatrix} \\ + (\mathbf{7.0747})^2 &\begin{bmatrix} \mathbf{0.5844} \\ \mathbf{0.5058} \\ \mathbf{0.6346} \end{bmatrix} \begin{bmatrix} \mathbf{0.5844} & \mathbf{0.5058} & \mathbf{0.6346} \end{bmatrix}. \end{aligned}$$

The final result is

$$\boldsymbol{B}^2 = \begin{bmatrix} 21 & 16 & 14 \\ 16 & 14 & 14 \\ 15 & 14 & 26 \end{bmatrix}.$$

1.5 Coordinate Transformation

The vectors and tensors used to describe physical quantities may be expressed as a column vector or a matrix representing the components of the tensor within a specified coordinate system. While the physical quantity does not depend on the coordinate system, the components of the matrix representation do. In this section, we will show that the components of the matrix representation within different coordinate systems are related via the rotation tensor that relates the two coordinate systems.

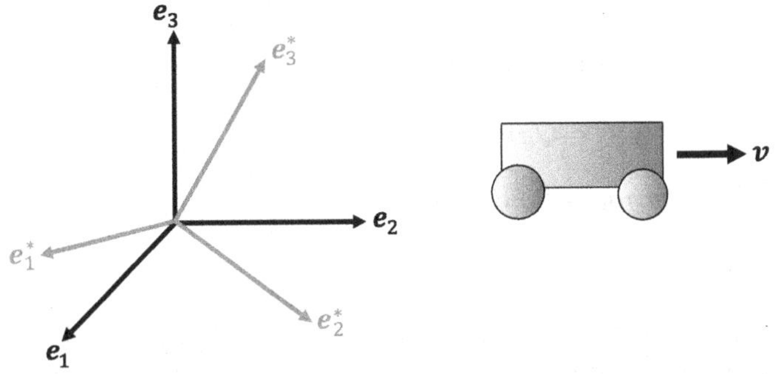

Figure 1.3. Illustration of a vector representation in multiple coordinate systems.

Imagine a race car traveling down a track with a velocity vector, $\boldsymbol{v}$. We select a coordinate system with basis vectors $\{\boldsymbol{e}_1, \boldsymbol{e}_2, \boldsymbol{e}_3\}$. The velocity vector can be expressed as a linear combination of these basis vectors such that

$$\boldsymbol{v} = v_i \boldsymbol{e}_i. \tag{1.27}$$

However, this coordinate system is not unique. We could have chosen a different coordinate system with basis vectors $\{\boldsymbol{e}_1^*, \boldsymbol{e}_2^*, \boldsymbol{e}_3^*\}$. The vector, $\boldsymbol{v}$, expressed in terms of components in this new coordinate system is given by

$$\boldsymbol{v} = v_i^* \boldsymbol{e}_i^*. \tag{1.28}$$

It is important to note that the vector itself has not changed. Only the components have changed. The physical interpretation of the velocity vector, $\boldsymbol{v}$, cannot change since the car is traveling down the same track at the same speed no matter which coordinate system we select. Therefore, Equation (1.27) and Equation (1.28) must be equal,

$$v_m \boldsymbol{e}_m = v_i^* \boldsymbol{e}_i^*.$$

If we dot both sides of this equation with the basis vector, $\boldsymbol{e}_j$, we obtain an equation that relates components of the vector in the * reference frame to those in the original reference frame

$$\begin{aligned} v_m \boldsymbol{e}_m \cdot \boldsymbol{e}_j &= v_i^* \boldsymbol{e}_i^* \cdot \boldsymbol{e}_j, \\ v_m \delta_{mj} &= v_i^* \left(\boldsymbol{e}_i^* \cdot \boldsymbol{e}_j\right), \\ v_j &= v_i^* \left(\boldsymbol{e}_i^* \cdot \boldsymbol{e}_j\right). \end{aligned} \tag{1.29}$$

The components are related to one another via the ***transformation matrix*** defined by

$$Q_{ij} \equiv \boldsymbol{e}_i^* \cdot \boldsymbol{e}_j.$$

Since the magnitude of the unit vectors is equal to one, the transformation matrix can be written in terms of the angles between the basis vectors as

$$\begin{aligned} Q_{ij} &= \cos\theta\left(\boldsymbol{e}_i^*, \boldsymbol{e}_j\right) \\ &= \boldsymbol{e}_i^* \cdot \boldsymbol{e}_j, \end{aligned}$$

where $\theta\left(e_i^*, e_j\right)$ is the angle between e_j and e_i^*. Therefore, the transformation of vector components, Equation (1.29), can be written as

$$v_j = Q_{ij} v_i^*.$$

In tensor notation, the ***vector transformation rules*** become

$$\boldsymbol{v}^* = \boldsymbol{Q} \cdot \boldsymbol{v} \quad \text{or} \quad \boldsymbol{v} = \boldsymbol{Q}^T \cdot \boldsymbol{v}^*. \tag{1.30}$$

The components of a tensor, $\boldsymbol{A}$, transform in a similar manner. Again, the physical quantity represented by the tensor does not change. Only the components of the tensor expressed within a given reference frame change. Using the results of Example 1.10, we can express the components of a tensor in the starred reference frame as

$$A_{ij}^* = \boldsymbol{e}_i^* \cdot (A \cdot \boldsymbol{e}_j^*). \tag{1.31}$$

The basis vectors in the unstarred coordinate system can be expressed as a linear combination of the starred basis vectors as

$$\begin{aligned}
\boldsymbol{e}_i &= (\boldsymbol{e}_1^* \cdot \boldsymbol{e}_i)\boldsymbol{e}_1^* + (\boldsymbol{e}_2^* \cdot \boldsymbol{e}_i)\boldsymbol{e}_2^* + (\boldsymbol{e}_3^* \cdot \boldsymbol{e}_i)\boldsymbol{e}_3^* \\
&= (\boldsymbol{e}_j^* \cdot \boldsymbol{e}_i)\boldsymbol{e}_j^* \\
&= (Q_{1i}\boldsymbol{e}_1^* + Q_{2i}\boldsymbol{e}_2^* + Q_{3i}\boldsymbol{e}_3^*) \\
&= Q_{ji}\boldsymbol{e}_j^*.
\end{aligned}$$

The coefficient before each of the starred basis vectors is simply the projection of the unstarred basis vector onto the starred basis vector. The basis vectors in the starred coordinate system may also be expressed as a linear combination of the unstarred basis vectors as

$$\begin{aligned}
\boldsymbol{e}_i^* &= (\boldsymbol{e}_1 \cdot \boldsymbol{e}_i^*)\boldsymbol{e}_1 + (\boldsymbol{e}_2 \cdot \boldsymbol{e}_i^*)\boldsymbol{e}_2 + (\boldsymbol{e}_3 \cdot \boldsymbol{e}_i^*)\boldsymbol{e}_3 \\
&= (\boldsymbol{e}_i^* \cdot \boldsymbol{e}_j)\boldsymbol{e}_j \\
&= Q_{i1}\boldsymbol{e}_1 + Q_{i2}\boldsymbol{e}_2 + Q_{i3}\boldsymbol{e}_3 \\
&= Q_{ij}\boldsymbol{e}_j.
\end{aligned}$$

Substituting the basis transformation, into Equation (1.31) gives

$$\begin{aligned}
A_{ij}^* &= Q_{ik}\boldsymbol{e}_k \cdot \left(A \cdot \left(Q_{jl}\boldsymbol{e}_l\right)\right) \\
&= Q_{ik}Q_{jl}\left(\boldsymbol{e}_k \cdot (A \cdot \boldsymbol{e}_l)\right) \\
&= Q_{ik}Q_{jl}A_{kl}.
\end{aligned}$$

In tensor notation, the ***tensor transformation rule*** can be written as

$$\boldsymbol{A}^* = \boldsymbol{Q} \cdot \boldsymbol{A} \cdot \boldsymbol{Q}^T \quad \text{or} \quad \boldsymbol{A} = \boldsymbol{Q}^T \cdot \boldsymbol{A}^* \cdot \boldsymbol{Q}. \tag{1.32}$$

Whereas all tensors must obey the tensor transformation rule, there exists a subclass of tensors named ***isotropic tensors*** whose components do not change with transformation. These tensors have the special property $A_{ij}^* = A_{ij}$ for any possible transformation matrix. Isotropic tensors have the form, $\mathbf{A} = \alpha\mathbf{I}$.

EXAMPLE 1.16. *Two reference frames are shown in the following figure. Find the transformation matrix relating the basis vectors in the starred reference frame with those in the unstarred reference frame. The vectors* $\mathbf{e}_3, \mathbf{e}_3^*, \mathbf{e}_2$, *and* $\mathbf{e}_2^*$ *lie within a plane.*

Solution:
The components of the transformation matrix are

$$\begin{aligned} Q_{ij} &= \cos\theta(\boldsymbol{e}_i^*, \boldsymbol{e}_j) \\ &= \boldsymbol{e}_i^* \cdot \boldsymbol{e}_j. \end{aligned}$$

The first component, Q_{11}, is given by the angle between $\boldsymbol{e}_1$ and $\boldsymbol{e}_1^*$ such that

$$\begin{aligned} Q_{11} &= \cos\theta(\boldsymbol{e}_1^*, \boldsymbol{e}_1) \\ &= \cos\ 0° = 1. \end{aligned}$$

We must similarly evaluate all of the components of the transformation matrix. The full matrix can be written as

$$\mathbf{Q} = \begin{bmatrix} \cos\theta(\boldsymbol{e}_1^*, \boldsymbol{e}_1) & \cos\theta(\boldsymbol{e}_1^*, \boldsymbol{e}_2) & \cos\theta(\boldsymbol{e}_1^*, \boldsymbol{e}_3) \\ \cos\theta(\boldsymbol{e}_2^*, \boldsymbol{e}_1) & \cos\theta(\boldsymbol{e}_2^*, \boldsymbol{e}_2) & \cos\theta(\boldsymbol{e}_2^*, \boldsymbol{e}_3) \\ \cos\theta(\boldsymbol{e}_3^*, \boldsymbol{e}_1) & \cos\theta(\boldsymbol{e}_3^*, \boldsymbol{e}_2) & \cos\theta(\boldsymbol{e}_3^*, \boldsymbol{e}_3) \end{bmatrix}.$$

Evaluation of each component gives

$$\mathbf{Q} = \begin{bmatrix} \cos 0 & \cos\frac{\pi}{2} & \cos\frac{\pi}{2} \\ \cos\frac{\pi}{2} & \cos\theta & \cos\left(\frac{\pi}{2} + \theta\right) \\ \cos\frac{\pi}{2} & \cos\left(\frac{\pi}{2} - \theta\right) & \cos\theta \end{bmatrix}.$$

And finally, we obtain the transformation matrix

$$\mathbf{Q} = \begin{bmatrix} 1 & 0 & 0 \\ 0 & \cos\theta & -\sin\theta \\ 0 & \sin\theta & \cos\theta \end{bmatrix}.$$

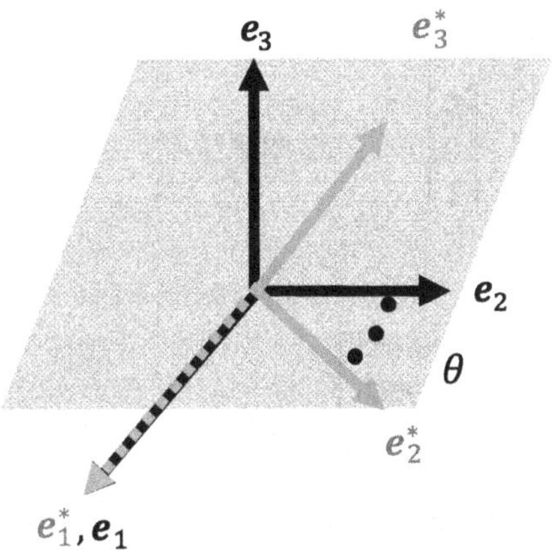

Figure 1.4. Coordinate rotation about the $\boldsymbol{e}_1$ axis

Note that the determinant of the transformation matrix is equal to one.

EXAMPLE 1.17. *The tensor* $\boldsymbol{B}$ *is expressed in component form within the xyz coordinate system with basis vectors* $\{\boldsymbol{e}_1, \boldsymbol{e}_2, \boldsymbol{e}_3\}$ *as*

$$\boldsymbol{B} = \begin{bmatrix} 2 & 1 & 4 \\ 1 & 1 & 3 \\ 4 & 3 & 5 \end{bmatrix}.$$

Find the orthogonal tensor which transforms from the current xyz coordinate system to the coordinate system defined by the normalized eigenvectors, $\{\hat{\boldsymbol{n}}^{(1)}, \hat{\boldsymbol{n}}^{(2)}, \hat{\boldsymbol{n}}^{(3)}\}$*, of the tensor* $\boldsymbol{B}$.

Solution:
We are searching for the orthogonal rotation tensor, $\boldsymbol{Q}$, which satisfies the following relationship:

$$\hat{\boldsymbol{n}}^{(i)} = \boldsymbol{Q} \cdot \boldsymbol{e}_i.$$

The components of $\boldsymbol{Q}$ are given by

$$Q_{ij} = \hat{\boldsymbol{n}}^{(i)} \cdot \boldsymbol{e}_i.$$

The eigenvalues and eigenvectors of the tensor, $\boldsymbol{B}$, are

$$\begin{aligned}
\lambda_1 &= -1.3560, & \hat{\boldsymbol{n}}^{(1)} &= \langle 0.5728 \quad 0.5410 \quad -0.6158 \rangle, \\
\lambda_2 &= 0.4123, & \hat{\boldsymbol{n}}^{(2)} &= \langle 0.6464 \quad -0.7601 \quad -0.0666 \rangle, \\
\lambda_3 &= 8.9438, & \hat{\boldsymbol{n}}^{(3)} &= \langle 0.5041 \quad 0.3599 \quad -0.7851 \rangle.
\end{aligned}$$

The components of the transformation can be computed as

$$Q_{11} = \hat{\boldsymbol{n}}^{(1)} \cdot \boldsymbol{e}_1 = \begin{bmatrix} 0.5728 \\ 0.5410 \\ -0.6158 \end{bmatrix} \cdot \begin{bmatrix} 1 \\ 0 \\ 0 \end{bmatrix} = 0.5728,$$

$$Q_{12} = \hat{\boldsymbol{n}}^{(1)} \cdot \boldsymbol{e}_2 = \begin{bmatrix} 0.5728 \\ 0.5410 \\ -0.6158 \end{bmatrix} \cdot \begin{bmatrix} 0 \\ 1 \\ 0 \end{bmatrix} = 0.5410,$$

$$Q_{13} = \hat{\boldsymbol{n}}^{(1)} \cdot \boldsymbol{e}_3 = \begin{bmatrix} 0.5728 \\ 0.5410 \\ -0.6158 \end{bmatrix} \cdot \begin{bmatrix} 0 \\ 0 \\ 1 \end{bmatrix} = -0.6158.$$

The resultant transformation matrix is

$$\boldsymbol{Q} = \begin{bmatrix} 0.5728 & 0.5410 & -0.6158 \\ 0.6464 & -0.7601 & -0.0666 \\ 0.5041 & 0.3599 & 0.7851 \end{bmatrix}.$$

In this case, each column of the rotation matrix consists of a normalized eigenvector.

EXAMPLE 1.18. *The tensor, $\boldsymbol{B}$, expressed in matrix form in the xyz coordinate system with basis vectors $\{\boldsymbol{e}_1, \boldsymbol{e}_2, \boldsymbol{e}_3\}$ is given by*

$$\boldsymbol{B} = \begin{bmatrix} 2 & 1 & 4 \\ 1 & 1 & 3 \\ 4 & 3 & 5 \end{bmatrix}.$$

Find the components of $\boldsymbol{B}$ in the $*$ coordinate system defined by its eigenvectors, $\{\hat{\boldsymbol{n}}^{(1)}, \hat{\boldsymbol{n}}^{(2)}, \hat{\boldsymbol{n}}^{(3)}\}$.

Solution:
In the previous example, we determined the transformation matrix, $\boldsymbol{Q}$, which relates the xyz coordinate system and that defined by the eigenvectors of tensor $\boldsymbol{B}$. The components in the $*$ coordinate system can be found using Equation (1.32) such that

$$\boldsymbol{B}^* = \boldsymbol{Q} \cdot \boldsymbol{B} \cdot \boldsymbol{Q}^T.$$

Using the transformation matrix from the previous example gives

$$\boldsymbol{B}^* = \begin{bmatrix} \mathbf{0.5728} & \mathbf{0.5410} & \mathbf{-0.6158} \\ \mathbf{0.6464} & \mathbf{-0.7601} & \mathbf{-0.0666} \\ \mathbf{0.5041} & \mathbf{0.3599} & \mathbf{0.7851} \end{bmatrix} \begin{bmatrix} \mathbf{2} & \mathbf{1} & \mathbf{4} \\ \mathbf{1} & \mathbf{1} & \mathbf{3} \\ \mathbf{4} & \mathbf{3} & \mathbf{5} \end{bmatrix} \begin{bmatrix} \mathbf{0.5728} & \mathbf{0.6464} & \mathbf{0.5041} \\ \mathbf{0.5410} & \mathbf{-0.7601} & \mathbf{0.3599} \\ \mathbf{-0.6158} & \mathbf{-0.0666} & \mathbf{0.7851} \end{bmatrix}.$$

Carrying out the matrix multiplication, we find that

$$\boldsymbol{B}^* = \begin{bmatrix} \mathbf{-1.3560} & \mathbf{0} & \mathbf{0} \\ \mathbf{0} & \mathbf{0.4123} & \mathbf{0} \\ \mathbf{0} & \mathbf{0} & \mathbf{8.9438} \end{bmatrix}.$$

Note that each component of this diagonal matrix is an eigenvalue of the tensor $\boldsymbol{B}$.

EXAMPLE 1.19. *Show that any tensor with the form, $\mathbf{A} = \alpha\mathbf{I}$, is an isotropic tensor.*

Solution:
If $\boldsymbol{A}$ is isotropic, then it has the property that the components of the tensor expressed in terms of any coordinate system are the same, $A^*_{ij} = A_{ij}$. The transformation of tensor components is given by Equation (1.32) as

$$\boldsymbol{A}^* = \boldsymbol{Q} \cdot \boldsymbol{A} \cdot \boldsymbol{Q}^T.$$

We substitute $\boldsymbol{A} = \boldsymbol{A}^* = \alpha\mathbf{I}$ into the tensor transformation rule giving

$$\boldsymbol{A}^* = \boldsymbol{Q} \cdot \boldsymbol{A}^* \cdot \boldsymbol{Q}^T$$

or

$$\alpha\mathbf{I} = \boldsymbol{Q} \cdot \alpha\mathbf{I} \cdot \boldsymbol{Q}^T.$$

This equation must be valid for all possible coordinate systems. Therefore, it must be true for all transformation matrices, $\boldsymbol{Q}$. Since $\boldsymbol{Q} \cdot \mathbf{I} = \boldsymbol{Q}$, we can write

$$\alpha\mathbf{I} = \alpha\boldsymbol{Q} \cdot \boldsymbol{Q}^T.$$

Finally, the orthogonal tensor, $\boldsymbol{Q}$, is its own inverse, which gives us

$$\alpha\mathbf{I} = \alpha\mathbf{I}.$$

Since the components of the tensor $\boldsymbol{A}$ remain invariant under all possible coordinate transformations, the tensor is an isotropic tensor.

1.6 Invariants

Even though the components of both tensors and vectors change with coordinate transformation, there are several combinations of components which do not change. For example, the magnitude of a vector does not change with coordinate transformation. Therefore, we say that the first invariant, I_1, of a vector is its magnitude

$$I_1(\boldsymbol{a}) = |\boldsymbol{a}| = a_1^2 + a_2^2 + a_3^2.$$

It is clear that if the magnitude of a vector is invariant, then the square of the magnitude must also be invariant. In fact, any function of the invariant will also be invariant. However, there is only one independent invariant of a vector.

A second-order tensor has three independent invariants, I_A, II_A, and III_A given by

$$\begin{aligned} I_A &= A_{ii} = tr\boldsymbol{A}, \\ II_A &= \frac{1}{2}(A_{ii}A_{jj} - A_{ji}A_{ij}) = \frac{1}{2}[(tr\boldsymbol{A})^2 - tr(\boldsymbol{A}^2)], \\ III_A &= \varepsilon_{ijk}A_{1i}A_{2j}A_{3k} = \det\boldsymbol{A}. \end{aligned}$$

Any function of these invariants, $\alpha = f(I_A, II_A, III_A)$, is also invariant. You will find that both the notation and the particular form of the invariants may differ from textbook to textbook. For example, one can refer to the invariants of tensor **A** using the notation $I_A = I_1(\mathbf{A})$, $II_{\mathbf{A}} = I_2(\mathbf{A})$, and $III_{\mathbf{A}} = I_3(\mathbf{A})$.

EXAMPLE 1.20. *Show that $\alpha(\mathbf{A}) = \det\boldsymbol{A}$ is invariant to coordinate transformation.*

Solution:
If the coordinate system is rotated, the components of the tensor **A** transform according to tensor transformation rule, $\boldsymbol{A}^* = \boldsymbol{Q}^T\boldsymbol{A}\boldsymbol{Q}$. When we suggest that this function is invariant to coordinate transformation, we are stating that the scalar value of this function, $\alpha(\mathbf{A})$, does not change when **A** is expressed in a different coordinate system. Therefore, we obtain the relation

$$\alpha(\mathbf{A}) = \alpha(\mathbf{A}^*).$$

Substituting the transformation rule into the preceding equation gives

$$\begin{aligned} \alpha(\mathbf{A}) &= \alpha(\boldsymbol{Q}^T\boldsymbol{A}\boldsymbol{Q}) \\ \det\boldsymbol{A} &= \det(\boldsymbol{Q}^T\boldsymbol{A}\boldsymbol{Q}) \\ \det\boldsymbol{A} &= \det\boldsymbol{Q}^T\det\boldsymbol{A}\det\boldsymbol{Q} \\ \det\boldsymbol{A} &= \ 1 \quad \det\boldsymbol{A} \quad 1. \end{aligned}$$

As we can see, this relation is satisfied. Therefore, the function is invariant to coordinate transformation.

1.7 Cayley-Hamilton Theorem

The Cayley-Hamilton theorem provides a useful relation between a tensor and its invariants. The theorem states that a second-order tensor will satisfy its own characteristic equation. Recall that the characteristic equation for a second-order tensor can be written as

$$\det(\mathbf{A} - \lambda\mathbf{I}) = 0.$$

This equation can be written in terms of the invariants of the tensor $\boldsymbol{A}$ as

$$\lambda^3 - I_A\lambda^2 + II_A\lambda - III_A = 0. \tag{1.33}$$

If we consider symmetric second-order tensors, spectral decomposition allows us to write $\boldsymbol{A} = \sum \lambda_i \boldsymbol{n}^{(i)} \otimes \boldsymbol{n}^{(i)}$, where λ_i and $\boldsymbol{n}^{(i)}$ are the eigenvalues and eigenvectors of the tensor $\boldsymbol{A}$, respectively. According to the Cayley-Hamilton theorem, the tensor $\boldsymbol{A}$ must satisfy its own characteristic equation. This statement can be written as

$$\mathbf{A}^3 - I_A\mathbf{A}^2 + II_A\mathbf{A} - III_A\mathbf{I} = 0. \tag{1.34}$$

Expanding this equation by using spectral decomposition gives

$$\sum \lambda_i^3 \boldsymbol{n}^{(i)} \otimes \boldsymbol{n}^{(i)} - I_A\left(\sum \lambda_i^2 \boldsymbol{n}^{(i)} \otimes \boldsymbol{n}^{(i)}\right) + II_A\left(\sum \lambda_i \boldsymbol{n}^{(i)} \otimes \boldsymbol{n}^{(i)}\right)$$
$$- III_A\left(\sum \boldsymbol{n}^{(i)} \otimes \boldsymbol{n}^{(i)}\right) = 0.$$

Rearranging the equation, we find that

$$\sum_{i=1}^{3}\left(\lambda_i^3 - I_A\lambda_i^2 + II_A\lambda_i - III_A\right)\boldsymbol{n}^{(i)} \otimes \boldsymbol{n}^{(i)} = 0.$$

This equation must be true since each of the eigenvalues satisfies the characteristic equation given as Equation (1.33). While the proof here applies only to second-order symmetric tensors, the Cayley-Hamilton theorem applies to *all* second-order tensors.

The Cayley-Hamilton theorem can also be used to find the inverse of a tensor. If we multiply Equation (1.34) by A^{-1}, we obtain

$$\mathbf{A}^2 - I_A\mathbf{A}^1 + II_A\mathbf{I} - III_A\mathbf{A}^{-1} = 0.$$

Therefore, we can write the inverse of a tensor in terms of higher powers and the invariants of the tensor itself such that

$$\mathbf{A}^{-1} = \frac{1}{III_A}(\mathbf{A}^2 - I_A\mathbf{A}^1 + II_A\mathbf{I}).$$

This can prove useful when numerically determining the inverse of a second-order tensor.

1.8 Scalar, Vector, and Tensor Functions and Fields

A function is classified by both its arguments and the nature of the resultant value. For example, a ***scalar-valued function***, a ***vector-valued function***, or a ***tensor-valued function*** returns a scalar, vector, or tensor respectively. We augment this description with terminology denoting the highest order argument. For example, a ***scalar-valued tensor function*** takes as an argument at least one tensor while it may also include vector and/or scalar arguments. A ***scalar-valued vector function*** takes as an argument at least one vector while it may also include a scalar argument. As an example, $\mathbf{B}(\boldsymbol{A},\boldsymbol{a},t) = 2t\boldsymbol{A} + \boldsymbol{a}\otimes\boldsymbol{a}$ is a tensor-valued tensor function while $\alpha(\boldsymbol{A},\boldsymbol{a},t) = \boldsymbol{A}:\boldsymbol{A} + 2t(\boldsymbol{a}\cdot\boldsymbol{a})$ is a scalar valued tensor function.

Within the text, we will encounter functions that explicitly depend on time and whose arguments also depend on time. When taking the derivative of these functions with respect to time, we must employ the chain rule. For example, the time derivative of a scalar-valued tensor function, $\alpha(\boldsymbol{A},\boldsymbol{a},t)$, whose arguments also depend on time, $\boldsymbol{A} = \boldsymbol{A}(t)$, and $\boldsymbol{a} = \boldsymbol{a}(t)$, can be written as

$$\begin{aligned}\frac{d\alpha(\boldsymbol{A}(t),\boldsymbol{a}(t),t)}{dt} &= \frac{\partial\alpha}{\partial\boldsymbol{A}}:\frac{d\boldsymbol{A}}{dt} + \frac{\partial\alpha}{\partial\boldsymbol{a}}\cdot\frac{d\boldsymbol{a}}{dt} + \frac{\partial\alpha}{\partial t}\\ &= \frac{\partial\alpha}{\partial A_{ij}}\frac{dA_{ij}}{dt} + \frac{\partial\alpha}{\partial a_i}\frac{da_i}{dt} + \frac{\partial\alpha}{\partial t}.\end{aligned}$$

Notice that we are taking the partial derivative of the function with respect to each of the components of the tensor and each component of the vector and summing each of these. Similarly, we can write the derivative for a tensor-valued tensor function, $\mathbf{B}(\boldsymbol{A}(t),\boldsymbol{a}(t),t)$, as

$$\frac{d\boldsymbol{B}(\boldsymbol{A}(t),\boldsymbol{a}(t),t)}{dt} = \frac{\partial\boldsymbol{B}}{\partial\boldsymbol{A}}:\frac{d\boldsymbol{A}}{dt} + \frac{\partial\boldsymbol{B}}{\partial\boldsymbol{a}}\cdot\frac{d\boldsymbol{a}}{dt} + \frac{\partial\boldsymbol{B}}{\partial t}.$$

We can write this in component form as

$$\frac{dB_{ij}(\boldsymbol{A}(t),\boldsymbol{a}(t),t)}{dt} = \frac{\partial B_{ij}}{\partial A_{kl}}\frac{dA_{kl}}{dt} + \frac{\partial B_{ij}}{\partial a_k}\frac{da_k}{dt} + \frac{\partial B_{ij}}{\partial t}.$$

When describing the behavior of materials, we use scalar, vector, and tensor functions to describe relevant forces and fields. A ***scalar field***, ***vector field***, or ***tensor field*** are scalar, vector, or tensor functions, respectively, which are defined for every point within a region of space. The region of space might be the material body, the space around the experimental apparatus, or the earth. The field might be the temperature of a material body, or the velocity field of the points within an object. Once we have defined a field, we can examine how the field varies at a given point with time (which is given by the time derivative of the field), or we can examine how the field varies at a given time with respect to space (which is given by the gradient of the field).

The gradient, divergence, Laplacian, and curl of vectors, tensors, and scalars are all measures of spatial field variation at a fixed time. The ***gradient*** of a field is a measure of the rate of change along each coordinate direction for each of the components of the field. Therefore, the result of the gradient of a field is always one order higher than the field itself. The gradient of a scalar is a vector; the gradient of a vector is a second-order tensor, and so on. The gradient of any order field, $grad(\cdot)$, is defined as the linear operator which gives the ***directional derivative*** of the field along the basis vectors, $\boldsymbol{e}_i$, according to

$$grad(\cdot)\cdot\boldsymbol{e}_i = \lim_{\alpha\to 0}\left[\frac{(\cdot)(\boldsymbol{x}+\alpha\,\boldsymbol{e}_i,t) - (\cdot)(\boldsymbol{x})}{\alpha}\right] = (\cdot)_{,i} \tag{1.35}$$

where $(\cdot)$ could be a scalar, vector, or higher-order tensor.

The gradient, divergence, Lapalacian, and curl can each be expressed in terms of a vector differential operator named the ***Del operator***, which is defined in Cartesian coordinates as either a right-differential operator, $\overleftarrow{\nabla}$,

$$\overleftarrow{\nabla} \equiv \boldsymbol{e}_i \frac{\overleftarrow{\partial}\,()}{\partial x_i},$$

or a left-differential operator, $\overrightarrow{\nabla}$,

$$\overrightarrow{\nabla} \equiv \nabla \equiv \boldsymbol{e}_i \frac{\overleftarrow{\partial}\,()}{\partial x_i}.$$

The direction of the arrow signifies the direction in which the operator acts. For example, the gradient of a scalar field, $grad\,\Phi$, is given by

$$grad\,\boldsymbol{\Phi} = \overrightarrow{\nabla}\boldsymbol{\Phi} = \boldsymbol{e}_i \frac{\partial}{\partial x_i}(\boldsymbol{\Phi}) = \boldsymbol{\Phi}_{,i}\boldsymbol{e}_i.$$

We can see that this result is consistent with the definition of the gradient in Equation (1.35). The direction of the gradient of the scalar field, $\boldsymbol{n} = \frac{\nabla\Phi}{|\nabla\Phi|}$, points in the direction of maximum increase in the scalar field, while the magnitude of the gradient is a measure of how rapidly the scalar field is changing in this direction.

Similarly, the gradient of a vector field produces a second-order tensor and can be denoted as the dyadic product of the vector with the right-differential operator, $\boldsymbol{a} \otimes \overleftarrow{\nabla}$. Many texts will also use the notation $\nabla\boldsymbol{a}$ to denote the gradient of a vector. The gradient of a vector is given by

$$grad\,\boldsymbol{a} = \boldsymbol{a} \otimes \overleftarrow{\nabla} = \boldsymbol{a}\overleftarrow{\nabla} = (a_i\boldsymbol{e}_i) \otimes \left(\boldsymbol{e}_j \frac{\overleftarrow{\partial}\,()}{\partial x_j}\right) = \frac{\partial a_i}{\partial x_j}\boldsymbol{e}_i \otimes \boldsymbol{e}_j.$$

Again, this is consistent with the definition of the gradient given in Equation (1.35). To verify this, consider

$$\frac{\partial a_i}{\partial x_j}\boldsymbol{e}_i \otimes \boldsymbol{e}_j \cdot \boldsymbol{e}_k = \frac{\partial a_i}{\partial x_j}(\boldsymbol{e}_i(\boldsymbol{e}_j \cdot \boldsymbol{e}_k)) = \frac{\partial a_i}{\partial x_k}\boldsymbol{e}_i = \boldsymbol{a}_{,k}.$$

In addition, the transpose of the gradient can be found using the left-differential operator as

$$\begin{aligned}(grad\,\boldsymbol{a})^{\boldsymbol{T}} &= \overrightarrow{\nabla} \otimes \boldsymbol{a} \\ &= \left(\boldsymbol{e}_i \frac{\overrightarrow{\partial}\,()}{\partial x_i}\right) \otimes (a_j\boldsymbol{e}_j) \\ &= \frac{\partial a_j}{\partial x_i}\boldsymbol{e}_i \otimes \boldsymbol{e}_j.\end{aligned}$$

The gradient of a second-order tensor, $grad\boldsymbol{A}$, produces a third-order tensor according to

$$grad\,\boldsymbol{A} = \boldsymbol{A} \otimes \overleftarrow{\nabla} = \boldsymbol{A}\overleftarrow{\nabla} = \left(\frac{\partial A_{ij}}{\partial x_k}\right)\boldsymbol{e} \otimes \boldsymbol{e}_j \otimes \boldsymbol{e}_k.$$

For the rest of this section, the left-differential operator is used and is denoted using the nabla without an arrow, ∇. The ***divergence*** of a field reduces the order of the field by one. The divergence of a vector field gives a scalar

$$\begin{aligned}\nabla\cdot\boldsymbol{a} &= div\,\boldsymbol{a}\\ &= \left(\boldsymbol{e}_i\frac{\partial()}{\partial x_i}\right)\cdot(a_j\boldsymbol{e}_j)\\ &= \frac{\partial a_j}{\partial x_i}(\boldsymbol{e}_i\cdot\boldsymbol{e}_j)\\ &= \frac{\partial a_i}{\partial x_i}.\end{aligned}$$

The divergence of the velocity vector of a material body is physically interpreted as the rate of expansion of the material.

The divergence of a tensor gives a vector

$$\begin{aligned}\nabla\cdot\boldsymbol{A} &= div\,\boldsymbol{A}\\ &= \left(\boldsymbol{e}_i\frac{\partial()}{\partial x_i}\right)\cdot(A_{\mathbf{jk}}\boldsymbol{e}_j\otimes\boldsymbol{e}_k)\\ &= \frac{\partial A_{jk}}{\partial x_i}(\boldsymbol{e}_i\cdot\boldsymbol{e}_j\otimes\boldsymbol{e}_k)\\ &= \frac{\partial A_{ik}}{\partial x_i}(\boldsymbol{e}_i\cdot\boldsymbol{e}_j)\boldsymbol{e}_k\\ &= A_{ij,i}\boldsymbol{e}_j.\end{aligned}$$

The curl of a vector is a vector expressed as

$$\begin{aligned}\nabla\times\boldsymbol{a} &= curl\,\boldsymbol{a}\\ &= \boldsymbol{e}_i\frac{\partial()}{\partial x_i}\times a_j\boldsymbol{e}_j = \frac{\partial a_j}{\partial x_i}(\boldsymbol{e}_i\times\boldsymbol{e}_j)\\ &= \varepsilon_{ijk}a_{j,i}\boldsymbol{e}_k.\end{aligned}$$

The ***curl*** of the velocity vector of a material body is a measure of the rotation within the material.

The ***Laplacian*** of a scalar is a scalar given by

$$\begin{aligned}\nabla^2\phi &= \nabla\cdot\nabla\phi\\ &= \boldsymbol{e}_i\frac{\partial()}{\partial x_i}\cdot\boldsymbol{e}_j\frac{\partial()}{\partial x_j}\phi\\ &= \frac{\partial}{\partial x_i}\left(\frac{\partial\phi}{\partial x_j}\right)(\boldsymbol{e}_i\cdot\boldsymbol{e}_j)\\ &= \frac{\partial^2\phi}{\partial x_i^2} = \phi_{,ii}.\end{aligned}$$

The Laplacian of a vector is given by

$$\begin{aligned}\nabla^2 \boldsymbol{v} &= \nabla \cdot \nabla \boldsymbol{v}\\ &= \boldsymbol{e}_k \frac{\partial}{\partial x_k} \cdot \boldsymbol{v}_{j,i} \boldsymbol{e}_i \otimes \boldsymbol{e}_j\\ &= \frac{\partial v_{j,i}}{\partial x_k} (\boldsymbol{e}_k \cdot \boldsymbol{e}_i) \boldsymbol{e}_j\\ &= v_{j,ii} \boldsymbol{e}_j.\end{aligned}$$

The gradient, divergence, curl, and Laplacian can all be computed in cylindrical coordinates or spherical coordinates by inserting the appropriate differential operator into the definitions provided above. The left-differential operator in cylindrical coordinates is defined as

$$\nabla \equiv \boldsymbol{e}_r \frac{\partial ()}{\partial x_r} + \boldsymbol{e}_\theta \frac{1}{r} \frac{\partial ()}{\partial x_\theta} + \boldsymbol{e}_z \frac{\partial ()}{\partial x_z}.$$

The left-differential operator in spherical coordinates is given by

$$\nabla \equiv \boldsymbol{e}_r \frac{\partial ()}{\partial x_r} + \boldsymbol{e}_\theta \frac{1}{r} \frac{\partial ()}{\partial x_\theta} + \boldsymbol{e}_\phi \frac{1}{r \sin\theta} \frac{\partial ()}{\partial x_\phi}.$$

EXAMPLE 1.21. *Find the gradient of the scalar field* $\alpha(\boldsymbol{x}) = 3.2x_1^2 + x_2 + x_3^3$.

Solution:
The gradient is given by $\vec{\nabla}\alpha = \alpha_i \boldsymbol{e}_i = 6.4x_1\boldsymbol{e}_1 + 1\boldsymbol{e}_2 + 3x_3^2\boldsymbol{e}_3$.

EXAMPLE 1.22. Find the divergence of the vector field $\boldsymbol{a}(\boldsymbol{x}) = 3x_1^2\boldsymbol{e}_1 + x_2^3\boldsymbol{e}_2 + (2x_3^2 - 1)\boldsymbol{e}_3$.

Solution:
The divergence is given by $\nabla \cdot \boldsymbol{a} = \boldsymbol{a}_{i,i} = 6x_1 + 3x_2^2 + 4x_3$.

EXAMPLE 1.23. *Use Matlab® to plot the gradient of the scalar-valued vector function* $\alpha(\boldsymbol{x}) = 1.5x_1 \cdot x_2 + 2.0x_3$ *for the values* $0 \le x_1 \le 1, 0 \le x_2 \le 1$, *and* $x_3 = 0$.

Solution:
The following Matlab® script can be used to generate a plot of the gradient of the given scalar field. Recall that the gradient of a scalar field is a vector. Therefore, the results are plotted as a vector field using the quiver function. For reference, a contour plot of the scalar field is superposed on the same plot.

```
>> x1range = [0:0.05:1]
>> x2range = [0:0.05:1];
>> x3range = [0];
>> [x1, x2, x3] = meshgrid(x1range, x2range, x3range);
>> alpha = 1.5 * x1.*x2+2.0*x3;
>> [px, py] = gradient(alpha, 0.2, 0.2)
>> contour(x1range, x2range, alpha);
>> hold on;
>> quiver(x1range, x2range, px, py)
```

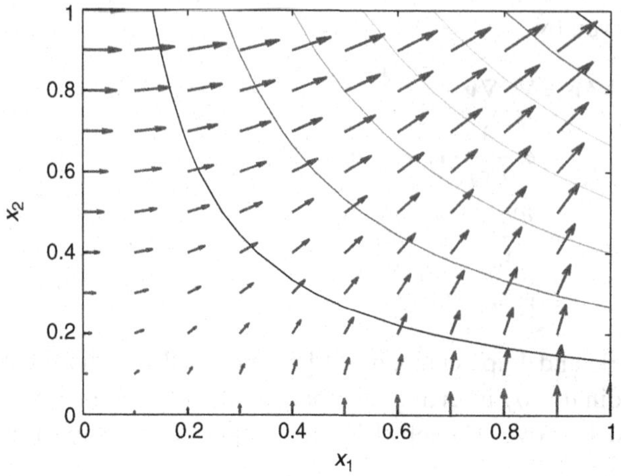

Figure 1.5. Vector plot illustrating the gradient of the scalar.

EXAMPLE 1.24. *Prove that* $\dfrac{\partial \boldsymbol{A}^{-1}}{\partial \alpha} = -\boldsymbol{A}^{-1} \cdot \dfrac{\partial \boldsymbol{A}}{\partial \alpha} \cdot \boldsymbol{A}^{-1}$.

Solution:

We being with the identity tensor

$$\mathbf{I} = \boldsymbol{A}^{-1} \cdot \boldsymbol{A}.$$

Taking the derivative of the identity tensor with respect to α gives

$$\frac{\partial (\mathbf{A}^{-1} \cdot \mathbf{A})}{\partial \alpha} = \frac{\partial (\mathbf{I})}{\partial \alpha},$$

$$\frac{\partial (\mathbf{A}^{-1})}{\partial \alpha} \cdot \mathbf{A} + \mathbf{A}^{-1} \cdot \frac{\partial (\mathbf{A})}{\partial \alpha} = 0.$$

Multiplying by $\boldsymbol{A}^{-1}$ from the right-hand side gives

$$\frac{\partial (\mathbf{A}^{-1})}{\partial \alpha} \cdot \mathbf{A} \cdot \mathbf{A}^{-1} + \mathbf{A}^{-1} \cdot \frac{\partial (\mathbf{A})}{\partial \alpha} \cdot \mathbf{A}^{-1} = 0,$$

$$\frac{\partial (\mathbf{A}^{-1})}{\partial \alpha} \cdot \mathbf{I} + \mathbf{A}^{-1} \cdot \frac{\partial (\mathbf{A})}{\partial \alpha} \cdot \mathbf{A}^{-1} = 0.$$

Rearranging this equation gives

$$\frac{\partial (\mathbf{A}^{-1})}{\partial \alpha} = -\mathbf{A}^{-1} \cdot \frac{\partial (\mathbf{A})}{\partial \alpha} \cdot \mathbf{A}^{-1}.$$

1.9 Integral Theorems

The ***divergence theorem*** states that the integration of the scalar product of a continuous differentiable field, and the surface normal over a closed surface, $\partial\Omega$, is equal to the integration of the divergence of the field over the enclosed volume, Ω. For a vector field, the divergence theorem states that

$$\int_{\Omega} \nabla \cdot \boldsymbol{b}\, dV = \int_{\partial\Omega} \boldsymbol{n} \cdot \boldsymbol{b}\, dS. \tag{1.36}$$

For a tensor field, we have

$$\int_{\Omega} \nabla \cdot \boldsymbol{A}\, dV = \int_{\partial\Omega} \boldsymbol{n} \cdot \boldsymbol{A}\, dS.$$

The ***gradient theorem*** states that the integration of the scalar product of a continuous differentiable scalar field and the surface normal over a closed surface, $\partial\Omega$, is equal to the integration of the gradient of the scalar field over the enclosed volume, Ω, such that

$$\int_{\Omega} \nabla\phi\, dV = \int_{\partial\Omega} \phi \boldsymbol{n}\, dS.$$

EXAMPLE 1.25. *Prove that* $\int_{\Omega} \nabla \cdot \boldsymbol{b}\, dV = \int_{\partial\Omega} \boldsymbol{b} \cdot \boldsymbol{n}\, dS$ *where* $\boldsymbol{b} = \boldsymbol{b}(x_1, x_2, x_3)$.

Solution:
Expanding the equation in terms of the components of $\boldsymbol{b}$ gives

$$\int_{\Omega} div\, \boldsymbol{b}\, dV = \int_{\partial\Omega} \boldsymbol{n} \cdot \boldsymbol{b}\, dS,$$

$$\int_{\Omega} \left(\frac{\partial b_1}{\partial x_1} + \frac{\partial b_2}{\partial x_2} + \frac{\partial b_3}{\partial x_3} \right) dV = \int_{\partial\Omega} \left(b_1 (\boldsymbol{e}_1 \cdot \boldsymbol{n}) + b_2 (\boldsymbol{e}_2 \cdot \boldsymbol{n}) + b_3 (\boldsymbol{e}_3 \cdot \boldsymbol{n}) \right) dS.$$

Let us look more closely at the first term on the right side of the equation. We split the object into a top and bottom part as shown in the figure. The surface integral can then be broken up into two parts

$$\int_{\partial\Omega} b_1 (\boldsymbol{e}_1 \cdot \boldsymbol{n})\, dS = \int_{\partial\Omega_a} b_1 (\boldsymbol{e}_1 \cdot \boldsymbol{n})\, dS + \int_{\partial\Omega_b} b_1 (\boldsymbol{e}_1 \cdot \boldsymbol{n})\, dS.$$

Instead of integrating on the curved surface $\partial\Omega_a$, we can carry out the integration over the projected area, $\partial\Omega_p$, by changing variables. The incremental surface, dS, becomes

$$dA = |e_1 \cdot \boldsymbol{n}|\, dS.$$

On the $\partial\Omega_b$ surface, $(\boldsymbol{e}_1 \cdot \boldsymbol{n}) > 0$, while $(\boldsymbol{e}_1 \cdot \boldsymbol{n}) < 0$ on the $\partial\Omega_a$ surface. In addition, for any position on the projected surface, x_2 and x_3, the x_1 coordinate of the top surface is given by the function $x_1 = f_b(x_2, x_3)$ and the x_1 coordinate of the bottom

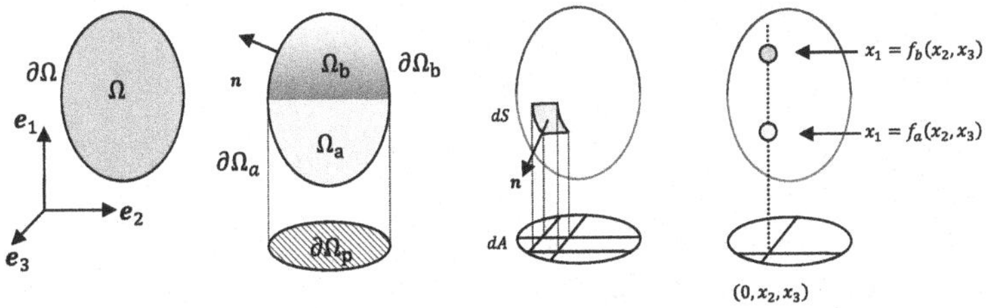

Figure 1.6. Illustration showing a volume, $\boldsymbol{\Omega}$, the surrounding surface, $\boldsymbol{\partial\Omega}$, and the outward normal, $\boldsymbol{n}$.

surface is given by the function $x_1 = f_a(x_2,x_3)$. This gives

$$\int_{\partial\Omega} b_1(\boldsymbol{e}_1 \cdot \boldsymbol{n})\,dS = \int_{\partial\Omega_p} \frac{b_1(\boldsymbol{e}_1 \cdot \boldsymbol{n})}{|\boldsymbol{e}_1 \cdot \boldsymbol{n}|}dA + \int_{\partial\Omega_p} \frac{b_1(\boldsymbol{e}_1 \cdot \boldsymbol{n})}{|\boldsymbol{e}_1 \cdot \boldsymbol{n}|}dA$$

$$= \int_{\partial\Omega_p} b_1(f_b(x_2,x_3),x_2,x_3)\,dA - \int_{\partial\Omega_p} b_1(f_a(x_2,x_3),x_2,x_3)\,dA$$

$$= \int_{\partial\Omega_p} \{b_1(f_b(x_2,x_3),x_2,x_3) - b_1(f_a(x_2,x_3),x_2,x_3)\}\,dA.$$

We will note that we can write the integrand as an integral over the derivative of $b_1$as

$$b_1(f_b(x_2,x_3),x_2,x_3) - b_1(f_a(x_2,x_3),x_2,x_3) = \int_{f_a(x_2,x_3)}^{f_b(x_2,x_3)} \frac{\partial b_1}{\partial x_1}dx_1.$$

Substituting this into the original surface integral gives

$$\int_{\partial\Omega} b_1(\boldsymbol{e}_1 \cdot \boldsymbol{n})\,dS = \int_{\partial\Omega_p}\int_{f_a(x_2,x_3)}^{f_b(x_2,x_3)} \frac{\partial b_1}{\partial x_1}dx_1\,dA.$$

The term on the right is a volume integral

$$\int_{\partial\Omega} b_1(\boldsymbol{e}_1 \cdot \boldsymbol{n})\,dS = \int_{\Omega} \frac{\partial b_1}{\partial x_1}dV.$$

Repeating the identical analysis in the x_2 and x_3 directions gives us the final result

$$\int_{\Omega}\left(\frac{\partial b_1}{\partial x_1} + \frac{\partial b_2}{\partial x_2} + \frac{\partial b_3}{\partial x_3}\right)dV = \int_{\partial\Omega}(b_1(\boldsymbol{e}_1 \cdot \boldsymbol{n}) + b_2(\boldsymbol{e}_2 \cdot \boldsymbol{n}) + b_3(\boldsymbol{e}_3 \cdot \boldsymbol{n}))\,dS,$$

$$\int_{\Omega} \nabla \cdot \boldsymbol{b}dV = \int_{\partial\Omega} \boldsymbol{n} \cdot \boldsymbol{b}dS.$$

EXERCISES

1. Show that

   a. $\delta_{ij}\delta_{ij} = 3$.
   b. $\varepsilon_{ijk}\varepsilon_{ijk} = 6$.
   c. $\delta_{ij}\varepsilon_{ijk} = 0$.
   d. $\varepsilon_{ijk}a_j a_k = 0$.
   e. $\varepsilon_{iks}\varepsilon_{mks} = 2\delta_{im}$.
   f. $\varepsilon_{amn}\varepsilon_{ars} + \varepsilon_{ams}\varepsilon_{anr} = \varepsilon_{amr}\varepsilon_{ans}$.

2. Using index notation, show that

   a. $(\boldsymbol{a}\times\boldsymbol{b})\cdot(\boldsymbol{c}\times\boldsymbol{d}) = (\boldsymbol{a}\cdot\boldsymbol{c})(\boldsymbol{b}\cdot\boldsymbol{d}) - (\boldsymbol{a}\cdot\boldsymbol{d})(\boldsymbol{b}\cdot\boldsymbol{c})$.
   b. $(\boldsymbol{a}\times\boldsymbol{b})\times(\boldsymbol{c}\times\boldsymbol{d}) = \boldsymbol{b}(\boldsymbol{a}\cdot(\boldsymbol{c}\times\boldsymbol{d})) - \boldsymbol{a}(\boldsymbol{b}\cdot(\boldsymbol{c}\times\boldsymbol{d}))$.
   c. $\boldsymbol{a}\times(\boldsymbol{b}\times\boldsymbol{c}) + \boldsymbol{b}\times(\boldsymbol{c}\times\boldsymbol{a}) + \boldsymbol{c}\times(\boldsymbol{a}\times\boldsymbol{b}) = \boldsymbol{0}$.
   d. $(\boldsymbol{a}\times\boldsymbol{b})\cdot(\boldsymbol{b}\times\boldsymbol{c})\times(\boldsymbol{c}\times\boldsymbol{a}) = (\boldsymbol{a}\cdot(\boldsymbol{b}\times\boldsymbol{c}))^2$.
   e. $tr(\boldsymbol{A}\cdot\boldsymbol{B}\cdot\boldsymbol{C}) = tr(\boldsymbol{C}\cdot\boldsymbol{A}\cdot\boldsymbol{B})$.

3. If $\boldsymbol{v}$ is a vector and $\xi_r \equiv \frac{1}{2}\varepsilon_{rst}v_{t,s}$, show that

   a. $\xi_{r,r} = 0$.
   b. $v_{i,k} - v_{k,i} = 2\varepsilon_{ijk}\xi_j$.
   c. $4\xi_j\xi_j = v_{j,k}(v_{j,k} - v_{k,j})$.

4. Any tensor, $\boldsymbol{A}$, may be written as the sum of a symmetric part, $\boldsymbol{A}^s$, and an antisymmetric part, $\boldsymbol{A}^a$, where the symmetric part is given by $\boldsymbol{A}^s = \frac{1}{2}(\boldsymbol{A} + \boldsymbol{A}^T)$ and the antisymmetric part is given by $\boldsymbol{A}^s = \frac{1}{2}(\boldsymbol{A} - \boldsymbol{A}^T)$. Show that

   a. $\boldsymbol{A}^s\boldsymbol{B}^a = 0$.
   b. $\boldsymbol{A}^s\boldsymbol{B}^s + \boldsymbol{A}^a\boldsymbol{B}^a = \boldsymbol{AB}$.

5. For any vector,$\boldsymbol{v}$, and scalar, α, show that

   a. $\nabla \cdot (\nabla \times \boldsymbol{v}) = 0$.
   b. $(\nabla \times \boldsymbol{v}) \times \boldsymbol{v} = \boldsymbol{v} \cdot \nabla\boldsymbol{v} - \nabla\boldsymbol{v} \cdot \boldsymbol{v}$.
   c. $\nabla \cdot (\alpha\boldsymbol{v}) = \alpha(\nabla \cdot \boldsymbol{v}) + \boldsymbol{v} \cdot (\nabla\boldsymbol{v})$.

6. Given the equation, $\boldsymbol{B} = (tr\boldsymbol{A})\mathbf{I} + 3\mathbf{A}$, show that

   a. $tr(\boldsymbol{B}) = 6tr(\boldsymbol{A})$.
   b. $\mathbf{A} = \frac{1}{3}(\boldsymbol{B} - \frac{tr(\boldsymbol{B})}{6}\mathbf{I})$.

7. Compute the eigenvalues and eigenvectors of the matrix

   a. $\boldsymbol{A} = \begin{bmatrix} 8 & 3 & 0 \\ 3 & 8 & 0 \\ 0 & 0 & 1 \end{bmatrix}$.
   b. $\boldsymbol{B} = \begin{bmatrix} 5 & 3 & 2 \\ 3 & 4 & 7 \\ 2 & 7 & 1 \end{bmatrix}$.
   c. $\boldsymbol{C} = \begin{bmatrix} 1 & 1 & 2 \\ 1 & 2 & 1 \\ 2 & 1 & 1 \end{bmatrix}$.

8. Find the three invariants for the tensors given in Problem 7.
9. Show that the tensors in Problem 7 can be expressed as the sum of the eigenvalues multiplied by the dyadic product of the eigenvectors such that $\boldsymbol{A} = \sum_{i=1}^{3} \lambda_i \hat{\boldsymbol{n}}^{(i)} \otimes \hat{\boldsymbol{n}}^{(i)}$.
10. Using spectral decomposition, find the square root of the tensors in Problem 7.
11. Show that the Cayley-Hamilton theorem is valid for the tensors in Problem 7.

MATLAB® EXERCISES

12. Using Matlab® compute the determinant, eigenvalues, and eigenvectors for the matrices in Problem 7.
13. Using Matlab® find the inverse for each of the matrices in Problem 7.
14. Using Matlab® and taking the values of $\boldsymbol{A}$, $\boldsymbol{B}$, and $\boldsymbol{C}$ from Problem 7, compute

    a. $\boldsymbol{D} = \boldsymbol{A} \cdot \boldsymbol{B}$.
    b. $\boldsymbol{D} = \boldsymbol{B} \cdot \boldsymbol{A} \cdot \boldsymbol{B}^T$.

c. $\boldsymbol{D} = sqrt(\boldsymbol{A}) \cdot \boldsymbol{B}^2 \cdot \boldsymbol{C}^{-1}$.
d. $\boldsymbol{D} = (\boldsymbol{A} : \boldsymbol{B})\boldsymbol{C} + tr(\mathbf{A})\mathbf{I}$.

15. Given the values of $\boldsymbol{A}$, $\boldsymbol{B}$, and $\boldsymbol{C}$ in Problem 7, plot the scalar-valued scalar function $\alpha(\beta) = (\boldsymbol{A} : \boldsymbol{C})\beta + (\boldsymbol{A} \cdot \boldsymbol{B}) : \boldsymbol{C}$ for $0 \leq \beta \leq 10$.
16. Plot the gradient of the scalar-valued vector function $\alpha(\boldsymbol{x}) = sqrt(x_1 \cdot x_2)$ for $0 \leq x_1 \leq 1$ and $0 \leq x_2 \leq 1$.

2 Kinematics

Kinematics is the study of motion without regards to the forces responsible for that motion. Intuitively, we know that the application of a force can lead to the movement of an object. The equations characterizing this movement are called the equations of motion. Perhaps, we might compute the displacement of an object by measuring how far it has moved from its initial location. In this chapter, we will build on these intuitive concepts to explore the kinematics of deformable continua. We will show how simple geometric relations allow us to compute the deformation and strain from the equations of motion. Similarly, the velocity and acceleration fields may be determined by differentiation the equations of motion.

2.1 Configurations

We know that matter consists of atoms, which consist of protons, electrons, and neutrons, all of which consist of quarks. However, this level of detail can often be ignored when mathematically modeling a macroscopic object's response to external fields. The true discreet nature of material can be modeled as a continuous distribution of mass and the atomic or subatomic structure can be ignored. Within this representation, an object is no longer made up of a finite set of atoms each with its own mass or charge but instead consists of an infinite number of ***material points*** or ***particles***. Instead of defining atomic mass or charge, we define a density and a charge density field.

Let us consider a ***continuum body*** made up of a collection of material points. A ***configuration*** refers to the simultaneous position of all particles within the body. The ***reference configuration*** consists of the position of each material point, $\boldsymbol{X}$, which lies within the real continuous space $\Omega_o$which defines our object at some instance in time

$$\boldsymbol{X} \in \Omega_o$$

We will reserve the capital letter, $\boldsymbol{X}$, contrary to our naming convention, for the vector position of material points within the reference configuration. The reference configuration might be the configuration of the undeformed object at time $t = 0$ before forces have been applied, or it might a theoretical configuration that is never realized by the object. As the name implies, the reference configuration is the configuration relative to which displacement or deformation will be measured.

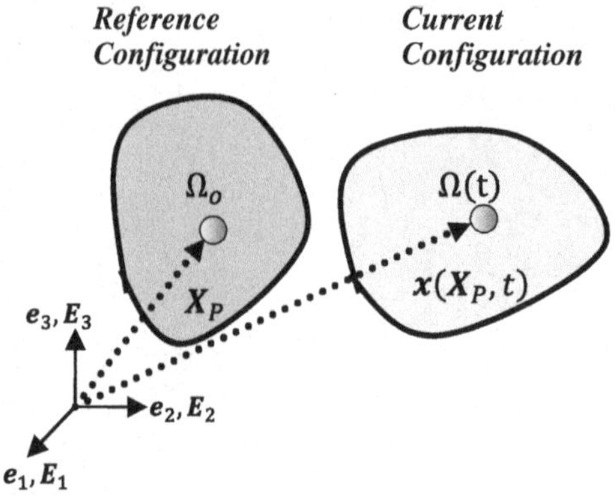

Figure 2.1. Illustration of the reference and current configuration of a continuum body.

When we apply an external force to the continuum body, the object ***deforms*** meaning the position of each material point may change. We need to determine the ***current configuration***, also referred to as the ***deformed configuration***, which is the new position, $\boldsymbol{x}$, of each of the material points originally occupying the reference configuration at the current time, t. This can be written as

$$\boldsymbol{x} = \kappa(\boldsymbol{X}, t). \tag{2.1}$$

This function is a ***deformation map***, which finds the position within the current configuration of the particle that originally occupied a given position within the reference configuration. If the deformation map in Equation (2.1) is a continuous function in time, it provides the ***equations of motion*** for the body. The equation need not be continuous in time since the deformation map may be a map between two configurations that are discontinuous in time.

Each material point within the reference configuration must uniquely occupy a position within the current configuration. Hence, there exists a one-to-one correspondence between the reference position of a material point and its current point. There must exist a mapping that determines the original reference position given the current position of the material particle and the current time

$$\boldsymbol{X} = \kappa^{-1}(\boldsymbol{x}, t). \tag{2.2}$$

This function can be used to find the position within the reference configuration of the material point which occupies a given position within the current configuration.

Continuum mechanics can be used to describe the response of gaseous, fluid, and solid systems. Whereas the underlying physics governing the behavior of these is the same, we usually formulate descriptions of solid systems using the ***Lagrangian description*** and for fluid systems we use an ***Eulerian description***. Within the Lagrangian description also known as the ***material description***, the reference position and the time are independent variables. The equations of motion are given in the form $\boldsymbol{x} = \kappa(\boldsymbol{X}, t)$, and the property fields describing the body are expressed as $\phi = \phi(\boldsymbol{X}, t)$. The material description is common in solid mechanics where the reference configuration is easily defined and material points are easily tracked. The position vectors in the reference configuration are referred to as ***material coordinates***, or ***referential coordinates***, and are measured with respect to the $\boldsymbol{E}_I$ basis vectors such that $\boldsymbol{X} = X_I \boldsymbol{E}_I$. Uppercase letters are reserved for indices denoting material coordinates.

The Eulerian description also known as the ***spatial description*** is commonly used in fluid mechanics where a reference configuration is difficult to describe. In addition, following specific material points is not as important as understanding the instantaneous motion of material points. The current position and the time are the independent variables in the Eulerian description. The property fields describing the body are expressed as $\phi = \phi(\boldsymbol{x}, t)$. A description of the motion of the continuum body is commonly provided or experimentally determined in the form of a velocity field, $\boldsymbol{v}(\boldsymbol{x}, t)$, defined at each position within the current reference frame. The position vectors in the spatial coordinate system are given by $\boldsymbol{x} = x_i \boldsymbol{e}_i$. Lowercase letters are reserved for indices denoting spatial coordinates. In most cases, the material coordinate frame and spatial coordinate frame will be coincide such that $\boldsymbol{E}_i = \boldsymbol{e}_i$.

EXAMPLE 2.1. *Given the equations of motion*

$$x_1(\boldsymbol{X}, t) = \alpha X_1 + \beta X_2,$$

$$x_2(\boldsymbol{X}, t) = \beta X_1 + \alpha X_2,$$

$$x_3(\boldsymbol{X}, t) = X_3,$$

find the inverse relation, $\boldsymbol{X} = f^{-1}(\boldsymbol{x}, t)$.

Solution:
We can determine the X_2 component by subtracting αx_2 from βx_1 such that

$$\beta x_1 - \alpha x_2 = \alpha\beta X_1 + \beta^2 X_2 - \alpha\beta X_1 - \alpha^2 X_2$$

$$X_2\left(\beta^2 - \alpha^2\right) = \beta x_1 - \alpha x_2$$

$$X_2 = \frac{\beta x_1 - \alpha x_2}{\left(\beta^2 - \alpha^2\right)}.$$

Similarly, we can determine the X_1 by subtracting βx_2 from αx_1

$$\alpha x_1 - \beta x_2 = \alpha^2 X_1 + \alpha\beta X_2 - \beta^2 X_1 - \alpha\beta X_2$$

$$X_1\left(\alpha^2 - \beta^2\right) = \alpha x_1 - \beta x_2$$

$$X_1 = \frac{\alpha x_1 - \beta x_2}{\left(\alpha^2 - \beta^2\right)}.$$

Finally, X_3 is given by

$$X_3 = x_3$$

Written together, this set of equations becomes

$$X_1(\boldsymbol{x}, t) = \frac{\alpha x_1 - \beta x_2}{\left(\alpha^2 - \beta^2\right)},$$

$$X_2(\boldsymbol{x}, t) = \frac{\beta x_1 - \alpha x_2}{\left(\beta^2 - \alpha^2\right)},$$

$$X_3(\boldsymbol{x}, t) = x_3.$$

2.2 Velocity and Acceleration

Any property field including the velocity may be expressed using a material or spatial description. However, the velocity, $\boldsymbol{v}_P$, of a material point, $\boldsymbol{X}_p$, is defined as the time rate of change of its current position, $\boldsymbol{x}(\boldsymbol{X}_p,t)$, such that

$$\boldsymbol{v}_P(\boldsymbol{X},t) = \frac{\partial \boldsymbol{x}(\boldsymbol{X}_P,t)}{\partial t}. \tag{2.3}$$

This equation provides a material description of the velocity vector with the reference position and time as the independent variables.

Similarly, the acceleration is given by the second derivative of the current position with respect to time such that

$$\boldsymbol{a}_P(\boldsymbol{x},t) = \frac{\partial^2 \boldsymbol{x}(\boldsymbol{X}_P,t)}{\partial t^2}. \tag{2.4}$$

Imagine that we have a solid which is being displaced and is possibly deforming with time as shown in Figure 2.2. We identify a specific material point, $\boldsymbol{X}_P$, by painting a small dot at its location within the reference configuration. The ***particle trajectory*** is a trace of the current position of the material point $\boldsymbol{X}_P$ as a function of time. The trajectory is that found by plotting the following equation:

$$\boldsymbol{x}_p = \boldsymbol{x}(\boldsymbol{X}_p,t)$$

for times $t = 0$ to t_2. We can see that for a solid material subject to small deformations, following the motion of particular material points makes sense since material points are mostly confined to a limited region of space, and their relative positions do not change much.

However, in the case of fluids, the spatial description may be more useful.Take the case of fluid flowing through a pipe (Figure 2.3). We might be more interested in determining the velocity of the fluid that is currently at the outlet of the pipe than we are in following the motion of a particular fluid particle. In this case, we know the spatial location, $\boldsymbol{x}$, where we need the velocity as a function of time. From instance to instance, the material point occupying this spatial location is different. In order to

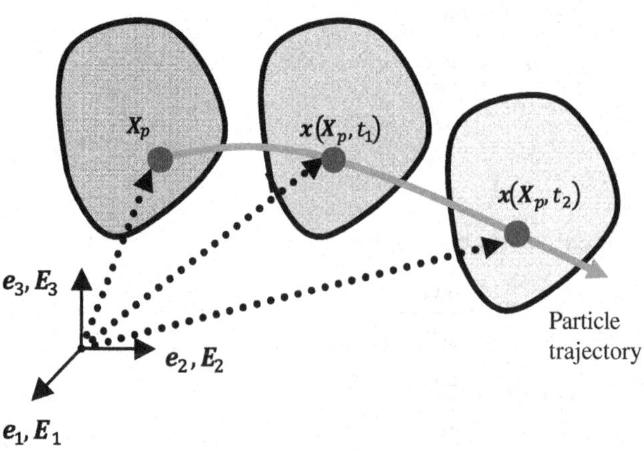

Figure 2.2. Illustration of an object in motion. The position of a material point, X_p, is shown for three distinct configurations at time $t = 0, t_1$, and t_1.

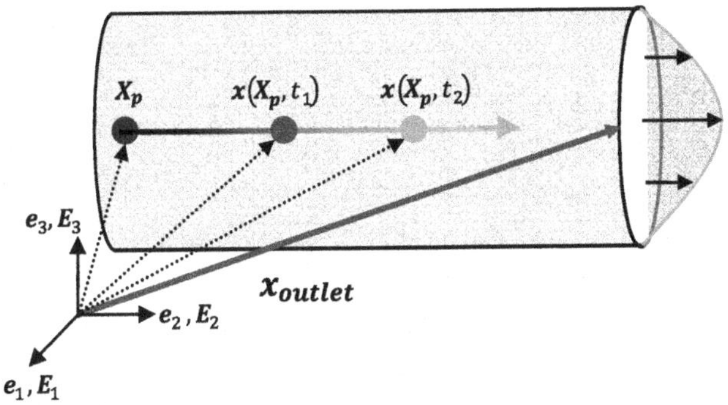

Figure 2.3. Illustration of fluid flow through a pipe.

find the spatial description of the velocity, we can substitute the inverted equations of motion, Equation (1.2), into the velocity field determined in Equation (1.3) after taking the time derivative such that

$$\boldsymbol{v}(\boldsymbol{x},t) = \left.\frac{\partial \boldsymbol{x}(\boldsymbol{X}_P,t)}{\partial t}\right|_{\boldsymbol{X}_P=\kappa^{-1}(\boldsymbol{x},t)}. \tag{2.5}$$

In this case, we have passed the reference position $\boldsymbol{X}_P = \kappa^{-1}(\boldsymbol{x},t)$ of the material point, which is currently located at $\boldsymbol{x}$ at time t into the velocity field.

Similarly, the spatial acceleration can be found by taking the material acceleration field and substituting the inverted equations of motion such that

$$\boldsymbol{a}(\boldsymbol{x},t) = \left.\frac{\partial^2 \boldsymbol{x}(\boldsymbol{X}_P,t)}{\partial t^2}\right|_{\boldsymbol{X}_P=\kappa^{-1}(\boldsymbol{x},t)}. \tag{2.6}$$

EXAMPLE 2.2. *A two-dimensional unit square is deformed according to the following equations of motion:*

$$x_1(\boldsymbol{X},t) = X_1 + 2t^2 X_2,$$
$$x_2(\boldsymbol{X},t) = 2t^2 X_1 + X_2.$$

1. *Draw the deformed unit square at $t = 0.5$ seconds.*
2. *Plot the material velocity field, $\boldsymbol{v}(\boldsymbol{X},t)$ at $t = 0.5$ seconds.*
3. *Plot the spatial velocity field, $\boldsymbol{v}(\boldsymbol{X},t)$ at $t = 0.5$ seconds.*

Solution:
Part 1

The positions of each of the points within the unit square move in time. At $t = 2$ s, the new positions are given by

$$x_1(\boldsymbol{X}, 0.5\,\mathrm{s}) = X_1 + 0.5X_2,$$
$$x_2(\boldsymbol{X}, 0.5\,\mathrm{s}) = 0.5X_1 + X_2.$$

We can plot both the reference configuration and the deformed configuration using Matlab®.

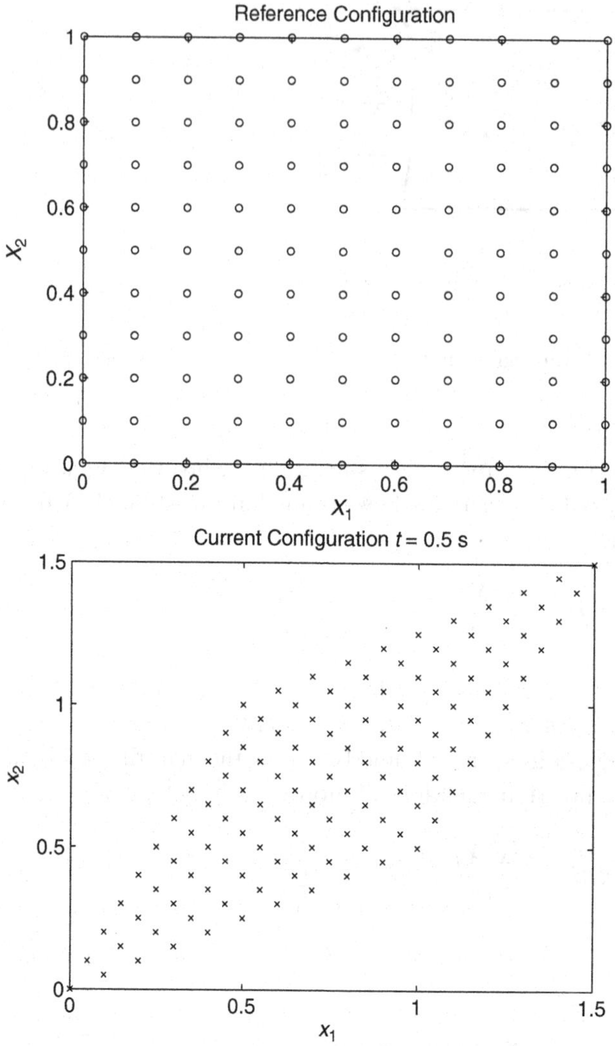

Figure 2.4. (Top) The positions of a set of reference points within the continuum body at $t = 0$. (Bottom) The positions of the reference points at $t = 0.5$ s.

```
>> [X1, X2] = meshgrid([0:0.1:1]);   % Create a grid of material points
>> plot(X1,X2, `ro');                % Reference configuration
>> x1 = X1 + 0.5*X2;                 % Calculate new position of each material point
>> x2 = 0.5*X1 + X2;
>> plot(x1, x2, `x', `bx');          % Plot current configuration positions
```

The reference and current configurations are shown below.

Part 2

The velocity is given by Equation (2.3). Taking the time derivative of the equations of motion gives

$$v_1(\boldsymbol{X}, t) = 4tX_2,$$

$$v_2(\boldsymbol{X}, t) = 4tX_1.$$

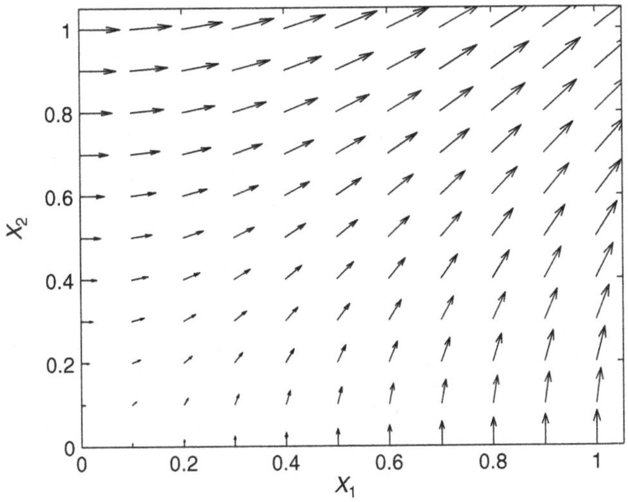

Figure 2.5. Material velocity field at $t = 0.5$ s.

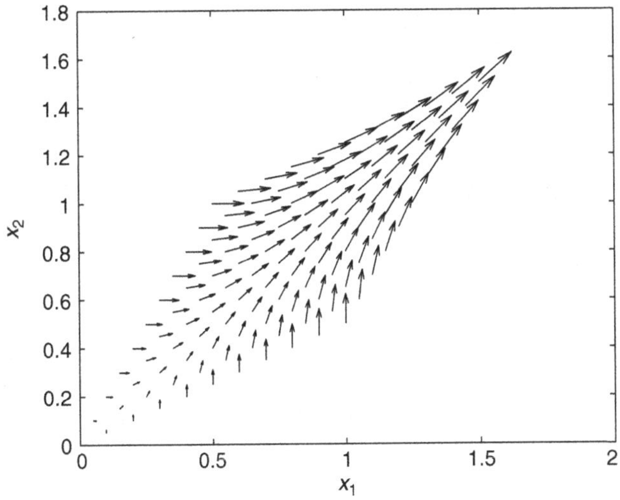

Figure 2.6. Spatial velocity field at $t = 0.5$ s.

Substituting the value $t = 2$ s gives

$$v_1(\boldsymbol{X}, 2s) = 2X_2,$$
$$v_2(\boldsymbol{X}, 2s) = 2X_1.$$

We may visualize this velocity field using Matlab®.

```
>> [X1, X2] = meshgrid([0:0.1:1]);    % Create a grid of material points
>> v1 = 2*X2;                         % Compute the velocity for each material point
>> v2 = 2*X1;
>> quiver(X1,X2, v1, v2);             % Plot the vector field
```

The resulting velocity field is shown in the figure below.

Part 3

Inverting the equations of motion gives

$$X_1 = \frac{x_1 - 2t^2 x_2}{1 - 4t^4},$$

$$X_2 = \frac{x_2 - 2t^2 x_1}{1 - 4t^4}.$$

The spatial description of the velocity can be found using Equation (2.5) as

$$v_1(\boldsymbol{x}(\boldsymbol{X},t),t) = \frac{4t(x_2 - 2t^2 x_1)}{1 - 4t^4},$$

$$v_2(\boldsymbol{x}(\boldsymbol{X},t),t) = \frac{4t(x_1 - 2t^2 x_2)}{1 - 4t^4}.$$

Substituting $t = 0.5$ s gives

$$v_1(\boldsymbol{x}(\boldsymbol{X},t),t) = \frac{8}{3}\left(x_2 - 2t^2 x_1\right)$$

$$v_2(\boldsymbol{x}(\boldsymbol{X},t),t) = \frac{8}{3}\left(x_1 - 2t^2 x_2\right)$$

This vector field may be plotted in Matlab® using the following code:

```
>> [X1, X2] = meshgrid([0:0.1:1]);
>> x1 = X1 + 0.5*X2;
>> x2 = 0.5*X1 + X2;
>> v1 = 8/3*(x2 - 0.5*x1);
>> v2 = 8/3*(x1 - 0.5*x2);
>> quiver(x1, x2, v1, v2);
```

The resulting velocity field is shown in the figure below.

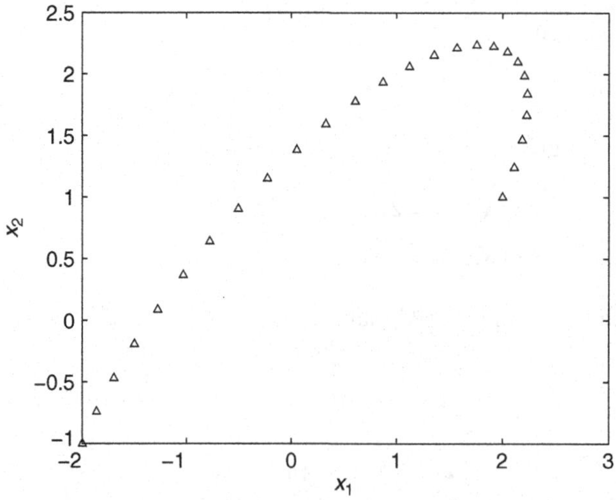

Figure 2.7. Illustration of the particle trajectory for $0\,\text{s} < t < 0.5\,\text{s}$.

EXAMPLE 2.3. *Plot the trajectory of the material particle* $\boldsymbol{X} = \langle 1,2 \rangle$ *for times* $t = 0$ s *to* $t = 0.5$ s *given the equations of motion*

$$x_1(\boldsymbol{X},t) = \sin(2\pi t) X_1 + \cos(2\pi t) X_2,$$
$$x_2(\boldsymbol{X},t) = \cos(2\pi t) X_1 + \sin(2\pi t) X_2.$$

Solution:
The trajectory can be found by substituting the material point into the equations of motion to give

$$x_1(\langle 1,2 \rangle,t) = \sin(2\pi t) + 2\cos(2\pi t),$$
$$x_2(\langle 1,2 \rangle,t) = \cos(2\pi t) + 2\sin(2\pi t).$$

These are now the equations of motion for a particular material point. The trajectory can be plotted using Matlab®.

```
>> t = [0:0.02:0.5];
>> x1 = sin(2*pi*t)+2*cos(2*pi*t);
>> x2 = cos(2*pi*t)+2*sin(2*pi*t);
>> plot(x1, x2);
```

The resulting trajectory is shown in the following figure.

2.3 Displacement

Imagine that within a continuum body, you identify a material point, $\boldsymbol{X}_p$, in the reference configuration as shown in Figure 2.8. With the application of a external force, the body deforms and the material point moves to a new location. The material point has been displaced. The ***displacement*** of this material point is the current position, $\boldsymbol{x}(\boldsymbol{X}_P,t)$, minus the original position, $\boldsymbol{X}_P$. This gives the equation

$$\boldsymbol{u}(\boldsymbol{X}_P,t) = \boldsymbol{x}(\boldsymbol{X}_P,t) - \boldsymbol{X}_P. \tag{2.7}$$

Notice that this gives a ***material description*** of the displacement vector field.

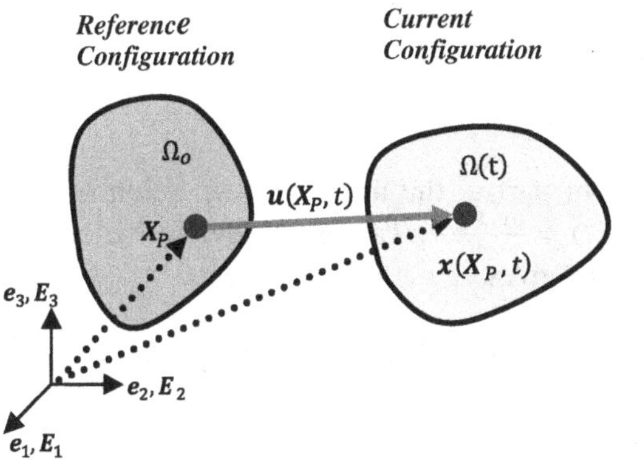

Figure 2.8. Illustration of the displacement of a material point.

The spatial description of the displacement field will give the displacement of the particle that currently occupies a particular spatial point, $\boldsymbol{x}_p$, as a function of time. In this case, we identify a location in space, $\boldsymbol{x}_p$, and determine the reference position of that particle, $\boldsymbol{X}\left(\boldsymbol{x}_p,t\right)$. The displacement of the particle which currently occupies that location is given by

$$\boldsymbol{u}\left(\boldsymbol{x}_p,t\right)=\boldsymbol{x}_p-\boldsymbol{X}\left(\boldsymbol{x}_p,t\right). \tag{2.8}$$

EXAMPLE 2.4. **One-dimensional stretch of a cube.** *A unit cube is stretched homogenously along the* $\mathbf{e}_1$ *axis as shown in the figure. The dimensions in the* $\mathbf{e}_2$*, and* $\mathbf{e}_3$ *directions remain unchanged.*

1. *Find the current position,* $\boldsymbol{x}$*, of a material point,* $\boldsymbol{X}$*, at time t.*
2. *Find the material position,* $\boldsymbol{X}$*, of the particle which currently occupies the spatial position* $\boldsymbol{x}$ *at time t.*
3. *Find the material description of the velocity field as a function of time.*
4. *Find the spatial description of the velocity field as a function of time.*
5. *Find the material description of the displacement field as a function of time.*
6. *Find the spatial description of the displacement field as a function of time.*

Solution:

1. Since there is no change in length along 2 or 3 axis, the x_2 and x_3 position of each particle will not change. The new x_1 position must be determined from the stretch. Let us look at two limiting cases. A material point that initially has $X_1=0$ will remain on the left face of the cube and not move during the deformation. However, a point that initially has $X_1=1$ will move along with the right face of the cube. The motion of the right face is given as $L_1(t)=1+a^2t^2$. The equations of motion become

$$x_1(\boldsymbol{X},t)=X_1\left(1+a^2t^2\right),$$
$$x_2(\boldsymbol{X},t)=X_2,$$
$$x_3(\boldsymbol{X},t)=X_3.$$

2. We invert the equations of motion to find

$$X_1(\boldsymbol{x},t)=\frac{x_1}{\left(1+a^2t^2\right)},$$
$$X_2(\boldsymbol{x},t)=x_2,$$
$$X_3(\boldsymbol{x},t)=x_3.$$

3. Given the displacement field from part a, the material description of the velocity is found by using $\boldsymbol{v}_P(\boldsymbol{X},t)=\frac{\partial \boldsymbol{x}(\boldsymbol{X}_P,t)}{\partial t}$. Taking the derivative of the equations of motion found in part a gives

$$v_1(\boldsymbol{X},t)=2a^2X_1t,$$
$$v_2(\boldsymbol{X},t)=0,$$
$$v_3(\boldsymbol{X},t)=0.$$

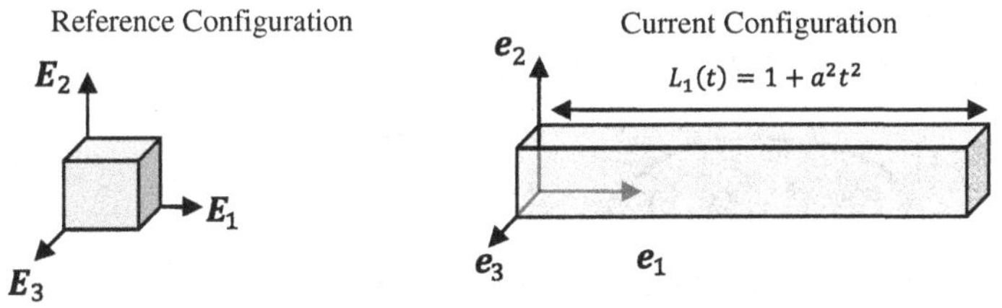

Figure 2.9. Illustration of the reference and current configuration for a cube being stretched along the e_1 direction.

4. The spatial description of the velocity field is found by substituting the inverse equations in part b into the velocity found in part c giving

$$v_1(\boldsymbol{x},t) = \left(\frac{2a^2tx_1}{1+a^2t^2}\right),$$
$$v_2(\boldsymbol{x},t) = 0,$$
$$v_3(\boldsymbol{x},t) = 0.$$

5. The displacement of a material point is given by $\boldsymbol{u}(\boldsymbol{X},t) = \boldsymbol{x}(\boldsymbol{X},t) - \boldsymbol{X}$. We have already found the function $\boldsymbol{x}(\boldsymbol{X},t)$ in part a. Therefore, we can write the following:

$$u_1(\boldsymbol{X},t) = X_1\left(1+a^2t^2\right) - X_1 = X_1a^2t^2,$$
$$u_2(\boldsymbol{X},t) = X_2 - X_2 = 0,$$
$$u_3(\boldsymbol{X},t) = X_3 - X_3 = 0.$$

6. The spatial displacement field is given by $\boldsymbol{u}(\boldsymbol{X},t) = \boldsymbol{x} - \boldsymbol{X}(\boldsymbol{x},t)$. We have found $\boldsymbol{X}(\boldsymbol{x}t)$ in part b. The spatial displacement field becomes

$$u_1(\boldsymbol{x},t) = x_1 - \frac{x_1}{1+a^2t^2} = \frac{x_1a^2t^2}{1+a^2t^2},$$
$$u_2(\boldsymbol{x},t) = x_2 - x_2 = 0,$$
$$u_3(\boldsymbol{x},t) = x_3 - x_3 = 0.$$

2.4 Deformation Gradient

The ***deformation gradient*** is a measure of the local relative displacement occurring between material points. More specifically, the deformation gradient transforms ***material line elements*** into ***spatial line elements***. A material line element, $\boldsymbol{dX}$, is a vector that connects two *infinitesimally* close points, P and Q, within the reference configuration such that $\boldsymbol{dX}_{QP} = \boldsymbol{X}_Q - \boldsymbol{X}_P$. A spatial line element, $\boldsymbol{dx}$, is the vector which connects points P and Q in the current configuration such that $\boldsymbol{dx}_{QP} = \boldsymbol{x}\left(\boldsymbol{X}_Q,t\right) - \boldsymbol{x}(\boldsymbol{X}_P,t)$. We can write the current position of point Q as the current position of point P, $\boldsymbol{x}(\boldsymbol{X}_P,t)$, plus the spatial line element, $\boldsymbol{dx}_{QP}$, such that

$$\boldsymbol{x}\left(\boldsymbol{X}_Q,t\right) = \boldsymbol{x}(\boldsymbol{X}_P,t) + \boldsymbol{dx}_{QP}.$$

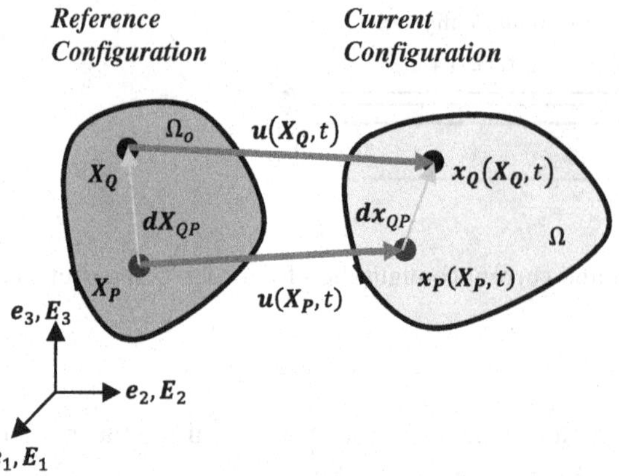

Figure 2.10. Illustration of a material and spatial line element.

Similarly, we may write the current position of point Q as the current position of the reference point $\boldsymbol{X}_P + \boldsymbol{dX}_{QP}$ such that

$$\boldsymbol{x}(\boldsymbol{X}_Q,t) = \boldsymbol{x}(\boldsymbol{X}_P + \boldsymbol{dX}_{QP},t).$$

Setting these two representations equation gives

$$\boldsymbol{x}(\boldsymbol{X}_P,t) + \boldsymbol{dx}_{QP} = \boldsymbol{x}(\boldsymbol{X}_P + \boldsymbol{dX}_{QP},t).$$

Performing a Taylor expansion of the function on the right side of the equation gives

$$\boldsymbol{x}(\boldsymbol{X}_P,t) + \boldsymbol{dx}_{QP} = \boldsymbol{x}(\boldsymbol{X}_P,t) + \left.\frac{\partial \boldsymbol{x}(\boldsymbol{X},t)}{\partial \boldsymbol{X}}\right|_{\boldsymbol{X}_P} \cdot \boldsymbol{dX}_{QP} + \mathrm{O}\left(|\boldsymbol{dX}_{QP}|^2\right).$$

The higher order, $\mathrm{O}\left(|\boldsymbol{dX}_{QP}|^2\right)$, terms can be neglected because we are considering an infinitesimal material line element which has an infinitesimally small magnitude $|\boldsymbol{dX}_{QP}| \ll 1$. Canceling the current position of point P gives the resulting relation between the material and spatial line elements as

$$\boldsymbol{dx}_{QP} = \left.\frac{\partial \boldsymbol{x}(\boldsymbol{X},t)}{\partial \boldsymbol{X}}\right|_{\boldsymbol{X}_P} \cdot \boldsymbol{dX}_{QP}$$

The deformation gradient is defined as the second-order tensor given by

$$\boldsymbol{F}(\boldsymbol{X},t) \equiv \frac{\partial \boldsymbol{x}(\boldsymbol{X},t)}{\partial \boldsymbol{X}} = \frac{\partial x_i}{\partial X_J} \boldsymbol{e}_i \otimes \boldsymbol{E}_J. \tag{2.9}$$

As stated earlier, the lowercase index, i, refers to the spatial Cartesian coordinate and the capital index, J, refers to the material Cartesian coordinate. The deformation gradient is a ***two-point tensor*** due to the fact the tensor is formed from dyadic products between basis vectors in the current configuration and basis vectors in the reference configuration. It transforms material line elements from the reference coordinate system into line elements in the current coordinate system. Notice that the deformation gradient is the gradient of the equations of motion with respect to the reference Cartesian coordinates such that

$$\boldsymbol{F}(\boldsymbol{X},t) = \boldsymbol{x}(\boldsymbol{X},t) \otimes \overleftarrow{\nabla}_X = Grad(\boldsymbol{x}(\boldsymbol{X},t)).$$

The capital *Grad* is reserved for the gradient taken with respect to the material coordinates, X_I, whereas the lowercase *grad* is reserved for the gradient taken with respect to the spatial coordinates, x_i. Written compactly, the deformation gradient and its transpose are given by

$$\boldsymbol{F} = \boldsymbol{x} \overleftarrow{\nabla}_X \quad \text{and} \quad \boldsymbol{F}^T = \overrightarrow{\nabla}_X \boldsymbol{x}.$$

The spatial line element, $\boldsymbol{dx}$, can be related to the material line element, $\boldsymbol{dX}$, according to

$$\boldsymbol{dx} = \boldsymbol{F} \cdot \boldsymbol{dX} \quad \text{and} \quad \boldsymbol{dX} = \boldsymbol{F}^{-1} \cdot \boldsymbol{dx}. \tag{2.10}$$

The inverse of the deformation gradient is given by

$$\boldsymbol{F}^{-1}(\boldsymbol{x},t) = \frac{\partial \boldsymbol{X}}{\partial \boldsymbol{x}} = \frac{\partial X_I}{\partial x_j} \boldsymbol{E}_J \otimes \boldsymbol{e}_i. \tag{2.11}$$

Therefore, the inverse of the deformation gradient is given by the spatial gradient of the inverse mapping such that

$$\boldsymbol{F}^{-1}(\boldsymbol{x},t) = \boldsymbol{X}(\boldsymbol{x},t) \otimes \overleftarrow{\nabla}_x = grad\,(X(x,t)).$$

Rearranging Equation (2.7), the relation between the current position and the displacement can be written as

$$\boldsymbol{x}(\boldsymbol{X},t) = \boldsymbol{u}(\boldsymbol{X},t) + \boldsymbol{X}.$$

Taking the gradient of this expression gives the deformation gradient in terms of the displacement field as

$$\boldsymbol{F} = (\boldsymbol{u} + \boldsymbol{X}) \overleftarrow{\nabla}_X = \boldsymbol{u} \overleftarrow{\nabla}_X + \mathbf{I}. \tag{2.12}$$

Similarly, the transpose of the deformation gradient may be written as

$$\boldsymbol{F}^T = \overrightarrow{\nabla}_X \boldsymbol{u} + \mathbf{I}. \tag{2.13}$$

The inverse of the deformation gradient and its transpose are given by

$$\boldsymbol{F}^{-1} = \overrightarrow{\nabla}_x \boldsymbol{u} + \mathbf{I} \quad \text{and} \quad \boldsymbol{F}^{-T} = \boldsymbol{u} \overleftarrow{\nabla}_x + \mathbf{I}. \tag{2.14}$$

EXAMPLE 2.5. *Given the following equations of motion for a deformable object,*

$$x_1(\boldsymbol{X},t) = (t+1)^2 X_1 + t X_2,$$
$$x_2(\boldsymbol{X},t) = t^2 X_1 + (t+1) X_2,$$
$$x_3(\boldsymbol{X},t) = X_3 + t X_3^2.$$

for $t \geq 0$, find the deformation gradient.

Solution:

The components of the deformation gradient, $F_{ij} = \frac{\partial x_i}{\partial X_j}$, are given by

$$\begin{array}{lll} F_{11} = x_{1,1} = (t+1)^2, & F_{12} = x_{1,2} = t, & F_{13} = x_{1,3} = 0, \\ F_{21} = x_{2,1} = t^2, & F_{22} = x_{2,2} = (t+1), & F_{23} = x_{2,3} = 0, \\ F_{31} = x_{3,1} = 0, & F_{32} = x_{3,2} = 0, & F_{33} = x_{3,3} = 1 + 2tX_3. \end{array}$$

In matrix form, this becomes

$$\boldsymbol{F} = \begin{bmatrix} (t+1)^2 & t & 0 \\ t^2 & (t+1) & 0 \\ 0 & 0 & 1+2tX_3 \end{bmatrix}.$$

Notice that for $t = 0$, we obtain the initial shape of the object and the deformation gradient becomes the identity. This is our reference configuration, and deformation is measured relative to this configuration. Also, the deformation is heterogeneous since F_{33} depends on the position, X_3, within the object.

EXAMPLE 2.6. *For a two-dimensional unit square, with the equations of motion,* $\boldsymbol{x}(\boldsymbol{X},t)$*, given by*

$$x_1(\boldsymbol{X},t) = X_1 + 2tX_1X_2^2,$$
$$x_2(\boldsymbol{X},t) = X_2.$$

1. *Find the deformation gradient and its inverse. Write the inverse in terms of the spatial coordinates.*
2. *Draw the deformed object at* $t = 1$ *second.*
3. *An infinitesimal material line element with magnitude* dS_o *at a position,* $\boldsymbol{X}^a = (1,1)$*, has the direction* $\left(\frac{1}{\sqrt{5}}, \frac{2}{\sqrt{5}}\right)$*. What is its direction and magnitude in the current configuration at time* $t = 0.5$ *seconds and at* $t = 1$ *seconds?*
4. *An infinitesimal spatial line element with magnitude* dS *at a position,* $\boldsymbol{x}^b = (1,0.5)$*, at time* $t = 2$ *second, has direction* $\left(\frac{1}{\sqrt{5}}, \frac{2}{\sqrt{5}}\right)$*. What was its direction and magnitude in the reference configuration?*

Solution:

1. The deformation gradient is found using Equation (2.9) giving

$$\boldsymbol{F}(\boldsymbol{X},t) = \begin{bmatrix} 1+2tX_2^2 & 4tX_1X_2 \\ 0 & 1 \end{bmatrix}.$$

   Again we notice that at $t = 0$ s, the deformation becomes the identity. The deformation is heterogeneous since it depends on the position within the object.
   The inverse of the deformation gradient is given by

$$\boldsymbol{F}^{-1}(\boldsymbol{X},t) = \frac{1}{1+2tX_2^2}\begin{bmatrix} 1 & -4tX_1X_2 \\ 0 & 1+2tX_2^2 \end{bmatrix}.$$

   Inverting the equations of motion gives $X_2 = x_2$, and $X_1 = x_1/(1+2tx_2^2)$. Substituting into the inverse of the deformation gradient gives the following spatial representation

$$\boldsymbol{F}^{-1}(\boldsymbol{x}(\boldsymbol{X},t),t) = \frac{1}{1+2tx_2^2}\begin{bmatrix} 1 & -\dfrac{4tx_1x_2}{1+2tx_2^2} \\ 0 & 1+2tx_2^2 \end{bmatrix}.$$

2. The reference shape is the unit square. The deformed shapes may be drawn by determining the current position of the material points within the unit

square. This gives a rhombus at $t = 0.5$ seconds and $t = 1.0$ seconds. For clarity, the points $\boldsymbol{X}^a$ and $\boldsymbol{x}^b$ are also shown.

3. The equation of the line in the reference configuration is given by

$$\boldsymbol{dX} = dS_o \begin{bmatrix} \frac{1}{\sqrt{5}} \\ \frac{2}{\sqrt{5}} \end{bmatrix}.$$

The direction and magnitude of this line in the current configuration at time $t = 0.5$ seconds is given by

$$\boldsymbol{dx} = \boldsymbol{F}\left(\boldsymbol{X}^a, t\right) \cdot \boldsymbol{dX}.$$

Inserting the position of the line and the time gives

$$\boldsymbol{dx} = dS_o \begin{bmatrix} 2 & 2 \\ 0 & 1 \end{bmatrix} \begin{bmatrix} \frac{1}{\sqrt{5}} \\ \frac{2}{\sqrt{5}} \end{bmatrix}.$$

Carrying out the matrix multiplication gives

$$\boldsymbol{dx} = dS_o \begin{bmatrix} \frac{6}{\sqrt{5}} \\ \frac{2}{\sqrt{5}} \end{bmatrix}.$$

The magnitude of the line in the current configuration is given by

$$\left|\boldsymbol{dx}\right| = 2\sqrt{2}\, dS_o.$$

And the current direction is given by

$$\frac{\boldsymbol{dx}}{\left|\boldsymbol{dx}\right|} = \begin{bmatrix} \frac{3}{\sqrt{10}} \\ \frac{1}{\sqrt{10}} \end{bmatrix}.$$

Notice that the vector in the current configuration is more than twice the length of the original vector in the reference configuration and the direction of the infinitesimal line element has changed.

4. The equation of the line in the current configuration is given by

$$\boldsymbol{dx} = dS \begin{bmatrix} \frac{1}{\sqrt{5}} \\ \frac{2}{\sqrt{5}} \end{bmatrix}.$$

The direction and magnitude of this line in the reference configuration is given by

$$\boldsymbol{dX} = \boldsymbol{F}^{-1}\left(\boldsymbol{x}^b, t\right) \cdot \boldsymbol{dx}.$$

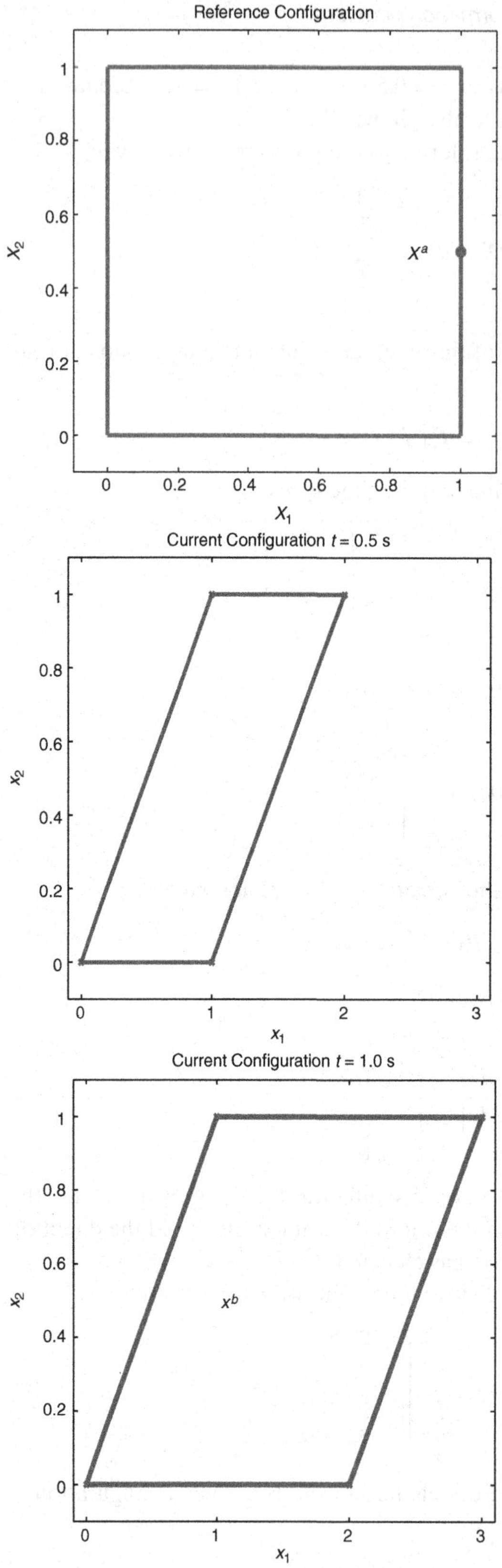

Figure 2.11. Illustration showing the reference configuration (top), current configuration at $t = 0.5$s (center), and the current configuration at $t = 1.0$s (bottom).

Written in terms of the inverse deformation gradient found earlier gives

$$dX = dS\frac{1}{1+2tx_2^2}\begin{bmatrix} 1 & -\dfrac{4tx_1x_2}{1+2tx_2^2} \\ 0 & 1+2tx_2^2 \end{bmatrix}\begin{bmatrix} \dfrac{1}{\sqrt{5}} \\ \dfrac{2}{\sqrt{5}} \end{bmatrix}.$$

Substituting time and position into this equation gives

$$dX = dS\begin{bmatrix} \dfrac{1}{2} & -1 \\ 0 & 1 \end{bmatrix}\begin{bmatrix} \dfrac{1}{\sqrt{5}} \\ \dfrac{2}{\sqrt{5}} \end{bmatrix}.$$

Carrying out the matrix multiplication gives

$$dX = dS\begin{bmatrix} -\dfrac{3}{2\sqrt{5}} \\ \dfrac{2}{\sqrt{5}} \end{bmatrix}.$$

The magnitude of the line in the reference configuration is given by

$$|dX| = \frac{\sqrt{5}}{2}dS.$$

And the direction is given by

$$\frac{dX}{|dX|} = \begin{bmatrix} -\dfrac{3}{5} \\ \dfrac{4}{5} \end{bmatrix}.$$

We see both magnitude and direction have changed for this infinitesimal line element. The magnitude of the line is approximately 1.12 times larger in the reference than the current configuration.

2.5 Jacobian

As we deform an object, it may locally compress or dilate causing the local density to vary with time. The changes in local volume may be computed directly from the deformation gradient.

Let us consider an infinitesimal cube of material within the reference configuration, Figure 2.12. When we deform the object globally, this infinitesimal cube will change shape and orientation. The size and orientation of the cube can be defined by the three vectors defining its sides. If we specify that the sides of the cube are given in the reference configuration by the infinitesimally small vectors $dX^{(1)}$, $dX^{(2)}$, and $dX^{(3)}$, then the volume of the cube in the reference configuration, dV_X, is given by

$$dV_X = \left(dX^{(1)} \times dX^{(2)}\right) \cdot dX^{(3)}. \tag{2.15}$$

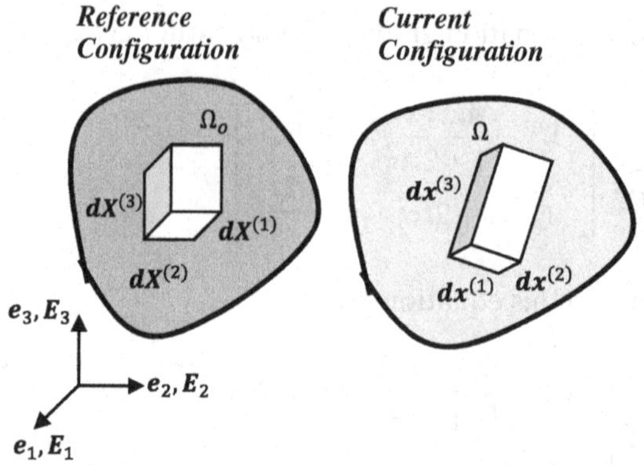

Figure 2.12. Change in volume of an infinitesimal volume element during deformation.

Due to the deformation, both the size and direction of the vectors defining the cube may change. The new volume of the cube in the current configuration, dV_x, is given by

$$dV_x = \left(\boldsymbol{dx}^{(1)} \times \boldsymbol{dx}^{(2)}\right) \cdot \boldsymbol{dx}^{(3)}, \tag{2.16}$$

where $\boldsymbol{dx}^{(1)}$, $\boldsymbol{dx}^{(2)}$, and $\boldsymbol{dx}^{(3)}$ are the vectors defining the edges of the cube in the current configuration. These infinitesimally small edge vectors in the reference and current configuration are related by Equation (2.10). Substituting $\boldsymbol{dx}^{(1)} = \boldsymbol{F} \cdot \boldsymbol{dX}^{(1)}$, $\boldsymbol{dx}^{(2)} = \boldsymbol{F} \cdot \boldsymbol{dX}^{(2)}$, and $\boldsymbol{dx}^{(3)} = \boldsymbol{F} \cdot \boldsymbol{dX}^{3}$ into Equation (2.16) gives

$$dV_x = \left(\left(\boldsymbol{F} \cdot \boldsymbol{dX}^{(1)}\right) \times \left(\boldsymbol{F} \cdot \boldsymbol{dX}^{(2)}\right)\right) \cdot \boldsymbol{F}\boldsymbol{dX}^{(3)}$$

Substituting the result of Example 1.9, $\det \boldsymbol{F} = \frac{\left(\left(\boldsymbol{F} \cdot \boldsymbol{dX}^{(1)}\right) \times \left(\boldsymbol{F} \cdot \boldsymbol{dX}^{(2)}\right)\right) \cdot \boldsymbol{F}\boldsymbol{dX}^{(3)}}{\left(\boldsymbol{dX}^{(1)} \times \boldsymbol{dX}^{(2)}\right) \cdot \boldsymbol{dX}^{(3)}}$, and using Equation (2.15), we obtain

$$\begin{aligned} dV_x &= \det \boldsymbol{F} \left(\boldsymbol{dX}^{(1)} \times \boldsymbol{dX}^{(2)}\right) \cdot \boldsymbol{dX}^{(3)} \\ &= \det \boldsymbol{F}\, dV_X. \end{aligned}$$

Therefore, the ratio of the current deformed volume to the original reference volume of the cube is given by the determinate of the deformation gradient

$$\frac{dV_x}{dV_X} = \det \boldsymbol{F}. \tag{2.17}$$

This ratio is known as the ***Jacobian***, *J*,

$$J\left(\boldsymbol{X}, t\right) \equiv \det \boldsymbol{F}\left(\boldsymbol{X}, t\right).$$

2.6 Nanson's Formula

Having related local current volume, dV_x, to the local reference volume, dV_X, in Equation (2.17), let us now relate changes in local surface area between the reference and current configurations. Let us define the local area on the surface of a continuum

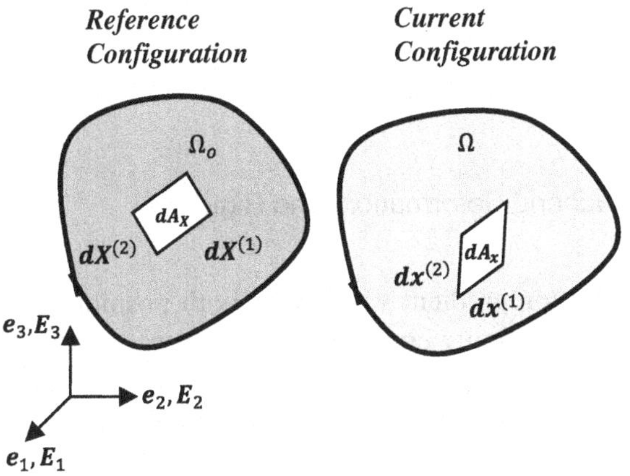

Figure 2.13. Change in area of an infinitesimal surface element during deformation.

body in the reference configuration, dA_X, as the area spanned by the two vectors $\boldsymbol{dX}^{(1)}$ and $\boldsymbol{dX}^{(2)}$ as shown in Figure 2.13. During deformation, the size and direction of these two vectors changes as does the enclosed area. In the current configuration, we will label the area of this surface element as dA_x.

Let us begin by recalling the results from the previous section. Combining Equations (2.17) and (2.16), we may write

$$\left(\boldsymbol{dx}^{(1)} \times \boldsymbol{dx}^{(2)}\right) \cdot \boldsymbol{dx}^{(3)} = J \left(\boldsymbol{dX}^{(1)} \times \boldsymbol{dX}^{(2)}\right) \cdot \boldsymbol{dX}^{(3)}.$$

Substituting Equation (2.10) for $\boldsymbol{dx}^{(3)}$ gives

$$\left(\boldsymbol{dx}^{(1)} \times \boldsymbol{dx}^{(2)}\right) \cdot \boldsymbol{F} \cdot \boldsymbol{dX}^{(3)} = J \left(\boldsymbol{dX}^{(1)} \times \boldsymbol{dX}^{(2)}\right) \cdot \boldsymbol{dX}^{(3)}. \tag{2.18}$$

Let us define the current oriented area element, $\boldsymbol{dA}_x$, and the reference oriented area element, $\boldsymbol{dA}_X$, as

$$\boldsymbol{dA}_x = dA_x \boldsymbol{n}_x = \boldsymbol{dx}^{(1)} \times \boldsymbol{dx}^{(2)}$$

and

$$\boldsymbol{dA}_X = dA_X \boldsymbol{n}_X = \boldsymbol{dX}^{(1)} \times \boldsymbol{dX}^{(2)},$$

where $\boldsymbol{n}_x$ and $\boldsymbol{n}_X$ are the unit vectors normal to the current area element and the reference area element respectively. Equation (2.18) can then be written as

$$\boldsymbol{dA}_x \cdot \boldsymbol{F} \cdot \boldsymbol{dX}^{(3)} = J\, \boldsymbol{dA}_X \cdot \boldsymbol{dX}^{(3)}. \tag{2.19}$$

Noting that $\boldsymbol{dA}_x$ is a vector, we may rearrange this equation using Equation (12.1) to write

$$\left(\boldsymbol{F}^T \cdot \boldsymbol{dA}_x\right) \cdot \boldsymbol{dX}^{(3)} = J\, \boldsymbol{dA}_X \cdot \boldsymbol{dX}^{(3)}. \tag{2.20}$$

Since Equation (2.20) must hold for all possible values of $\boldsymbol{dX}^{(3)}$, we obtain Nanson's formula as

$$\boldsymbol{F}^T \cdot \boldsymbol{dA}_x = J\, \boldsymbol{dA}_X \tag{2.21}$$

or equivalently

$$dA_x \boldsymbol{n}_x = J\, dA_X \boldsymbol{F}^{-T} \cdot \boldsymbol{n}_X. \tag{2.22}$$

2.7 Homogenous Deformation, Isochoric Deformation, and Rigid Body Rotation

In its most general form, the deformation gradient varies with both position and time, $\boldsymbol{F} = \boldsymbol{F}(\boldsymbol{X},t)$. There are several commonly encountered special cases where functional variation of the deformation is more limited.

Homogenous deformation is the term used to describe a deformation process in which the deformation gradient varies with time but does not vary with position within the object, $\boldsymbol{F} = \boldsymbol{F}(t)$. The deformation gradient has the same value at each point within the continuum body. In order to have homogenous deformation, the equations of motion for a deformation process must be reducible to the form

$$x(\boldsymbol{X},t) = \boldsymbol{F}(t) \cdot \boldsymbol{X} + \boldsymbol{c}(t),$$

where $\boldsymbol{c}(t)$ represents the vector translation of the entire body, and $\boldsymbol{F}(t)$ depends only on time. Notice that for a homogenous deformation, the inverse map can easily be determined as

$$X(\boldsymbol{x},t) = \boldsymbol{F}^{-1}(t) \cdot (\boldsymbol{x} - \boldsymbol{c}(t)).$$

The inverse mapping is much harder, and sometimes impossible to specify in a simple analytical form for nonhomogenous deformation.

Isochoric deformation or ***volume-preserving deformation*** is used to describe a deformation process in which the deformation field does not alter the local density (or local volume) of the material. Therefore, at any given point within the continuum body, the local volume in the reference configuration must equal the local volume in the current configuration, $dV_X = dV_x$. Any isochoric deformation process must have a Jacobian equal to one at all points within the material body such that

$$J(\boldsymbol{X},t) = \frac{dV_x}{dV_X} = 1.$$

There are deformation processes which by their nature do not change the local density of a material. For example, simple shear deformation which has a deformation gradient of the form

$$[F] = \begin{bmatrix} 1 & \gamma & 0 \\ 0 & 1 & 0 \\ 0 & 0 & 1 \end{bmatrix}$$

has a Jacobian equal to one. There are also cases where the material itself is assumed to be incompressible. If the material is incompressible, its density cannot change and only isochoric deformation is permissible. In the case of incompressible materials, the requirement that the Jacobian equal one becomes a constraint placed on the deformation field.

Rigid body motion involves the rotation and/or translation of an object with no internal deformation. Rigid body motion is both isochoric and homogenous. The generalized equations of motion for rigid body rotation are

$$\boldsymbol{x}(\boldsymbol{X},t) = \boldsymbol{Q}(t)\cdot\boldsymbol{X} + \boldsymbol{c}(t), \tag{2.23}$$

where $\boldsymbol{Q}$ is an orthogonal tensor, and $\boldsymbol{c}$ is the vector describing the translation of the center of mass of the object.

EXAMPLE 2.7. *Prove that for rigid body motion,* $\boldsymbol{Q}$ *must be an orthogonal tensor.*

Solution:
If a structure is rigid, the distance between any two points on the body remains constant We can state this mathematically as

$$|\boldsymbol{x}(\boldsymbol{X}_2,t) - \boldsymbol{x}(\boldsymbol{X}_1,t)| = |\boldsymbol{X}_2 - \boldsymbol{X}_1|.$$

Substituting the equations of motion into this relation gives

$$|\{\boldsymbol{Q}(t)\cdot\boldsymbol{X}_2 + \boldsymbol{c}(t)\} - \{\boldsymbol{Q}(t)\cdot\boldsymbol{X}_1 + \mathbf{c}(t)\}| = |\boldsymbol{X}_2 - \boldsymbol{X}_1|.$$

Simplifying the expression gives

$$|\boldsymbol{Q}\cdot(\boldsymbol{X}_2 - \boldsymbol{X}_1)| = |\boldsymbol{X}_2 - \boldsymbol{X}_1|.$$

The square magnitude of the vectors may be written as

$$(\boldsymbol{Q}\cdot(\boldsymbol{X}_2 - \boldsymbol{X}_1))\cdot(\boldsymbol{Q}\cdot(\boldsymbol{X}_2 - \boldsymbol{X}_1)) = (\boldsymbol{X}_2 - \boldsymbol{X}_1)\cdot(\boldsymbol{X}_2 - \boldsymbol{X}_1).$$

Using tensor identity 12.1, this equation becomes

$$((\boldsymbol{X}_2 - \boldsymbol{X}_1))\cdot\boldsymbol{Q}^T\boldsymbol{Q}\cdot(\boldsymbol{X}_2 - \boldsymbol{X}_1) = (\boldsymbol{X}_2 - \boldsymbol{X}_1)\cdot(\boldsymbol{X}_2 - \boldsymbol{X}_1).$$

This must be true for all points within the object measured using any coordinate system. The only way for this to be true is if $\boldsymbol{Q}^T\boldsymbol{Q} = \mathbf{I}$. This condition requires that $\boldsymbol{Q}$ is an orthogonal tensor. The deformation gradient for these equations of motion is $\boldsymbol{F} = \boldsymbol{Q}$.

2.8 Material and Spatial Derivatives

A material point or particle is an infinitesimal but identifiable segment of a continuum body. If we are interested in the properties of a particular material point, we could follow the motion of that particle by plotting the particle trajectory, or we could plot a property, such as the temperature, as a function of time for that particle. The ***material derivative*** also termed the ***convective derivative*** or ***substantial derivative*** is the rate of change of a property as recorded by an observer fixed to and traveling along with a material point. It is the rate of change experienced by this particular material point as it moves during deformation.

If the material description of a property field, $\phi(\boldsymbol{X},t)$, is provided, then the material derivative of this field, $\frac{D\phi}{Dt}$, is given by

$$\frac{D\phi(\boldsymbol{X},t)}{Dt} \equiv \dot{\phi}(\boldsymbol{X}t) \equiv \frac{\partial\phi(\boldsymbol{X},t)}{\partial t} \tag{2.24}$$

The material derivative is equal to the partial derivative of the material description of the field with respect to time while keeping the reference coordinate of the material point constant. Both the derivative with a capital D and the raised dot above the variable are reserved for the material derivative.

If a spatial description of a property field, $\phi(\boldsymbol{x},t)$, is provided, then the material derivative must be obtained by using the chain rule such that

$$\frac{D\phi(\boldsymbol{x},t)}{Dt} \equiv \frac{\partial\phi(\boldsymbol{x},t)}{\partial\boldsymbol{x}} \cdot \frac{d\boldsymbol{x}}{dt} + \frac{\partial\phi(\boldsymbol{x},t)}{\partial t}. \tag{2.25}$$

Recall that the special coordinate varies with time such that $\boldsymbol{x} = \boldsymbol{x}(\boldsymbol{X}, t)$. The material derivative of a field is taken while keeping the material coordinate constant not the spatial coordinate. Therefore, the implicit dependence of the spatial coordinates on time must be accounted for through the chain rule. Notice that Equation (2.25) gives a spatial description of the material derivative of a property field. We can convert this to a material description of the property field by substituting the equations of motion into special description such that

$$\frac{D\phi(\boldsymbol{X},t)}{Dt} \equiv \left[\frac{\partial\phi(\boldsymbol{x},t)}{\partial\boldsymbol{x}} \cdot \frac{d\boldsymbol{x}}{dt} + \frac{\partial\phi(\boldsymbol{x},t)}{\partial t}\right]_{\boldsymbol{x}=\boldsymbol{x}(\boldsymbol{X},t)}. \tag{2.26}$$

Practically speaking, if the equations of motion were available, one could convert the field to a material description before evaluating the material derivative. More commonly, the material derivative of a spatial field is computed due to a lack of equations of motion. We can also rewrite Equation (2.25) in terms of the gradient of the spatial property field and the velocity field as

$$\frac{D\phi(\boldsymbol{x},t)}{Dt} \equiv grad\,\phi(\boldsymbol{x},t) \cdot \boldsymbol{v}(\boldsymbol{x},t) + \frac{\partial\phi(\boldsymbol{x},t)}{\partial t}.$$

In contrast, the ***spatial derivative*** is the rate of change of a property as recorded by an observer fixed at a spatial coordinate. If you imagine fluid flowing through a pipe, the spatial derivative of the temperature at the outlet would be the temperature measured by an ideal thermocouple that was placed at the outlet. The spatial derivative gives the temperature change at that specific point in space regardless of the particle currently occupying the space. The spatial derivative of a property field given with a spatial description is simply

$$\phi' \equiv \frac{\partial\phi(\boldsymbol{x},t)}{\partial t} \tag{2.27}$$

The spatial derivative is the partial derivative of the field with respect to time while keeping the spatial position constant. Notice that the spatial derivative of the property field is the last term in Equation (2.25).

EXAMPLE 2.8. *The material description of the temperature field for a unit cube subject to the equations of motion*

$$x_1(\boldsymbol{X},t) = (t+1)^2 X_1,$$

$$x_2(\boldsymbol{X},t) = (t+1)^2 X_2,$$

$$x_3(\boldsymbol{X},t) = X_3,$$

has been found to be

$$\theta(\boldsymbol{X},t) = 1.1X_1 + 2.3tX_2 \ [^\circ\mathrm{C}].$$

1. *Find the material derivative of the temperature field for the material point,* $\boldsymbol{X}^P = (0.1,\, 0.1, 0)$.
2. *Find the time rate of change of the temperature of the particle which currently occupies the spatial position* $\boldsymbol{x}^p = (0.1, 0.1, 0)$ *at time t.*

Solution:

1. Given the temperature field is written in terms of material coordinates, this can easily be computed for all material points as

$$\dot{\theta}(\boldsymbol{X},t) = 2.3X_2 [^\circ\mathrm{C/s}].$$

At the point of interest, we have

$$\dot{\theta}\left(\boldsymbol{X}^P,t\right) = 2.3 \times 0.1 = 0.23 [^\circ\mathrm{C/s}].$$

2. We must find the material derivative of the temperature field for the particle which currently occupies the spatial location $\boldsymbol{x}^p$. We have already found the material derivative in part a. However, we must either rewrite the material derivative in terms of the spatial coordinates of the particles, or find the reference position of the particle which currently occupies $\boldsymbol{x}^P$. In both cases, we need to invert the equations of motion:

$$X_1 = (t+1)^{-2}\, x_1$$

$$X_2 = (t+1)^{-2} x_2$$

$$X_3 = x_3$$

The spatial form of the material derivative can be found by substituting the inverted equations of motion into the material derivative found in part a. This gives

$$\dot{\theta}(\boldsymbol{x},t) = 2.3\left((t+1)^{-2}\, x_2\right) [^\circ\mathrm{C/s}].$$

For the particular spatial point of interest, we obtain

$$\dot{\theta}\left(\boldsymbol{x}^p,t\right) = 0.23(t+1)^{-2} \ [^\circ\mathrm{C/s}].$$

EXAMPLE 2.9. *A pipe carries hot water into a tank. Using a thermocouple, you measure the temperature field within the pipe as*

$$\theta(\boldsymbol{x},t) = \frac{20.3t}{2+\left(x_1^2+x_2^2\right)} - 1.3x_3 \ [^\circ\mathrm{C}].$$

The temperature is highest at the center of the cross section and drops linearly as the fluid travels along the length of the pipe. You also measure the velocity field in the pipe to be

$$v_1(\boldsymbol{x},t) = 0 \,[\mathrm{m/s}],$$

$$v_2(\boldsymbol{x},t) = 0 \,[\mathrm{m/s}],$$

$$v_3(\boldsymbol{X},t) = 3.0 - \left(x_1^2 + x_2^2\right) [\mathrm{m/s}].$$

1. *What is the material derivative of the fluid temperature at the pipe outlet,* $\boldsymbol{x}^p = (0, 0.01, 1)\,[\mathrm{m}]$?
2. *If you stuck the thermocouple back into the stream at ,* $\boldsymbol{x}^p = (0, 0.01, 1)\,[\mathrm{m}]$, *what would be the rate of change of your temperature reading?*

Solution:

1. The material derivative of the spatial temperature field is given by

$$\dot{\theta} = (grad\,\theta)\cdot \boldsymbol{v} + \theta'$$

$$= \begin{bmatrix} -40.3\dfrac{x_1 t}{\left(2+(x_1^2+x_2^2)\right)^2} \\ -40.3\dfrac{x_2 t}{\left(2+(x_1^2+x_2^2)\right)^2} \\ -1.3 \end{bmatrix} \cdot \begin{bmatrix} 0 \\ 0 \\ 3.0-(x_1^2+x_2^2) \end{bmatrix} + \frac{20.3}{2+(x_1^2+x_2^2)}.$$

This simplifies to give

$$\dot{\theta} = -1.3\left(3.0-\left(x_1^2+x_2^2\right)\right) + \frac{20.3}{2+(x_1^2+x_2^2)}.$$

Inserting the point of interest gives

$$\dot{\theta} = -1.3(3.0-(0+(0.1)^2)) + \frac{20.3}{2+\left(0+0.1^2\right)}$$

$$= -1.3\,(2.99) + \frac{20.3}{2.01}\,[^\circ\mathrm{C/s}].$$

2. The thermocouple simply measures the current temperature at a given point in the fluid. The rate of change of your temperature reading is given by

$$\theta' = \frac{\partial\theta(\boldsymbol{x},t)}{\partial t} = \frac{20.3}{2+(x_1^2+x_2^2)}.$$

Inserting the point of interest gives

$$\theta' = \frac{\partial\theta(\boldsymbol{x},t)}{\partial t} = \frac{20.3}{2+\left(0+0.1^2\right)} = \frac{20.3}{2.01}\,[^\circ\mathrm{C/s}].$$

2.9 Polar Decomposition of the Deformation Gradient

Strain measures are used to quantify local rearrangements in a material. While there are many strain measures, they are constructed with the constraint that there should be no strain if a continuum body undergoes only rigid body rotation and/or translation. The deformation gradient, while characterizing local rearrangement to some degree, does not satisfy this condition since its value also depends on the rotation of an object. In this section, we will use the polar decomposition theorem to separate the deformation gradient into tensor which depends on the rotation of an object and a tensor whose value is independent of rotation.

The polar decomposition theorem states that the deformation gradient, $\boldsymbol{F}(\boldsymbol{X},t)$, can be written as the product of an orthogonal rotation tensor, $\boldsymbol{R}$, and the symmetric positive definite tensors $\boldsymbol{U}$ and $\boldsymbol{V}$ such that

$$\boldsymbol{F} = \boldsymbol{R}\cdot\boldsymbol{U} = \boldsymbol{V}\cdot\boldsymbol{R}. \tag{2.28}$$

Each of these tensors is important in the development of strain measures and each is given a name. $\boldsymbol{R}$ is the ***rotation tensor***, $\boldsymbol{U}$ is the ***right stretching tensor***, $\boldsymbol{V}$ is the ***left stretching tensor***.

The proof begins with the statement that if $\boldsymbol{F}$ is nonsingular (since the $\det\boldsymbol{F} \neq 0$), and positive definite since $\boldsymbol{x}\cdot\boldsymbol{F}^T\cdot\boldsymbol{F}\cdot\boldsymbol{x} = (\boldsymbol{F}\cdot\boldsymbol{x})\cdot(\boldsymbol{F}\cdot\boldsymbol{x}) > 0$, then $\boldsymbol{F}\cdot\boldsymbol{F}^T$ and $\boldsymbol{F}^T\cdot\boldsymbol{F}$ are both symmetric and positive definite. The spectral decomposition theorem says that the square root of $\boldsymbol{F}\cdot\boldsymbol{F}^T$ and $\boldsymbol{F}^T\cdot\boldsymbol{F}$ will be symmetric real tensors and can be found using the eigenvalues and eigenvectors of these tensors. We define $\boldsymbol{U}$ and $\boldsymbol{V}$ such that

$$\boldsymbol{U}^2 \equiv \boldsymbol{F}^T\cdot\boldsymbol{F},$$
$$\boldsymbol{V}^2 \equiv \boldsymbol{F}\cdot\boldsymbol{F}^T.$$

Rearranging Equation (1.28), we may write $\boldsymbol{R} = \boldsymbol{F}\cdot\boldsymbol{U}^{-1}$. Recall that for an orthogonal tensor, $\boldsymbol{Q}\cdot\boldsymbol{Q}^T = \mathbf{I}$. Evaluating $\boldsymbol{R}^T\cdot\boldsymbol{R}$ gives

$$\boldsymbol{R}^T\cdot\boldsymbol{R} = \left(\boldsymbol{F}\cdot\boldsymbol{U}^{-1}\right)^T\cdot\left(\boldsymbol{F}\cdot\boldsymbol{U}^{-1}\right) = \boldsymbol{U}^{-T}\cdot\boldsymbol{F}^T\cdot\boldsymbol{F}\cdot\boldsymbol{U}^{-1} = \boldsymbol{U}^{-T}\cdot\boldsymbol{U}^T\cdot\boldsymbol{U}\cdot\boldsymbol{U}^{-1} = \mathbf{I}.$$

This result proves that $\boldsymbol{R}$ is an orthogonal tensor since $\boldsymbol{R}^T = \boldsymbol{R}^{-1}$.

The terminology used to describe these tensors is quite informative. For simplicity, we will consider a homogenous deformation process, such that the continuum body deforms uniformly at each point within the continuum body, Figure 2.14. A material line element in the reference configuration, $\boldsymbol{dX}$, is both stretched and rotated, finally becoming the spatial line element $\boldsymbol{dx}$ in the current configuration. The relation between the material and spatial line elements is given by

$$\boldsymbol{dx} = \boldsymbol{F}\cdot\boldsymbol{dX}.$$

Substituting Equation (2.28) into this relation gives

$$\boldsymbol{dx} = \boldsymbol{R}\cdot\boldsymbol{U}\cdot\boldsymbol{dX}.$$

This is equivalent to first stretching the material fiber by $\boldsymbol{U}$, then rotating the stretched fiber by $\boldsymbol{R}$ (see Figure 2.14). Note that we will show in the next section that the tensor $\boldsymbol{U}$ captures the stretch of material fibers. The stretched fiber, $\boldsymbol{dy}$, is given by $\boldsymbol{dy} = \boldsymbol{U}\cdot\boldsymbol{dX}$. The stretched fiber is then rotated to obtain the current fiber configuration, $\boldsymbol{x} = \boldsymbol{R}\cdot\boldsymbol{dy}$.

We could equivalently express our equations of motion in terms of $\boldsymbol{R}$ and $\boldsymbol{V}$ as

$$\mathbf{dx} = \boldsymbol{V}\cdot\boldsymbol{R}\cdot\boldsymbol{dX}.$$

In this case, the material fiber is first rotated into an intermediate state and then stretched. Notice in the following figure that intermediate configuration 1 is not equivalent to intermediate configuration 2.

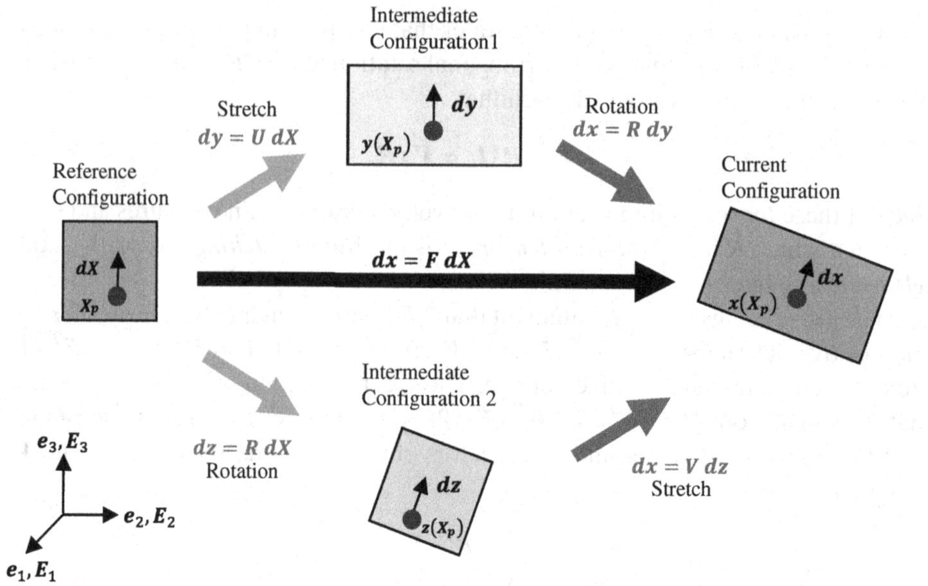

Figure 2.14. Decomposition of a homogenous deformation gradient into rotation and stretch.

EXAMPLE 2.10. *The deformed configuration for a unit cube is given by*

$$x_1(\boldsymbol{X},t) = X_1 + 3.0X_2,$$
$$x_2(\boldsymbol{X},t) = X_2,$$
$$x_3(\boldsymbol{X},t) = X_3.$$

1. *Find the deformation gradient,* $\boldsymbol{F}$.
2. *Find the left stretch tensor,* $\boldsymbol{U}$.
3. *Find* $\boldsymbol{U}^{-1}$.
4. *Find the right stretch tensor,* $\boldsymbol{V}$.
5. *Find the rotation tensor,* $\boldsymbol{R}$.
6. *Draw the reference and current configurations. Label a material fiber at point,* $X^a = \left(\frac{1}{2}, \frac{1}{2}, 0\right)$, *with magnitude* dS_o *and direction* (0,1,0). *Find its new magnitude, direction, and position in the current configuration.*

Solution:

1. From the equations of motion, we can find the deformation gradient

$$\boldsymbol{F}(\boldsymbol{X},t) = \begin{bmatrix} 1 & 3 & 0 \\ 0 & 1 & 0 \\ 0 & 0 & 1 \end{bmatrix}.$$

2. We start by finding $\boldsymbol{U}^2$

$$\boldsymbol{U}^2 = \boldsymbol{F}^T \cdot \boldsymbol{F} = \begin{bmatrix} 1 & 0 & 0 \\ 3 & 1 & 0 \\ 0 & 0 & 1 \end{bmatrix} \begin{bmatrix} 1 & 3 & 0 \\ 0 & 1 & 0 \\ 0 & 0 & 1 \end{bmatrix} = \begin{bmatrix} 1 & 3 & 0 \\ 3 & 10 & 0 \\ 0 & 0 & 1 \end{bmatrix}.$$

Then using spectral decomposition, or Matlab®, we find $\boldsymbol{U}$ to be

$$\boldsymbol{U} = \frac{1}{\sqrt{13}} \begin{bmatrix} 2 & 3 & 0 \\ 3 & 11 & 0 \\ 0 & 0 & 1 \end{bmatrix}.$$

3. Again using Matlab® or spectral decomposition gives

$$\boldsymbol{U}^{-1} = \frac{1}{\sqrt{13}} \begin{bmatrix} 11 & -3 & 0 \\ -3 & 2 & 0 \\ 0 & 0 & 1 \end{bmatrix}.$$

4. Start by finding $\boldsymbol{V}^2$

$$\boldsymbol{V}^2 = \boldsymbol{F} \cdot \boldsymbol{F}^T = \begin{bmatrix} 1 & 3 & 0 \\ 0 & 1 & 0 \\ 0 & 0 & 1 \end{bmatrix} \begin{bmatrix} 1 & 0 & 0 \\ 3 & 1 & 0 \\ 0 & 0 & 1 \end{bmatrix} = \begin{bmatrix} 10 & 3 & 0 \\ 3 & 1 & 0 \\ 0 & 0 & 1 \end{bmatrix}.$$

Then finding the square root of the tensor gives

$$\boldsymbol{V} = \frac{1}{\sqrt{13}} \begin{bmatrix} 11 & 3 & 0 \\ 3 & 2 & 0 \\ 0 & 0 & 1 \end{bmatrix}.$$

5. The rotation tensor is found through matrix multiplication

$$\boldsymbol{R} = \boldsymbol{F} \cdot \boldsymbol{U}^{-1} = \frac{1}{\sqrt{13}} \begin{bmatrix} 1 & 3 & 0 \\ 0 & 1 & 0 \\ 0 & 0 & 1 \end{bmatrix} \begin{bmatrix} 11 & 3 & 0 \\ 3 & 2 & 0 \\ 0 & 0 & 1 \end{bmatrix} = \frac{1}{\sqrt{13}} \begin{bmatrix} 20 & 9 & 0 \\ 3 & 2 & 0 \\ 0 & 0 & 1 \end{bmatrix}.$$

6. The current position, $\boldsymbol{x}^a$, of the material point, $\boldsymbol{X}^a$, is given by the equations of motion

$$\boldsymbol{x}^a \left(\boldsymbol{X}^a \right) = \begin{bmatrix} 2 \\ 0.5 \\ 0 \end{bmatrix}.$$

The current orientation of the material fiber is given by

$$\boldsymbol{dx} = \boldsymbol{F} \cdot \boldsymbol{dX} = dS_o \begin{bmatrix} 1 & 3 & 0 \\ 0 & 1 & 0 \\ 0 & 0 & 1 \end{bmatrix} \begin{bmatrix} 0 \\ 1 \\ 0 \end{bmatrix} = dS_o \begin{bmatrix} 3 \\ 1 \\ 0 \end{bmatrix}.$$

2.10 Stretch Ratios

In the previous section, we decomposed the deformation gradient, $\boldsymbol{F}$, into a rotation tensor, $\boldsymbol{R}$, and the right stretch tensor, $\boldsymbol{U}$. We stated without proof that the deformation was equivalent to stretching the material fibers by $\boldsymbol{U}$ and then rotating them by $\boldsymbol{R}$. The intermediate configuration, Figure 2.14, contains the stretched line element, $\boldsymbol{dy}$, given by

$$\boldsymbol{dy} = \boldsymbol{U} \cdot \boldsymbol{dX}, \tag{2.29}$$

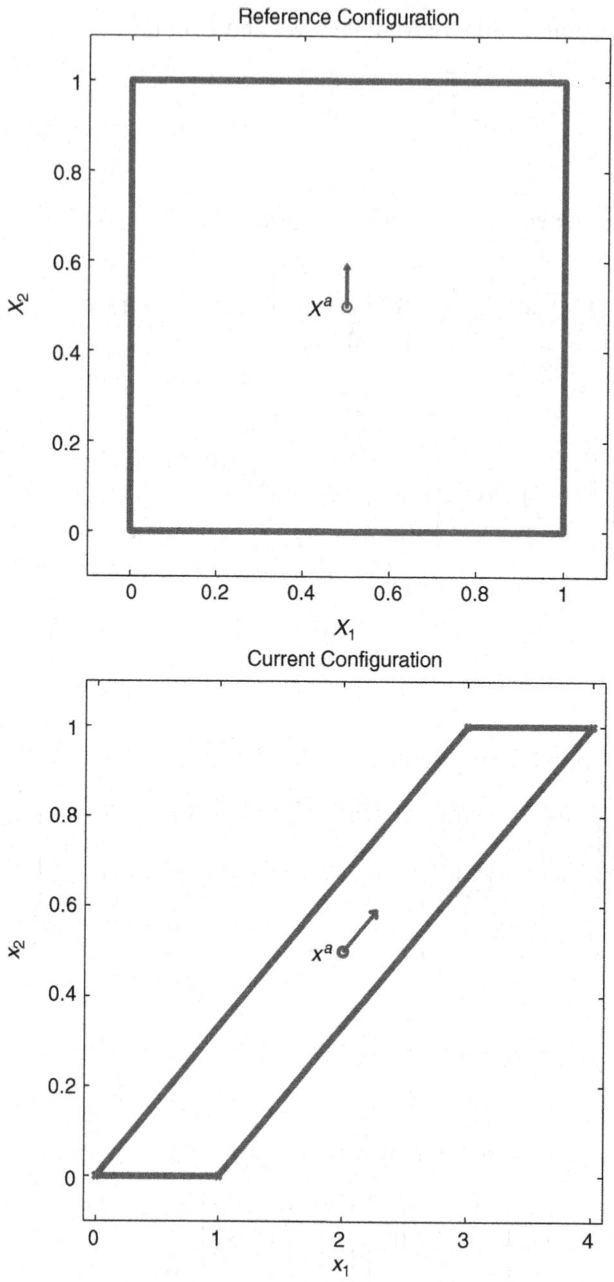

Figure 2.15. Illustration showing the reference configuration (top), and current configuration (bottom).

where $\boldsymbol{dX}$ is the material line element. Since $\boldsymbol{U}$ is a real symmetric tensor, it has real eigenvalues and three linearly independent eigenvectors. The eigenvectors must satisfy the relation

$$\boldsymbol{U} \cdot \boldsymbol{r}_i = \lambda_i \boldsymbol{r}_i,$$

where λ_i are the eigenvalues, and $\boldsymbol{r}_i$ are the eigenvectors of the tensor $\boldsymbol{U}$. If we look at a material fiber that is parallel to one of the eigenvectors such that $\boldsymbol{dX} = \alpha \boldsymbol{r}_i$, then Equation (2.29) becomes

$$\boldsymbol{dy} = \boldsymbol{U} \cdot \alpha \boldsymbol{r}_i = \lambda_i \alpha \boldsymbol{r}_i. \tag{2.30}$$

This implies that any material fiber that is oriented parallel to an eigenvector of the tensor $\boldsymbol{U}$ changes in length but not direction. In this case, the stretched fiber, $\boldsymbol{dy}$, is parallel to the original material fiber, $\boldsymbol{dX}$. Since $\boldsymbol{U}$ has three orthogonal eigenvectors, fibers in these three directions stretch but don't rotate. Therefore, $\boldsymbol{U}$ is said to capture the stretching of material fibers.

The ***stretch ratio*** is defined as the final length of a material fiber divided by its original length. By taking the magnitude of the vectors in Equation (1.30), we find that the material line element parallel to an eigenvector, $\boldsymbol{r}_i$, has a stretch ratio equal to the corresponding eigenvalue, λ_i,

$$\lambda_i = \frac{|\boldsymbol{U} \cdot \alpha \boldsymbol{r}_i|}{|\alpha \boldsymbol{r}_i|} = \frac{\text{final length}}{\text{original length}}.$$

Recall that the rotation tensor does not change the length of a material fiber on which it acts. Therefore, even taking into account the rotation, the eigenvalue gives the stretch of a material fiber parallel to the eigenvector of $\boldsymbol{U}$,

$$\lambda_i = \frac{|\boldsymbol{R} \cdot \boldsymbol{U} \cdot \alpha \boldsymbol{r}_i|}{|\alpha \boldsymbol{r}_i|} = \frac{|\boldsymbol{F} \cdot \alpha \boldsymbol{r}_i|}{|\alpha \boldsymbol{r}_i|}.$$

2.11 Left and Right Cauchy Deformation Tensor

The right Cauchy deformation tensor, $\boldsymbol{C}$, and the left Cauchy deformation tensor, $\boldsymbol{B}$, defined as

$$\boldsymbol{C} = \boldsymbol{U}^2 = \boldsymbol{F}^T \cdot \boldsymbol{F} \quad \text{and} \quad \boldsymbol{B} = \boldsymbol{V}^2 = \boldsymbol{F} \cdot \boldsymbol{F}^T, \tag{2.31}$$

provide a measure of the change in the size of material and spatial line elements. In particular, if one selects a material line element, $\boldsymbol{dX}$, the magnitude of the corresponding line element in the current configuration, $|\boldsymbol{dx}|$, is given by a transformation involving the right Cauchy deformation tensor such that

$$\begin{aligned} \boldsymbol{dx} \cdot \boldsymbol{dx} &= (\boldsymbol{F} \cdot \boldsymbol{dX}) \cdot (\boldsymbol{F} \cdot \boldsymbol{dX}) = \boldsymbol{dX} \cdot \boldsymbol{F}^T \cdot \boldsymbol{F} \cdot \boldsymbol{dX}, \\ |\boldsymbol{dx}| &= \sqrt{\boldsymbol{dX} \cdot \boldsymbol{C} \cdot \boldsymbol{dX}}. \end{aligned} \tag{2.32}$$

If one selects a spatial line element, $\boldsymbol{dx}$, the magnitude of the corresponding material line element, $|\boldsymbol{dX}|$, is given by a transformation involving the left Cauchy deformation tensor

$$\begin{aligned} \boldsymbol{dX} \cdot \boldsymbol{dX} &= \left(\boldsymbol{F}^{-1} \cdot \boldsymbol{dx}\right) \cdot \left(\boldsymbol{F}^{-1} \cdot \boldsymbol{dx}\right) = \boldsymbol{dx} \cdot \boldsymbol{F}^{-T} \cdot \boldsymbol{F}^{-1} \cdot \boldsymbol{dx}, \\ |\boldsymbol{dX}| &= \sqrt{\boldsymbol{dx} \cdot \boldsymbol{B}^{-1} \cdot \boldsymbol{dx}}. \end{aligned} \tag{2.33}$$

2.12 Green Strain Tensor

The ***Green strain tensor*** also known as the ***Lagrangian strain tensor*** is defined as

$$\boldsymbol{dx} \cdot \boldsymbol{dx} - \boldsymbol{dX} \cdot \boldsymbol{dX} = 2\boldsymbol{dX} \cdot \boldsymbol{E} \cdot \boldsymbol{dX}. \tag{2.34}$$

Substituting Equation (2.31) into Equation (2.34) gives the equation

$$\boldsymbol{dX} \cdot \boldsymbol{C} \cdot \boldsymbol{dX} - \boldsymbol{dX} \cdot \mathbf{I} \cdot \boldsymbol{dX} = 2\,\boldsymbol{dX} \cdot \boldsymbol{E} \cdot \boldsymbol{dX}.$$

The Green strain can then be expressed as

$$\boldsymbol{E} = \frac{1}{2}(\boldsymbol{C} - \mathbf{I}). \tag{2.35}$$

Substitution of Equation (2.12) and (2.13) gives the Green strain in terms of the displacement vector, $\boldsymbol{u}$, as

$$\begin{aligned} \boldsymbol{E} &= \frac{1}{2}\left(\boldsymbol{F}^T\boldsymbol{F} - \mathbf{I}\right) \\ &= \frac{1}{2}(\left(\mathbf{I} + \overrightarrow{\nabla}_X \boldsymbol{u}\right) \cdot \left(\mathbf{I} + \boldsymbol{u}\overleftarrow{\nabla}_X\right) - \mathbf{I}) \\ &= \frac{1}{2}\left(\overrightarrow{\nabla}_X \boldsymbol{u} + \boldsymbol{u}\overleftarrow{\nabla}_X + \overleftarrow{\nabla}_X \boldsymbol{u} \cdot \boldsymbol{u}\overleftarrow{\nabla}_X\right). \end{aligned} \tag{2.36}$$

Written in index notation this becomes

$$E_{ij} = \frac{1}{2}\left(\frac{\partial u_j}{\partial X_i} + \frac{\partial u_i}{\partial X_j} + \frac{\partial u_k}{\partial X_i}\frac{\partial u_k}{\partial X_j}\right). \tag{2.37}$$

The diagonal components of the Green strain are measures of the change in length of material fibers oriented parallel to the $\boldsymbol{E}_i$ basis vectors relative to their original length. This can be seen by substituting $\boldsymbol{dX} = (dS)\boldsymbol{E}_1$ where dS is the magnitude of the infinitesimal material fiber oriented parallel to $\boldsymbol{E}_1$, and $\boldsymbol{dx} = (ds)\hat{\boldsymbol{m}}$ where ds is the magnitude of the corresponding spatial fiber and $\hat{\boldsymbol{m}}$ is the unit vector parallel to $\boldsymbol{dx}$. Substituting this into Equation (1.34) gives

$$\begin{aligned} |\boldsymbol{dx}|^2 - |\boldsymbol{dX}|^2 &= 2dS^2\boldsymbol{E}_1 \cdot \boldsymbol{E} \cdot \boldsymbol{E}_1, \\ E_{11} &= \frac{ds^2 - dS^2}{2\,dS^2} = \frac{1}{2}(\lambda_1^2 - 1). \end{aligned} \tag{2.38}$$

2.13 Almansi Strain Tensor

The ***Almansi strain tensor*** also known as the ***Euler strain tensor*** is defined as

$$\boldsymbol{dx} \cdot \boldsymbol{dx} - \boldsymbol{dX} \cdot \boldsymbol{dX} = 2\boldsymbol{dx} \cdot \eta \cdot \boldsymbol{dx}. \tag{2.39}$$

Substituting Equation (2.33) into (2.39) gives

$$\boldsymbol{dx} \cdot \boldsymbol{dx} - \boldsymbol{dx} \cdot \boldsymbol{B}^{-1} \cdot \boldsymbol{dx} = 2\boldsymbol{dx} \cdot \eta \cdot \boldsymbol{dx}.$$

The Almansi strain tensor can then be expressed as

$$\eta = \frac{1}{2}(\mathbf{I} - \mathbf{B}^{-1}). \tag{2.40}$$

Expressed in terms of the displacement vector, $\boldsymbol{u}$, this becomes

$$\begin{aligned}\eta &= \frac{1}{2}\left(\mathbf{I}-\mathbf{F}^{-\mathrm{T}}\cdot\mathbf{F}^{-1}\right)\\ &= \frac{1}{2}\left(\mathbf{I}-\left(\mathbf{I}+\overrightarrow{\nabla}_x\boldsymbol{u}\right)\cdot\left(\mathbf{I}+\boldsymbol{u}\overleftarrow{\nabla}_x\right)\right)\\ &= \frac{1}{2}\left(\boldsymbol{u}\overleftarrow{\nabla}_x+\overrightarrow{\nabla}_x\boldsymbol{u}-\left(\overrightarrow{\nabla}_x\boldsymbol{u}\right)\cdot\left(\boldsymbol{u}\overleftarrow{\nabla}_x\right)\right).\end{aligned} \tag{2.41}$$

Written in index notation, this becomes

$$\eta_{ij} = \frac{1}{2}\left(\frac{\partial u_j}{\partial x_i}+\frac{\partial u_i}{\partial x_j}-\frac{\partial u_k}{\partial x_i}\frac{\partial u_k}{\partial x_j}\right). \tag{2.42}$$

The diagonal components of the Almansi strain are a measure of the change in length of the spatial line elements, which are parallel to the $\boldsymbol{e}_i$ basis vectors relative to their final length. This can be seen by assuming we have an infinitesimal spatial line element given by $\boldsymbol{dx} = (ds)\,\boldsymbol{e}_1$, where ds is the magnitude of the line element and $\boldsymbol{dX} = (dS)\hat{\boldsymbol{m}}$ where dS is the magnitude of the corresponding material line element and $\hat{\boldsymbol{m}}$ is the unit vector parallel to $\boldsymbol{dX}$. Substituting this into Equation (2.39) gives

$$\begin{aligned} ds^2 - dS^2 &= 2\,ds^2\boldsymbol{e}_1\cdot\eta\cdot\boldsymbol{e}_1,\\ \eta_{11} &= \frac{1}{2}\frac{\left(ds^2-dS^2\right)}{ds^2} = \frac{1}{2}\left(1-\lambda_1^{-2}\right).\end{aligned} \tag{2.43}$$

2.14 Infinitesimal Strain Tensor

Both the components of the Green strain tensor, Equations (2.37), and the components of the Almansi strain tensor, Equation (2.42), are nonlinear in terms of the displacement gradients. If both the displacements $\|\boldsymbol{u}\|$ and the material displacement gradient are small, $\left\|\boldsymbol{u}\overleftarrow{\nabla}\right\| \ll 1$, then the spatial gradient would also be small, $\left\|\boldsymbol{u}\overleftarrow{\nabla}_x\right\| \ll 1$, and the distinction between spatial and material gradients can be ignored, $\boldsymbol{u}\overleftarrow{\nabla}_X \cong \boldsymbol{u}\overleftarrow{\nabla}_x$. Linearizing either Equations (2.37) or Equation (2.42) gives

$$\boldsymbol{\varepsilon} = \frac{1}{2}\left(\boldsymbol{u}\overleftarrow{\nabla}_X+\overrightarrow{\nabla}_X\boldsymbol{u}\right) \quad \text{or} \quad \varepsilon_{ij} = \frac{1}{2}\left(\frac{\partial u_j}{\partial X_i}+\frac{\partial u_i}{\partial X_j}\right), \tag{2.44}$$

where $\boldsymbol{\varepsilon}$ is the ***infinitesimal strain tensor***.

2.15 Velocity Gradient, Rate of Deformation, Vorticity

The ***velocity gradient tensor***, $\boldsymbol{L}$, is defined as the spatial gradient of the velocity vector such that

$$\boldsymbol{L} = \boldsymbol{v}(\boldsymbol{X},t)\,\overleftarrow{\nabla}_x = grad\ \boldsymbol{v}(\boldsymbol{X},t) \quad \text{or} \quad L_{ij} = \frac{\partial v_i}{\partial x_j}.$$

The velocity gradient can be related to the material derivative of the deformation gradient. The material derivative of the deformation gradient is given by

$$\dot{\boldsymbol{F}} = \frac{\boldsymbol{D}}{\boldsymbol{Dt}} = \left(\boldsymbol{x}\overleftarrow{\nabla}_X\right) \quad \text{or} \quad \dot{F}_{ij} = \frac{d}{dt}\left(\frac{\partial x_i}{\partial X_J}\right).$$

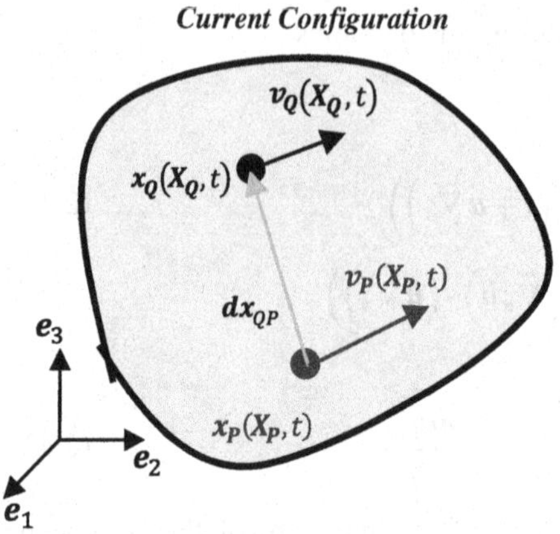

Figure 2.16. Illustration of the spatial velocity gradient.

Switching the order of differentiation and rewriting the equation in terms of the velocity vector gives

$$\dot{F}_{ij} = \frac{\partial}{\partial X_J}\left(\frac{dx_i}{dt}\right) = \frac{\partial v_i}{\partial X_J} = \frac{\partial v_i}{\partial x_m}\frac{\partial x_m}{\partial X_J} = L_{im}F_{mJ}.$$

In tensor notation, this can be written as

$$\dot{\boldsymbol{F}} = \boldsymbol{L}\cdot\boldsymbol{F} \quad \text{or} \quad \boldsymbol{L} = \dot{\boldsymbol{F}}\cdot\boldsymbol{F}^{-1}. \tag{2.45}$$

The velocity gradient describes the relative velocity of neighboring particles in the current configuration. Specifically, the relative velocity vector, $\delta\boldsymbol{v}$, of two infinitesimally close particles, P and Q, is given by

$$\boldsymbol{dv} = \boldsymbol{v}_Q - \boldsymbol{v}_P = \boldsymbol{L}\cdot\boldsymbol{dx}_{QP},$$

where $\boldsymbol{v}_Q$ is the velocity of particle Q, $\boldsymbol{v}_P$ is the velocity of particle P, and $\boldsymbol{dx}_{QP}$ is the spatial line element connecting the two points. In fact, the material derivative of a spatial line element is equal to the relative velocity between its two endpoints such that

$$\begin{aligned}\frac{D}{Dt}(\boldsymbol{dx}) &= \frac{D}{Dt}(\boldsymbol{F}\cdot\boldsymbol{dX})\\ &= \dot{\boldsymbol{F}}\cdot d\boldsymbol{X} + \boldsymbol{F}\cdot\frac{D}{Dt}(d\boldsymbol{X})\\ &= \boldsymbol{L}\cdot\boldsymbol{F}\cdot d\boldsymbol{X}\\ &= \boldsymbol{L}\cdot d\boldsymbol{x}\\ &= d\boldsymbol{v}.\end{aligned} \tag{2.46}$$

Note that in this derivation, we have made use of the fact that the material derivative of the material line element $\boldsymbol{dX}$ is zero, $\frac{D}{Dt}(d\boldsymbol{X}) = 0$.

The velocity gradient tensor can be decomposed into a symmetric tensor and a skew-symmetric tensor. The symmetric ***rate of deformation tensor*** or the ***stretching***

tensor, $\boldsymbol{D}$, captures the rate at which spatial line elements are stretched. The skew-symmetric ***spin tensor*** or the ***vorticity tensor***, $\boldsymbol{W}$, captures the rate at which spatial line elements are rotated. These tensors are related via the following equations:

$$\begin{aligned} \boldsymbol{L} &= \boldsymbol{D}+\boldsymbol{W}, \\ \boldsymbol{D} &= \frac{1}{2}\left(\boldsymbol{L}+\boldsymbol{L}^T\right), \\ \boldsymbol{W} &= \frac{1}{2}\left(\boldsymbol{L}-\boldsymbol{L}^T\right). \end{aligned} \tag{2.47}$$

The diagonal components of the rate of deformation tensor are related to the rate of stretch of spatial line elements. If we identify a spatial line element, $d\boldsymbol{x} = (ds)\,\hat{\boldsymbol{m}}$, where ds is the magnitude of the infinitesimal line element and $\hat{\boldsymbol{m}}$ is the unit vector parallel to $d\boldsymbol{x}$, then the material derivative of the magnitude squared of the spatial line element may be written as

$$\begin{aligned} \frac{D}{Dt}\left(\left(ds^2\right)\right) &= \frac{D}{Dt}\left(d\boldsymbol{x}\cdot d\boldsymbol{x}\right) \\ 2ds\frac{D}{Dt}(ds) &= 2\,d\boldsymbol{x}\cdot\frac{D}{Dt}(d\boldsymbol{x}) \\ ds\frac{D}{Dt}(ds) &= d\boldsymbol{x}\cdot\boldsymbol{L}\cdot d\boldsymbol{x}. \end{aligned} \tag{2.48}$$

Substituting Equation (2.47) into Equation (2.48) gives

$$ds\frac{D}{Dt}(ds) = d\boldsymbol{x}\cdot(\boldsymbol{D}+\boldsymbol{W})\cdot d\boldsymbol{x}.$$

For a skew-symmetric tensor such as the vorticity, $d\boldsymbol{x}\cdot\boldsymbol{W}\cdot d\boldsymbol{x} = 0$. Therefore, we have the relation

$$\begin{aligned} ds\frac{D}{Dt}(ds) &= d\boldsymbol{x}\cdot(\boldsymbol{D})\cdot d\boldsymbol{x} = (ds)\,\hat{\boldsymbol{m}}\cdot\boldsymbol{D}\cdot(ds)\hat{\boldsymbol{m}}, \\ \frac{\dot{ds}}{ds} &= \hat{\boldsymbol{m}}\cdot\boldsymbol{D}\cdot\hat{\boldsymbol{m}}. \end{aligned} \tag{2.49}$$

If we select a spatial line element which is parallel to $\boldsymbol{e}_1$, such that $\hat{\boldsymbol{m}} = \boldsymbol{e}_1$, Equation (2.49) gives

$$D_{11} = \frac{\dot{ds}}{ds}.$$

As you can see, the component, D_{11}, gives the normalized rate of change of the magnitude of the spatial line element parallel to the $\boldsymbol{e}_1$ basis vector. Similarly, the components, D_{22} and D_{33}, give the normalized stretch rates in the $\boldsymbol{e}_2$ and $\boldsymbol{e}_3$ directions respectively.

The shear rate components of the rate of deformation tensor are related to the rate of change in the angle between spatial line elements at a given point in the current configuration. Let us consider two spatial line elements $d\boldsymbol{x}^{(1)}$ and $d\boldsymbol{x}^{(2)}$ located at a spatial point $\boldsymbol{x}_p$. We can write the dot product of these two vectors as $d\boldsymbol{x}^{(1)}\cdot d\boldsymbol{x}^{(2)} = \left|d\boldsymbol{x}^{(1)}\right|\left|d\boldsymbol{x}^{(2)}\right|\cos\theta$, where θ is the angle between vectors 1 and 2. Taking the material derivative of $\left|d\boldsymbol{x}^{(1)}\right|\left|d\boldsymbol{x}^{(2)}\right|\cos\theta$ gives

$$\frac{D}{Dt}\left(\left|d\boldsymbol{x}^{(1)}\right|\left|d\boldsymbol{x}^{(2)}\right|\cos\theta\right) = \left|d\dot{\boldsymbol{x}}^{(1)}\right|\left|d\boldsymbol{x}^{(2)}\right|\cos\theta + \left|d\dot{\boldsymbol{x}}^{(1)}\right|\left|d\dot{\boldsymbol{x}}^{(2)}\right|\cos\theta - \left|d\boldsymbol{x}^{(1)}\right|\left|d\boldsymbol{x}^{(2)}\right|\dot{\theta}\sin\theta.$$

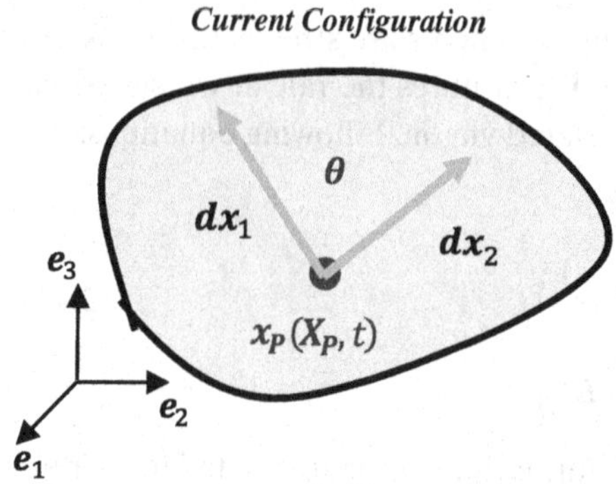

Figure 2.17. Illustration of two spatial line elements in the current configuration at point x_P.

The material derivative of $d\boldsymbol{x}^{(1)} \cdot d\boldsymbol{x}^{(2)}$ is given by

$$\begin{aligned}
\frac{D}{Dt}\left(d\boldsymbol{x}^{(1)} \cdot d\boldsymbol{x}^{(2)}\right) &= \frac{D\left(d\boldsymbol{x}^{(1)}\right)}{Dt} \cdot d\boldsymbol{x}^{(2)} + d\boldsymbol{x}^{(1)} \cdot \frac{D\left(d\boldsymbol{x}^{(2)}\right)}{Dt} \\
&= \boldsymbol{L} \cdot d\boldsymbol{x}^{(1)} \cdot d\boldsymbol{x}^{(2)} + d\boldsymbol{x}^{(1)} \cdot \boldsymbol{L} \cdot d\boldsymbol{x}^{(2)} \\
&= d\boldsymbol{x}^{(1)} \cdot \left(\boldsymbol{L}^T + \boldsymbol{L}\right) \cdot d\boldsymbol{x}^{(2)} \\
&= 2\, d\boldsymbol{x}^{(1)} \cdot \boldsymbol{D} \cdot d\boldsymbol{x}^{(2)}.
\end{aligned}$$

Equating these two results gives

$$2\, d\boldsymbol{x}^{(1)} \cdot \boldsymbol{D} \cdot d\boldsymbol{x}^{(2)} = \left|d\dot{\boldsymbol{x}}^{(1)}\right|\left|d\boldsymbol{x}^{(2)}\right| \cos\theta + \left|d\boldsymbol{x}^{(1)}\right|\left|d\dot{\boldsymbol{x}}^{(2)}\right| \cos\theta - \left|d\boldsymbol{x}^{(1)}\right|\left|d\boldsymbol{x}^{(2)}\right| \dot{\theta} \sin\theta. \tag{2.50}$$

Selecting any two perpendicular spatial line elements, we obtain $\cos\theta = 0$ and $\sin\theta = 1$. If we specifically select a vector parallel to $\boldsymbol{e}_1$ and another parallel to $\boldsymbol{e}_2$, we can write $d\boldsymbol{x}^{(1)} = ds^{(1)}\boldsymbol{e}_1$ and $d\boldsymbol{x}^{(2)} = ds^{(2)}\boldsymbol{e}_2$, where $ds^{(1)}$ and $ds^{(2)}$ are the magnitudes of the first and second vectors, respectively. Substituting these vectors into Equation (2.50) gives the relation

$$D_{12} = -\frac{1}{2}\dot{\theta}$$

The rate of deformation tensor is symmetric so that $D_{12} = D_{21}$. We see clearly, that the off-diagonal terms of the stretching tensor give the rate of change of the angles between fibers oriented along the basis vectors. This is a measure of the rate of shear within the system.

2.16 Reynolds' Transport Theorem

Throughout this text, we will be asked to find the material derivative of an integral computed over a volume in the current configuration. The Reynolds' transport theorem allows us to simplify these expressions and exchange the order of integration and differentiation.

The integral of a scalar field, $\phi = \phi(\boldsymbol{x}, t)$, computed over an enclosed volume in the current configuration, $\zeta(t)$, is given by

$$\zeta(t) = \int_{\Omega(t)} \phi(\boldsymbol{x}, t) dV_x,$$

where $\Omega(t)$ is the volume over which the integral is computed, and dV_x is the incremental spatial volume element. For example, if we integrated the density field over a volume, $\Omega(t)$, in the current configuration, we would obtain the total mass enclosed within the volume defined by $\Omega(t)$. Returning to the general case and taking the material time derivative of the integral, $\dot{\zeta}(t)$ we obtain

$$\dot{\zeta}(t) = \frac{D}{Dt}\zeta(t) = \frac{D}{Dt}\int_{\Omega(t)} \phi(\boldsymbol{x}, t) dV_x.$$

We would like to reverse the order of differentiation and integration, but the differential and integral operators do not commute since the limits of integration, $\Omega(t)$, change with time. Instead, we will first convert our integral over the current configuration into an integral over the reference configuration, we will then reverse the order of the operators and transform the integral back to the current configuration.

The first step is to convert the integral over the spatial volume to an integral over the material volume. The incremental volume in the current configuration, dV_x, is related to the incremental volume in the reference configuration, dV_X by the Jacobian such that

$$dV_x = J(\boldsymbol{X}, t) dV_X.$$

We can write the spatial scalar field in terms of the reference configuration by finding the current position of the reference material point

$$\phi(\boldsymbol{X}, t) = \phi(\boldsymbol{x}(\boldsymbol{X}, t), t).$$

The spatial integral becomes

$$\frac{D}{Dt}\int_{\Omega(t)} \phi(\boldsymbol{x}, t)\, dV_x = \frac{D}{Dt}\int_{\Omega_0} \phi(\boldsymbol{x}(X, t), t)\, J(X, t)\, dV_X.$$

Notice that the limits of integration have changed from an enclosed volume in the current configuration, $\Omega(t)$, to the corresponding enclosed volume in the reference configuration, Ω_0. Therefore, we can either integrate the spatial description of the scalar function over a volume in the current configuration or integrate the material description of the scalar function over the volume in the reference configuration. We now benefit from the fact that the limits of integration no longer depend on time. The integral and differential operators can be reversed giving

$$\begin{aligned}\dot{\zeta}(t) &= \frac{D}{Dt}\int_{\Omega_0} \phi(\boldsymbol{x}(X, t), t) J(X, t) dV_X \\ &= \int_{\Omega_0} \frac{D}{Dt}[\phi J] dV_X\end{aligned}$$

$$= \int_{\Omega_0} (\dot{\phi}J + \phi\dot{J})\, dV_X \tag{2.51}$$
$$= \int_{\Omega_0} \left(\dot{\phi} + \phi\frac{\dot{J}}{J}\right) J\, dV_X.$$

The material derivative of the Jacobian can be expanded using the chain rule to give

$$\dot{J} = \frac{D}{Dt}J = \frac{\partial J}{\partial F_{ij}}\frac{\partial F_{ij}}{\partial t} = \frac{\partial J}{\partial \boldsymbol{F}} : \dot{\boldsymbol{F}}.$$

The partial derivative of the Jacobian with respect to the deformation gradient is given by

$$\frac{\partial J}{\partial \boldsymbol{F}} = \frac{\partial}{\partial \boldsymbol{F}} det\, \boldsymbol{F} = J\boldsymbol{F}^{-T}.$$

This gives us

$$\begin{aligned}\dot{J} &= J\boldsymbol{F}^{-T} : \dot{\boldsymbol{F}} = J\boldsymbol{F}^{-T} : \boldsymbol{L}\boldsymbol{F} \\ &= J\boldsymbol{F}^{-T}\boldsymbol{F}^{T} : \boldsymbol{L} \\ &= J(I : \boldsymbol{L}) \\ &= J\, tr(\boldsymbol{L}) \\ &= J\, div(\boldsymbol{v}).\end{aligned} \tag{2.52}$$

Substitution of Equation (2.52) into Equation (2.51) gives

$$\dot{\varsigma}(t) = \int_{\Omega_o} (\dot{\phi} + \phi\, div\, (\boldsymbol{v}))J\, dV_X.$$

Now that we have exchanged the differential and integral operators, we will convert back to an integral over the current configuration. Substituting the inverted incremental volume relation, $\frac{dV_x}{J} = dV_X$, while converting back to a spatial description gives

$$\dot{\varsigma}(t) = \int_{\Omega(t)} (\dot{\phi}(\boldsymbol{x},t) + \phi(\boldsymbol{x},t) div(\boldsymbol{v}(\boldsymbol{x},t)))\, dV_x.$$

This relation may be written in several equivalent forms. Note that expanding the material derivative such that $\frac{D\phi}{Dt} = \frac{\partial\phi}{\partial t} + grad\,\phi \cdot \boldsymbol{v}$, we can also write

$$\begin{aligned}\frac{D}{Dt}\int_{\Omega(t)} \phi\, dV_x &= \int_{\Omega(t)} \left(\frac{D\phi}{Dt} + \phi\, div\,(\boldsymbol{v})\right) dV_x \\ &= \int_{\Omega(t)} \left(\frac{\partial\phi}{\partial t} + grad\,\phi \cdot \boldsymbol{v} + \phi\, div\,(\boldsymbol{v})\right) dV_x \\ &= \int_{\Omega(t)} \left(\frac{\partial\phi}{\partial t} + div\,(\phi\boldsymbol{v})\right) dV_x.\end{aligned} \tag{2.53}$$

In this last step, we made use of the fact that $grad\,\phi \cdot \boldsymbol{v} + \phi\, div\,(\boldsymbol{v}) = div\,(\phi\boldsymbol{v})$. In addition, using the results of the divergence theorem, Equation (1.36), we can convert

Table 2.1. *Summary of Reynolds' transport theorem*

$\frac{D}{Dt}\int_{\Omega(t)}\phi\, dV_x = \int_{\Omega(t)}(\dot{\phi}+\phi(\nabla\cdot\boldsymbol{v}))\, dV_x$	2.54
$\frac{D}{Dt}\int_{\Omega(t)}\phi\, dV_x = \int_{\Omega(t)}\left(\frac{\partial\phi}{\partial t}+\nabla\cdot(\phi\boldsymbol{v})\right) dV_x$	2.55
$\frac{D}{Dt}\int_{\Omega(t)}\phi\, dV_x = \int_{\Omega(t)}\left(\frac{\partial\phi}{\partial t}\right) dV_x + \int_{\partial\Omega(t)}\phi\,(\boldsymbol{v}\cdot\boldsymbol{n})\, dS_x$	2.56
$\frac{D}{Dt}\int_{\Omega(t)}\boldsymbol{a}\, dV_x = \int_{\Omega(t)}\left(\frac{\partial\boldsymbol{a}}{\partial t}\right) dV_x + \int_{\partial\Omega(t)}\boldsymbol{a}\,(\boldsymbol{v}\cdot\boldsymbol{n})\, dS_x$	2.57

the integral to a surface integral such that

$$\frac{D}{Dt}\int_{\Omega(t)}\phi\, dV_x = \int_{\Omega(t)}\left(\frac{\partial\phi}{\partial t}\right) dV_x + \int_{\partial\Omega(t)}\phi\,(\boldsymbol{v}\cdot\boldsymbol{n})\, dS_x.$$

This equation is particularly interesting since it says that we can write the material derivative of a scalar field as the addition of the local time rate of change of the spatial scalar field within a region Ω (first term) and the rate of transport in or out of the bounding surface $\partial\Omega$ (second term).

The Reynolds' transport theorem applied to vectors gives a similar result,

$$\frac{D}{Dt}\int_{\Omega(t)}\boldsymbol{a}\, dV_x = \int_{\Omega(t)}\left(\frac{\partial\boldsymbol{a}}{\partial t}\right) dV_x + \int_{\partial\Omega(t)}\boldsymbol{a}\,(\boldsymbol{v}\cdot\boldsymbol{n})\, dS_x.$$

EXERCISES

1. Given the following equations of motion for an infinite continuum body for $t > 0$:

$$x_1 = (t+1)^2 X_1 - tX_2,$$
$$x_2 = tX_1 - (t+1)^2 X_2,$$
$$x_3 = X_3,$$

   a. Find the Lagrangian description of the velocity and acceleration vectors.
   b. Find the Eulerian description of the velocity and acceleration vectors.
   c. Find the velocity as a function of time for the particle initially found at the reference location $\boldsymbol{X} = (1,2,0)$.
   d. Find the velocity as a function of time of the particle which occupies the position $\boldsymbol{x} = (1,2,0)$ at the current time t.

2. A two-dimensional continuum body consisting of a unit square is deformed such that the deformation is given by

$$x_1 = 4 - 2X_1 - X_2,$$
$$x_2 = 2 + \frac{3}{2}X_1 - \frac{1}{2}X_2.$$

   a. Find the deformation gradient and its inverse.
   b. Draw an illustration of the current configuration of the continuum body.

c. Find the spatial line element that corresponds to the material line element $\boldsymbol{a}_o = \frac{1}{\sqrt{2}}(1,1)$.
d. Find the material line element that transforms into the spatial line element $\boldsymbol{b} = (1,0)$.

3. Given the following equations of motion for an infinite continuum body for $t > 0$:

$$x_1 = t^2 X_2^2 + X_1,$$
$$x_2 = t X_1^2 + X_2,$$
$$x_3 = (t+1) X_3.$$

a. Find the left and right Cauchy deformation tensors.
b. Find the Green, Almansi, and infinitesimal strain tensors.
c. Find the velocity gradient and the vorticity.

4. A unit cube is subject to the following equations of motion:

$$x_1(\boldsymbol{X},t) = \left(\sqrt{t+1}\right) X_1,$$
$$x_2(\boldsymbol{X},t) = \left(\sqrt{t+1}\right) X_2,$$
$$x_3(\boldsymbol{X},t) = \left(\sqrt{t+1}\right) X_3.$$

The material description of the temperature field is found to be

$$\theta(\boldsymbol{X},t) = 1.3(t+1) X_1 X_2 X_3 \ [°\mathrm{C}].$$

a. Find the material derivative of the temperature field for the material point, $\boldsymbol{X}^P = (0.5, 0.5, 0.5)$.
b. Find the time rate of change of the temperature of the particle that currently occupies the spatial position $\boldsymbol{x}^P = (0.5, 0.5, 0.5)$ at time t.

5. The velocity of the fluid within a pipe of radius $R = 10\,\mathrm{cm}$, is found to be

$$v_1(\boldsymbol{x},t) = 0,$$
$$v_2(\boldsymbol{x},t) = 0,$$
$$v_3(\boldsymbol{x},t) = c_o * (R^2 - (x_1^2 + x_2^2)),$$

where $c_o = 5$ per meter per second. The origin is located at the center of the pipe, and the e_3 axis is oriented along the axis of the pipe. The spatial temperature field is found to be $\theta(\boldsymbol{x},t) = c_1(1 - \sin(c_2 t))\sqrt{x_3}$.

a. Find the spatial derivative of the temperature field.
b. Find the material derivative of the temperature field.

MATLAB® EXERCISES

1. A two-dimensional unit circle is deformed according to the following equations of motion:

$$x_1(X,t) = tX_1^2 + X_1,$$
$$x_2(X,t) = \sqrt{t}X_1 + X_2.$$

The coordinate system is located at the center of the unit circle such that $0 \leq X_1^2 + X_2^2 \leq 1$. For each of the parts below, remember that the functions of the reference coordinates have values within the unit circle but are not defined outside of the unit circle.

a. Use Matlab® to plot a set of points within the reference configuration.
b. Plot the new location of these points within the current configuration at $t = 2.0$ seconds.
c. Plot the material velocity field, $\boldsymbol{v}(\boldsymbol{X},t)$ at $t = 2.0$ seconds.
d. Plot the spatial velocity field, $\boldsymbol{v}(\boldsymbol{x},t)$ at $t = 2.0$ seconds.
e. Plot the norm of the deformation gradient at $t = 2.0$ seconds as a contour plot. Note that the deformation gradient is naturally a function of the reference positions. Therefore, you should plot the norm $\alpha(\boldsymbol{X},t) = \sqrt{(\boldsymbol{F}:\boldsymbol{F})}$.
f. Plot the norm of the Green strain tensor at $t = 2.0$ seconds as a contour plot. Note that the Green strain is naturally a function of the reference positions.

3 The Stress Tensor

In any material system, discrete forces may be transmitted between the atomic and subatomic particles in a nonuniform manner in response to external forces and fields. Depending on the nature of these internal forces, a material may yield or fracture. For a continuum which does not have discreet particles exerting discrete forces on one another, we need some measure of the intensity and orientation of the internally transmitted forces within a material. For this, we define the stress tensor which has units of force per unit area. In this chapter, we will introduce several measures of stress and show how they are related to one another.

3.1 Mass, Density, and Forces

Excluding relativistic corrections, the mass, force, and acceleration of a point mass are related via Newton's Second Law, $\sum \boldsymbol{F} = m\boldsymbol{a}$. While discreetly distributed in real materials, the mass, m, which is the measure of an objects resistance to acceleration, is assumed to be continuously distributed over the continuum body. We assume that there exists a density field, $\rho = \rho(\boldsymbol{x}, t)$, such that the total mass, m_x, within an enclosed region, Ω, may be written as

$$m_x = \int_\Omega \rho(\boldsymbol{x}, t)\, dV_x.$$

The total volume, V_x, within the enclosed region is given by

$$V_x = \int_\Omega dV_x.$$

The density, $\rho(\boldsymbol{x}, t)$, at any point in the continuum body is found by dividing the mass by the volume as the enclosed region, Ω, is shrunk to a point such that

$$\rho(\boldsymbol{x}, t) = \lim_{\Omega \to 0} \frac{m_x(\boldsymbol{x}, t)}{V_x(\boldsymbol{x}, t)}.$$

The forces described in this text will be split into ***body forces***, ***surface forces***, ***body couples***, and ***surface couples***. Surface forces are due to physical contact between materials and act on the bounding surface of the region of interest. Body forces such as gravity, electric fields, and magnetic fields may exert forces from a distance on material points within a region of interest. The dimensions of the body forces are typically given by force per unit mass or force per unit volume. The distributed body and surface couples represent the transfer of angular momentum at a distance. Contact couples and body couples represent the transfer of angular momentum at a surface and a volume, respectively.

3.2 Traction Vector

The ***traction vector*** or ***stress vector*** is equal to the resultant force acting on a surface divided by the surface area over which it acts in the limit as the surface area goes to zero. We will use the traction vector to define internal stress measures within a continuum body. However, there are several different ways of defining traction vectors. Each leads to a different measure of stress.

Let us begin by introducing the geometric and force quantities used to define various traction vectors. If we apply an ***external*** force to an object, it responds through some combination of acceleration and deformation. The object remains cohesive as the atoms of the material interact with one another resulting in an ***internal force*** distribution, which balances the applied external force. Imagine that we apply both a set of ***surface forces*** and ***body forces*** to a continuum body, Figure 3.1. Forces applied to the exterior surface of the object are termed surface forces or surface traction. Forces that act throughout the interior of a material are termed body forces. The

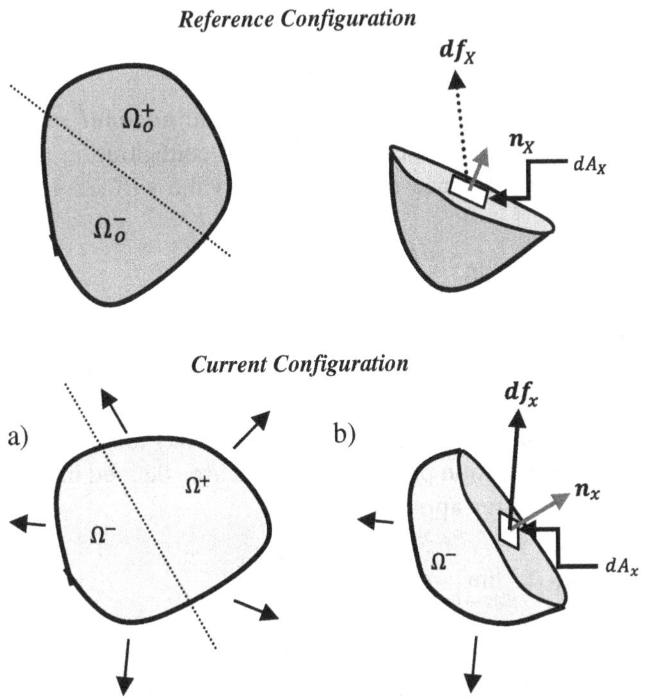

Figure 3.1. (a) Illustration of the reference and current configuration of a continuum body with applied forces. (b) Isolated segment of a continuum body cut along an internal plane with internal forces, surface normal, and infinitesimal area shown.

force exerted by gravity would be a common example of a body force. Furthermore, let us make an imaginary cut through the continuum body along the dashed line shown in Figure 3.1 exposing an internal surface. There are internal forces acting along this cut surface in the current configuration that are responsible for keeping the object in one piece. In other words, one segment of the object, Ω^+, exerts some force across this surface on the other segment, Ω^-, and vice versa. The instantaneous force vector, $d\boldsymbol{f}_x$, acting on an infinitesimal area, dA_x, with outward unit normal vector, $\boldsymbol{n}_x$, is shown in the current configuration. The corresponding infinitesimal area, dA_X, and surface normal, $\boldsymbol{n}_X$, in the reference configuration can be found using Nanson's formula

$$\boldsymbol{n}_x dA_x = J\boldsymbol{F}^{-T} \cdot \boldsymbol{n}_X dA_X. \tag{3.1}$$

The instantaneous force vector in the current configuration can be transformed back into the reference configuration as

$$d\boldsymbol{f}_X = \boldsymbol{F}^{-1} \cdot d\boldsymbol{f}_x. \tag{3.2}$$

Notice that the transformed instantaneous force, $d\boldsymbol{f}_X$, does not truly exist in the reference configuration. It is a mathematical construction, a projection of the instantaneous force back into the reference configuration that will be used later to define a particular traction vector.

The ***Cauchy traction vector,*** $\boldsymbol{t}_x$, also known as the ***true traction vector***, is defined as the instantaneous force in the current configuration per unit of surface area in the current configuration as the surface area approaches zero such that

$$\boldsymbol{t}_x = \lim_{dA_x \to 0} \frac{d\boldsymbol{f}_x}{dA_x}. \tag{3.3}$$

The ***first Piola-Kirchhoff traction vector,*** $\boldsymbol{t}_X$, also known as the ***nominal traction vector***, is defined as the instantaneous force in the current configuration per unit of surface area defined in the reference configuration as the surface area approaches zero

$$\boldsymbol{t}_X = \lim_{dA_x \to 0} \frac{d\boldsymbol{f}_x}{dA_X}. \tag{3.4}$$

Note that the force vector, $d\boldsymbol{f}_x$, in the definition of the Cauchy and the first Piola-Kirchhoff traction vector is the same. Therefore, these traction vectors are parallel to one another.

The ***second Piola-Kirchhoff traction vector***, is the instantaneous force transformed back into the reference configuration per unit of surface area defined in the reference configuration as the surface area approaches zero

$$\boldsymbol{t}_X^* = \lim_{dA_x \to 0} \frac{d\boldsymbol{f}_X}{dA_X}. \tag{3.5}$$

Depending on the angle at which the surface is cut, the traction on a surface that passes through the same point varies. Therefore, we must specify the surface on which the traction is being computed. This will be accomplished by using a superscript such that $\boldsymbol{t}_x^{(\boldsymbol{n}_x)}$ is the traction acting on a surface, which has an outward normal $\boldsymbol{n}_x$. Since the force exerted by Ω^- on Ω^+ is equal and opposite to the force exerted by Ω^+ on

Ω^-, the Cauchy traction vector on the surface with outward normal, $\boldsymbol{n}_x$, is equal and opposite to the Cauchy traction vector with outward normal, $-\boldsymbol{n}_x$, such that

$$\boldsymbol{t}_x^{(\boldsymbol{n}_x)} = -\boldsymbol{t}_x^{(-\boldsymbol{n}_x)}.$$

Each of these traction vectors can be decomposed into a component that is normal to the surface on which it is acting, $\boldsymbol{t}_x^n$, and a vector that is tangential to the surface on which it is acting, $\boldsymbol{t}_x^s$, such that $\boldsymbol{t}_x^{(\boldsymbol{n}_x)} = \boldsymbol{t}_x^n + \boldsymbol{t}_x^s$. If $\boldsymbol{n}_x$ is the outward normal, then the normal traction vector, $\boldsymbol{t}_x^n$, is found by taking the projection of the traction along the outward normal and assigning it the vector direction of the outward normal

$$\boldsymbol{t}_x^n = \left(\boldsymbol{t}_x^{(\boldsymbol{n}_x)} \cdot \boldsymbol{n}_x\right)\boldsymbol{n}_x.$$

The tangential traction vector is found by subtracting the normal traction vector from the total traction vector

$$\boldsymbol{t}_x^s = \boldsymbol{t}_x^{(\boldsymbol{n}_x)} - \boldsymbol{t}_x^n.$$

The magnitudes of the normal and tangential traction vectors are related via the Pythagorean theorem such that

$$\left(\boldsymbol{t}_x^n\right)^2 + \left(\boldsymbol{t}_x^s\right)^2 = \left(\boldsymbol{t}_x^{(\boldsymbol{n}_x)}\right)^2.$$

3.3 Cauchy Stress Tensor

The traction vector passing through a point in a material is a function of the outward normal to the surface cutting through the point. At a single point, P, let us define the traction vector acting on the surface with outward normal $\boldsymbol{n}_x = \boldsymbol{e}_1$ as $\boldsymbol{t}_x^{(\boldsymbol{e}_1)}$. Similarly, the traction vector, $\boldsymbol{t}_x^{(\boldsymbol{e}_2)}$, acts on the surface with outward normal $\boldsymbol{n}_x = \boldsymbol{e}_2$ and $\boldsymbol{t}_x^{(\boldsymbol{e}_3)}$ acts on the surface with outward normal $\boldsymbol{n}_x = \boldsymbol{e}_3$. Expanding the vector traction vector in terms of the basis vectors, we can write

$$\begin{aligned}
\boldsymbol{t}_x^{(\boldsymbol{e}_1)} &= \left(\boldsymbol{t}_x^{(\boldsymbol{e}_1)} \cdot \boldsymbol{e}_1\right)\boldsymbol{e}_1 + \left(\boldsymbol{t}_x^{(\boldsymbol{e}_1)} \cdot \boldsymbol{e}_2\right)\boldsymbol{e}_2 + \left(\boldsymbol{t}_x^{(\boldsymbol{e}_1)} \cdot \boldsymbol{e}_3\right)\boldsymbol{e}_3,\\
\boldsymbol{t}_x^{(\boldsymbol{e}_2)} &= \left(\boldsymbol{t}_x^{(\boldsymbol{e}_2)} \cdot \boldsymbol{e}_1\right)\boldsymbol{e}_1 + \left(\boldsymbol{t}_x^{(\boldsymbol{e}_2)} \cdot \boldsymbol{e}_2\right)\boldsymbol{e}_2 + \left(\boldsymbol{t}_x^{(\boldsymbol{e}_2)} \cdot \boldsymbol{e}_3\right)\boldsymbol{e}_3,\\
\boldsymbol{t}_x^{(\boldsymbol{e}_3)} &= \left(\boldsymbol{t}_x^{(\boldsymbol{e}_3)} \cdot \boldsymbol{e}_1\right)\boldsymbol{e}_1 + \left(\boldsymbol{t}_x^{(\boldsymbol{e}_3)} \cdot \boldsymbol{e}_2\right)\boldsymbol{e}_2 + \left(\boldsymbol{t}_x^{(\boldsymbol{e}_3)} \cdot \boldsymbol{e}_3\right)\boldsymbol{e}_3.
\end{aligned}$$

We define the ***Cauchy stress tensor***, $\boldsymbol{\sigma}$, as

$$\sigma_{ij} = \boldsymbol{t}_x^{(\boldsymbol{e}_i)} \cdot \boldsymbol{e}_j,$$

such that the three resultant traction vectors can be written in component form as

$$\begin{aligned}
\boldsymbol{t}_x^{(\boldsymbol{e}_1)} &= \sigma_{11}\boldsymbol{e}_1 + \sigma_{12}\boldsymbol{e}_2 + \sigma_{13}\boldsymbol{e}_3,\\
\boldsymbol{t}_x^{(\boldsymbol{e}_2)} &= \sigma_{21}\boldsymbol{e}_1 + \sigma_{22}\boldsymbol{e}_2 + \sigma_{23}\boldsymbol{e}_3,\\
\boldsymbol{t}_x^{(\boldsymbol{e}_3)} &= \sigma_{31}\boldsymbol{e}_1 + \sigma_{32}\boldsymbol{e}_2 + \sigma_{33}\boldsymbol{e}_3.
\end{aligned}$$

This set of equations may be written compactly as

$$\boldsymbol{t}_x^{(\boldsymbol{e}_i)} = \sigma_{ij}\boldsymbol{e}_j. \tag{3.6}$$

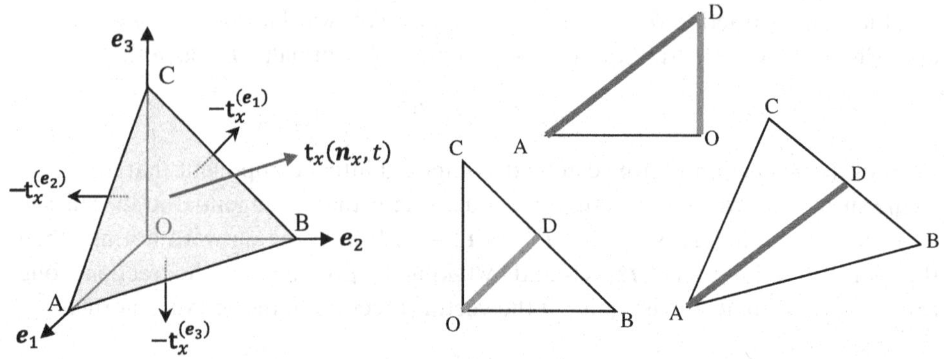

Figure 3.2. Illustration of a tetrahedral volume element. The triangles OCB, ADO, and ACB that make up the tetrahedral volume element are also shown.

Once we have the components of the Cauchy stress tensor, we may find the traction vector acting on any arbitrary surface. To show this, we select an infinitesimal tetrahedral volume element. We locate the origin of our coordinate system at point O, and define the tetrahedron as shown in the preceding figure. Let us assume that our object is subjected to a body force per unit mass, $\boldsymbol{b}_x$. Conservation of linear momentum requires that the sum of forces acting on the tetrahedron equal the mass of the tetrahedron times the acceleration of the center of mass, $\boldsymbol{a}$, such that $\sum \boldsymbol{F} = m\boldsymbol{a}$. The forces acting on the tetrahedral volume element are the body force acting on the internal volume and the traction vectors acting on the surfaces of the infinitesimal tetrahedral volume element. If we note that the mass of enclosed in the volume element is given by $m = \rho\,\delta V$, we can write

$$\sum \boldsymbol{F} = \rho \boldsymbol{a}\,\delta V.$$

The resultant force acting on the volume element due to the body force, $\boldsymbol{b}_x$, is given by $\rho\,\mathbf{b}_x\,\delta V$. Recall that this is an infinitesimal volume element and the surface traction may be assumed to be approximately constant over each of the surfaces of the element. In addition, let us define the surface area of surface OCB, OAC, and OBA as $\delta S^{(1)}$, $\delta S^{(2)}$, and $\delta S^{(3)}$ respectively. The resultant forces due to the surface traction on surface OCB, OAC, and OBA are then given by $-\boldsymbol{t}_x^{(e_1)}\,\delta S^{(1)}$, $-\boldsymbol{t}_x^{(e_2)}\delta S^{(2)}$, and $-\boldsymbol{t}_x^{(e_3)}\delta S^{(3)}$ respectively. The sum of forces acting on the volume element can then be written as

$$-\boldsymbol{t}_x^{(e_1)}\,\delta S^{(1)} - \boldsymbol{t}_x^{(e_2)}\delta S^{(2)} - \boldsymbol{t}_x^{(e_3)}\delta S^{(3)} + \boldsymbol{t}_x \delta S^{(n)} + \rho \boldsymbol{b}_x \delta V = \rho\,\boldsymbol{a}\,\delta V.$$

The surface areas on the side faces of the tetrahedron, as $\delta S^{(1)}$, $\delta S^{(2)}$, and $\delta S^{(3)}$, can be related to the surface area of the inclined surface, δS. Since each surface is triangular, the surface area of each face is given by one-half the base times the height of the surface. We can write the surface area of the face OBC which has an outward normal of $-\boldsymbol{e}_1$, as

$$\delta S^{(1)} = \frac{1}{2}\mathrm{CB} \cdot \mathrm{OD},$$

where CB and OD are the lengths of the respective line segments shown in Figure 3.2. Similarly, the surface area of the inclined surface ABC is given by

$$\delta S = \frac{1}{2} \mathrm{CB} \cdot \mathrm{AD},$$

where AD is the length of the line segment shown in Figure 3.2. Notice that OD and AD are related via the projection

$$\mathrm{OD} = \mathrm{AD} \cos \theta$$

where θ is the angle between OD and AD. Noting that both $\boldsymbol{n}_x$ and $\boldsymbol{e}_1$ are unit vectors, the angle, θ, can be found using the relation

$$\cos \theta = \boldsymbol{n}_x \cdot \boldsymbol{e}_1.$$

Therefore, the surface area of the inclined plane may be related to the surface area of a face such that

$$\delta S^{(1)} = \frac{1}{2} \mathrm{CB} \cdot \mathrm{OD} = \frac{1}{2} \mathrm{CB} \cdot \mathrm{AD} \left(\boldsymbol{n}_x \cdot \boldsymbol{e}_1\right) = \delta S \left(\boldsymbol{n}_x \cdot \boldsymbol{e}_1\right).$$

Similarly, we obtain the following set of relations

$$\begin{aligned} \delta S^{(1)} &= \delta S \left(\boldsymbol{n}_x \cdot \boldsymbol{e}_1\right) = \delta S\, n_1, \\ \delta S^{(2)} &= \delta S \left(\boldsymbol{n}_x \cdot \boldsymbol{e}_2\right) = \delta S\, n_2, \\ \delta S^{(3)} &= \delta S \left(\boldsymbol{n}_x \cdot \boldsymbol{e}_3\right) = \delta S\, n_3. \end{aligned}$$

Substitution of these relations into the equation for conservation of linear momentum gives

$$-\boldsymbol{t}_x^{(\boldsymbol{e}_1)} n_1 - \boldsymbol{t}_x^{(\boldsymbol{e}_2)} n_2 - \boldsymbol{t}_x^{(\boldsymbol{e}_3)} n_3 + \boldsymbol{t}_x = \frac{\delta V}{\delta S_n} \left(\rho \boldsymbol{a} - \rho \boldsymbol{b}_x\right).$$

The volume of the tetrahedron is related to the base area and the perpendicular height from origin to the angled face, h. Therefore, $\delta V = \frac{h}{3} \delta S$ and as the tetrahedron shrinks to a point and $h \to 0$, we obtain

$$\begin{aligned} \boldsymbol{t}_x &= \boldsymbol{t}_x^{(\boldsymbol{e}_1)} n_1 + \boldsymbol{t}_x^{(\boldsymbol{e}_2)} n_2 + \boldsymbol{t}_x^{(\boldsymbol{e}_3)} n_3 \\ &= \boldsymbol{t}_x^{(\boldsymbol{e}_i)} n_i. \end{aligned} \tag{3.7}$$

We substitute Equation (3.6) into Equation (3.7) to obtain

$$\boldsymbol{t}_x = \left(\sigma_{ij} \boldsymbol{e}_j\right) n_i.$$

Taking the dot product of this equation with $\boldsymbol{e}_k$ gives

$$\begin{aligned} \boldsymbol{t}_x \cdot \boldsymbol{e}_k &= \left[\left(\sigma_{ij} \boldsymbol{e}_j\right) n_i \cdot \boldsymbol{e}_k\right] \\ (t_x)_k &= \sigma_{ij} n_i \left(\boldsymbol{e}_j \cdot \boldsymbol{e}_k\right) \\ (t_x)_k &= \sigma_{ik} n_i. \end{aligned}$$

In tensor notation, this equation can be written

$$\boldsymbol{t}_x = \boldsymbol{n}_x \cdot \boldsymbol{\sigma} = \boldsymbol{\sigma}^T \cdot \boldsymbol{n}_x. \tag{3.8}$$

Equation (3.8) is known as the ***Cauchy stress formula***.

3.4 First Piola-Kirchhoff Stress Tensor

The first Piola-Kirchhoff stress tensor, $\boldsymbol{P}$, is a measure of the stress per unit referential area. Following a similar derivation as that provided for the Cauchy stress tensor, the ***first Piola-Kirchhoff stress tensor***, $\boldsymbol{P}$, can be written in terms of the first Piola-Kirchhoff traction vector as

$$\boldsymbol{t}_X = \boldsymbol{n}_X \cdot \boldsymbol{P}^T = \boldsymbol{P} \cdot \boldsymbol{n}_X. \tag{3.9}$$

Let us examine the relation which relates the first Piola-Kirchhoff stress tensor to the Cauchy stress tensor. Combining Equations (3.3) and (3.4), the instantaneous force in the current configuration can be written as either the product of the Cauchy traction vector times the area in the reference configuration or the first Piola-Kirchhoff traction vector times the area in the reference configuration such that

$$d\boldsymbol{f} = \boldsymbol{t}_x dA_x = \boldsymbol{t}_X dA_X. \tag{3.10}$$

Substitution of the Cauchy stress formula, Equation (3.8), and the first Piola-Kirchhoff stress formula, Equation (3.9) gives

$$\boldsymbol{\sigma}^T \cdot \boldsymbol{n}_x dA_x = \boldsymbol{P} \cdot \boldsymbol{n}_X dA_X.$$

Using Nanson's formula, Equation (3.1), to convert the current infinitesimal area to the corresponding reference area gives

$$\boldsymbol{\sigma}^T \cdot J \boldsymbol{F}^{-T} \cdot \boldsymbol{n}_X dA_X = \boldsymbol{P} \cdot \boldsymbol{n}_X dA_X.$$

Since this equation must hold for all possible surfaces passing through a given point, it must be true for any selection of $\boldsymbol{n}_X$. Therefore, the Piola-Kirchhoff stress tensor can be written in terms of the symmetric Cauchy stress tensor as

$$\boldsymbol{P} = J \boldsymbol{\sigma} \cdot \boldsymbol{F}^{-T} \quad or \quad P_{iJ} = J \sigma_{ik} F_{Jk}^{-1}. \tag{3.11}$$

The inverse relation is given by

$$\boldsymbol{\sigma} = J^{-1} \boldsymbol{P} \cdot \boldsymbol{F}^T \quad or \quad \sigma_{ij} = J^{-1} P_{iK} F_{jK}. \tag{3.12}$$

Notice that the first Piola-Kirchhoff stress tensor is a mixed tensor with one index in the current configuration and one index in the reference configuration. Unlike the Cauchy stress tensor, the Piola-Kirchhoff stress tensor is ***not*** a symmetric tensor.

3.5 Second Piola-Kirchhoff Stress Tensor

The ***second Piola-Kirchhoff stress tensor***, $\boldsymbol{S}$, is a measure of the transformed force per unit referential area. The second Piola-Kirchhoff stress tensor can be written in terms of the second Piola-Kirchhoff traction vector, $\boldsymbol{t}_X^*$, acting on a surface with outward normal, $\boldsymbol{n}_X$, in the reference configuration such that

$$\boldsymbol{t}_X^* = \boldsymbol{n}_X \cdot \boldsymbol{S}^T = \boldsymbol{S} \cdot \boldsymbol{n}_X.$$

We can obtain a relation between the second Piola-Kirchhoff stress tensor and the Cauchy stress tensor by making use of Equations (3.2), (3.3), and (3.5) to write

$$\begin{aligned} d\boldsymbol{f}_X &= \boldsymbol{F}^{-1} \cdot d\boldsymbol{f}_x \\ \boldsymbol{t}_X^* \, dA_X &= \boldsymbol{F}^{-1} \cdot \boldsymbol{t}_x^n \, dA_x \\ \boldsymbol{S} \cdot \boldsymbol{n}_X dA_X &= \boldsymbol{F}^{-1} \cdot \boldsymbol{\sigma}^T \cdot \boldsymbol{n}_x \, dA_x. \end{aligned}$$

Using Nanson's formula, Equation (3.1), to convert the current infinitesimal area to the corresponding reference area gives

$$\boldsymbol{S} \cdot \boldsymbol{n}_X dA_X = \boldsymbol{F}^{-1} \cdot \boldsymbol{\sigma}^T \cdot J \boldsymbol{F}^{-T} \cdot \boldsymbol{n}_X \, dA_X.$$

The second Piola-Kirchhoff stress tensor is then written as

$$\boldsymbol{S} = J \boldsymbol{F}^{-1} \cdot \boldsymbol{\sigma} \cdot \boldsymbol{F}^{-T} \quad or \quad S_{IJ} = J F_{Ik}^{-1} \sigma_{km} F_{Jm}^{-1}. \tag{3.13}$$

The inverse relation is given by

$$\boldsymbol{\sigma} = J^{-1} \boldsymbol{F} \cdot \boldsymbol{S} \cdot \boldsymbol{F}^T \quad or \quad \sigma_{ij} = J F_{iK} S_{KM} F_{jM}. \tag{3.14}$$

The second Piola-Kirchhoff stress tensor is symmetric.

3.6 Maximum Normal and Shear Stress

Earlier in this chapter, we learned that the components of the stress tensor are the projections of the traction vector onto the surfaces of an infinitesimal volume element. When the coordinate system changes and the volume element rotates, the components of the stress change. The maximum normal component of the stress is obtained when the traction vector is perpendicular to the surface on which it acts. For the Cauchy traction vector, this condition can be written as

$$\boldsymbol{t}_x = \alpha \boldsymbol{n},$$

where α is a scalar, and $\boldsymbol{n}$ is the outward normal of the surface. Substituting the Cauchy stress formula, Equation (3.8), for the traction vector gives

$$\boldsymbol{\sigma}^T \cdot \boldsymbol{n} = \alpha \boldsymbol{n}.$$

As you can see, this becomes an eigenvalue problem. The maximum true normal stress components are given by the principal values of the Cauchy stress tensor. The principal values of the Cauchy stress tensor will be denoted as σ_1, σ_2, and σ_3, and the cooresponding normalized eigenvectors are denoted as $\hat{\boldsymbol{m}}_1$, $\hat{\boldsymbol{m}}_2$, and $\hat{\boldsymbol{m}}_3$. Each eigenvector is normal to the plane on which the normal stress is maximum. Since the traction vector is perpendicular to the surface, there is no shear stress component on this surface. When considering only body and surface forces, the Cauchy stress tensor is symmetric and will have three orthonormal eigenvectors.

In order to determine the maximum shear stress, we find the outward normal that maximizes the shear traction vector. This is accomplished by expressing the stress and traction vector in terms of components in the coordinate system defined by the eigenvectors of the stress tensor. The components of the stress tensor in the basis defined by the three eigenvectors is given by

$$\boldsymbol{\sigma} = \begin{bmatrix} \sigma_1 & 0 & 0 \\ 0 & \sigma_2 & 0 \\ 0 & 0 & \sigma_3 \end{bmatrix}.$$

Similarly, given an outward normal, $\boldsymbol{n}_x = n_1\boldsymbol{e}_1 + n_2\boldsymbol{e}_2 + n_3\boldsymbol{e}_3$, the traction vector can be written as

$$\boldsymbol{t}_x = \boldsymbol{\sigma}^T \cdot \boldsymbol{n}_x = \sigma_1 n_1 \boldsymbol{e}_1 + \sigma_2 n_2 \boldsymbol{e}_2 + \sigma_3 n_3 \boldsymbol{e}_3.$$

The normal component of the traction is given by

$$t_x^n = \boldsymbol{t}_x \cdot \boldsymbol{n}_x = \left(\boldsymbol{\sigma}^T \cdot \boldsymbol{n}_x\right) \cdot \boldsymbol{n}_x = \sigma_1 n_1^2 + \sigma_2 n_2^2 + \sigma_3 n_3^2.$$

Therefore, the magnitude of the tangential component of the traction vector can be written as

$$\left|\boldsymbol{t}_x^s\right|^2 = |\boldsymbol{t}_x|^2 - \left|\boldsymbol{t}_x^n\right|^2 = \sum_{i=1}^{3} (\sigma_i n_i)^2 - \left(\sum_{i=1}^{3} \sigma_i n_i^2\right)^2.$$

In order to find the maximum shear stress, we must find the maxima of the function $\left|\boldsymbol{t}_x^s\right|^2$ subject to the constraint that the magnitude of the normal is one such that $|\boldsymbol{n}_x| = 1$. This is accomplished by introducing a Lagrange multiplier, λ, and finding the minimum of the function

$$f(n_i,\lambda) = \sum_{i=1}^{3} (\sigma_i n_i)^2 - \left(\sum_{i=1}^{3} \sigma_i n_i^2\right)^2 - \lambda\left(n_i^2 - 1\right).$$

Maximizing the function $f(n_i,\lambda)$, requires the simultaneous solution of the equations

$$\frac{\partial f}{\partial n_1} = 2\left(\sigma_1^2 n_1\right) - 4\left(\sigma_1 n_1^2 + \sigma_2 n_2^2 + \sigma_3 n_3^2\right)\sigma_1 n_1 - 2\lambda n_1 = 0,$$

$$\frac{\partial f}{\partial n_2} = 2\left(\sigma_2^2 n_2\right) - 4\left(\sigma_1 n_1^2 + \sigma_2 n_2^2 + \sigma_3 n_3^2\right)\sigma_2 n_2 - 2\lambda n_2 = 0,$$

$$\frac{\partial f}{\partial n_3} = 2\left(\sigma_3^2 n_3\right) - 4\left(\sigma_1 n_1^2 + \sigma_2 n_2^2 + \sigma_3 n_3^2\right)\sigma_3 n_3 - 2\lambda n_3 = 0,$$

$$\frac{\partial f}{\partial \lambda} = -\left(n_1^2 + n_2^2 + n_3^2 - 1\right) = 0.$$

The maximum shear stress is occurs for values of the outward normal

$$\boldsymbol{n} = \sqrt{\frac{1}{2}}\langle 0,1,1\rangle, \quad \sqrt{\frac{1}{2}}\langle 1,0,1\rangle, \quad \text{and} \quad \sqrt{\frac{1}{2}}\langle 1,1,0\rangle.$$

This indicates that the outward normal of the plane containing the maximum shear stress is oriented at a 45° angle to the principal axes. Substituting these values of the outward normal into the above equation gives

$$\left|\boldsymbol{t}_x^s\right| = \frac{1}{2}|\sigma_2 - \sigma_3|, \quad \frac{1}{2}|\sigma_1 - \sigma_3|, \quad \text{and} \quad \frac{1}{2}|\sigma_1 - \sigma_2|.$$

3.7 Decomposition of the Stress Tensor

While we can decompose any tensor into an ***isotropic*** and ***deviatoric*** part, the decomposition of the stress tensor has particular significance in constitutive modeling. Decomposition of the stress is often performed to isolate the driving force causing some process. For example, the density of an isotropic compressible fluid is a function of only the hydrostatic part of the stress tensor, whereas the condition for plastic deformation in a metal might depend only on the deviatoric part of the stress tensor.

We begin with a discussion of pure shear stress and hydrostatic stress. The deviatoric part of the stress tensor represents a state of pure shear. A state of pure shear stress is one in which a coordinate system can be found such that the three normal stresses, σ_{11}, σ_{22}, and σ_{33} are each zero. Therefore, there exists a coordinate system for which the stress tensor may be represented in matrix form as

$$\boldsymbol{\sigma}_{deviatoric} = \begin{bmatrix} 0 & \sigma_{12} & \sigma_{13} \\ \sigma_{21} & 0 & \sigma_{23} \\ \sigma_{31} & \sigma_{32} & 0 \end{bmatrix}.$$

Recall that the trace of the tensor is invariant under coordinate transformation. Therefore, if the trace of a tensor is zero in one coordinate system, then the trace is zero in any other coordinate system. In fact, having a zero trace is a necessary and sufficient condition for a state of pure shear stress.

The ***hydrostatic*** or isotropic stress state is the equivalent of a mechanical pressure. The components of the stress are same in all coordinate systems and can be represented in matrix form as a scalar multiplied by the identity matrix such that

$$\boldsymbol{\sigma}_{iso} = \alpha \mathbf{I} = \begin{bmatrix} \alpha & 0 & 0 \\ 0 & \alpha & 0 \\ 0 & 0 & \alpha \end{bmatrix}.$$

We can decompose any generic stress tensor, $\boldsymbol{\sigma}$, into a hydrostatic, $\boldsymbol{\sigma}_{iso}$, and deviatoric part, $\mathcal{S}$, such that

$$\boldsymbol{\sigma} = \boldsymbol{\sigma}_{\mathbf{iso}} + \mathcal{S}.$$

The isotropic part is determined by taking one third the trace of the stress tensor such that

$$\boldsymbol{\sigma}_{iso} = \frac{1}{3} tr(\boldsymbol{\sigma})\, \mathbf{I}.$$

The deviatoric part of the stress is found by subtracting the isotropic component from the total stress tensor such that

$$\mathcal{S} = \boldsymbol{\sigma} - \frac{1}{3} tr(\boldsymbol{\sigma}).$$

EXERCISES

1. The Cauchy stress tensor is given at a point within the material body by

$$\boldsymbol{\sigma} = \begin{bmatrix} 4 & 2 & 3 \\ 2 & 3 & -6 \\ 3 & -6 & 5 \end{bmatrix} [\text{MPa}].$$

   a. Find the principal stresses and their directions.
   b. Find the Cauchy traction vector on the plane with unit normal $\boldsymbol{n} = \frac{1}{\sqrt{14}}(2, 3, 1)$. Find the normal and shear components of the traction vector.

2. The Cauchy stress tensor field for a deformed body is given by

$$\sigma = \begin{bmatrix} 2X_1 & 3X_2 & 0 \\ 3X_2 & 3X_1X_2 & 0 \\ 0 & 0 & 0 \end{bmatrix} \text{[MPa]}.$$

   a. Find the resultant force vector on the surface of a unit cube in the current configuration, which has unit normal $n = (1,0,0)$.
   b. Find the resultant force vector on the surface of a unit cube in the current configuration, which has unit normal $n = (0,1,0)$.
   c. Find the resultant force vector on the surface of a unit cube in the current configuration, which has unit normal $n = (0,0,1)$.

3. A unit cube is deformed according to the following equations of motion

$$x_1 = \alpha X_1 + \beta X_2,$$

$$x_2 = \beta X_1 + \alpha X_2,$$

$$x_3 = X_3,$$

   where $\alpha = 1.2$, and $\beta = 0.3$. The Cauchy stress for this unit cube is given by

$$\sigma = \begin{bmatrix} 4 & 2 & 3 \\ 2 & 3 & -6 \\ 3 & -6 & 5 \end{bmatrix} \text{[MPa]}.$$

   a. Find the first Piola-Kirchhoff stress tensor.
   b. Find the second Piola-Kirchhoff stress tensor.
   c. Decompose the Cauchy stress tensor into a deviatoric part, $\boldsymbol{S}$, and an isotropic part, $p\mathbf{I}$, such that $\boldsymbol{\sigma} = -p\mathbf{I} + \boldsymbol{S}$. Note that p is a scalar and $\mathbf{I}$. is the identity tensor.

4 Introduction to Material Modeling

As engineers, we seek to develop mathematical models that allow us to predict a system's response to external stimuli. For example, one might want to predict the strain that results when an object is subjected to a set of prescribed forces. In this text, we will discuss the development of a set of equations that describe the relationship between applied forces, thermodynamic variables, and deformation. The procedure presented for building a practical model of a material system consists of four major steps. First, we must identify the forces, fields, and thermodynamic variables that we would like to model. For example, we might be interested in modeling the material's response to changes in temperature and electric field. In nature, there are many forces and fields which influence the behavior of materials. A model that captures the coupling between all of these fields would be exceedingly complex. Instead, we must select the forces and field which are of primary interest or restrict the applicability of the mathematical model to a narrow range of external forces to simplify the model. Second, the balance laws and constitutive model must be formulated given the relevant variables and material characteristics determined in step one. The result of this second step is a set of mathematical equations describing the connections between the selected forces and fields in the given material system. Third, a strategy for parameterizing the constitutive model must be developed. This involves selection of an appropriate experimental method and the derivation of working equations particular to that experiment. Fourth, a numerical solution must be obtained in order to apply the mathematical model to conditions mimicking real-world systems. The mathematical equations derived in step two are often complex and an analytical solution may be difficult to obtain. In addition, the geometry and boundary conditions found in real world material applications make analytical solution for some constitutive models impossible.

The final result of this procedure consists of a set of balance equations and a parameterized constitutive model that can be used to solve for the behavior of structures or machines incorporating the modeled material. Often analytical solution of even the most elementary constitutive model is impossible given the complex geometries encountered in engineering applications. Typically, the finite element method is used to solve the set of partial differential equations which make up the constructed material model.

Perhaps the most critical step in material modeling is the specification of the characteristics which are to be captured by the model. In this chapter we limit ourselves to thermo mechanical materials. Before we begin, let us review some basic terminology which will be used to classify our material models. An ***isothermal material model*** is used to describe the behavior of a material that is held at constant temperature. Heat flux, and thermo-mechanical coupling need not be included in the formulation of an isothermal material model. A ***thermal material model*** will include thermo-mechanical coupling and must include heat flux within the formulation. An ***elastic material*** exhibits no hysteresis. If one were to stretch and then release an elastic material, the load and unload curves would coincide. Models may include ***finite deformation*** or be limited to small strains. Models may include ***internal variables***, which may be used to account for processes such as damage, plasticity, or even growth of materials.

4.1 Forces and Fields

When developing a material model, the first, most important, step is to identify the forces and fields that will be incorporated in the model. Including extraneous forces or fields leads to an overly complex and unwieldy model. As a simple example, consider a machine component made of structural steel. If one intends to model the response of this part under the application of very small static loads in an application where the temperature varies, one would include the temperature field in the model formulation. However, if the temperature of the part remains constant, and the loading is very slow, then one might omit the temperature field without sacrificing the ability to model the system. On the other hand, if you would like to capture the piezoelectric behavior in a system, you would have to include the electric field within your formulation.

In this chapter, we will examine the general development of a ***thermo-mechanical model***. We consider only mechanical and thermal fields and forces neglecting fields such as but not limited to the electric field, magnetic field, and chemical reactions. Within this thermo-mechanical model, we must account for the velocity vector $\boldsymbol{v}$, displacement vector $\boldsymbol{u}$, temperature θ, heat flux vector $\boldsymbol{q}$, stress tensor $\boldsymbol{\sigma}$, density ρ, internal energy e, and entropy η at each point within the continuum body. In later chapters, we will add additional variables to this list which account for interactions between material phases, growth of a material, and other phenomena. In addition to listing the relevant variables, we will also list important assumptions made concerning this material system. We will assume that there are no material sources or sinks, and we assume that only body forces and traction forces are transmitted to the material.

Note that the general thermo-mechanical model captures the mechanical and thermal response of a system. We have not yet specified whether the system is elastic, or viscoelastic. We have made no mention of the material phase. In fact, in this textbook, the models for the ideal gas, Newtonian, and non-Newtonian fluids, elastic, and hyperelastic materials are all examples of thermo-mechanical models.

4.2 Balance Laws

The balance laws also known as the conservation laws are physical principles that must be satisfied for all materials systems. They consist of conservation of mass, conservation of linear momentum, conservation of angular momentum, conservation of energy, and the second law of thermodynamics. The specific form of the balance laws depends on the forces and fields included within the model. For example, the balance laws for a piezoelectric material will include terms not found in the thermo-mechanical materials. In this section, we will derive both the Eulerian and Lagrangian forms of the balance equations for thermo-mechanical materials. In addition, we will derive the integral and local forms of each equation.

4.2.1 Conservation of Mass

The principle of conservation of mass states that mass is neither created nor destroyed within a system. Even though we can have terms which represent sinks or sources of mass, these are mathematical constructs that allow us to transfer mass to or from the system surroundings. In this chapter, we will consider the statement of conservation of mass in both spatial and material forms, neglecting the presence of mass sources or sinks.

4.2.1.1 Spatial Form

The principle of conservation of mass states that the material derivative of the integral of a continuous spatial density field over a volume in the current configuration must equal the rate of change of the total mass of the system. This can be written as

$$\frac{D}{Dt}[m(t)] = \frac{D}{Dt}\int_{\Omega(t)} \rho(\boldsymbol{x}, t)\, dV_x,$$

where m is the mass within the volume, ρ is the spatial density field, and dV_x is a volume element within the current configuration. Having previously assumed that there are no mass sources or sinks in the material being modeled, we can state that $\dot{m}(t) = 0$. Using the Reynolds transport theorem in the form given in Equation (2.54), the integral form of the conservation of mass becomes

$$0 = \int_{\Omega(t)} \left[\frac{D\rho}{Dt} + \rho\, div(\boldsymbol{v})\right] dV_x.$$

Since this relation must hold for any arbitrary choice of $\Omega(t)$, the integrand must vanish at all points. If the integrand were nonzero at even a single point, we could always select a volume that included only that single point and violate the relation. Therefore, the ***local form*** or ***differential form*** of the conservation of mass can be written as

$$0 = \frac{D\rho}{Dt} + \rho\, div(\boldsymbol{v}), \tag{4.1}$$

or equivalently as

$$0 = \frac{\partial \rho}{\partial t} + div(\rho\boldsymbol{v}).$$

These are referred to as local forms due to the fact that they must be true at each and every point within the material.

As we proceed through the derivation of the remaining conservation laws, we will often encounter terms that have the form $\frac{D}{Dt}\int_{\Omega} \rho(\boldsymbol{x}, t)\, \phi(\boldsymbol{x}, t)\, dV_x$, where ϕ is a scalar field and ρ is the density field. When there are no mass sources or sinks, the Reynolds transport theorem to the product of any field, ϕ, and the scalar density, ρ, may be greatly simplified. The Reynolds transport theorem applied to this product gives

$$\frac{D}{Dt}\int_{\Omega}\rho(\boldsymbol{x},t)\,\phi(\boldsymbol{x},t)\,dV_x = \int_{\Omega}\left[\rho\frac{D\phi}{Dt}+\phi\frac{D\rho}{Dt}+(\rho\,\phi)\,div\,(\boldsymbol{v})\right]dV_x$$
$$= \int_{\Omega}\left[\rho\frac{D\phi}{Dt}+\phi\left(\frac{D\rho}{Dt}+\rho\,div\,(\boldsymbol{v})\right)\right]dV_x.$$

Combining this with the local form of conservation of mass, Equation (4.1), gives

$$\frac{D}{Dt}\int_{\Omega}\rho\,\phi\,dV_x = \int_{\Omega}\rho\frac{D\phi}{Dt}\,dV_x. \tag{4.2}$$

The equivalent result holds for any vector field. We will use this result extensively in the derivative of the remaining conservation equations.

4.2.1.2 Material Form

The material form of the conservation of mass equation is found by following a set of material points through time. The particles of interest initially occupy a volume Ω_o at $t=0$, and occupy a volume $\Omega(t)$ at time t. We have already assumed that there are no sinks or sources of mass. The mass of the collection of particles can then be written as an integral of the referential density over the reference configuration or an integral of the spatial density field over the current configuration

$$\int_{\Omega_o}\rho_o(\boldsymbol{X})\,dV_X = \int_{\Omega(t)}\rho(\boldsymbol{x},t)\,dV_x.$$

The infinitesimal volume element in the current configuration, dV_x, is related to the infinitesimal volume element in the reference configuration, dV_X, by the Jacobian according to

$$dV_x = J\,dV_X.$$

The statement of conservation of mass then becomes

$$\int_{\Omega_o} J(\boldsymbol{X},t)\,\rho(\boldsymbol{x}(\boldsymbol{X},t),t)\,dV_X = \int_{\Omega_o}\rho_o(\boldsymbol{X})\,dV_X.$$

Notice that the limits of integration have changed, and the spatial density field is expressed in terms of reference coordinates by employing the equations of motion. Once again, this integral equation must be true for any arbitrary choice of initial volume Ω_o. This requires that the integrands be equal such that

$$\rho_o = \rho J. \tag{4.3}$$

Therefore, the ratio of the density in the reference configuration to that in the current configuration is given by the Jacobian.

4.2.2 Conservation of Linear Momentum

The principle of conservation of linear momentum also known as Newton's second law states the rate of change of the linear momentum of a body is equal to the sum of the forces acting on the body. For a rigid body, this can be written as

$$\frac{D}{Dt}(m\,\boldsymbol{v}_{cm}) = \sum_i \mathbf{f}^{(i)},$$

where $\mathbf{f}^{(i)}$ are the applied forces, m is the mass of the object, and $\boldsymbol{v}_{cm}$ is the velocity of the center of mass of the object.

4.2.2.1 Spatial Form

In spatial form, we begin with the statement that the rate of change of the linear momentum within a system is equal to the sum of the forces acting on that system. Writing the conservation of linear momentum as an integral over a volume of space, the change in linear momentum is equal to the sum of forces acting on the surface of the body plus the sum of forces acting in the interior of the body

$$\frac{D}{Dt}\int_{\Omega(t)} \rho(\boldsymbol{x},t)\,\boldsymbol{v}(\boldsymbol{x},t)\,dV_x = \int_{\partial\Omega(t)} \boldsymbol{t}_x(\boldsymbol{x},t)\,dS_x + \int_{\Omega(t)} \rho(\boldsymbol{x},t)\,\boldsymbol{b}_x(\boldsymbol{x},t)\,dV_x, \quad (4.4)$$

where $\boldsymbol{t}_x$ is the Cauchy traction, $\boldsymbol{b}_x$ is the body force per unit mass, ρ is the density, and $\boldsymbol{v}$ is the velocity vector. Combining Equation (4.4), the modified Reynolds transport theorem given by Equation (4.2), and the Cauchy stress formula, Equation (3.8), gives

$$\int_{\Omega(t)} \rho\,\dot{\boldsymbol{v}}\,dV_x = \int_{\partial\Omega(t)} \boldsymbol{\sigma}^T\cdot\boldsymbol{n}_x\,dS_x + \int_{\Omega(t)} \rho\,\boldsymbol{b}_x\,dV_x.$$

Using the divergence theorem, Equation (1.36), to convert the surface integral to a volume integral and combining the integrands gives

$$\int_{\Omega(t)} \left(\rho\,\dot{\boldsymbol{v}} - div\,\boldsymbol{\sigma}^T - \rho\boldsymbol{b}_x\right) dV_x = \mathbf{0}.$$

This equation is valid for any arbitrary choice of volume, $\Omega(t)$. Therefore, the integrand must vanish at each point within the continuum body giving

$$\rho\,\dot{\boldsymbol{v}} - div\,\boldsymbol{\sigma}^T - \rho\,\boldsymbol{b}_x = \mathbf{0}. \quad (4.5)$$

4.2.2.2 Material Form

We can find the material form of the principle of conservation of linear momentum by starting with Equation (4.4) and converting the integral of spatial quantities over a volume in the current configuration into the integral over the corresponding volume in the reference configuration. We begin by restating Equation (4.4) for convenience

$$\frac{D}{Dt}\int_{\Omega(t)} \rho(\boldsymbol{x},t)\,\boldsymbol{v}(\boldsymbol{x},t)\,dV_x = \int_{\partial\Omega(t)} \boldsymbol{t}_x(\boldsymbol{x},t)\,dS_x + \int_{\Omega(t)} \rho(\boldsymbol{x},t)\,\boldsymbol{b}_x(\boldsymbol{x},t)\,dV_x. \quad (4.6)$$

We note that an integral over the volume $\Omega(t)$ can be written in terms of the corresponding reference volume Ω_o as

$$\int_{\Omega(t)} dV_x = \int_{\Omega_o} J(\boldsymbol{X},t)\,dV_X. \quad (4.7)$$

where J is the Jacobian. We also make note of Equation (3.10), which tells us that $\boldsymbol{t}_x^n\, dS_x = \boldsymbol{t}_X^n\, dS_X$:

$$\int_{\partial\Omega(t)} \boldsymbol{t}_x(\boldsymbol{x}, t)\, dS_x = \int_{\partial\Omega_0} \boldsymbol{t}_X(\boldsymbol{X}, t)\, dS_X. \tag{4.8}$$

Combining Equation (4.7) and Equation (4.8) with Equation (4.6) gives

$$\frac{D}{Dt}\int_{\Omega_0} \rho(\boldsymbol{x}(\boldsymbol{X},t),t)\, \boldsymbol{v}(\boldsymbol{x}(\boldsymbol{X},t),t) J(\boldsymbol{X},t)\, dV_X$$

$$= \int_{\partial\Omega_0} \boldsymbol{t}_X(\boldsymbol{X},t)\, dS_X + \int_{\Omega_0} \rho(\boldsymbol{x}(\boldsymbol{X},t),t)\, \boldsymbol{b}_x(\boldsymbol{x}(\boldsymbol{X},t),t)\, J(\boldsymbol{X},t)\, dV_X. \tag{4.9}$$

We introduce a new variable for the material description of the body force field, $\boldsymbol{b}_o$ such that

$$\boldsymbol{b}_o(\boldsymbol{X}, t) = \boldsymbol{b}_x(\boldsymbol{x}(\boldsymbol{X}, t), t). \tag{4.10}$$

The material description of the velocity field will be noted as $\boldsymbol{v}_o$ such that

$$\boldsymbol{v}_o(\boldsymbol{X}, t) = \boldsymbol{v}_x(\boldsymbol{x}(\boldsymbol{X}, t), t). \tag{4.11}$$

Combining Equation (4.3), which provides a relation between the referential and spatial descriptions of the density field, with Equations (4.10) and (4.11) gives

$$\frac{D}{Dt}\int_{\Omega_0} \rho_o \boldsymbol{v}_o\, dV_X = \int_{\partial\Omega_0} \boldsymbol{t}_X\, dS_X + \int_{\Omega_0} \rho_o \boldsymbol{b}_o\, dV_X. \tag{4.12}$$

Substituting Equation (3.9) which gives the first Piola-Kirchoff traction vector, $\boldsymbol{t}_X$, in terms of the first Piola-Kirchoff stress tensor, $\boldsymbol{P}$, and outward surface normal, and using the divergence theorem, Equation (1.36), to convert the surface integral into a volume integral gives

$$\int_{\partial\Omega_0} \boldsymbol{t}_X\, dS_X = \int_{\partial\Omega_0} \boldsymbol{P}\cdot\boldsymbol{n}_X\, dS_X = \int_{\Omega_0} Div(\boldsymbol{P})\, dV_X.$$

Equation (4.12) becomes

$$\int_{\Omega_0} \left(\rho_o \dot{\boldsymbol{v}}_o - Div(\boldsymbol{P}) - \rho_o \boldsymbol{b}_o\right) dV_X = \boldsymbol{0}.$$

This equation must be satisfied for any arbitrary choice of volume, Ω_o. Therefore, the integrand must be zero at each point within the continuum body. The local material form of the conservation of linear momentum becomes

$$\rho_o \dot{\boldsymbol{v}}_o - Div(\boldsymbol{P}) - \rho_o \boldsymbol{b}_o = \boldsymbol{0}. \tag{4.13}$$

4.2.3 Conservation of Angular Momentum

The principle of conservation of angular momentum states that the time rate of change of the total angular momentum of a body is equal to the sum of the moments acting on the body. For a rigid body, this can be written as

$$\frac{D}{Dt}(\boldsymbol{I}\,\boldsymbol{\omega}) = \sum_i \mathbf{m}^{(i)},$$

where $\boldsymbol{I}$ is the moment of intertia, $\boldsymbol{\omega}$ is the angular velocity, and $\mathbf{m}^{(i)}$ are the applied moments.

4.2.3.1 Spatial Form

The spatial integral form of the principle of conservation of angular momentum expressed about a fixed point, $\boldsymbol{x}_o$, for a collection of particles making up a continuum body is given by

$$\frac{D}{Dt}\int_{\Omega(t)} ((\boldsymbol{x}-\boldsymbol{x}_o)\times\rho(\boldsymbol{x},t)\,\boldsymbol{v}(\boldsymbol{x},t))\,dV_x$$
$$=\int_{\partial\Omega(t)} ((\boldsymbol{x}-\boldsymbol{x}_o)\times(\boldsymbol{t}_x(\boldsymbol{x},t)))\,dS_x+\int_{\Omega(t)}\left((\boldsymbol{x}-\boldsymbol{x}_o)\times\rho(\boldsymbol{x},t)\,\boldsymbol{b}_x(\boldsymbol{x},t)\right)dV_x. \quad (4.14)$$

The Reynolds transport theorem given by Equation (2.54) can be used to simplify the first term

$$\frac{D}{Dt}\int_{\Omega(t)} ((\boldsymbol{x}-\boldsymbol{x}_o)\times\rho\,\boldsymbol{v})\,dV_x$$
$$=\int_{\Omega(t)}\left(\overline{(\boldsymbol{x}-\boldsymbol{x}_o)}^{\cdot}\times\rho\,\boldsymbol{v}+(\boldsymbol{x}-\boldsymbol{x}_o)\times(\dot{\rho}\,\boldsymbol{v}+\rho\dot{\boldsymbol{v}}+(\rho\,\boldsymbol{v})\,div\,(\boldsymbol{v}))\right)dV_x.$$

Notice that $\overline{(\boldsymbol{x}-\boldsymbol{x}_o)}^{\cdot}\times\boldsymbol{v}=\dot{x}\times\boldsymbol{v}=\boldsymbol{v}\times\boldsymbol{v}=0$ and that the principle of conservation of mass can be used to state that $(\dot{\rho}\,\boldsymbol{v}+\rho\dot{\boldsymbol{v}}+(\rho\,\boldsymbol{v})\,div\,(\boldsymbol{v}))=\rho\dot{\boldsymbol{v}}$. Combining these relations with the Cauchy stress formula, Equation (3.8), the principle of conservation of angular momentum can be written as

$$\int_{\Omega(t)} (\boldsymbol{r}\times\rho\,\dot{\boldsymbol{v}})\,dV_x=\int_{\partial\Omega(t)}\left(\boldsymbol{r}\times\left(\boldsymbol{\sigma}^T\cdot\boldsymbol{n}_x\right)\right)dS_x+\int_{\Omega(t)}\left(\boldsymbol{r}\times\rho\,\boldsymbol{b}_x\right)dV_x, \quad (4.15)$$

where $\boldsymbol{r}=\boldsymbol{x}-\boldsymbol{x}_o$. The divergence theorem, Equation (1.36), can be applied to convert the surface integral into a volume integral. Written in index notation, the first term on the right of Equation (4.15) becomes

$$\int_{\partial\Omega(t)}\left(\varepsilon_{ijk}r_j\sigma_{mk}n_m\right)dS_x=\int_{\Omega(t)}\frac{\partial}{\partial x_m}\left(\varepsilon_{ijk}r_j\sigma_{mk}\right)dV_x$$
$$=\int_{\Omega(t)}\varepsilon_{ijk}\left(r_{j,m}\sigma_{mk}+r_j\sigma_{mk,m}\right)dV_x$$
$$=\int_{\Omega(t)}\varepsilon_{ijk}\left(\delta_{jm}\sigma_{mk}+r_j\sigma_{mk,m}\right)dV_x$$
$$=\int_{\Omega(t)}\left(\varepsilon_{ijk}\sigma_{jk}+r_j\sigma_{mk,m}\right)dV_x.$$

Written in tensor notation, this becomes

$$\int_{\partial\Omega(t)}\left(\boldsymbol{r}\times\left(\boldsymbol{\sigma}^T\cdot\boldsymbol{n}_x\right)\right)dS_x=\int_{\Omega(t)}\left(\boldsymbol{r}\times div\,\boldsymbol{\sigma}^T+\boldsymbol{\varepsilon}:\boldsymbol{\sigma}\right)dV_x. \quad (4.16)$$

Equation (4.15) becomes

$$\int_{\Omega(t)}\left(\boldsymbol{r}\times\left(\rho\dot{\boldsymbol{v}}-\boldsymbol{div}\,\boldsymbol{\sigma}^T-\rho\boldsymbol{b}_x\right)-\boldsymbol{\varepsilon}:\boldsymbol{\sigma}\right)dV_x=0.$$

Combining this with the principle of conservation of linear momentum given by Equation (4.5), this becomes

$$\int_{\Omega(t)}\boldsymbol{\varepsilon}:\boldsymbol{\sigma}\,dV_x=0.$$

Since this must be true for any arbitrary choice of volume, $\Omega(t)$, the integrand must vanish at each point within the continuum body. Therefore, we have the condition that

$$\varepsilon_{ijk}\sigma_{jk} = 0.$$

This is satisfied if and only if the stress tensor is symmetric

$$\boldsymbol{\sigma} = \boldsymbol{\sigma}^T.$$

4.2.3.2 Material Form

The material form of the principle of conservation of angular momentum is found by converting the integral of spatial quantities over a volume in the current configuration into the integral over the corresponding volume in the reference configuration in Equation (4.14). Combining Equations (4.14), (4.7), (4.8), (4.3), (4.10), and (4.11) along with the relation that $\boldsymbol{r}(\boldsymbol{X},t) = (\boldsymbol{x}(\boldsymbol{X},t) - \boldsymbol{x}_o)$ gives

$$\frac{D}{Dt}\int_{\Omega_0} (\boldsymbol{r}\times\rho_o\boldsymbol{v}_o)\,dV_X = \int_{\partial\Omega_0} (\boldsymbol{r}\times\boldsymbol{t}_X)\,dS_X + \int_{\Omega_0} \left(\boldsymbol{r}\times\rho_o\boldsymbol{b}_o\right)dV_X. \tag{4.17}$$

Using Equation (3.9) to write the first Piola-Kirchhoff traction vector in terms of the first Piola-Kirchhoff stress tensor and the outward surface normal, and the divergence theorem, Equation (1.36), the second term in Equation (4.17) becomes

$$\begin{aligned}
\int_{\partial\Omega_o} \left(\varepsilon_{ijK} r_j P_{KM} n_M\right) dS_X &= \int_{\Omega_o} \frac{\partial}{\partial X_M}\left(\varepsilon_{ijK} r_j P_{KM}\right) dV_X \\
&= \int_{\Omega_o} \varepsilon_{ijK}\left(r_{j,M} P_{KM} + r_j P_{KM,M}\right) dV_X \\
&= \int_{\Omega_0} \varepsilon_{ijK}\left(x_{j,M} P_{KM} + r_j P_{KM,M}\right) dV_X \\
&= \int_{\Omega_o} \left(\varepsilon_{ijK} F_{jM} P_{KM} + r_j P_{KM,M}\right) dV_X.
\end{aligned}$$

Written in tensorial notation, this becomes

$$\int_{\partial\Omega_0} (\boldsymbol{r}\times(\boldsymbol{P}\cdot\boldsymbol{n}_X))\,dS_X = \int_{\Omega_0} \left(\boldsymbol{r}\times(Div\,\boldsymbol{P}) + \boldsymbol{\varepsilon} : \left(\boldsymbol{F}\cdot\boldsymbol{P}^T\right)\right) dV_X. \tag{4.18}$$

Substituting Equation (4.18) into Equation (4.17) gives

$$\int_{\Omega_0} \left(\boldsymbol{r}\times(\rho_o\dot{\boldsymbol{v}}_o - div\,\boldsymbol{P} - \rho_o\boldsymbol{b}_o) + \boldsymbol{\varepsilon} : \left(\boldsymbol{F}\cdot\boldsymbol{P}^T\right)\right) dV_X = 0. \tag{4.19}$$

Combining this with the material form of the principle of conservation of linear momentum, Equation (4.15), gives

$$\int_{\Omega_0} \left(\boldsymbol{\varepsilon} : \left(\boldsymbol{F}\cdot\boldsymbol{P}^T\right)\right) dV_X = 0. \tag{4.20}$$

Since this equation holds for any arbitrary choice of Ω_o, the integrand must vanish at each point within the continuum body. The local material form of the principle of angular momentum is then

$$\boldsymbol{\varepsilon} : \left(\boldsymbol{F}\cdot\boldsymbol{P}^T\right) = 0.$$

This gives the three equations

$$\varepsilon_{123}F_{2l}P_{l3} - \varepsilon_{132}F_{3l}P_{l2} = 0,$$
$$\varepsilon_{231}F_{3l}P_{l1} - \varepsilon_{213}F_{1l}P_{l3} = 0,$$
$$\varepsilon_{312}F_{1l}P_{l2} - \varepsilon_{321}F_{2l}P_{l1} = 0.$$

Ultimately this requires

$$\boldsymbol{F}\cdot\boldsymbol{P}^T = \boldsymbol{P}\cdot\boldsymbol{F}^T.$$

This can be rewritten in terms of the second Piola-Kirchhoff stress tensor to show that

$$\boldsymbol{S}^T = \boldsymbol{S}.$$

Therefore, the second Piola-Kirchhoff stress tensor is symmetric, but the first Piola-Kirchhoff stress tensor is not.

4.2.4 Conservation of Energy

The principle of conservation of energy states that the time rate of change of the total energy of a continuum body is equal to the sum of the work done by the surroundings on the body and the heat added to the body. The total energy of a particle is the sum of the kinetic and potential energy. The principle of conservation of energy can then be written as

$$\frac{D}{Dt}(\mathcal{K}+\mathfrak{U}) = \sum \mathfrak{E}_i,$$

where $\mathcal{K}$ is the kinetic energy per unit mass, $\mathfrak{U}$ is the potential energy per unit mass, and $\mathfrak{E}_i$ represents the transfer of energy per unit mass due to, but not limited to, mechanical, thermal, electromechanical, and chemical forces. Having limited ourselves within this chapter to a thermo-mechanical model, we will consider only mechanical work and thermal energy in the following analysis.

4.2.4.1 Spatial Form

The spatial integral form of the conservation of energy can be written as the integral of the energy density over the spatial volume occupied by the object. For the case of the thermo-mechanical model, we have work done by body forces, surface tractions, we have heat added through conduction and radiation or absorption and the energy density of the material can be broken up into kinetic and potential energy densities.

The principle of conservation of energy for a thermo-mechanical material can then be written as

$$\frac{D}{Dt}(\mathcal{K}+\mathfrak{U}) = \mathfrak{E}_b + \mathfrak{E}_t + \mathfrak{E}_q + \mathfrak{E}_r,$$

where $\mathfrak{E}_b$, $\mathfrak{E}_t$, $\mathfrak{E}_q$, and $\mathfrak{E}_r$ are defined in Table 4.1. Equivalently, we have

$$\frac{D}{Dt}\int_{\Omega(t)}\left(\rho e + \frac{1}{2}\rho\,(\boldsymbol{v}\cdot\boldsymbol{v})\right)dV_x = \int_{\partial\Omega(t)}\left(\boldsymbol{t}_x\cdot\boldsymbol{v} - \boldsymbol{q}_x\cdot\boldsymbol{n}_x\right)dS_x + \int_{\Omega(t)}\left(\rho\boldsymbol{b}_x\cdot\boldsymbol{v} + \rho\,r_x\right)dV_x, \tag{4.21}$$

where ρ is the density, e is the internal energy per unit mass, $\boldsymbol{v}$ is the velocity, $\boldsymbol{t}_x$ is the Cauchy traction vector, $\boldsymbol{q}_x$ is the heat flux vector, $\boldsymbol{n}_x$ is the outward surface normal on

Table 4.1. *Energy and work terms relevant to a thermo-mechanical model*

	Symbol	Integral equation
Kinetic energy	$\mathcal{K}$	$\int_{\Omega(t)} \left(\frac{1}{2}\rho \boldsymbol{v}\cdot\boldsymbol{v}\right) dV_x$
Potential energy	$\mathfrak{U}$	$\int_{\Omega(t)} (\rho e)\, dV_x$
Rate of work done by body forces	$\mathfrak{E}_b$	$\int_{\Omega(t)} (\rho \boldsymbol{b}_x\cdot\boldsymbol{v})\, dV_x$
Rate of work done by surface traction	$\mathfrak{E}_t$	$\int_{\partial\Omega(t)} (\boldsymbol{t}_x\cdot\boldsymbol{v})\, dS_x$
Heat flux	$\mathfrak{E}_q$	$\int_{\Omega(t)} -(\boldsymbol{q}_x\cdot\boldsymbol{n}_x)\, dS_x$
Heat radiation or absorption	$\mathfrak{E}_r$	$\int_{\Omega(t)} (\rho r_x)\, dV_x$

the surface defined by $\partial\Omega(t)$, $\boldsymbol{b}_x$ is the body force, and r_x is the heat supply per unit mass. Using the Reynolds transport theorem in the form given by Equation (4.2), the Cauchy stress formula, Equation (3.8), the principle of conservation of energy becomes

$$\int_{\Omega(t)} (\rho\dot{e}+\rho\boldsymbol{v}\cdot\dot{\boldsymbol{v}})\, dV_x = \int_{\partial\Omega(t)} (\boldsymbol{n}_x\cdot\boldsymbol{\sigma}\cdot\boldsymbol{v}-\boldsymbol{q}_x\cdot\boldsymbol{n}_x)\, dS_x + \int_{\Omega(\mathrm{t})} (\rho\boldsymbol{b}_x\cdot\boldsymbol{v}+\rho r_x)\, dV_x.$$

Using the divergence theorem, Equation (1.36), to convert the surface integral into a volume integral gives the equation

$$\int_{\Omega(t)} \left(\rho\dot{e}+\rho\boldsymbol{v}\cdot\dot{\boldsymbol{v}}-div(\boldsymbol{\sigma}^T\cdot\boldsymbol{v})\right)+div\,(\boldsymbol{q}_x)-\rho\boldsymbol{b}_x\cdot\boldsymbol{v}-\rho r_x)\, dV_x=0.$$

Note that since $div\,(\boldsymbol{\sigma}^T\cdot\boldsymbol{v}) = \frac{\partial}{\partial x_k}(\sigma_{jk}v_j) = \sigma_{jk,k}v_j+\sigma_{jk}v_{j,k} = div\,\boldsymbol{\sigma}^T\cdot\boldsymbol{v}+\boldsymbol{\sigma}:\left(\boldsymbol{v}\overleftarrow{\nabla}\right)$ and the velocity gradient is given by $\boldsymbol{L}=\boldsymbol{v}\overleftarrow{\nabla}$, we can write

$$\int_{\Omega(t)} \left(\rho\dot{e}+\boldsymbol{v}\cdot\left(\rho\dot{\boldsymbol{v}}-div\left(\boldsymbol{\sigma}^T\right)-\rho\boldsymbol{b}_x\right)-\boldsymbol{\sigma}:\boldsymbol{L}+div\,(\boldsymbol{q}_x)-\rho r_x\right) dV_x=0.$$

Combining this with the principle of conservation of linear momentum given by Equation (4.5), this becomes

$$\int_{\Omega(t)} (\rho\dot{e}-\boldsymbol{\sigma}:\boldsymbol{L}+div\,(\boldsymbol{q}_x)-\rho r_x)\, dV_x=0.$$

Since this must be true for any arbitrary volume, $\Omega(t)$, the integrand must vanish at each point leading to the equation

$$\rho\dot{e}-\boldsymbol{\sigma}:\boldsymbol{L}+div\,(\boldsymbol{q}_x)-\rho r_x=0.$$

Since $\boldsymbol{\sigma}$ is a symmetric tensor, we note that $\boldsymbol{\sigma}:\boldsymbol{L}=\boldsymbol{\sigma}:\boldsymbol{D}$, where $\boldsymbol{D}$ is the rate of deformation tensor. Therefore, we can write the local form of the principle of conservation of energy as

$$\rho\dot{e}-\boldsymbol{\sigma}:\boldsymbol{D}+div\,(\boldsymbol{q}_x)-\rho r_x=0. \tag{4.22}$$

4.2.4.2 Material Form

The material form of the principle of conservation of energy can be found by converting the integral of spatial quantities over a volume in the current configuration into the integral over the corresponding volume in the reference configuration in Equation (4.21). We begin by restating Equation (4.21) for convenience

$$\frac{D}{Dt}\int_{\Omega(t)}\left(\rho e+\frac{1}{2}\rho\left(\boldsymbol{v}\cdot\boldsymbol{v}\right)\right)dV_x=\int_{\partial\Omega(t)}\left(\boldsymbol{t}_x\cdot\boldsymbol{v}-\boldsymbol{q}_x\cdot\boldsymbol{n}_x\right)dS_x$$
$$+\int_{\Omega(t)}\left(\rho\boldsymbol{b}_x\cdot\boldsymbol{v}+\rho\, r_x\right)dV_x.$$

The material description of the scalar heat supply per unit mass, r_o, may be written as

$$r_o\left(\boldsymbol{X},t\right)=r_x\left(\boldsymbol{x}\left(\boldsymbol{X},t\right),t\right). \tag{4.23}$$

Introducing the material description of the heat flux vector, $\boldsymbol{q}_o$, such that the heat flux across a surface may be written as

$$\boldsymbol{q}_o\cdot\boldsymbol{n}_X\,dS_X=\boldsymbol{q}_x\cdot\boldsymbol{n}_x\,dS_x,$$

where $\boldsymbol{n}_X$ is the outward normal to the surface in the reference configuration, dS_X, and $\boldsymbol{n}_x$ is the outward normal to the same surface in its current configuration, dS_x. Substituting Nanson's formula, Equation (2.22), we obtain

$$\boldsymbol{q}_o\left(\boldsymbol{X},t\right)=J\boldsymbol{F}^{-1}\left(\boldsymbol{X},t\right)\cdot\boldsymbol{q}_x\left(\boldsymbol{x}\left(\boldsymbol{X},t\right),t\right). \tag{4.24}$$

Combining Equations (4.21), (4.7), (4.8), (4.3), (4.10), and (4.11) gives

$$\int_{\Omega_0}\left(\rho_o\dot{e}+\rho_o\left(\boldsymbol{v}_o\cdot\dot{\boldsymbol{v}}_o\right)\right)dV_X=\int_{\partial\Omega_0}\left(\boldsymbol{t}_X\cdot\boldsymbol{v}_o-\boldsymbol{q}_o\cdot\boldsymbol{n}_X\right)dS_X$$
$$+\int_{\Omega_0}\left(\rho_o\boldsymbol{b}_o\cdot\boldsymbol{v}_o+\rho_o\, r_o\right)dV_X. \tag{4.25}$$

Using Equation (3.9) to write the first Piola-Kirchhoff traction vector in terms of the first Piola-Kirchhoff stress tensor and the outward surface normal, and the divergence theorem, Equation (1.36), Equation (4.25) becomes

$$\int_{\Omega_0}\left(\rho_o\dot{e}+\rho_o\left(\boldsymbol{v}_o\cdot\dot{\boldsymbol{v}}_o\right)-\rho_o\boldsymbol{b}_o\cdot\boldsymbol{v}_o-\rho_o\, r_o+Div\left(\boldsymbol{q}_o\right)-Div\left(\boldsymbol{P}^T\cdot\boldsymbol{v}_o\right)\right)dV_X=0. \tag{4.26}$$

We note that

$$Div\left(\boldsymbol{P}^T\cdot\boldsymbol{v}_o\right)=\frac{\partial}{\partial X_I}\left(P_{jI}v_j\right)=P_{jI,I}v_j+P_{jI}v_{j,I}=Div\boldsymbol{P}\cdot\boldsymbol{v}_o+\boldsymbol{P}:\dot{\boldsymbol{F}}. \tag{4.27}$$

Combining Equations (4.26) and (4.27) gives

$$\int_{\Omega_0}\left(\rho_o\dot{e}+\left(\boldsymbol{v}_o\cdot\left(\rho_o\dot{\boldsymbol{v}}_o-\boldsymbol{b}_o-Div\boldsymbol{P}\right)\right)-\boldsymbol{P}:\dot{\boldsymbol{F}}-\rho_o\, r_o+Div\left(\boldsymbol{q}_o\right)\right)dV_X=0. \tag{4.28}$$

Combining this with the material form of the principle of conservation of linear momentum, Equation (4.15), gives

$$\int_{\Omega_0}\left(\rho_o\dot{e}-\boldsymbol{P}:\dot{\boldsymbol{F}}-\rho_o\, r_o+Div\left(\boldsymbol{q}_o\right)\right)dV_X=0. \tag{4.29}$$

Since this is true for any arbitrary choice of volume, we obtain the local material form of the principle of conservation of energy is

$$\rho_o\dot{e}-\boldsymbol{P}:\dot{\boldsymbol{F}}-\rho_o\, r_o+Div\left(\boldsymbol{q}_o\right)=0. \tag{4.30}$$

4.2.5 The Second Law of Thermodynamics

The second law of thermodynamics states that the time rate of change of the total entropy of an isolated system must be equal to or greater than zero. In general, we must also account for the entropy exchanged between a system and its surroundings due to heat transfer, $\int \frac{\delta Q}{\theta}$. Thus, the full statement becomes the time rate of change of the total entropy of a system must be greater than or equal to the entropy transfer of entropy into the system.

4.2.5.1 Spatial Form

The second law of thermodynamics can be stated in spatial integral form as

$$\frac{D}{Dt}\int_{\Omega(t)} \rho\,\eta\, dV_x + \int_{\partial\Omega(t)} \frac{\boldsymbol{q}_x \cdot \boldsymbol{n}_x}{\theta}\, dS_x - \int_{\Omega(t)} \frac{\rho\, r_x}{\theta}\, dV_x \geq 0, \tag{4.31}$$

where η is the entropy per unit mass, $\boldsymbol{q}_x$ is the heat flux vector, and r_x is the heat supply. The second and third term account for the entropy transferred into the system in the form of heat. Using the Reynolds transport theorem in the form given by Equation (4.2), and the divergence theorem, Equation (1.36), to convert the surface integral into a volume integral, we obtain

$$\int_{\Omega(t)} \left(\rho\,\dot{\eta} + div\left(\frac{\boldsymbol{q}_x}{\theta}\right) - \frac{\rho r_x}{\theta}\right) dV_x \geq 0.$$

Once again, this must be true for each portion of the object, and hence the integrand must be greater than or equal to zero at each point within the continuum body. The local form of the second law of thermodynamics becomes

$$\rho\,\dot{\eta} + div\left(\frac{\boldsymbol{q}_x}{\theta}\right) - \frac{\rho r_x}{\theta} \geq 0.$$

Expanding the divergence term and multiplying the entire equation by temperature gives

$$\theta\,\rho\,\dot{\eta} - \frac{1}{\theta}\boldsymbol{q}_x \cdot grad\,\theta + div\left(\boldsymbol{q}_x\right) - \rho r_x \geq 0.$$

Note that the temperature is measured in Kelvin and must be a positive number. Rearranging the local form of the conservation of energy equation, Equation (4.22), we find that $div\left(\boldsymbol{q}_x\right) + \rho r_x = \boldsymbol{\sigma}^T : \boldsymbol{D} - \rho\dot{e}$. Substitution into the entropy equation gives us the ***Clausius-Duhem inequality***

$$\theta\,\rho\,\dot{\eta} - \frac{1}{\theta}\boldsymbol{q}_x \cdot grad\,\theta + \left(\boldsymbol{\sigma}^T : \boldsymbol{D} - \rho\dot{e}\right) \geq 0. \tag{4.32}$$

We introduce a new variable named the ***internal dissipation***, $\mathfrak{D}$, which is defined as

$$\mathfrak{D} = \rho\,\theta\,\dot{\eta} + div\left(\boldsymbol{q}_x\right) - \rho r_x.$$

Substitution into Equation (4.32), gives the second law in the form

$$\mathfrak{D} - \frac{1}{\theta}\boldsymbol{q}_x \cdot grad\,\theta \geq 0.$$

For a reversible process, the dissipation is zero. For an irreversible process, the dissipation is greater than zero.

Finally, we can rewrite the Clausius-Duhem inequality in terms of the Helmholtz free energy density, ψ, which is measured per unit mass

$$\psi(\boldsymbol{x},t) = e(\boldsymbol{x},t) - \eta(\boldsymbol{x},t)\,\theta(\boldsymbol{x},t).$$

The Helmholtz free energy is a measure of the energy available to do work at constant temperature. The material time derivative of the Helmholtz free energy is given by

$$\dot{\psi} = \dot{e} - \dot{\eta}\theta - \eta\dot{\theta}. \tag{4.33}$$

Substitution of Equation (4.33) into Equation (4.32) gives

$$-\rho\dot{\psi} - \rho\eta\dot{\theta} - \frac{1}{\theta}\boldsymbol{q}_x \cdot grad\,\theta + \boldsymbol{\sigma}^T : \boldsymbol{D} \geq 0. \tag{4.34}$$

4.2.5.2 Material Form

The material form of the second law of thermodynamics is found by converting the integral of spatial quantities over a volume in the current configuration into the integral over the corresponding volume in the reference configuration in Equation (4.31). We begin by restating Equation (4.31) for convenience

$$\frac{D}{Dt}\int_{\Omega(t)} \rho\,\eta\, dV_x + \int_{\partial\Omega(t)} \frac{\boldsymbol{q}_x \cdot \boldsymbol{n}_x}{\theta}\, dS_x - \int_{\Omega(t)} \frac{\rho\, r_x}{\theta}\, dV_x \geq 0. \tag{4.35}$$

Equation (4.7) allows us to convert the spatial volume integral into a material volume integral, Equation (4.3) allows us to convert the spatial density field to the material density field, and Equations (4.23) and (4.24) define the material heat flux vector and the material heat generation field. When combined with Equation (4.35), we obtain

$$\frac{D}{Dt}\int_{\Omega_0} \rho_o\,\eta_o\, dV_X + \int_{\partial\Omega_0} \frac{\boldsymbol{q}_o \cdot \boldsymbol{n}_X}{\theta}\, dS_X - \int_{\Omega_0} \frac{\rho_o\, r_o}{\theta}\, dV_X \geq 0. \tag{4.36}$$

Using the Reynolds transport theorem in the form given by Equation (4.2) and the divergence theorem, Equation (1.36), to convert the surface integral into a volume integral, we obtain

$$\int_{\Omega_0} \left(\rho_o\,\dot{\eta}_o + Div\left(\frac{\boldsymbol{q}_o}{\theta_o}\right) - \frac{\rho_o r_o}{\theta_o} \right) dV_X \geq 0.$$

Since this condition must hold for any arbitrary choice of volume, the integrand must vanish at each point within the continuum body, giving the local form of the second law of thermodynamics

$$\rho_o\,\dot{\eta}_o + Div\left(\frac{\boldsymbol{q}_o}{\theta_o}\right) - \frac{\rho_o r_o}{\theta_o} \geq 0.$$

Expanding the divergence term and multiplying by the temperature gives

$$\theta_o\,\rho_o\,\dot{\eta}_o - \frac{1}{\theta_o}\boldsymbol{q}_o \cdot Grad\,\theta_o + Div\left(\boldsymbol{q}_o\right) - \rho_o r_o \geq 0.$$

Substituting the material form of the principle of conservation of energy gives

$$\theta\,\rho_o\,\dot{\eta} - \frac{1}{\theta_o}\boldsymbol{q}_o \cdot Grad\,\theta_o + \left(\boldsymbol{P} : \dot{\boldsymbol{F}} - \rho_o\dot{e}_o\right) \geq 0.$$

Table 4.2. *List of unknown variables within the thermo-mechanical model*

Spatial form		Material form			Unknowns
Variable	Name	Variable	Name	Conversion equation	(Material, spatial)
$\boldsymbol{x}(\boldsymbol{X}, t)$	Equations of motion	$\boldsymbol{x}(\boldsymbol{X}, t)$	Equations of motion		3,3
$\rho(\boldsymbol{x}, t)$	Spatial density field	$\rho_o(\boldsymbol{X})$	Material density field	$\rho_o = \rho J$	1,0
$\boldsymbol{v}(\boldsymbol{x}, t)$	Spatial velocity field	$\boldsymbol{v_o}(\boldsymbol{X}, t)$	Material velocity field	$\boldsymbol{v_o}(\boldsymbol{X}, t) = \boldsymbol{v}(\boldsymbol{\kappa}(\boldsymbol{X}, t), t)$	3,3
$\boldsymbol{\sigma}(\boldsymbol{x}, t)$	Cauchy stress tensor	$\boldsymbol{P}(\boldsymbol{X}t)$	First Piola-Kirchhoff stress tensor	$\boldsymbol{P} = J\boldsymbol{\sigma} \cdot \boldsymbol{F}^{-T}$	9,9
$\boldsymbol{q}(\boldsymbol{x}, t)$	Spatial heat flux	$\boldsymbol{q}_o(\boldsymbol{X}, t)$	Material heat flux vector	$\boldsymbol{q}_o = J\boldsymbol{F}^{-1} \cdot \boldsymbol{q}_x$	3,3
$e(\boldsymbol{x}, t)$	Spatial internal energy density field	$e_o(\boldsymbol{X}, t)$	Material internal energy density field	$e_o(\boldsymbol{X}, t) = e(\boldsymbol{\kappa}(\boldsymbol{X}, t), t)$	1,1
$\theta(\boldsymbol{x}, t)$	Spatial temperature field	$\theta_o(\boldsymbol{X}, t)$	Material temperature field	$\theta_o(\boldsymbol{X}, t) = \theta(\boldsymbol{\kappa}(\boldsymbol{X}, t), t)$	1,1
$\eta(\boldsymbol{x}, t)$	Spatial entropy density field	$\eta_o(\boldsymbol{X}, t)$	Material entropy density field	$\eta_o(\boldsymbol{X}, t) = \eta(\boldsymbol{\kappa}(\boldsymbol{X}, t), t)$	1,1
$\boldsymbol{b}(\boldsymbol{x}, t)$	Spatial body force field	$\boldsymbol{b}_o(\boldsymbol{X}, t)$	Spatial body force field	$\boldsymbol{b}_o(\boldsymbol{X}, t) = \boldsymbol{b}(\boldsymbol{\kappa}(\boldsymbol{X}, t), t)$	0,0

Introducing the material form of the Helmholtz free energy gives the material form of the Clausius-Duhem inequality

$$-\rho_o \dot{\psi}_o - \rho_o \eta \dot{\theta}_o - \frac{1}{\theta_o} \boldsymbol{q}_o \cdot Grad\, \theta_o + \boldsymbol{P} : \dot{\boldsymbol{F}} \geq 0.$$

4.2.6 Summary of the Field Equations

In this chapter, we began the development of a material model by specifying the general forces and fields that were to be included. Specifically, we selected a thermo-mechanical model that included both thermal and mechanical variables. We have identified 22 unknown quantities in the spatial formulation and 21 unknown quantities in the material formulation of this model assuming that the material density field and both the spatial and material body force fields are prescribed. Therefore, these do not add to the count of unknown variables. We have used the kinematical relations and the basic laws of physics to develop 11 independent equations in the spatial formulation and 10 independent equations in the material formulation. Therefore, we require an additional 11 equations to form a complete mathematical model, Tables 4.2 and 4.3.

Specifying simply the general forces and fields being modeled has provided significant and useful information but has fallen short of providing a complete set of

Table 4.3. *List of the spatial and material forms of the balance equations*

Balance law	Spatial form	Material form	Independent equations (material, spatial)
Kinematical relations	$\boldsymbol{v} = \frac{\partial \boldsymbol{x}}{\partial t}$	$v_o = \left.\frac{\partial \boldsymbol{x}}{\partial t}\right\|_{\boldsymbol{x}=\kappa(\boldsymbol{X},t)}$	3,3
Conservation of Mass	$0 = \frac{\partial \rho}{\partial t} + div\,(\rho\,\boldsymbol{v})$	$\rho_o = \rho J$	1,0
Conservation of linear momentum	$\rho\,\dot{\boldsymbol{v}} - div\,\boldsymbol{\sigma}^T - \rho\,\boldsymbol{b}_x = 0$	$\rho_o\,\dot{\boldsymbol{v}}_o - Div\,(\boldsymbol{P}) - \rho_o \boldsymbol{b_o} = 0$	3,3
Conservation of angular momentum	$\boldsymbol{\sigma} = \boldsymbol{\sigma}^T$	$\boldsymbol{F}\cdot\boldsymbol{P}^T = \boldsymbol{P}\cdot\boldsymbol{F}^T$	3,3
Conservation of energy	$\rho\dot{e} - \boldsymbol{\sigma} : \boldsymbol{D} + div\,(\boldsymbol{q}_x)$ $-\rho\, r_x = 0$	$\rho_o\dot{e}_o - \boldsymbol{P} : \dot{\boldsymbol{F}} - \rho_o r_o$ $+Div\,(\boldsymbol{q}_o) = 0$	1,1
Entropy inequality	$-\rho\,\dot{\psi} - \rho\eta\dot{\theta} - \frac{1}{\theta}\boldsymbol{q}_x \cdot grad\,\theta$ $+\boldsymbol{\sigma}^T : \boldsymbol{D} \geq 0$	$-\rho_o\,\dot{\psi}_o - \rho_o\eta\dot{\theta}_o - \frac{1}{\theta_o}\boldsymbol{q}_o \cdot Grad\,\theta_o$ $+\boldsymbol{P} : \dot{\boldsymbol{F}} \geq 0$	

equations. Additional equations will be determined by creating a constitutive model for the material. For example, we will form constitutive models for ideal gases, Newtonian fluids, and elastic solids. These material models share the same set of unknowns and the same kinematical and balance equations. However, gases, fluids, and elastic solids exhibit radically different behavior and their constitutive models are quite dissimilar.

4.3 Stress Power

The rate of work done by the internal stresses within an object is termed the ***stress power***. In this section, we will use the concept of stress power to define conjugate stress-strain pairs. The derivation of this quantity proceeds by considering the rate of work, $\dot{\mathcal{W}}$, done by the applied surface tractions and body forces on an object such that

$$\dot{\mathcal{W}} = \int_{\partial\Omega(t)} (\boldsymbol{t}_x \cdot \boldsymbol{v})\, dS_x + \int_{\Omega(t)} \left(\rho \boldsymbol{b}_x \cdot \boldsymbol{v}\right) dV_x.$$

Substituting the Cauchy stress for the traction vector, using the divergence theorem, Equation (1.36), to eliminate the surface integral, substituting $div\left(\boldsymbol{\sigma}^T \cdot \boldsymbol{v}\right) = div\,\boldsymbol{\sigma}^{\boldsymbol{T}} \cdot \boldsymbol{v} + \boldsymbol{\sigma} : \left(\boldsymbol{v}\overleftarrow{\nabla}\right)$, and $\boldsymbol{L} = \boldsymbol{v}\overleftarrow{\nabla}$ gives

$$\dot{\mathcal{W}} = \int_{\Omega(t)} \left(\boldsymbol{v} \cdot \left(div\left(\boldsymbol{\sigma}^T\right) + \rho \boldsymbol{b}_x\right) + \boldsymbol{\sigma} : \boldsymbol{L}\right) dV_x.$$

Finally, substitution of the conservation of linear momentum in the form $div\,\boldsymbol{\sigma}^T + \rho\,\boldsymbol{b}_x = \rho\,\dot{\boldsymbol{v}}$, gives

$$\dot{\mathrm{W}} = \int_{\Omega(t)} (\rho \boldsymbol{v} \cdot \dot{\boldsymbol{v}} + \boldsymbol{\sigma} : \boldsymbol{L})\, dV_x.$$

Therefore, we have the relation

$$\int_{\partial\Omega(t)} (\boldsymbol{t}_x \cdot \boldsymbol{v})\, dS_x + \int_{\Omega(t)} (\rho \boldsymbol{b}_x \cdot \boldsymbol{v})\, dV_x = \int_{\Omega(t)} (\rho \boldsymbol{v} \cdot \dot{\boldsymbol{v}} + \boldsymbol{\sigma} : \boldsymbol{L})\, dV_x,$$

which implies that the rate of work done by the surface traction and the body force is equal to the change in the kinetic energy of the object plus the work done by the internal stresses. This last term, $\boldsymbol{\sigma} : \boldsymbol{L}$, is the stress power. Noting that the Cauchy stress is symmetric, we can write

$$\boldsymbol{\sigma} : \boldsymbol{L} = \boldsymbol{\sigma} : \boldsymbol{D}.$$

Written in this form, the stress power is the rate of work done per unit spatial volume in the spatial configuration.

We can equivalently derive an equation for the stress power per unit reference volume by converting the integral to an integral over the reference volume and substituting the first Piola-Kirchhoff stress tensor for the Cauchy stress tensor such that, $\boldsymbol{\sigma} = J^{-1}\boldsymbol{P} \cdot \boldsymbol{F}^T$, and

$$\int_{\Omega(t)} \boldsymbol{\sigma} : \boldsymbol{L}\, dV_x = \int_{\Omega_0} J^{-1}\boldsymbol{P} \cdot \boldsymbol{F}^T : \boldsymbol{L} J\, dV_X.$$

The order of the tensor contraction can be switched by noting that $\boldsymbol{P} \cdot \boldsymbol{F}^T : \boldsymbol{L} = \boldsymbol{P} : \boldsymbol{L} \cdot \boldsymbol{F}$ and the derivative of the deformation gradient is given by $\dot{\boldsymbol{F}} = \boldsymbol{L} \cdot \boldsymbol{F}$, giving the result

$$\int_{\Omega(t)} \boldsymbol{\sigma} : \boldsymbol{L}\, dV_x = \int_{\Omega_0} \boldsymbol{P} : \dot{\boldsymbol{F}}\, dV_X.$$

Therefore, the deformation gradient is the conjugate strain measure for the first Piola-Kirchhoff stress tensor.

If we write the stress power in terms of the second Piola-Kirchhoff stress tensor, where $\boldsymbol{S} = J^{-1}\boldsymbol{F} \cdot \boldsymbol{S} \cdot \boldsymbol{F}^T$, we obtain

$$\int_{\Omega(t)} \boldsymbol{\sigma} : \boldsymbol{D}\, dV_x = \int_{\Omega_0} J^{-1}\boldsymbol{F} \cdot \boldsymbol{S} \cdot \boldsymbol{F}^T : \boldsymbol{D} J\, dV_X.$$

Shifting the order of the contracted terms gives $\boldsymbol{F} \cdot \boldsymbol{S} \cdot \boldsymbol{F}^T : \boldsymbol{D} = \boldsymbol{S} : \boldsymbol{F}^T \cdot \boldsymbol{D} \cdot \boldsymbol{F}$, and substituting $\dot{\boldsymbol{E}} = \boldsymbol{F}^T \cdot \boldsymbol{D} \cdot \boldsymbol{F}$, we have

$$\int_{\Omega(t)} \boldsymbol{\sigma} : \boldsymbol{D}\, dV_x = \int_{\Omega_0} \boldsymbol{S} : \dot{\boldsymbol{E}}\, dV_X.$$

As you can see, the choice of stress and strain measure is arbitrary. However, in each case, the stress measure is matched with the time derivative of its *conjugate* strain measure. In the case of the Cauchy stress, the deformation rate tensor, $\boldsymbol{D}$, is not the time derivative of a strain measure. Therefore, it does not have a conjugate strain measure, Table 4.4.

4.4 Jump Conditions

The field equations summarized in the previous section assume that the stress, density, energy, and velocity fields are all continuous. However, there are many cases

Table 4.4. *Stress-strain conjugate pairs*

Stress	Strain
σ, Cauchy stress tensor	Not properly defined
$\boldsymbol{P}$, First Piola-Kirchhoff stress tensor	$\boldsymbol{F}$, deformation gradient
$\boldsymbol{S}$, Second Piola-Kirchhoff stress tensor	$\boldsymbol{E}$, Green strain tensor

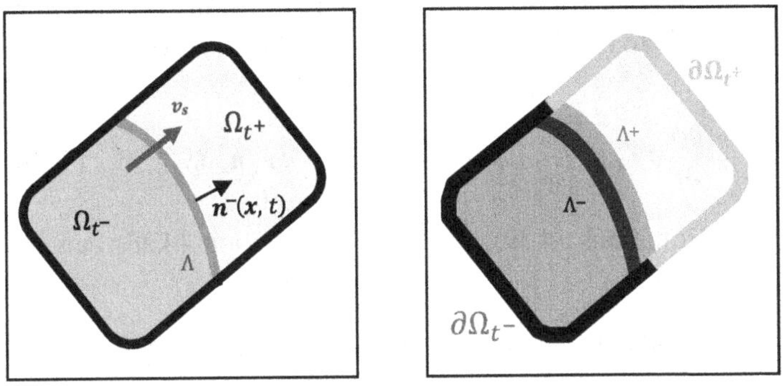

Figure 4.1. (Left) Illustration of a region of space that contains singular surface (a discontinuity) where $\boldsymbol{v}_s$ is the velocity of the singular surface and n is the outward unit normal to the singular surface. The outward surface normal, $\boldsymbol{n}^-$, points out of the volume $\boldsymbol{\Omega}_{t^-}$ while $\boldsymbol{n}^+$ points out of the volume $\boldsymbol{\Omega}_{t^+}$. (Right) $\boldsymbol{\Lambda}^+$ is the surface of the singularity on the + side. $\boldsymbol{\partial\Omega}_{t^+}$ is the part of the surface surrounding $\boldsymbol{\Omega}_{t^+}$ excluding $\boldsymbol{\Lambda}^+$. $\boldsymbol{\Lambda}^-$ is the surface of the singularity on the - side. $\boldsymbol{\partial\Omega}_{t^-}$ is the part of the surface surrounding $\boldsymbol{\Omega}_{t^-}$ excluding $\boldsymbol{\Lambda}^-$.

where these fields may have sharp interfaces, which can be approximated as discontinuous interfaces. Take for example the case of a shock wave. The velocity and density of material on one side of the infinitesimally thin shock wave may be significantly different than the velocity and density on the other side.

If we integrate a property field, ϕ, over a region, Ω_t, which includes a singularity along the surface given by Λ, such as that shown in Figure 4.1, we must break the integral into two continuous parts

$$\int_{\Omega_t} \phi\, dV_x = \int_{\Omega_{t^-}} \phi\, dV_x + \int_{\Omega_{t^+}} \phi\, dV_x, \tag{4.37}$$

where $\Omega_t = \Omega_{t^-} + \Omega_{t^+}$ and the scalar field, ϕ, is individually continuous within Ω_{t^-} and Ω_{t^+}. If we take the material time derivative of Equation (4.37), we obtain

$$\frac{D}{Dt}\int_{\Omega_t} \phi\, dV_x = \frac{D}{Dt}\int_{\Omega_{t^-}} \phi\, dV_x + \frac{D}{Dt}\int_{\Omega_{t^+}} \phi\, dV_x. \tag{4.38}$$

Since each of the two integrals on the right side of Equation (4.38) is continuous, we can use the Reynolds transport theorem on each individually. Note that we cannot use the Reynolds transport theorem on the first term because there is a discontinuity within this volume defined by Ω_t. Applying the Reynolds transport theorem in the form given by Equation (2.56) to the second term in Equation (4.38), we obtain

$$\begin{aligned}\frac{D}{Dt}\int_{\Omega_{t^-}}\phi\, dV_x &= \int_{\Omega_{t^-}}\frac{\partial\phi}{\partial t}dV_x + \int_{(\partial\Omega_{t^-}+\Lambda^-)}\phi\,(\boldsymbol{v}\cdot\boldsymbol{n}_x)\, dS_x\\ &= \int_{\Omega_{t^-}}\frac{\partial\phi}{\partial t}dV_x + \int_{\partial\Omega_{t^-}}\phi\,(\boldsymbol{v}\cdot\boldsymbol{n}_x)\, dS_x + \int_{\Lambda^-}\phi\,(\boldsymbol{v}\cdot\boldsymbol{n}_x)\, dS_x,\end{aligned} \tag{4.39}$$

where $\partial\Omega_{t^-}+\Lambda^-$ is the total surface surrounding the region Ω_{t^-}. We have split the surface integral into an integral over $\partial\Omega_{t^-}$ and one over Λ^-. Similarly, applying the Reynolds transport theorem to the third term in Equation (4.38) and splitting the integral into two parts gives

$$\frac{D}{Dt}\int_{\Omega_{t^+}}\phi\, dV_x = \int_{\Omega_{t^+}}\frac{\partial\phi}{\partial t}dV_x + \int_{\partial\Omega_{t^+}}\phi\boldsymbol{v}\cdot\boldsymbol{n}_x\, dS_x + \int_{\Lambda^+}\phi\boldsymbol{v}\cdot\boldsymbol{n}_x\, dS_x. \tag{4.40}$$

Combining the results of Equations (4.39) and (4.40), Equation (4.38) can be written as

$$\begin{aligned}\frac{D}{Dt}\int_{\Omega_t}\phi\, dV_x = \int_{\Omega_t}\frac{\partial\phi}{\partial t}dV_x &+ \int_{\partial\Omega_t}\phi\,(\boldsymbol{v}\cdot\boldsymbol{n}_x)\, dS_x + \int_{\Lambda^+}\phi\left(\boldsymbol{v}\cdot\boldsymbol{n}_x^+\right) dS_x\\ &+ \int_{\Lambda^-}\phi\left(\boldsymbol{v}\cdot\boldsymbol{n}_x^-\right) dS_x,\end{aligned}$$

where $\partial\Omega_t = \partial\Omega_{t^-}+\partial\Omega_{t^+}$. This resembles the Reynolds transport theorem with the addition of an integral over the surface singularity captured in the last two terms. These terms may be simplified when we realize the outward normal on the + and − side of the singularity are equal and opposite

$$\begin{aligned}\frac{D}{Dt}\int_{\Omega_t}\phi\, dV_x = \int_{\Omega_t}\frac{\partial\phi}{\partial t}dV_x &+ \int_{\partial\Omega_t}\phi(\boldsymbol{v}\cdot\boldsymbol{n}_x)\, dS_x - \int_{\Lambda^+}\phi\left(\boldsymbol{v}\cdot\boldsymbol{n}_x^-\right) dS_x\\ &+ \int_{\Lambda^-}\phi\left(\boldsymbol{v}\cdot\boldsymbol{n}_x^-\right) dS_x.\end{aligned}$$

We can further simplify this equation if we know the velocity of the singular surface, $\boldsymbol{v}_s$,

$$\begin{aligned}\frac{D}{Dt}\int_{\Omega_t}\phi\, dV_x = \int_{\Omega_t}\frac{\partial\phi}{\partial t}dV_x &+ \int_{\partial\Omega_t}\phi(\boldsymbol{v}\cdot\boldsymbol{n}_x)\, dS_x - \int_{\Lambda^+}\phi\left(\boldsymbol{v}_s\cdot\boldsymbol{n}_x^-\right) dS_x\\ &+ \int_{\Lambda^-}\phi\left(\boldsymbol{v}_s\cdot\boldsymbol{n}_x^-\right) dS_x.\end{aligned}$$

Often in the literature, you will find that $\boldsymbol{v}_s\cdot\boldsymbol{n}_x^-$ is written as the component of the velocity normal to the singular surface, v_s^n,

$$\frac{D}{Dt}\int_{\Omega_t}\phi\, dV_x = \int_{\Omega_t}\frac{\partial\phi}{\partial t}dV_x + \int_{\partial\Omega_t}\phi(\boldsymbol{v}\cdot\boldsymbol{n}_x)\, dS_x - \int_{\Lambda^+}\phi v_s^n\, dS_x + \int_{\Lambda^-}\phi v_s^n\, dS_x.$$

The equation can be consolidated in the form

$$\frac{D}{Dt}\int_{\Omega_t}\phi\, dV_x = \int_{\Omega_t}\frac{\partial\phi}{\partial t}dV_x + \int_{\partial\Omega_t}\phi(\boldsymbol{v}\cdot\boldsymbol{n}_x)\, dS_x - \int_{\Lambda^+}[\phi]_-^+\, v_s^n\, dS_x, \tag{4.41}$$

where$[\phi]_-^+ = \phi^+ - \phi^-$.

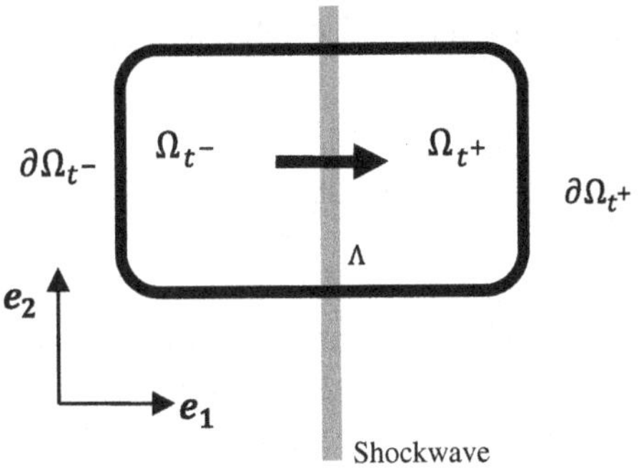

Figure 4.2. Illustration of a passing shockwave.

If we repeat this analysis for a vector field, we can obtain

$$\frac{D}{Dt}\int_{\Omega_t} \boldsymbol{a}\, dV_X = \int_{\Omega_t} \frac{\partial \boldsymbol{a}}{\partial t} dV_X + \int_{\partial\Omega_t} \boldsymbol{a}\,(\boldsymbol{v}\cdot\boldsymbol{n}_X)\, dS_X - \int_{\Lambda^+} [\boldsymbol{a}]_-^+\, v_s^n\, dS_X. \tag{4.42}$$

EXAMPLE 4.1. *One-dimensional shockwave traveling through a fluid*
Let us take the example of a shockwave moving across a fluid in the positive x direction with a velocity of v_s^n. The fluid in front of the shockwave is stationary, and the fluid behind the shockwave moves at a velocity, v_{final}. We would like to determine the relation between the density behind the shockwave, ρ^-, and the velocity in front of the shockwave, ρ^+.

Solution:
The basic approach follows:

1. Formulate the conservation of mass equation for a volume that includes a portion of the shockwave.
2. Since the shockwave represents a discontinuity in the density and velocity fields, we must use the discontinuous Reynolds transport theorem, Equation (4.41), to evaluate the conservation of mass equation.
3. In the limit as the volume approaches zero, the integral is evaluated only on the surface of shockwave.
4. Using the velocity of the shockwave, we can find the relative density of the fluid behind and in front of the shockwave.

Step 1.
The principle of conservation of mass in the spatial form for a volume, Ω_t, which includes a portion of the shockwave, can be written as

$$\frac{D}{Dt}\int_{\Omega_t} \rho(\boldsymbol{x},t)\, dV_X = 0. \tag{4.43}$$

Step 2.
Since this volume includes a discontinuity in the density field, we must use the Reynolds transport theorem in the form given by Equation (4.41), to evaluate material derivative. Equation (4.43) becomes

$$\frac{D}{Dt}\int_{\Omega_t} \rho\, dV_X = \int_{\Omega_t} \frac{\partial \rho}{\partial t} dV_X + \int_{\partial\Omega_t} \rho\,(\boldsymbol{v}\cdot\boldsymbol{n}_X)\, dS_X - \int_{\Lambda} [\rho]_-^+\, v_s^n\, dS_X.$$

We assume there is no creation or destruction of mass involved in this process, $\frac{D}{Dt}\int_{\Omega_t}\rho\, dV_x = 0$. Therefore, we can write

$$0 = \int_{\Omega_t}\frac{\partial\rho}{\partial t}dV_x + \int_{\partial\Omega_t}\rho\,(\boldsymbol{v}\cdot\boldsymbol{n}_x)\,dS_x - \int_{\Lambda}[\rho]_-^+\,v_s^n\,dS_x.$$

Step 3.
Next, we shrink the volume of the region of integration while keeping the shockwave in the integrated region (i.e., we let Ω_{t^-} and Ω_{t^+} go to zero simultaneously). The integral over volume approaches zero

$$\lim_{\substack{\Omega_{t^-}\to 0\\ \Omega_{t^+}\to 0}}\int_{\Omega_t}\frac{\partial\rho}{\partial t}dV_x = 0,$$

and the surface $\partial\Omega_t$ collapses onto the singular surface and becomes equal to $\Lambda_+ + \Lambda_-$. We can then write

$$\begin{aligned}0 &= \int_{\Lambda^+}\rho^+\left(\boldsymbol{v}^+\cdot\boldsymbol{n}_x^+\right)dS_x + \int_{\Lambda^-}\rho^-\left(\boldsymbol{v}^-\cdot\boldsymbol{n}_x^-\right)dS_x - \int_{\Lambda}[\rho]_-^+\,v_s^n\,dS_x\\ &= \int_{\Lambda}[\rho\,(\boldsymbol{v}\cdot\boldsymbol{n}_x) + \rho\,v_s^n]_-^+\,dS_x.\end{aligned}$$

This must hold for any arbitrary choice of surface segment, Λ, along the discontinuity. Therefore, we have a differential equation that must be satisfied at each point on the discontinuity

$$0 = [\rho\,(\boldsymbol{v}\cdot\boldsymbol{n}_x) + \rho\,v_s^n]_-^+. \tag{4.44}$$

Step 4.
Notice that the velocity of the shockwave relative to the fluid is given by $\boldsymbol{v}_{rel} = \boldsymbol{v}\cdot\boldsymbol{n}_x + v_s^n$. Therefore, Equation (4.44) can be written as

$$\begin{aligned}0 &= \left[\rho\,v_{rel}^n\right]_-^+\\ &= \rho^+v_{rel}^+ - \rho^-v_{rel}^-.\end{aligned} \tag{4.45}$$

This equation states that the product of the density and the relative velocity in front of the shockwave is equal to the product of the density and the relative velocity behind the shockwave. By inserting the relative velocity given in the problem, Equation (4.45) becomes

$$\rho^+v_s^n = \rho^-\left(-v_{final} + v_s^n\right).$$

Rearranging this gives the final solution

$$\rho^+ = \rho^-\left(1 - \frac{v_{final}}{v_s^n}\right).$$

4.5 Constitutive Modeling

As stated in Section 4.2.6, the material model for a thermo-mechanical material requires 11 additional equations beyond the balance laws. In fact, it is impossible to develop a single model that captures the behavior of all thermo-mechanical materials. These 11 additional equations will vary from material to material and may be valid for only a limited range of temperatures, stresses, and so on. For example, the model for a linear elastic material will describe the behavior of a structural steel quite reasonably at small strains. However, as the applied stress approaches the yield strength, this linear elastic material model cannot account for the nonlinearity of the stress-strain relation and the plastic deformation in the material. The ***constitutive model*** consists of a set of mathematical equations that relate the thermodynamic and kinematic quantities for a specific material within some finite range of applicability. When combined, the balance laws and the constitutive model form a complete set of equations that can theoretically be solved to determine the response of the system to imposed boundary conditions. Development of a constitutive model proceeds via a three-step process.

First, the general form of the constitutive model is formulated by selection the dependent and independent variables for the equations required to complete the material model. In this section, we will specify the general form for a thermoelastic model, which is an example of a thermo-mechanical model. We need 11 equations beyond the balance laws. We chose in this example to provide constitutive equations for the Helmholtz free energy, ψ, Cauchy stress tensor, $\boldsymbol{\sigma}$, the spatial heat flux vector, $\boldsymbol{q}_x$, and entropy, η, as independent variables. Due to the symmetry of the Cauchy stress tensor, that gives $1+6+3+1=11$ additional equations. The dependent variables are selected based on the material characteristics. For example, a thermoelastic material exhibits no dissipation. Energy is stored during deformation and is released when the external stress is removed. Therefore, the internal stress depends only on the current deformation and not on the history of previous deformation. The mechanical response is therefore dictated by some function of the deformation gradient, $\boldsymbol{F}$. Since we are including thermal effects, the material response will depend on temperature, θ, and the heat flux will depend on the gradient of temperature, $grad\,\theta$. Applying the principle of equi-presence (which will be discussed in more detail later in this section), the general form of the constitutive model is then

$$\boldsymbol{\sigma} = \boldsymbol{\sigma}\left(\boldsymbol{F}, \theta, grad\,\theta\right),$$

$$\psi = \psi\left(\boldsymbol{F}, \theta, grad\,\theta\right),$$

$$\eta = \eta\left(\boldsymbol{F}, \theta, grad\,\theta\right),$$

$$\boldsymbol{q} = \boldsymbol{q}\left(\boldsymbol{F}, \theta, grad\,\theta\right).$$

Second, constraints on the specific form of the constitutive model are developed by applying the principles of constitutive model development to the general form of the equations. Later in this chapter we will show that the principal of dissipation requires that the specific form of any thermoelastic model must satisfy the condition $\sigma^T = -\rho \frac{\partial \psi}{\partial \boldsymbol{F}} \boldsymbol{F}^T$. The constraints generated in this step may help to simplify the required form of the constitutive model, or may be used to find the range of permissible values for a material constant within a specific form of the constitutive model.

Third, the specific form of the constitutive model, which is consistent with the constraints developed in step 2, is postulated based on available experimental trends for the specific material or group of materials. For example, in a later chapter, we will discuss the neo-Hookean model, which is a hyperelastic model. The neo-Hookean model can be written as

$$\psi = G(I_B - 3),$$

where G is a material parameter, I_B are the invariants of the Cauchy deformation tensor and is a material parameter.

Fourth, we must parameterize the model through experimentation for a specific material. The neo-Hookean model has a single material parameter G. The value of this parameter must be found through experimentation. For example, we could subject a bar of polyethylene to a uniaxial tension experiment in which stress and strain are measured. The value of G specific to this type of polyethylene would be extracted from the stress-strain curve.

In summary, the thermo-mechanical class of material models includes ideal gases, Newtonian fluids, nonlinear elastic solids, and other broad classes of materials. The general balance laws and the form of the constitutive model for a thermoelastic solid derived in this chapter applies to the neo-Hookean model but also applies to the Mooney-Rivlin model, and hundreds of other nonlinear elastic models. The material parameters in these models would vary from material to material. For example, material parameters for a polymer vary for polyethylene and polycarbonate. Given the type of polymer, they also depend on the molecular chain length, the crosslink density, impurities in the material, and a host of other parameters. It is only when we parameterize the constitutive model for a specific material with specific chemical structure and microstructure that we have a model that can be used to study structures built from this specific material.

4.5.1 Constitutive Modeling Principles

When creating constitutive models, we are creating a complete set of equations with which we can determine the response of a material to various forces and fields. We seek a set of physically admissible equations which describe the specific material of interest. By physically admissible, we mean that this set of equations should not contradict any the laws of thermodynamics and that the set should obey the following principles of constitutive modeling.

1. ***Principle of causality:*** As applied to thermo-mechancial models, the principle of causality states that the motion and temperature of all particles within the continuum body are observable causal variables. All other quantities such as free energy, entropy, and heat flux are dependent constitutive variables. These dependent quantities may be functions of temperature, motion, and all other quantities that are directly derivable from these such as the deformation gradient, velocity, velocity gradient, and the time rate of change of the temperature:

   $$\{\psi, \eta, \boldsymbol{q}, \sigma\} = f\left(\boldsymbol{x}(\boldsymbol{X}, t), \theta(\boldsymbol{X}, t), \boldsymbol{F}, \dot{\boldsymbol{F}}, \dot{\theta}, \nabla\theta, \ldots\right).$$

   It is important to note that the exact form of the causality principle depends on the class of the material model. For example, thermo-electro-mechanical systems require the addition of charge to the causal variables.

2. ***Principle of determinism:*** The present state of the independent constitutive variables can be completely determined from its kinematic and temperature history.
3. ***Principle of local action:*** The interaction strength between particles decays with increasing distance, and particles separated by very large distances have insignificant interactions.
4. ***Principle of fading memory:*** The dependence of the current state of a material particle on its history decays with increasing separation in time, and states in the distant past have insignificant impact on the current state.
5. ***Principle of equi-presence:*** When initially formulating constitutive models, each constitutive equation should have the same independent variables until they are explicitly excluded by one of the other principles. The principle of equi-presence ensures coupled phenomena are correctly accounted for in the initial stages of constitutive model development.
6. ***Principle of material frame indifference or objectivity:*** The material response must be independent of the observer's reference frame. Therefore, the constitutive equations must be invariant to rigid translations or rigid rotations of the spatial reference frame.
7. ***Principle of material symmetry:*** The material response should be invariant under rotations belonging to the material symmetry group.
8. ***Principle of dissipation:*** The material model must satisfy the second law of thermodynamics for any thermodynamic process that is compatible with the governing equations.

4.5.2 Principle of Dissipation

The principle of dissipation states that for any admissible process, the material model must satisfy the Clausius-Duhem (CD) inequality. Taking the proposed thermoelastic material model, we will derive the constraints on the constitutive equations imposed by the second law of thermodynamics. Following similar steps, one may derive the constraints for any general constitutive model. The functional from of the thermoelastic model previously proposed is

$$\begin{aligned}
\psi &= \psi\left(\boldsymbol{F}, \theta, grad\,\theta\right), \\
\boldsymbol{\sigma} &= \boldsymbol{\sigma}\left(\boldsymbol{F}, \theta, grad\,\theta\right), \\
\eta &= \eta\left(\boldsymbol{F}, \theta, grad\,\theta\right), \\
\boldsymbol{q} &= \boldsymbol{q}\left(\boldsymbol{F}, \theta, grad\,\theta\right).
\end{aligned}$$

To simplify notation, we will introduce a new variable, $\boldsymbol{g} = grad\,\theta$, The spatial form of the CD inequality was given in Equation (4.34) but is repeated here for convenience

$$-\rho\dot{\psi} - \rho\eta\dot{\theta} - \frac{1}{\theta}\boldsymbol{q}_x \cdot \boldsymbol{g} + \boldsymbol{\sigma}^T : \boldsymbol{D} \geq 0. \tag{4.46}$$

Using the chain rule to expand the Helmholtz free energy in terms of the independent variables gives

$$\dot{\psi}\left(\boldsymbol{F}, \theta, grad\,\theta\right) = \frac{\partial\psi}{\partial\boldsymbol{F}} : \frac{D\boldsymbol{F}}{Dt} + \frac{\partial\psi}{\partial\theta}\frac{D\theta}{Dt} + \frac{\partial\psi}{\partial\boldsymbol{g}} \cdot \frac{D\boldsymbol{g}}{Dt}. \tag{4.47}$$

Substitution of Equation (4.54) into the CD inequality gives

$$-\rho\left(\frac{\partial\psi}{\partial \boldsymbol{F}}:\dot{\boldsymbol{F}}+\frac{\partial\psi}{\partial\theta}\dot{\theta}+\frac{\partial\psi}{\partial \boldsymbol{g}}\cdot\dot{g}\right)-\rho\eta\dot{\theta}-\frac{1}{\theta}\boldsymbol{q}\cdot\boldsymbol{g}+\boldsymbol{\sigma}^T:\boldsymbol{D}\geq 0.$$

Using the relation $\dot{\boldsymbol{F}}=\boldsymbol{L}\cdot\boldsymbol{F}$ and the fact that the Cauchy stress tensor is symmetric,

$$\boldsymbol{\sigma}^T:\boldsymbol{D}=\boldsymbol{\sigma}^T:\boldsymbol{L}=\boldsymbol{\sigma}^T:\dot{\boldsymbol{F}}\cdot\boldsymbol{F}^{-1}.$$

Using the tensor identity $\boldsymbol{A}:\boldsymbol{B}\cdot\boldsymbol{C}=\boldsymbol{A}\cdot\boldsymbol{C}^T:\boldsymbol{B}$, we obtain

$$\left(-\rho\frac{\partial\psi}{\partial \boldsymbol{F}}+\boldsymbol{\sigma}\cdot\boldsymbol{F}^{-T}\right):\dot{\boldsymbol{F}}-\rho\left(\frac{\partial\psi}{\partial\theta}-\eta\right)\dot{\theta}-\rho\frac{\partial\psi}{\partial \boldsymbol{g}}\cdot\dot{\boldsymbol{g}}-\frac{1}{\theta}\boldsymbol{q}\cdot\boldsymbol{g}\geq 0.$$

This equation must hold for any and all physically admissible processes. An admissible process is any time-dependent solution of the field variables that satisfies the balance laws and the constitutive relations. Since we can vary both the applied body force field and heat supplied to our continuum body, every deformation-temperature path can be realized in the process. The values of $\{\boldsymbol{F},\dot{\boldsymbol{F}},\theta,\dot{\theta},g,\dot{g}\}$ can be independently varied.

For example, we may select a thermodynamic process that has no change in the deformation gradient with respect to time, $\dot{\boldsymbol{F}}=0$, a steady-state temperature that gives $\dot{\boldsymbol{g}}=0$ and $\dot{\theta}=0$ but that has a nonzero temperature gradient, $\boldsymbol{g}\neq 0$. This corresponds to steady-state heat conduction through a segment of the material with no change in the deformation. For this process, the CD inequality becomes

$$-\frac{\boldsymbol{q}\cdot\boldsymbol{g}}{\boldsymbol{\theta}}\geq 0.$$

Since the temperature is always positive, we can write

$$-\boldsymbol{q}\cdot\boldsymbol{g}\geq 0. \tag{4.48}$$

This is known as ***Fourier's inequality***. It states that the heat flux must be opposite to the direction of the temperature gradient. In simple terms, heat must flow from hot to cold.

There is no requirement that a process be steady state, or even that it persist for more than an instant. In addition, the CD inequality is applied locally to each individual material particle. To determine the constraints imposed by the CD inequality, we need only consider admissible processes at any given material point. For example, we can imagine a process for a material particle with no change in deformation, $\dot{F}=0$, with a fixed temperature, $\dot{\theta}=0$, which at a given instance in time has no temperature gradient, $\boldsymbol{g}=0$, but whose temperature gradient is changing with respect to time, $\dot{g}\neq 0$. Substitution of this process into the CD inequality gives

$$-\rho\frac{\partial\psi}{\partial \boldsymbol{g}}\cdot\dot{\boldsymbol{g}}\geq 0.$$

Since the direction of $\dot{\boldsymbol{g}}$ is arbitrary and independent of $\boldsymbol{F}$, θ, and $\boldsymbol{g}$, the only way for the CD inequality to be satisfied is for all admissible processes for this term to go to zero. This condition requires that the Helmholtz free energy does not vary with the temperature gradient

$$\frac{\partial\psi}{\partial \boldsymbol{g}}=0. \tag{4.49}$$

Similarly, we can find a thermodynamic process for which only $\dot{\boldsymbol{F}} \neq 0$. Namely, an isothermal process for which $\dot{\theta} = 0$, $\boldsymbol{g} = 0$, and $\dot{\boldsymbol{g}} = 0$, gives the condition

$$\left(-\rho \frac{\partial \psi}{\partial \boldsymbol{F}} + \boldsymbol{\sigma} \cdot \boldsymbol{F}^{-T}\right) : \dot{\boldsymbol{F}} \geq 0.$$

Since the direction of the change in the deformation gradient can be changed arbitrarily and independently of $\boldsymbol{F}$, θ, and $\boldsymbol{g}$, the term in brackets must vanish. This gives the condition

$$\sigma = \rho \frac{\partial \psi}{\partial \boldsymbol{F}} \cdot \boldsymbol{F}^T. \tag{4.50}$$

The final constraint is obtained by substituting a process for which the only nonzero term is $\dot{\theta}$ into the CD inequality. This final process gives

$$\eta = \frac{\partial \psi}{\partial \theta}. \tag{4.51}$$

Gathering the constraints, we may simplify our constitutive model formulation. Equation (4.49) states that for the thermoelastic material, the Helmholtz free energy must not depend on the temperature gradient. Equations (4.50) and (4.51) give the Cauchy stress and the entropy in terms of the Helmholtz free energy, respectively. Therefore, neither the Cauchy stress nor the entropy may be functions of the temperature gradient. Finally, Equation (4.48) gives the condition that the heat must travel from hot to cold. We conclude that only two constitutive response functions which are consistent with the constraints from the CD inequality are needed to characterize a thermoelastic material. The constitutive model can now be written as

$$\psi = \psi\left(\boldsymbol{F},\theta\right),$$
$$\boldsymbol{q} = \boldsymbol{q}\left(\boldsymbol{F},\theta,\boldsymbol{g}\right),$$

with the stress and entropy determined from the Helmholtz free energy as

$$\begin{aligned} \sigma &= \rho \frac{\partial \psi}{\partial \boldsymbol{F}} \cdot \boldsymbol{F}^T, \\ \eta &= \frac{\partial \psi}{\partial \theta}. \end{aligned} \tag{4.52}$$

4.5.3 Principle of Material Frame Indifference

The principle of material frame indifference states that the material response must be independent of the observer's reference frame. Practically, this means that if we have a scalar function for the temperature, θ, for a material particle, its value should be the same for all observers regardless of the reference frame which they inhabit. If we have one observer in the starred reference frame and another observer in the unstarred reference frame, then objective scalars, vectors, and tensors have the respective properties

$$\begin{aligned} \alpha^* &= \alpha, \\ \boldsymbol{a}^* &= \boldsymbol{Q} \cdot \boldsymbol{a}, \\ \boldsymbol{A}^* &= \boldsymbol{Q} \cdot \boldsymbol{A} \cdot \boldsymbol{Q}^T. \end{aligned} \tag{4.53}$$

Similarly, an ***objective scalar function*** will have the property

$$\alpha^*\left(\zeta_1^*,\zeta_2^*,\ldots\zeta_n^*\right)=\alpha\left(\zeta_1,\zeta_2,\ldots\zeta_n\right). \tag{4.54}$$

where $\boldsymbol{Q}$ is the orthogonal rotation tensor relating the basis vectors of the starred reference frame to those in the unstarred reference frame, ζ_i are the parameters measured in the unstarred reference frame, whereas ζ_i^* are the parameters measured in the starred reference frame. The parameters may be scalars, vectors, and tensors. The components of an ***objective vector function***, $\boldsymbol{a}$, transform according to Equation (1.30) such that

$$\boldsymbol{a}^*\left(\zeta_1^*,\zeta_2^*,\ldots,\zeta_n^*\right)=\boldsymbol{Q}\cdot\boldsymbol{a}\left(\zeta_1,\zeta_2,\ldots,\zeta_n\right). \tag{4.55}$$

The components of an ***objection tensor function***, $\boldsymbol{A}$, transform according to Equation (1.32) such that

$$\boldsymbol{A}^*\left(\zeta_1^*,\zeta_2^*,\ldots,\zeta_n^*\right)=\boldsymbol{Q}\cdot\boldsymbol{A}\left(\zeta_1,\zeta_2,\ldots,\zeta_n\right)\cdot\boldsymbol{Q}^T. \tag{4.56}$$

We seek to determine the constraints imposed on our constitutive model from the principle of objectivity. In the next sections, we will look at how this is done for Eulerian and Lagrangian constitutive models.

Before looking at the constraints due to the principle of material frame indifference, let us look at the objectivity of several common kinematic tensors. Let us assume that the observer A is in the unstarred reference frame given by $\boldsymbol{x}$, and observer B is in the starred reference frame given by $\boldsymbol{x}^*$. The two spatial coordinate systems can be related via the equation

$$\boldsymbol{x}^*(\boldsymbol{X},t)=\boldsymbol{Q}(t)\cdot\boldsymbol{x}(\boldsymbol{X},t)+\boldsymbol{c}(t), \tag{4.57}$$

where $\boldsymbol{Q}$ is the relative rotation and $\boldsymbol{c}$ is the relative translation between the two reference frames. Let us also assume that at time $t=0$, the two frames coincide such that the two observers share the same material reference configuration and $\boldsymbol{x}(0)=\boldsymbol{X}=\boldsymbol{x}^*(0)=\boldsymbol{X}^*$.

Observer B and observer A measure the velocity of a material point as

$$\boldsymbol{v}^*(\boldsymbol{X},t)=\frac{\partial\boldsymbol{x}^*(\boldsymbol{X},t)}{\partial t}\quad\text{and}\quad\boldsymbol{v}(\boldsymbol{X},t)=\frac{\partial\boldsymbol{x}(\boldsymbol{X},t)}{\partial t}, \tag{4.58}$$

respectively. The relation between $\boldsymbol{v}$ and $\boldsymbol{v}^*$ can be found by substituting Equation (4.57) into Equation (4.58) giving

$$\boldsymbol{v}^*(\boldsymbol{X},t)=\frac{\partial}{\partial t}[\boldsymbol{Q}\cdot\boldsymbol{x}+\boldsymbol{c}]=\frac{\partial\boldsymbol{Q}}{\partial t}\cdot\boldsymbol{x}+\boldsymbol{Q}\cdot\boldsymbol{v}+\frac{\partial\boldsymbol{c}}{\partial t}. \tag{4.59}$$

In order to be objective, the velocity would have to transform according to $\boldsymbol{v}^*=\boldsymbol{Q}\cdot\boldsymbol{v}$. As you can see from Equation (4.59), this only occurs if the two reference frames are not rotating or translating relative to one another in time. When the two reference frames move relative to one another, the measured velocity is fundamentally different in the two reference frames. Therefore, we conclude that *the velocity vector is not an objective vector* unless the rotation $\boldsymbol{Q}$ and translation $\boldsymbol{c}$ are constant in time.

Next, let us examine the deformation gradient. The deformation gradients measured by observer A and observer B within the two reference frames are given by

$$\boldsymbol{F}(\boldsymbol{X},t)=\frac{\partial\boldsymbol{x}(\boldsymbol{X},t)}{\partial\boldsymbol{X}}\ \text{and}\ \boldsymbol{F}^*(\boldsymbol{X},t)=\frac{\partial\boldsymbol{x}^*(\boldsymbol{X},t)}{\partial\boldsymbol{X}^*}. \tag{4.60}$$

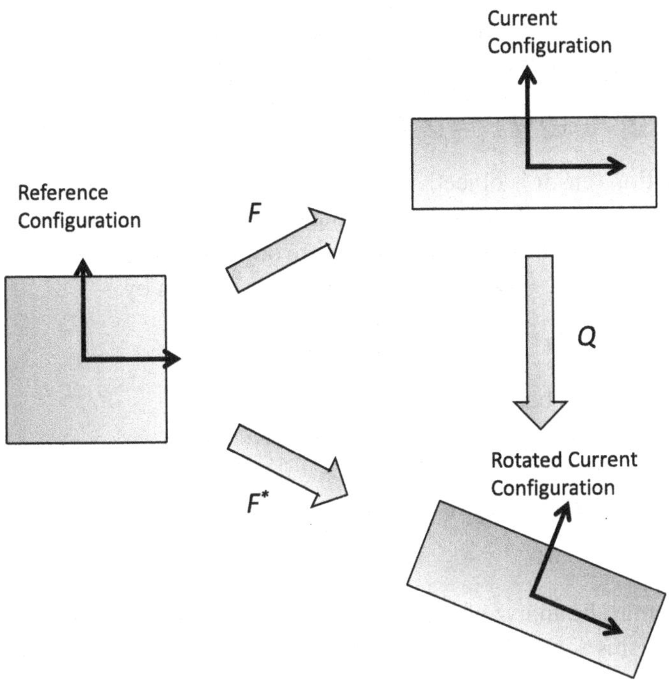

Figure 4.3. Illustration of the configurations used to derive the constraints dictated by material frame indifference. The coordinates of the current configuration are give by $\boldsymbol{x(X,t)}$, the coordinates in the rotated current configuration are given by $\boldsymbol{x}^{-}\boldsymbol{(X,t)}$, and the deformation gradients are related by $\boldsymbol{F}^{-} = \boldsymbol{Q} \cdot \boldsymbol{F}$.

We substitute Equation (4.57) into Equation (4.60) and make use of the fact that the two observers share the same reference configuration to obtain

$$\boldsymbol{F}^{*}(\boldsymbol{X},t) = \frac{\partial}{\partial \boldsymbol{X}}[\boldsymbol{Q}\cdot\boldsymbol{x}+\boldsymbol{c}] = \boldsymbol{Q}\cdot\frac{\partial \boldsymbol{x}}{\partial \boldsymbol{X}} = \boldsymbol{Q}\cdot\boldsymbol{F}.$$

Since the deformation gradient is a two-point tensor, it does not transform according to Equation (1.32). Instead, we write the tensor as

$$F_{iJ}^{*} = \boldsymbol{e}_i^{*}\cdot\left(\boldsymbol{F}\cdot\boldsymbol{E}_J^{*}\right). \tag{4.61}$$

The relation between basis vectors in the two reference frames depends only on the rotation and not on the relative translation of the two frames. Therefore, we can substitute $\boldsymbol{e}_i^{*} = \boldsymbol{Q}\cdot\boldsymbol{e}_i$ and $\boldsymbol{E}_j^{*} = \boldsymbol{E}_j$ into Equation (4.61) to get

$$\boldsymbol{F}^{*}(\boldsymbol{X},t) = \boldsymbol{Q}\cdot\boldsymbol{e}_i\cdot(\boldsymbol{F}\cdot\boldsymbol{E}_J) = \boldsymbol{Q}\cdot\boldsymbol{F}.$$

So you see that a two-point tensor will transform in a manner similar to an objective vector. Therefore, the deformation gradient is an objective two-point tensor.

Next we consider the deformation tensors. The right Cauchy deformation tensor measured by observer B is given by

$$\boldsymbol{C}^{*} = \left(\boldsymbol{F}^{*}\right)^{T}\cdot\boldsymbol{F}^{*} = (\boldsymbol{Q}\cdot\boldsymbol{F})^{T}\cdot(\boldsymbol{Q}\cdot\boldsymbol{F}) = \boldsymbol{F}^{T}\cdot\boldsymbol{Q}^{T}\cdot\boldsymbol{Q}\cdot\boldsymbol{F} = \boldsymbol{F}^{T}\cdot\boldsymbol{F} = \boldsymbol{C}.$$

This is clearly not an objective tensor under the spatial transformation. However, the left Cauchy deformation tensor measured by observer B is given by

$$\boldsymbol{B}^* = \boldsymbol{F}^* \cdot \left(\boldsymbol{F}^*\right)^T = (\boldsymbol{Q}\cdot\boldsymbol{F})\cdot(\boldsymbol{Q}\cdot\boldsymbol{F})^T = \boldsymbol{Q}\cdot\boldsymbol{F}\cdot\boldsymbol{F}^T\cdot\boldsymbol{Q}^T = \boldsymbol{Q}\cdot\boldsymbol{B}\cdot\boldsymbol{Q}^T. \tag{4.62}$$

The left Cauchy deformation tensor is objective under spatial transformation.

Similarly, we can find the transformation laws for the Cauchy stress tensor. Since both the traction vector and the outward surface normal are objective and are rotated according to $\left(\boldsymbol{t}_x^n\right)^* = \boldsymbol{Q}\cdot\left(\boldsymbol{t}_x^n\right)$ and $\boldsymbol{n}_x^* = \boldsymbol{Q}\cdot\boldsymbol{n}_x$, the Cauchy stress is given by

$$\boldsymbol{\sigma}^* = \boldsymbol{Q}\cdot\boldsymbol{\sigma}\cdot\boldsymbol{Q}^T.$$

The second Piola-Kirchhoff stress becomes

$$\begin{aligned}\boldsymbol{S}^* &= J^*\left(\boldsymbol{F}^*\right)^{-1}\cdot\boldsymbol{\sigma}^*\cdot\left(\boldsymbol{F}^*\right)^{-T}\\ &= J\,(\boldsymbol{QF})^{-1}\cdot\boldsymbol{Q}\cdot\boldsymbol{\sigma}\cdot\boldsymbol{Q}^T\cdot(\boldsymbol{QF})^{-T}\\ &= \boldsymbol{S}.\end{aligned}$$

Ultimately, each variable must be analyzed for its transformation under rigid motion. The analysis presented in this section has shown that the vector velocity and the tensors, $\boldsymbol{C}$ and $\boldsymbol{S}$, are not frame-indifferent, whereas the tensors, $\boldsymbol{B}$ and $\boldsymbol{\sigma}$, are frame-indifferent under rigid motion of the spatial reference frame.

The constraints arising from the principle of objectivity depend on the particular choice of independent variables used to formulate the constitutive model. For example, if we have a constitutive equation for the Cauchy stress tensor written in terms of the left Cauchy deformation tensor, $\boldsymbol{\sigma} = \boldsymbol{\sigma}(\boldsymbol{B})$, the principle of objectivity requires

$$\boldsymbol{\sigma}\left(\boldsymbol{B}^*\right) = \boldsymbol{Q}\cdot\boldsymbol{\sigma}(\boldsymbol{B})\cdot\boldsymbol{Q}^T.$$

We have determined the manner in which $\boldsymbol{B}^*$ transforms in Equation (4.62), giving the result

$$\boldsymbol{\sigma}\left(\boldsymbol{Q}\cdot\boldsymbol{B}\cdot\boldsymbol{Q}^T\right) = \boldsymbol{Q}\cdot\boldsymbol{\sigma}(\boldsymbol{B})\cdot\boldsymbol{Q}^T.$$

This must be true for each and every possible rotation $\boldsymbol{Q}$.

On the other hand, a constitutive relation for the Cauchy stress tensor as a function of the right Cauchy deformation tensor, $\boldsymbol{\sigma} = \boldsymbol{\sigma}(\boldsymbol{C})$, becomes

$$\begin{aligned}\boldsymbol{\sigma}\left(\boldsymbol{C}^*\right) &= \boldsymbol{Q}\cdot\boldsymbol{\sigma}(\boldsymbol{C})\cdot\boldsymbol{Q}^T,\\ \boldsymbol{\sigma}(\boldsymbol{C}) &= \boldsymbol{Q}\cdot\boldsymbol{\sigma}(\boldsymbol{C})\cdot\boldsymbol{Q}^T.\end{aligned}$$

This equation must be true for all possible rotations $\boldsymbol{Q}$. The only way for this to be true is if the function $\boldsymbol{\sigma}(\boldsymbol{C})$ is an isotropic function that depends only on the invariants of the right Cauchy deformation gradient.

Let us examine the thermoelastic material model we have proposed. The principle of material frame indifference requires that our constitutive equations be objective functions. This requires that they adhere to the condition

$$\begin{aligned}\boldsymbol{\sigma}\left(\boldsymbol{F}^*,\theta^*\right) &= \boldsymbol{Q}\cdot\boldsymbol{\sigma}(\boldsymbol{F},\theta)\cdot\boldsymbol{Q}^T,\\ \psi\left(\boldsymbol{F}^*,\theta^*\right) &= \psi(\boldsymbol{F},\theta),\end{aligned}$$

$$\eta\left(\boldsymbol{F}^*,\theta^*\right) = \eta\left(\boldsymbol{F},\theta\right),$$
$$\boldsymbol{q}\left(\boldsymbol{F}^*,\theta^*,\, grad\,\theta^*\right) = \boldsymbol{Q}\cdot\boldsymbol{q}\left(\boldsymbol{F},\theta,\, grad\,\theta\right),$$

where $\boldsymbol{F}^* = \boldsymbol{Q}\cdot\boldsymbol{F}$, $\theta^* = \theta$ since it is a scalar parameter, and

$$(grad\,\theta)^* = \frac{\partial\theta^*}{\partial\boldsymbol{x}^*} = \frac{\partial\theta}{\partial\boldsymbol{x}}\cdot\frac{\partial\boldsymbol{x}}{\partial\boldsymbol{x}^*} = \frac{\partial\theta}{\partial\boldsymbol{x}}\cdot\boldsymbol{Q}^T = \boldsymbol{Q}\cdot grad\,\theta.$$

Therefore, the constraints on the constitutive model can be written as

$$\psi\left(\boldsymbol{Q}\cdot\boldsymbol{F},\theta\right) = \psi\left(\boldsymbol{F},\theta\right),$$
$$\eta\left(\boldsymbol{Q}\cdot\boldsymbol{F},\theta\right) = \eta\left(\boldsymbol{F},\theta\right),$$
$$\boldsymbol{\sigma}\left(\boldsymbol{Q}\cdot\boldsymbol{F},\theta\right) = \boldsymbol{Q}\cdot\boldsymbol{\sigma}\left(\boldsymbol{F},\theta\right)\cdot\boldsymbol{Q}^T,$$
$$\boldsymbol{q}\left(\boldsymbol{Q}\cdot\boldsymbol{F},\theta\right) = \boldsymbol{Q}\cdot\boldsymbol{q}\left(\boldsymbol{F},\theta,\, grad\,\theta\right).$$

These constraints must be satisfied for each and every possible choice of $\boldsymbol{Q}$. Let us look at one of the infinite set of constraints that this set of equations represents. We are free to choose one particular value of $\boldsymbol{Q} = \boldsymbol{R}^T$ which gives

$$\psi\left(\boldsymbol{F},\theta\right) = \psi\left(\boldsymbol{U},\theta\right),$$
$$\eta\left(\boldsymbol{F},\theta\right) = \eta\left(\boldsymbol{U},\theta\right),$$
$$\boldsymbol{\sigma}\left(\boldsymbol{F},\theta\right) = \boldsymbol{R}^T\cdot\boldsymbol{\sigma}\left(\boldsymbol{U},\theta\right)\cdot\boldsymbol{R},$$
$$\boldsymbol{q}\left(\boldsymbol{F},\theta,\, grad\,\theta\right) = \boldsymbol{R}^T\cdot\boldsymbol{q}\left(\boldsymbol{U},\theta,\boldsymbol{R}^T\cdot grad\,\theta\right),$$

where we note that $\boldsymbol{F} = \boldsymbol{R}\cdot\boldsymbol{U}$ and therefore $\boldsymbol{R}^T\cdot\boldsymbol{F} = \boldsymbol{U}$, and we have inverted the resulting constraints. These constraints require that the form of the constitutive model be reducible to a function of the right stretch tensor. Notice that since $\boldsymbol{U}^2 = \boldsymbol{C}$ and since $\boldsymbol{C}^* = \boldsymbol{C}$, we have $\boldsymbol{U}^* = \boldsymbol{U}$ and any function of solely $\boldsymbol{U}$, and $\theta, \boldsymbol{f}\left(\boldsymbol{U},\theta\right)$ will return the same value independent of the rotation of the coordinate system.

EXAMPLE 4.2. *Show that the function for the internal energy* $\boldsymbol{e}\left(\boldsymbol{F}\right) = \boldsymbol{det F}$ *is consistent with the principal of material frame indifference.*

Solution:
The function must obey the rule:

$$e^*\left(\boldsymbol{F}^*\right) = e\left(\boldsymbol{F}\right),$$
$$e\left(\boldsymbol{Q}\cdot\boldsymbol{F}\right) = e\left(\boldsymbol{F}\right),$$
$$\det\left(\boldsymbol{Q}\cdot\boldsymbol{F}\right) = \det\left(\boldsymbol{F}\right),$$
$$\det\left(\boldsymbol{Q}\right)\det\left(\boldsymbol{F}\right) = \det\left(\boldsymbol{F}\right),$$
$$1\det\left(\boldsymbol{F}\right) = \det\left(\boldsymbol{F}\right),$$

which shows that this particular function does obey the principal of material frame indifference.

EXAMPLE 4.3. *Show that the function for the stress,* $\boldsymbol{\sigma}\left(\boldsymbol{F},t\right) = \alpha t\left(\boldsymbol{F}+\boldsymbol{F}^T\right)$, *is not consistent with the principal of material frame indifference.*

Solution:

First, we note that the scalar time, $t = t^*$. We can then substitute this into the relation:

$$\sigma^*\left(\boldsymbol{F}^*, t^*\right) = \boldsymbol{Q}\cdot\boldsymbol{\sigma}\left(\boldsymbol{F}, t\right)\cdot\boldsymbol{Q}^T,$$

$$\sigma\left(\boldsymbol{Q}\cdot\boldsymbol{F}, t\right) = \boldsymbol{Q}\cdot\boldsymbol{\sigma}\left(\boldsymbol{F}, t\right)\cdot\boldsymbol{Q}^T,$$

$$\alpha t\left(\boldsymbol{Q}\cdot\boldsymbol{F} + (\boldsymbol{Q}\cdot\boldsymbol{F})^T\right) = \boldsymbol{Q}\cdot\alpha t\left(\boldsymbol{F} + \boldsymbol{F}^T\right)\cdot\boldsymbol{Q}^T,$$

$$\left(\boldsymbol{Q}\cdot\boldsymbol{F} + \boldsymbol{F}^T\cdot\boldsymbol{Q}^T\right) \neq \boldsymbol{Q}\cdot\boldsymbol{F}\cdot\boldsymbol{Q}^T + \boldsymbol{Q}\cdot\boldsymbol{F}^T\cdot\boldsymbol{Q}^T.$$

4.6 Material Symmetry

Many materials exhibit similar behavior in multiple directions. For example, the material properties of an isotropic solid do not vary directionally. The stress-strain response of the material element along one axis is identical to the stress-strain response along any other axis. If one starts with a unit sphere of material, compresses the material, and then rotates the sphere and compresses again, the experimentally determined response is identical. In fact, what you are doing is rotating the reference configuration of the material and subjecting it to an identical deformation. An isotropic material must have the same response for all possible rotations of the reference configuration. Many fluids exhibit isotropic behavior. Other materials exhibit lesser symmetry. A multilayered laminated composite (such as plywood) might exhibit transverse isotropy. The material properties within the plane are invariant, but are distinct from the properties perpendicular to the ply.

If the material posses symmetry, there exists some set of rigid body rotations, $\{\boldsymbol{Q}\}$, of the reference configuration which do not change the material response. For the isotropic material, the set of rotations includes all possible rotations. For the transversely isotropic material, the set of rotations includes all rotations about the axis perpendicular to the ply.

Let us examine the constraints on the form of the constitutive equations, which arise due to material symmetry. We will compare two configurations. First, we deform the body according to the equations of motion $\boldsymbol{x} = \boldsymbol{x}(\boldsymbol{X}, t)$. Next we imagine rotating the reference configuration of the object. Since the reference configuration is rotated and translated with respect to the original reference configuration, we can write the new reference configuration, $\boldsymbol{X}^*$, in terms of the original reference configuration, $\boldsymbol{X}$, and the rotation relating the two, $\boldsymbol{Q}$, as

$$\boldsymbol{X}^* = \boldsymbol{Q}\cdot\boldsymbol{X} + \boldsymbol{c} \text{ or } X_i^* = Q_{ij}X_j + c_i.$$

Finally, we apply the same equations of motion to the rotated configuration such that $\boldsymbol{x}^* = \boldsymbol{x}(\boldsymbol{X}^*, t)$. If we assume for simplicity that both the deformation and material are homogenous, we can write the deformation gradient as

$$\boldsymbol{F} = \frac{\partial \boldsymbol{x}}{\partial \boldsymbol{X}} = \frac{\partial \boldsymbol{x}}{\partial \boldsymbol{X}^*}\cdot\frac{\partial \boldsymbol{X}^*}{\partial \boldsymbol{X}}.$$

Given $\frac{\partial \boldsymbol{X}^*}{\partial \boldsymbol{X}} = \boldsymbol{Q}$, and $\boldsymbol{F}^* = \frac{\partial \boldsymbol{x}}{\partial \boldsymbol{X}^*}$, we can write

$$\boldsymbol{F} = \boldsymbol{F}^*\cdot\boldsymbol{Q} \text{ and } \boldsymbol{F}^* = \boldsymbol{F}\cdot\boldsymbol{Q}^T.$$

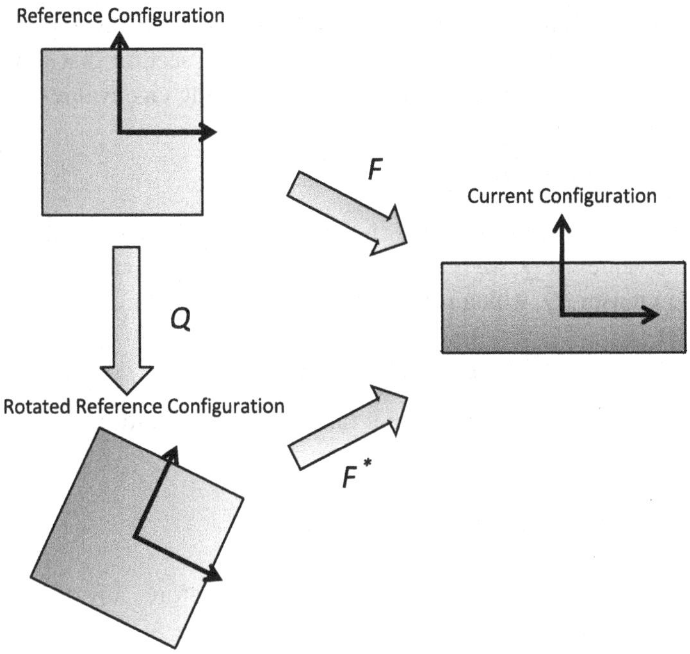

Figure 4.4. An illustration of the configurations used to derive the material symmetry condition. The positions of particles in the reference configuration are given by $\boldsymbol{X}$, the positions in the rotated reference configuration are given by $\boldsymbol{X}^{-} = \boldsymbol{Q} \cdot \boldsymbol{X} + \boldsymbol{c}$, and the deformation gradient $\boldsymbol{F}^{-} = \boldsymbol{F} \cdot \boldsymbol{Q}^{T}$.

The constitutive response must be the same for all rotations of the reference configuration that conform to the material symmetry. For the isotropic case, we have

$$\begin{aligned} \psi(\boldsymbol{F}) &= \psi\left(\boldsymbol{F}^{*}\right) \\ &= \psi\left(\boldsymbol{F} \cdot \boldsymbol{Q}^{T}\right), \end{aligned}$$

for *all possible rotations*. Similarly, we obtain a transformation rule for the right Cauchy deformation tensor given by

$$\boldsymbol{C}^{*} = \left(\boldsymbol{F}^{*}\right)^{T} \cdot \boldsymbol{F}^{*} = \boldsymbol{Q} \cdot \boldsymbol{F}^{T} \cdot \boldsymbol{F} \cdot \boldsymbol{Q}^{T} = \boldsymbol{Q} \cdot \boldsymbol{C} \cdot \boldsymbol{Q}^{T}.$$

Therefore, a constitutive model proposed in terms of the right Cauchy deformation tensor must obey the relation

$$\begin{aligned} \psi(\boldsymbol{C}) &= \psi\left(\boldsymbol{C}^{*}\right) \\ &= \psi\left(\boldsymbol{Q} \cdot \boldsymbol{C} \cdot \boldsymbol{Q}^{T}\right). \end{aligned}$$

4.6.1 Isotropic Scalar-Valued Functions

A scalar-valued tensor function, $\alpha(\boldsymbol{A})$, is called an isotropic scalar-valued function if

$$\alpha(\boldsymbol{A}) = \alpha\left(\boldsymbol{Q} \cdot \boldsymbol{A} \cdot \boldsymbol{Q}^{T}\right), \tag{4.63}$$

for every possible rotation, $\boldsymbol{Q}$. Clearly, these functions are common when developing constitutive models for isotropic materials. We will show in this section that any isotropic scalar-valued function may be written as a function of the eigenvalues of the tensor parameter, $\boldsymbol{A}$, such that

$$\alpha(\mathbf{A}) = \alpha(\lambda_1, \lambda_2, \lambda_3).$$

The proof is rather straightforward. We begin by noting that since Equation (4.63) must hold for all possible values of $\boldsymbol{Q}$, we may rewrite it for any specific value of $\boldsymbol{Q}$. Let us select the rotation tensor, $\overline{\boldsymbol{Q}}$, which rotates $\boldsymbol{A}$ into its principle directions. $\overline{\boldsymbol{Q}}$ is given by

$$\overline{Q}_{ij} = \hat{\boldsymbol{n}}^{(i)} \cdot \boldsymbol{e}_j,$$

where $\hat{\boldsymbol{n}}^{(i)}$ are the normalized eigenvectors of $\boldsymbol{A}$. For this selection of rotation tensor, we have

$$\boldsymbol{\lambda} = \overline{\boldsymbol{Q}} \cdot \boldsymbol{A} \cdot \overline{\boldsymbol{Q}}^T = \begin{bmatrix} \lambda_1 & 0 & 0 \\ 0 & \lambda_2 & 0 \\ 0 & 0 & \lambda_3 \end{bmatrix},$$

where λ_1, λ_2, and λ_3 are the eigenvalues of the tensor A. Therefore, any scalar function for an isotropic material must conform to

$$\alpha(\mathbf{A}) = \alpha\left(\boldsymbol{Q} \cdot \boldsymbol{A} \cdot \boldsymbol{Q}^T\right) = \alpha(\boldsymbol{\lambda}).$$

Since the three independent components of the tensor, $\boldsymbol{\lambda}$, are the eigenvalues of the tensor $\boldsymbol{A}$, we can write

$$\alpha(\mathbf{A}) = \alpha(\lambda_1, \lambda_2, \lambda_3).$$

Therefore, any scalar function for an isotropic material can be written in terms of the eigenvalues of the tensor parameter.

4.6.2 Isotropic Tensor-Valued Functions

A symmetric tensor-valued tensor function, $\boldsymbol{\sigma}$, with a symmetric tensor parameter, $\boldsymbol{A}$, is called an isotropic tensor-valued function if

$$\boldsymbol{\sigma}\left(\boldsymbol{Q} \cdot \boldsymbol{A} \cdot \boldsymbol{Q}^T\right) = \boldsymbol{Q} \cdot \boldsymbol{\sigma}(\boldsymbol{A}) \cdot \boldsymbol{Q}^T, \tag{4.64}$$

for every possible rotation $\boldsymbol{Q}$. These functions have the property that they can be written as a linear combination of the identity matrix, and the first and second powers of the tensor parameter $\boldsymbol{A}$ such that

$$\boldsymbol{\sigma}(\boldsymbol{A}) = \alpha_1 \mathbf{I} + \alpha_2 \boldsymbol{A} + \alpha_3 \boldsymbol{A}^2. \tag{4.65}$$

First, we show that the eigenvalues of $\boldsymbol{A}$ and $\boldsymbol{Q} \cdot \boldsymbol{\sigma}(\boldsymbol{A}) \cdot \boldsymbol{Q}^T$ must always be equal for all rotations $\boldsymbol{Q}$. Given an eigenpair, λ_1 and $\boldsymbol{n}^{(1)}$, for the tensor $\boldsymbol{A}$, we can write

$$A \cdot \boldsymbol{n}^{(1)} = \lambda_1 \boldsymbol{n}^{(1)}.$$

Multiplying on the left by $\boldsymbol{Q}$ and substituting $\boldsymbol{Q}^T \cdot \boldsymbol{Q} = \mathbf{I}$ gives

$$\boldsymbol{Q} \cdot A \cdot \boldsymbol{n}^{(1)} = \boldsymbol{Q} \cdot \lambda_1 \boldsymbol{n}^{(1)},$$

$$\left(\boldsymbol{Q} \cdot A \cdot \boldsymbol{Q}^T\right) \cdot \left(\boldsymbol{Q} \cdot \boldsymbol{n}^{(1)}\right) = \lambda_1 \left(\boldsymbol{Q} \cdot \boldsymbol{n}^{(1)}\right).$$

Therefore, $\boldsymbol{Q}\cdot\boldsymbol{n}^{(1)}$ is an eigenvector of the tensor product $\boldsymbol{Q}\cdot A\cdot\boldsymbol{Q}^T$, and λ_1 is an eigenvalue. Similarly, we can show that λ_2 and λ_3 are also eigenvalues of the tensor product $\boldsymbol{Q}\cdot A\cdot\boldsymbol{Q}^T$.

Next we can show that the only functions which satisfy Equation (4.64) are those which have the property that the eigenvectors of $\boldsymbol{A}$ are the same as $\boldsymbol{\sigma}(\boldsymbol{A})$. This can be demonstrated by selecting a particular rotation, $\overline{\boldsymbol{Q}}$, which represents the rotation of the coordinate system about an eigenvector of the tensor $\boldsymbol{A}$, $\boldsymbol{n}^{(1)}$ by an angle π. $\overline{\boldsymbol{Q}}$ is given by

$$\overline{\boldsymbol{Q}} = \boldsymbol{n}^{(1)}\otimes\boldsymbol{n}^{(1)} - \boldsymbol{n}^{(2)}\otimes\boldsymbol{n}^{(2)} - \boldsymbol{n}^{(3)}\otimes\boldsymbol{n}^{(3)}.$$

Notice that this rotation does not alter the direction of the eigenvector, $\boldsymbol{n}^{(1)}$, while the plane containing $\boldsymbol{n}^{(2)}$ and $\boldsymbol{n}^{(3)}$ eigenvectors will be rotated by 180° such that

$$\boldsymbol{n}^{(1)} = \hat{\boldsymbol{n}}^{(1)},\quad \boldsymbol{n}^{(2)} = -\hat{\boldsymbol{n}}^{(2)},\quad \text{and } \boldsymbol{n}^{(3)} = -\hat{\boldsymbol{n}}^{(3)}.$$

Given the particular rotation, $\overline{\boldsymbol{Q}}$, let us define the tensor $\boldsymbol{B}$ as

$$\boldsymbol{B} = \overline{\boldsymbol{Q}}\cdot\boldsymbol{A}\cdot\overline{\boldsymbol{Q}}^T. \tag{4.66}$$

Noting once again that $\overline{\boldsymbol{Q}}$ does not alter the direction of the eigenvector, $\boldsymbol{n}^{(1)}$, we can multiply this equation on the right by $\boldsymbol{n}^{(1)}$. This gives

$$\begin{aligned}
\boldsymbol{B}\cdot\,\boldsymbol{n}^{(1)} &= \overline{\boldsymbol{Q}}\cdot\boldsymbol{A}\cdot\overline{\boldsymbol{Q}}^T\cdot\boldsymbol{n}^{(1)},\\
\boldsymbol{B}\cdot\,\boldsymbol{n}^{(1)} &= \overline{\boldsymbol{Q}}\cdot\boldsymbol{A}\cdot\boldsymbol{n}^{(1)},\\
\boldsymbol{B}\cdot\,\boldsymbol{n}^{(1)} &= \overline{\boldsymbol{Q}}\cdot\lambda_1\cdot\boldsymbol{n}^{(1)},\\
\boldsymbol{B}\cdot\,\boldsymbol{n}^{(1)} &= \lambda_1\cdot\boldsymbol{n}^{(1)}.
\end{aligned}$$

Therefore, the eigenpair λ_1 and $\boldsymbol{n}^{(1)}$ for tensor $\boldsymbol{A}$ are also the eigenpair for tensor $\boldsymbol{B}$. Similarly, you can multiply Equation (4.66) on the right by $\boldsymbol{n}^{(2)}$ and $\boldsymbol{n}^{(3)}$ which shows that all of the eigenpairs for $\boldsymbol{A}$ are also eigenpairs for $\boldsymbol{B}$. In fact, the two tensors are equal, $\boldsymbol{A}=\boldsymbol{B}$, and we have the condition

$$\boldsymbol{A} = \overline{\boldsymbol{Q}}\cdot\boldsymbol{A}\cdot\overline{\boldsymbol{Q}}^T.$$

Armed with this condition, we can now examine the isotropic tensor-valued function

$$\begin{aligned}
\boldsymbol{\sigma}\left(\overline{\boldsymbol{Q}}\cdot\boldsymbol{A}\cdot\overline{\boldsymbol{Q}}^T\right) &= \overline{\boldsymbol{Q}}\cdot\boldsymbol{\sigma}(\boldsymbol{A})\cdot\overline{\boldsymbol{Q}}^T,\\
\boldsymbol{\sigma}(\boldsymbol{A}) &= \overline{\boldsymbol{Q}}\cdot\boldsymbol{\sigma}(\boldsymbol{A})\cdot\overline{\boldsymbol{Q}}^T,\\
\boldsymbol{\sigma}(\boldsymbol{A})\cdot\overline{\boldsymbol{Q}} &= \overline{\boldsymbol{Q}}\cdot\boldsymbol{\sigma}(\boldsymbol{A}).
\end{aligned}$$

Multiplying on the right by the eigenvector $\boldsymbol{n}^{(1)}$, we obtain

$$\begin{aligned}
\boldsymbol{\sigma}(\boldsymbol{A})\cdot\overline{\boldsymbol{Q}}\cdot\boldsymbol{n}^{(1)} &= \overline{\boldsymbol{Q}}\cdot\boldsymbol{\sigma}(\boldsymbol{A})\cdot\boldsymbol{n}^{(1)},\\
\boldsymbol{\sigma}(\boldsymbol{A})\cdot\boldsymbol{n}^{(1)} &= \overline{\boldsymbol{Q}}\cdot\boldsymbol{\sigma}(\boldsymbol{A})\cdot\boldsymbol{n}^{(1)}.
\end{aligned}$$

Substituting the variable, $\boldsymbol{v}=\boldsymbol{\sigma}(\boldsymbol{A})\cdot\boldsymbol{n}^{(1)}$, for clarity gives

$$\boldsymbol{v} = \overline{\boldsymbol{Q}}\cdot\boldsymbol{v}.$$

Since the only direction which remains unaltered by the rotation $\overline{\boldsymbol{Q}}$ is $\boldsymbol{n}^{(1)}$, then $\boldsymbol{\sigma}(\mathbf{A})\cdot\boldsymbol{n}^{(1)}$ must be parallel to $\boldsymbol{n}^{(1)}$. This gives the condition

$$\boldsymbol{\sigma}(\mathbf{A})\cdot\boldsymbol{n}^{(1)}=\beta_1\boldsymbol{n}^{(1)},$$

showing that $\boldsymbol{n}^{(1)}$ must be an eigenvector of the isotropic tensor-valued function $\boldsymbol{\sigma}(\mathbf{A})$. Similar analyses for $\boldsymbol{n}^{(2)}$ and $\boldsymbol{n}^{(3)}$ reveal that each of the eigenvectors of $\boldsymbol{\sigma}(\mathbf{A})$ must all be identical to those of the tensor $\mathbf{A}$.

Now we return to Equation (4.64) and use spectral decomposition to write

$$\boldsymbol{\sigma}(\mathbf{A})=\sum_{i=1}^{3}\beta_i(\mathbf{A})\left(\boldsymbol{n}^{(i)}\otimes\boldsymbol{n}^{(i)}\right),$$

where $\boldsymbol{n}^{(i)}$ are the eigenvectors of the tensor $\boldsymbol{A}$, and $\beta_i(\mathbf{A})$ are the distinct eigenvalues of the function $\boldsymbol{\sigma}(\mathbf{A})$. We have left them here as functions of the tensor parameter $\mathbf{A}$. Similarly, we can write

$$\boldsymbol{\sigma}\left(\boldsymbol{Q}\cdot\mathbf{A}\cdot\boldsymbol{Q}^T\right)=\sum_{i=1}^{3}\beta_i\left(\boldsymbol{Q}\cdot\mathbf{A}\cdot\boldsymbol{Q}^T\right)\ \left(\boldsymbol{Q}\cdot\left(\boldsymbol{n}^{(i)}\otimes\boldsymbol{n}^{(i)}\right)\cdot\boldsymbol{Q}^T\right),$$

since we have shown that the eigenvectors of $\boldsymbol{\sigma}\left(\boldsymbol{Q}\cdot\mathbf{A}\cdot\boldsymbol{Q}^T\right)$ are equal to $\boldsymbol{Q}\cdot\boldsymbol{n}^{(i)}$.

According to equation 4.64, we have

$$\boldsymbol{\sigma}\left(\boldsymbol{Q}\cdot\mathbf{A}\cdot\boldsymbol{Q}^T\right)=\boldsymbol{Q}\cdot\boldsymbol{\sigma}(\mathbf{A})\cdot\boldsymbol{Q}^T,$$

$$\sum_{i=1}^{3}\beta_i(\mathbf{A})\ \left(\boldsymbol{Q}\cdot\left(\boldsymbol{n}^{(i)}\otimes\boldsymbol{n}^{(i)}\right)\cdot\boldsymbol{Q}^T\right)=\boldsymbol{Q}\cdot\sum_{j=1}^{3}\beta_j\left(\boldsymbol{Q}\cdot\mathbf{A}\cdot\boldsymbol{Q}^T\right)\left(\boldsymbol{n}^{(j)}\otimes\boldsymbol{n}^{(j)}\right)\cdot\boldsymbol{Q}^T.$$

Ultimately, this gives the condition

$$\beta_i(\mathbf{A})\ =\beta_i\left(\boldsymbol{Q}\cdot\mathbf{A}\cdot\boldsymbol{Q}^T\right).$$

Therefore, the eigenvalues of the function $\boldsymbol{\sigma}(\mathbf{A})$ are isotropic scalar-valued functions and can be written in terms of the eigenvalues of the tensor $\mathbf{A}$ such that

$$\beta_i(\mathbf{A})=\beta_i(\lambda_1,\lambda_2,\lambda_3).$$

The spectral representation of the tensor parameter, $\mathbf{A}$, is given by

$$\mathbf{A}=\sum_{i=1}^{3}\lambda_i\left(\boldsymbol{n}^{(i)}\otimes\boldsymbol{n}^{(i)}\right).$$

Therefore, we can write the isotropic representation of the tensor function

$$\boldsymbol{\sigma}(\mathbf{A})=\alpha_1\mathbf{I}+\alpha_2\mathbf{A}+\alpha_3\mathbf{A}^2.$$

Substitution of the spectral decomposition for both $\boldsymbol{\sigma}$ and $\mathbf{A}$ gives

$$\sum_i\beta_i\left(\boldsymbol{n}^{(i)}\otimes\boldsymbol{n}^{(i)}\right)=\alpha_1\sum_i\left(\boldsymbol{n}^{(i)}\otimes\boldsymbol{n}^{(i)}\right)+\alpha_2\sum_i\lambda_i\left(\boldsymbol{n}^{(i)}\otimes\boldsymbol{n}^{(i)}\right)+\alpha_3\sum_i\lambda_i^2\left(\boldsymbol{n}^{(i)}\otimes\boldsymbol{n}^{(i)}\right),$$

where the coefficients α_i and β_i are isotropic scalar functions of $\boldsymbol{A}$.

By taking the dot product of this equation with each $\boldsymbol{n}^{(i)}$, we obtain a system of equations,

$$\begin{bmatrix} 1 & \lambda_1 & \lambda_1^2 \\ 1 & \lambda_2 & \lambda_2^2 \\ 1 & \lambda_3 & \lambda_3^2 \end{bmatrix} \begin{bmatrix} \alpha_1 \\ \alpha_2 \\ \alpha_3 \end{bmatrix} = \begin{bmatrix} \beta_1 \\ \beta_2 \\ \beta_3 \end{bmatrix},$$

for which we can always find a set of coefficients, α_i, that satisfy the relation. Therefore, any symmetric isotropic tensor-valued function must have the representation given by Equation (4.65).

4.7 Internal Variables

Occasionally, it is necessary to augment our continuum formulation with internal variables which describe the changes in microstructure that occur during time. We might use an internal variable to account for the damage done to a material during deformation. Perhaps, we have a growing tissue that changes its structure due to the application of stress. In order to use this concept of an internal variable within the thermodynamic framework we have developed, we will introduce the concept of affinities. Let us assume that we have a Helmholtz free energy function which describes the behavior of our material system as a function of the temperature, deformation gradient, and a set of internal variables, $\xi = \{\xi_1, \xi_2, \ldots, \xi_m\}$:

$$\psi = \psi\left(\boldsymbol{F}, \theta, \xi_1, \ldots, \xi_m\right) \tag{4.67}$$

If we take the material derivative of the Helmholtz free energy with respect to time, we obtain

$$\frac{D\psi\left(\boldsymbol{F}, \theta, \xi_1, \ldots, \xi_m\right)}{Dt} = \frac{\partial \psi}{\partial \boldsymbol{F}} : \frac{D\boldsymbol{F}}{Dt} + \frac{\partial \psi}{\partial \theta}\frac{D\theta}{Dt} + \frac{\partial \psi}{\partial \xi_i}\frac{D\xi_i}{Dt}. \tag{4.68}$$

The ***affinity*** for each internal variable is given by the derivative of the Helmholtz free energy with respect to that internal variable while holding all other variables constant:

$$\begin{aligned} \Lambda_1 &= \left.\frac{\partial \psi}{\partial \xi_1}\right|_{\boldsymbol{F},\ \theta, \xi_2, \ldots \xi_m}, \\ \Lambda_i &= \left.\frac{\partial \psi}{\partial \xi_i}\right|_{\boldsymbol{F},\ \theta, \xi_{j=1\ldots m, i\neq j}}. \end{aligned} \tag{4.69}$$

The power dissipated by the internal variable is given by the product of the affinity with the rate of change of the internal variable:

$$\mathfrak{E}_{\xi_i} = \Lambda_i \xi_l. \tag{4.70}$$

We then write equations that describe the evolution of the internal variables during deformation. In this case, we can write the rate of change of the internal variable in terms of the other variables:

$$\xi_i = f\left(\boldsymbol{F}, \theta, \xi_{i=1\ldots m}\right) \tag{4.71}$$

Table 4.5. *Definition and natural variables for thermodynamic potentials*

Thermodynamic potential	Formula	Natural variables
Internal energy	e	$\mathrm{n}_i,\ \boldsymbol{E},\ \eta$
Helmholtz free energy	$\psi = e - \eta\theta$	$\mathrm{n}_i,\ \boldsymbol{E},\ \theta$
Gibbs free energy	$\mathcal{G} = e - \eta\theta - \sigma^T : \boldsymbol{D}$	$\mathrm{n}_i,\ \boldsymbol{\sigma},\ \theta$
Enthalpy	$\mathcal{H} = e - \sigma^T : \boldsymbol{D}$	$\mathrm{n}_i,\ \boldsymbol{\sigma},\ \eta$
Landau potential	$\mathcal{L} = e - \eta\theta - \mu_i \mathrm{n}_i$	$\mu_i,\ \boldsymbol{E},\ \theta$

In this example, we have assumed the internal variable depends on stress, deformation, temperature, time, and the current value of the internal variable. This functional dependence will vary from problem to problem.

4.8 Thermodynamics of Materials

The thermodynamic state of a system consists of the stress, $\boldsymbol{\sigma}$, temperature, θ, entropy, η, and so on. As we act on the system by transferring mass, energy, momentum and the like, the thermodynamic state variables change value. A thermodynamic potential captures the relation between these thermodynamic variables in the presence of conservative forces. The thermodynamic equilibrium of the system can be found by identifying the thermodynamic variables that are held constant during relaxation and minimizing the appropriate thermodynamic potential. The variables held constant during the thermodynamic process are called the natural variables. For example, the minimum of the Gibbs free energy gives the equilibrium state of a material system at constant stress, temperature, and number of particles, Table 4.5.

4.9 Heat Transfer

The transfer of heat energy occurs both by direct molecular interaction and the transfer of electromagnetic radiation. These transfer mechanisms and the equations used to describe them are the same in all materials. It is necessary simply to identify the mechanisms that are active in the system being modeled and include the effects appropriately.

Conductive heat transfer is both the direct transfer of heat between molecules either through collisions, electron transfer, or molecular vibration, and the transfer of heat accompanying the diffusion of molecules. The conduction of heat is described using ***Fourier's law***,

$$\boldsymbol{q} = -\boldsymbol{k} \cdot \nabla\theta.$$

Convective heat transfer refers to the heat transferred within a fluid due to the motion of the molecules within the fluid. ***Natural convection*** refers to a process that induces fluid flow due to the temperature gradients while ***forced convection*** refers to a process in which the established fluid flow field is not influenced by the heat transfer process.

Radiative heat transfer refers to the indirect transfer of heat energy between molecules via electromagnetic radiation. Radiative heat transfer occurs when atoms are stimulated due to thermal energy and emit radiation. The heat transfer due to radiation between surfaces is given by the Stefan-Boltzmann law,

$$\boldsymbol{q} = \sigma\left(\theta^4 - \theta_o^0\right).$$

EXERCISES

1. The equations of motion for a unit cube that has an initial uniform density $\rho_o = 8\ \text{g/cm}^3$ are given by

$$x_1 = \alpha(t+1)X_1,$$
$$x_2 = \beta(t+1)X_2,$$
$$x_3 = X_3,$$

   where $0 \le X_1 \le 1, 0 \le X_2 \le 1, 0 \le X_3 \le 1$, and $0 \le t \le \infty$.

   a. Find the deformation gradient, $\boldsymbol{F}$, and describe in words the way in which this unit cube is deformed.
   b. Find the Jacobian, J.
   c. Noting that the deformation is homogenous, find the volume of the unit cube in the reference configuration, V_X, and the volume of the deformed cube in the current configuration, V_x.
   d. If the mass of the cube is conserved, find the current density, $\rho(\boldsymbol{x}, t)$.
   e. Evaluate the integral $\zeta = \int_{\Omega_o} \rho_o(\boldsymbol{X})\, dV_X$.
   f. Evaluate the integral $\zeta = \int_{\Omega(t)} \rho(\boldsymbol{x}, t)\, dV_x$.

2. Assume that a material is subject to (1) body forces, (2) surface traction, (3) body couples, and (4) surface couples.

   a. Derive the spatial form of the equation for conservation of mass.
   b. Derive the spatial form of the equation for conservation of linear momentum.
   c. Derive the material form of the equation for conservation of linear momentum.
   d. Derive the spatial form of the equation for conservation of angular momentum.
   e. Derive the material form of the equation for conservation of angular momentum.

3. Determine the objectivity of the

   a. Cauchy deformation tensor, $\boldsymbol{C}$.
   b. Cauchy deformation tensor, $\boldsymbol{B}$.
   c. Rate of deformation tensor, $\boldsymbol{D}$.
   d. Velocity gradient, $\boldsymbol{L}$.

5 Ideal Gas

We begin our survey of constitutive models with the development of a material model for ideal gases. The ideal gas can be modeled as a thermoelastic material which has a spherical stress tensor and no internal dissipation. In this chapter, the history and development of the constitutive law for an ideal gas are discussed, the field equations are derived from the balance laws, and the theory and numerical solution for linear acoustic wave propagation in an ideal gas are presented.

5.1 Historical Perspective

In 1845, James Prescott Joule conducted a simple though flawed experiment. Joule submerged two connected canisters separated by a valve within a thermally isolated water bath. One canister was filled with a gas; the other, a vacuum (Figure 5.1). Opening the valve, the gas expands into the empty canister, increasing its volume. Joule was unable to detect any change in the temperature of the water bath during the expansion. Recall that the first law of thermodynamics states that for a closed system, the change in internal energy between two equilibrium states, $\Delta\mathfrak{U}$, can be written as the heat transferred to the system from the surroundings, $\mathcal{Q}$, plus the work done by the surroundings on the system, $\mathcal{W}$, such that

$$d\mathfrak{U} = d\mathcal{Q} + d\mathcal{W}.$$

His conclusion was that since expansion of a gas into a vacuum that has zero pressure does no work and there was no transfer of energy between the water bath and the gas, the internal energy of the gas did not change. This implies that the internal energy density of the gas has the form $e = e(\theta)$. This result requires that there are no energetic interactions between the atoms within the gas. While this is the basic assumption of the ideal gas model, the atoms in a real gas do have weak energetic interactions. For real gases, there is a weak dependence of the internal energy on volume. Since the heat capacity of the gas canisters and the water bath are significantly higher than that of a gas, the resulting temperature change during expansion is very small. Joule's experiment apparatus was unable to resolve such small temperature changes. However, the result that the internal energy density does not depend on volume is true for the ideal gas.

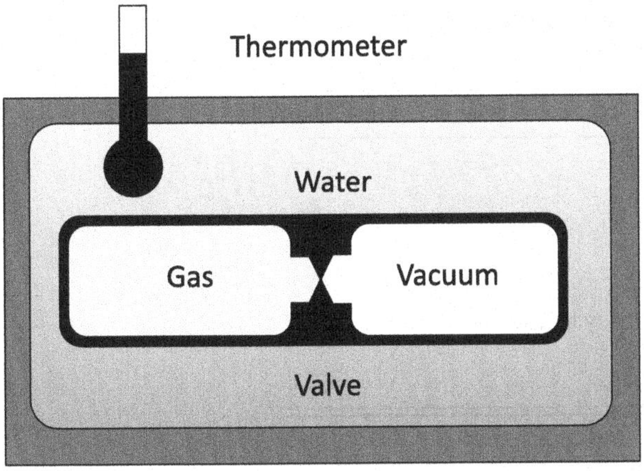

Figure 5.1. Insulated tank used for studying the energy absorbed by a gas during free expansion.

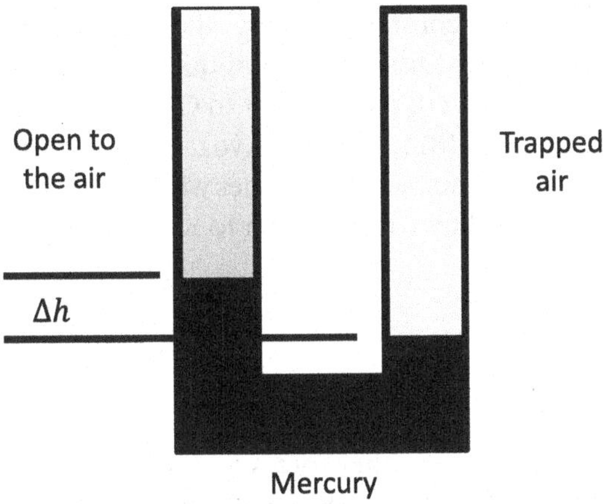

Figure 5.2. Illustration of Robert Boyle's experimental apparatus. As mercury is added, the trapped air is compressed within the manometer.

In 1834, Benoit Paul Emile Clapeyron stated the ***ideal gas law*** in its modern form

$$pV = nR_g\theta,$$

where p is the pressure, V is the volume, θ is the temperature, n is the number of moles of the gas, and R_g is the universal gas constant. It was based on earlier work demonstrating the relationship between the pressure, volume, temperature, and the number of molecules in a gas. A brief description of the early observations and experiments leading to the ideal gas law follows.

In 1662, Robert Boyle showed that at a constant temperature the pressure was inversely proportional to the volume of a gas, $p \propto \frac{1}{V}$. The experimental apparatus consisted of a manometer with air trapped on one end by mercury (Figure 5.2). The pressure at the open end is equal to atmospheric pressure and the pressure of the trapped gas, p, can be determined as $p - p_{atm} = \rho_m g \Delta h$, where ρ_m is the density of mercury, g is the acceleration of gravity, and Δh is the height difference of the

Table 5.1. *Field variables required to specify the finite deformation response of the thermoelastic material model*

Variable	Name	Number of unknowns
$\boldsymbol{x}(\boldsymbol{X},t)$	Equations of motion	3
$\rho(\boldsymbol{x},t)$	Spatial density field	1
$\boldsymbol{v}(\boldsymbol{x},t)$	Spatial velocity field	3
$\boldsymbol{\sigma}(\boldsymbol{x},t)$	Cauchy stress tensor	9
$\boldsymbol{q}(\boldsymbol{x},t)$	Spatial heat flux	3
$e(\boldsymbol{x},t)$	Spatial internal energy density field	1
$\theta(\boldsymbol{x},t)$	Spatial temperature field	1
$\eta(\boldsymbol{x},t)$	Spatial entropy density field	1

mercury within the tube. By adding mercury, the pressure, and volume of the trapped air changes. In 1702, Guillaume Amontons developed a thermometer based on the principle that that the pressure of a gas is proportional to the temperature of a gas, $p \propto \theta$. In 1787, Jacques Charles conducted unpublished experiments on balloons filled with various gasses, which showed that at constant pressure, the volume and temperature of a gas were related. In 1802, Gay-Lussac proposed that the volume was proportional to the temperature of a gas, $V \propto \theta$, giving credit to Charles for his earlier work and naming the law after Charles. In 1811, Amadeo Avogadro proposed that the volume of a gas was proportional to the number of molecules within it, $V \propto n$. It was the combination of these gas laws that allowed Clapeyron to write the ideal gas law.

5.2 Forces and Fields

The ideal gas is assumed to interact with the external universe via body forces, surface tractions, and heat transfer. We are neglecting other forces and fields such as electrical, magnetic, and chemical effects. The ideal gas material model is a specialized form of the thermo-mechanical material model developed in Chapter 4. The material model for an ideal gas has a total of 22 unknown fields that must be determined at each point within the material. These fields are listed in Table 5.1. Note that the Helmholtz free energy, ψ, is not listed since it can be computed from the internal energy, temperature, and entropy.

5.3 Balance Laws

Since the ideal gas model is an example of a thermo-mechanical material model, the balance laws derived in Chapter 4 remain unchanged. They are reproduced below for clarity in Table 5.2. The full derivation of these balance laws can be found in Chapter 4.

5.4 Constitutive Model

The gas phase is a form of matter that lacks both shape and volume. Unlike a solid or liquid, a gas will spontaneously expand to fill a container. The stress state of a gas depends only on the hydrostatic compression or expansion and not on the

Table 5.2. *Balance laws and kinematical relation for finite deformation of the thermo-mechanical material model*

		Number of equations
Kinematical relations	$\boldsymbol{v} = \dfrac{\partial \boldsymbol{x}}{\partial t}$	3
Conservation of mass	$0 = \dfrac{\partial \rho}{\partial t} + div\,(\rho\,\boldsymbol{v})$	1
Conservation of linear momentum	$\rho\,\dot{\boldsymbol{v}} - div\,\boldsymbol{\sigma}^T - \rho\,\boldsymbol{b}_x = \boldsymbol{0}$	3
Conservation of angular momentum	$\boldsymbol{\sigma} = \boldsymbol{\sigma}^T$	3
Conservation of energy	$\rho\dot{e} - \boldsymbol{\sigma} : \boldsymbol{D} + div\,(\boldsymbol{q}_x) - \rho\,r_x = 0$	1
Entropy inequality	$-\rho\,\dot{\psi} - \rho\eta\dot{\theta} - \dfrac{1}{\theta}\boldsymbol{q}_x \cdot grad\theta + \boldsymbol{\sigma}^T : \boldsymbol{D} \geq 0$	0

deviatoric deformation or the rate of deformation. This is another way of saying the gas lacks both a mechanism for viscous dissipation, and a mechanism to resist shear deformation. Molecular motion is chaotic and nearly uncorrelated as the molecular interactions are weak and the molecular separation is large. An ideal gas is one in which the molecules do not interact in any way. This includes a lack of both energetic interactions and excluded volume. Therefore, there are no internal mechanisms for dissipation or transfer of heat from one molecule to another. The internal energy density is simply a measure of the kinetic energy of the molecules at the given material point. Since the kinetic energy is proportional to the temperature, we expect the internal energy to be a monotonically increasing function of temperature.

For the sake of clarity, let us enumerate the assumptions within the ideal gas model.

1. The general material response depends only on the hydrostatic compression or rarefaction.
2. The thermodynamic pressure is given by the ideal gas law, $pV = nR_g\theta$.
3. There are no energetic or excluded volume interactions between molecules.
4. There are no dissipative or heat transfer mechanisms. The temperature of the gas must remain uniform.
5. The internal energy is a function of only temperature, $e = e\,(\theta)$.
6. The internal energy is a monotonically increasing function such that $\dfrac{\partial e\,(\theta)}{\partial \theta} > 0$ for all θ.
7. The molecular mass is small so that there are no variations in the pressure and density fields due to gravity.

We have shown in the previous chapter that the constitutive model for the thermoelastic material must have the form

$$\{\psi, \boldsymbol{\sigma}, \eta, \boldsymbol{q}\} = f\,(\boldsymbol{F}, \theta\ grad\ \theta)\,.$$

However, we have made additional simplifying assumptions that are specific to the ideal gas. For example, model assumption 4 requires that there is no heat transfer within the ideal gas and $\boldsymbol{q} = \boldsymbol{0}$ and the material response may not depend on the gradient of temperature. Note that a temperature gradient cannot be sustained in an ideal gas. We can then set the heat flux to zero and disregard it as a known quantity within the formulation leading to

$$\{\psi,\boldsymbol{\sigma},\eta\} = f(\boldsymbol{F},\theta).$$

Model assumption 1 says that the material response depends only on the hydrostatic compression or rarefaction. Therefore, we expect the constitutive model to depend not on the general deformation gradient but only on $\det \boldsymbol{F} = J$. Equivalently, we can have the constitutive model depend on ρ with no loss in generality since $\rho = \rho_o J$ and ρ_o is the constant reference density. This leads to a constitutive model with the form

$$\{\psi,\boldsymbol{\sigma},\eta\} = f(\rho,\theta).$$

Equivalently, we can postulate the model in terms of the internal energy instead of the Helmholtz free energy as

$$\{e,\boldsymbol{\sigma},\eta\} = f(\rho,\theta).$$

We will proceed with the development of a constitutive model with this form.

5.4.1 Constraints

Given the general form of the constitutive model, we may now apply the principle of material frame indifference and use the Clausius-Duhem inequality to simplify the constitutive equations or potentially remove inadmissible variables from the formulation.

5.4.1.1 Material Frame Indifference

The principle of material frame indifference requires that

$$\begin{aligned}
\boldsymbol{\sigma}\left(\rho^*,\theta^*\right) &= \boldsymbol{Q}\cdot\boldsymbol{\sigma}(\rho,\theta)\cdot\mathbf{Q}^{\mathrm{T}},\\
e\left(\rho^*,\theta^*\right) &= e(\rho,\theta),\\
\eta\left(\rho^*,\theta^*\right) &= \eta(\rho,\theta).
\end{aligned}$$

The scalar density and temperature are both invariant under observer coordinate transformations. Therefore, we obtain

$$\begin{aligned}
\boldsymbol{\sigma}(\rho,\theta) &= \boldsymbol{Q}\cdot\boldsymbol{\sigma}(\rho,\theta)\cdot\mathbf{Q}^{\mathrm{T}},\\
e(\rho,\theta) &= e(\rho,\theta),\\
\eta(\rho,\theta) &= \eta(\rho,\theta).
\end{aligned}$$

The second and third equations are naturally satisfied. Only the equation for the stress must be examined further. Since this condition requires that components of the stress tensor do not change under coordinate transformation, the stress tensor must be isotropic. The stress can then be written as the product of the negative of the thermodynamic pressure times the identity tensor

$$\boldsymbol{\sigma}(\rho,\theta) = -p(\rho,\theta)\mathbf{I}. \tag{5.1}$$

This is a natural outcome of the assumption that the ideal gas response depends only on the hydrostatic deformation.

5.4.1.2 Clausius-Duhem Inequality

We have chosen a constitutive model written in terms of the internal energy. Therefore, we will use the Clausius-Duhem inequality in the form

$$\rho\theta\,\dot{\eta} - \rho\dot{e} - \frac{1}{\theta}\boldsymbol{q}\cdot grad\,\theta + \boldsymbol{\sigma}^T : \boldsymbol{D} \geq 0.$$

Since there is no heat flux, $\boldsymbol{q} = \boldsymbol{0}$, this reduces to

$$\rho\theta\,\dot{\eta} - \rho\dot{e} + \boldsymbol{\sigma}^T : \boldsymbol{D} \geq 0.$$

We substitute the material time derivative of the dependent variables, which are the internal energy and the entropy, into this equation. Using the chain rule to expand each in terms of the independent variables, which are the density and temperature, we obtain

$$\dot{e} = \frac{\partial e}{\partial \rho}\frac{D\rho}{Dt} + \frac{\partial e}{\partial \theta}\frac{D\theta}{Dt},$$
$$\dot{\eta} = \frac{\partial \eta}{\partial \rho}\frac{D\rho}{Dt} + \frac{\partial \eta}{\partial \theta}\frac{D\theta}{Dt}.$$

Substitution back into the inequality gives

$$\rho\theta\left[\frac{\partial \eta}{\partial \rho}\dot{\rho} + \frac{\partial \eta}{\partial \theta}\dot{\theta}\right] - \rho\left[\frac{\partial e}{\partial \rho}\dot{\rho} + \frac{\partial e}{\partial \theta}\dot{\theta}\right] + \boldsymbol{\sigma}^T : \boldsymbol{D} \geq 0. \tag{5.2}$$

Since the stress tensor must be isotropic, Equation (5.1), we can write

$$\boldsymbol{\sigma}^T : \boldsymbol{D} = -p\boldsymbol{I} : \boldsymbol{D} = -p\,div\,\boldsymbol{v}.$$

Substituting this along with the equation for the conservation of mass, $\frac{D\rho}{Dt} + \rho\,div\,(\boldsymbol{v}) = 0$, into Equation (5.2) gives

$$\left[p - \rho^2\frac{\partial e}{\partial \rho} + \rho^2\theta\frac{\partial \eta}{\partial \rho}\right] div\,(\boldsymbol{v}) + \left[-\rho\frac{\partial e}{\partial \theta} + \rho\theta\frac{\partial \eta}{\partial \theta}\right]\dot{\theta} \geq 0. \tag{5.3}$$

The divergence of the velocity field can be varied independently of the rate of change of the temperature in a thermodynamic process. Since Equation (5.3) must be true for all possible thermodynamic processes, we obtain two constraint equations

$$p = \rho^2\frac{\partial e}{\partial \rho} - \rho^2\theta\frac{\partial \eta}{\partial \rho},$$
$$\frac{\partial e}{\partial \theta} = \theta\frac{\partial \eta}{\partial \theta}.$$

5.4.2 Constitutive Relations

We have two constitutive relations which are consistent with the experimental work outlined in Section 5.1 and the constraints imposed in Sections 5.4.1.1 and 5.4.1.2. First, the pressure is given by the ideal gas law

$$p = p\,(\rho,\theta) = \frac{nR_g\theta}{V} = \rho R_s T, \tag{5.4}$$

where $R_s = \frac{R_g}{\mathfrak{M}}$ is the specific gas constant for the ideal gas of interest, and $\mathfrak{M}$ is the molar mass of the gas. The stress can then be written as

$$\boldsymbol{\sigma}(\rho,\theta) = -\rho R_s \theta \mathbf{I}.$$

The resulting equation for the internal energy can be written as

$$e = e(\rho,\theta) = \left(\frac{\partial e}{\partial \theta}\right)_V \theta = c_V \theta, \tag{5.5}$$

where $c_V = \left(\frac{\partial e}{\partial \theta}\right)_V$ is the specific heat at constant volume. For an ideal monatomic gas, the specific heat per unit mass is given by $c_V = \frac{3}{2}R_s$.

The constitutive relation for the entropy can be found by examining the change in internal energy between two thermodynamic states given by the first law of thermodynamics,

$$d\mathfrak{U} = d\mathcal{Q} + d\mathcal{W}.$$

Rearranging this equation, the heat added to the system is given by

$$\begin{aligned} dQ &= de - dW \\ &= c_V d\theta + P dV \\ &= c_V d\theta + \frac{nR_g\theta}{V} dV. \end{aligned}$$

Dividing by the temperature, θ, gives

$$\frac{dQ}{\theta} = d\eta = \frac{c_V}{\theta} d\theta + \frac{nR_g}{V} dV.$$

Integrating the equation between two thermodynamic states gives

$$\int_{\eta_o}^{\eta} d\eta = c_V \int_{\theta_o}^{\theta} \frac{1}{\theta} d\theta + nR_g \int_{V_o}^{V} \frac{1}{V} dV.$$

The final result of this integration is

$$\begin{aligned} \eta - \eta_o &= c_V \ln\left(\frac{\theta}{\theta_o}\right) + nR_g \ln\left(\frac{V}{V_o}\right) \\ &= c_V \ln\left(\frac{\theta}{\theta_o}\right) + nR_g \ln\left(\frac{\rho_o}{\rho}\right). \end{aligned}$$

Assuming we are interested only in the relative entropy measured from a reference state, we can lump all of the reference values into a single constant η_*:

$$\eta = c_V \ln \theta - nR_g \ln \rho + \eta_*$$

Gathering equations, the constitutive model can be written as

$$\boxed{\begin{aligned} e(\rho,\theta) &= c_V \theta, \\ \boldsymbol{\sigma}(\rho,\theta) &= -\rho R_s \theta \boldsymbol{I}, \\ \eta(\rho,\theta) &= c_V \ln \theta - nR_g \ln \rho + \eta_*, \\ \boldsymbol{q}(\rho,\theta,\boldsymbol{g}) &= \mathbf{0}. \end{aligned}} \tag{5.6}$$

5.4.3 Molecular Model of an Ideal Gas

The molecular model for an ideal gas relates the molecular motion to the internal pressure and temperature. In the case of the ideal gas, a molecular model was first derived by Bernoulli in 1738. The behavior of the material is dictated by the motion and interaction of the atoms within the gas. Let us assume that the chamber of a piston is filled with a pure monatomic ideal gas (Figure 5.3). The temperature of the gas is a measure of the average velocity of the atoms. The force required to maintain the position of the piston, F, is equal and opposite to the momentum transferred from the atoms in the gas to the surface of the piston. In the case of the ideal gas, we have assumed that the particles do not interact with each other nor do they have excluded volume.

They are phantom atoms that can pass through one another without attracting or repulsing each other. The motion of one particle is independent of the motion of any other particle. The internal energy is the sum of the kinetic and potential energy of a system. However, the potential energy due to self-interaction is zero in the ideal gas, and we will neglect gravity in this analysis. Therefore, the internal energy density can be written as

$$e = \frac{\mathcal{K}}{m_{total}} = \frac{\frac{1}{2}\sum_{i=1}^{N} m^{(i)}\left(\boldsymbol{v}^{(i)} \cdot \boldsymbol{v}^{(i)}\right)}{\sum_{i=1}^{N} m^{(i)}}, \tag{5.7}$$

where $\mathcal{K} = \frac{1}{2}\sum_{i=1}^{N} m^{(i)}\left(\boldsymbol{v}^{(i)} \cdot \boldsymbol{v}^{(i)}\right)$ is the kinetic energy of the system, m_{total} is the total mass of the system, $m^{(i)}$ is the mass of atom i, and $\boldsymbol{v}^{(i)}$ is the velocity vector of atom i. Since the chamber is filled with a pure ideal gas, each molecule is identical, and the mass of each molecule is the same so that $m^{(i)} = m$. Equation (5.7) becomes

$$e = \frac{\frac{1}{2}m\sum_{i=1}^{N}\left(\boldsymbol{v}^{(i)} \cdot \boldsymbol{v}^{(i)}\right)}{Nm} = \frac{1}{2}\langle \boldsymbol{v} \cdot \boldsymbol{v} \rangle, \tag{5.8}$$

where $\langle \boldsymbol{v} \cdot \boldsymbol{v} \rangle$ denotes the average over all atoms of the atomic velocities squared such that $\frac{1}{N}\sum_{i=1}^{N}\left(\boldsymbol{v}^{(i)} \cdot \boldsymbol{v}^{(i)}\right)$. In general, the average of any discrete quantity $(\cdot)_i$ is given by

$$\langle (\cdot) \rangle = \frac{1}{N}\sum_{i=1}^{N} (\cdot)_i. \tag{5.9}$$

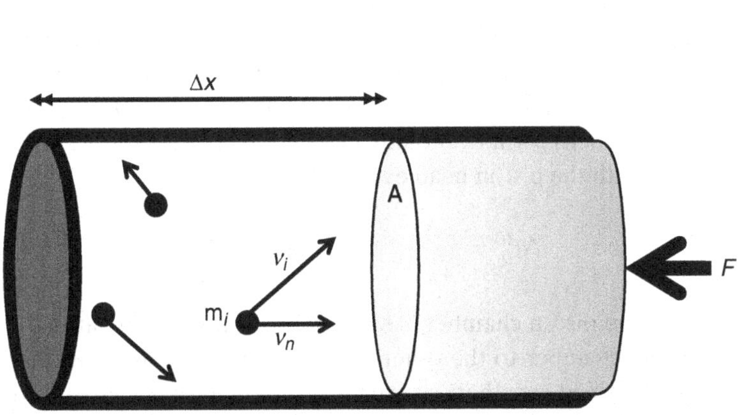

Figure 5.3. Illustration showing the molecules within an ideal gas trapped within a container.

Applying the equipartition theorem for a monatomic gas, the temperature of a collection of atoms is defined in terms of the kinetic energy as

$$\theta = \frac{2}{3}\frac{\mathcal{K}}{Nk_B} = \frac{1}{3}\frac{m\langle \boldsymbol{v}\cdot\boldsymbol{v}\rangle}{k_B}, \tag{5.10}$$

where k_B is the Boltzmann constant. Combining Equations (5.8) and (5.10), we obtain an expression for the internal energy in terms of the system temperature

$$e = \frac{3}{2}\frac{k_B}{m}\theta = \frac{3}{2}R_s\theta. \tag{5.11}$$

This gives us the first constitutive relation based on the molecular structure of the gas.

If we look at the piston head as an isolated body in equilibrium, the external applied force, f, must balance the force exerted by the ideal gas. In macroscopic terms, the force exerted by the ideal gas on the piston head is equal to the pressure of the gas times the cross-sectional area of the piston head. But from the molecular prospective, the force exerted by the gas is caused by individual collisions of atoms with the piston head. Each collision transfers some momentum from the gas to the piston head. The force exerted by a single atom on the wall, $\boldsymbol{f}^{(i)}$, is given by Newton's law, which states that the force exerted *by* a particle is equal to the negative of its change in momentum

$$\boldsymbol{f}^{(i)} = -\left(m^{(i)}\frac{d\boldsymbol{v}^{(i)}}{dt}\right).$$

For clarity, the traditional statement of Newton's law is that the force exerted *on* a particle is equal to its change in its momentum. Assuming we have completely elastic collisions between the atoms in the gas and the piston head, each collision will change the direction of the velocity vector, but not change its magnitude. The component of the velocity tangent to the piston head surface will not change. Only the normal component of the velocity vector will change. Therefore, the force exerted by the atom on the piston head can be written in terms of the normal components of the velocity vectors, $v_n^{(i)} = \boldsymbol{v}^{(i)}\cdot\boldsymbol{n}$, as

$$f_n^{(i)} = -\left(m^{(i)}\frac{dv_n^{(i)}}{dt}\right), \tag{5.12}$$

where $f_n^{(i)}$ is the component of the force normal to the piston head. An atom with an initial normal velocity of $-v_n$ before a collision will have a normal velocity of v_n after a collision. Therefore, $dv_n = -2v_n$. Every atom in the gas (with a nonzero component of the normal velocity) will eventually collide with the piston head. In fact, an atom i will collide with the piston head every Δt seconds:

$$\Delta t^{(i)} = \frac{2\,\Delta x}{v_n^{(i)}}, \tag{5.13}$$

where Δx is the length of the piston chamber. This is the time it takes the atom to travel from one side of the chamber to the other and back to its original position. Remember that even in a rarified gas, there are many atoms, and that these atoms are moving quite quickly. If we are simply interested in the average force exerted by a single atom on the piston, $\langle f_n \rangle$, is given by combining Equations (5.9), (5.12), and (5.13) to obtain

$$\langle f_n \rangle = \frac{1}{N}\sum_{i=1}^{N}\left(-m^{(i)}\frac{\left(-2v_n^{(i)}\right)}{\Delta t}\right) = \frac{1}{N}\sum_{i=1}^{N}\left(2m^{(i)}v_n^{(i)}\frac{v_n^{(i)}}{2\Delta x}\right) = \frac{m\langle v_n^2\rangle}{\Delta x}. \tag{5.14}$$

Construct a coordinate system with the x axis parallel to the normal of the piston head, $v_n = v_x$. Noting that the motion of any atom is random and uncorrelated with its neighbors, the average velocity along any coordinate direction will be the same such that

$$\langle v_x^2\rangle = \langle v_y^2\rangle = \langle v_z^2\rangle.$$

The magnitude of the velocity vector can then be written as

$$\langle \boldsymbol{v}\cdot\boldsymbol{v}\rangle = \langle v_x^2 + v_y^2 + v_z^2\rangle = \langle v_x^2\rangle + \langle v_y^2\rangle + \langle v_z^2\rangle = 3\langle v_x^2\rangle. \tag{5.15}$$

Combining Equations (5.10), (5.14), and (5.15), we obtain

$$\langle f_n \rangle = \frac{m\langle \boldsymbol{v}\cdot\boldsymbol{v}\rangle}{3\,\Delta x} = \frac{k_b\theta}{\Delta x}.$$

The pressure exerted on the piston head, p, is simply the total force exerted by the collection of atoms on the piston head divided by the cross-sectional area, which can be written as

$$p = \frac{N\langle f_n\rangle}{A} = \frac{Nk_b\theta}{\Delta x A} = \frac{Nk_b\theta}{V}.$$

Since $Nk_B = nR_g$, where N is the number of molecules, and n is the mole number of molecules,

$$Vp = nR_g\theta. \tag{5.16}$$

We have derived the two constitutive relations required for the ideal gas, Equations (5.11), and (5.16). Comparing to the empirical constitutive relations, Equations (5.4) and (5.5) shows that for a monatomic ideal gas, $c_v = \frac{3}{2}R_s$.

5.5 Governing Equations

The ideal gas model is a specialization of the thermoelastic material model. By combining the constitutive model in Section 5.4.2 with the governing principles described in Section 4.1, we obtain a set of equations specific to the ideal gas. They are generally referred to as field equations and for the ideal gas, these are the equations of gas dynamics.

The principle of conservation of mass, Equation (4.1), cannot be simplified any further and remains as

$$0 = \dot{\rho} + \rho\,(\boldsymbol{\nabla}\cdot\boldsymbol{v}).$$

The principle of conservation of linear momentum, Equation (4.5), can be simplified by combining it with the constitutive relation for the Cauchy stress, Equation (5.1). The equation for conservation of linear momentum is restated for clarity as

$$\rho\dot{\boldsymbol{v}} - \boldsymbol{\nabla}\cdot\boldsymbol{\sigma}^T - \rho\boldsymbol{b} = \boldsymbol{0}. \tag{5.17}$$

Employing Equation (5.1), the divergence of the stress can be written as

$$\nabla \cdot \boldsymbol{\sigma}^T = \nabla \cdot (-p\boldsymbol{I}) = \frac{\partial\left(-p\delta_{ij}\right)}{\partial x_j}\boldsymbol{e}_i = \frac{\partial(-p)}{\partial x_i}\boldsymbol{e}_i = -\nabla p. \tag{5.18}$$

Substitution of Equation (5.18) into Equation (5.17) gives

$$\rho\dot{\boldsymbol{v}} + \nabla p - \rho\,\boldsymbol{b} = \boldsymbol{0}.$$

The principle of conservation of energy, Equation (4.22), can be simplified in a similar way. We restate the principle of conservation of energy for clarity as

$$\rho\dot{e} - \boldsymbol{\sigma}^T : \boldsymbol{D} + \nabla \cdot \boldsymbol{q} - \rho\, r = 0. \tag{5.19}$$

Equation (5.1) allows us to write

$$\boldsymbol{\sigma}^T : \boldsymbol{D} = -p\boldsymbol{I} : \boldsymbol{D} = -p\,\delta_{ij}D_{ij} = -p\,tr\,\boldsymbol{D}. \tag{5.20}$$

Recall that $\boldsymbol{D} = \frac{1}{2}\left(\boldsymbol{v}\overleftarrow{\nabla} + \overrightarrow{\nabla}\boldsymbol{v}\right)$, and we can write

$$tr\left(\left(\boldsymbol{v}\overleftarrow{\nabla}\right)\right) = tr\left(\frac{\partial v_i}{\partial x_j}\boldsymbol{e}_j\right) = \frac{\partial v_i}{\partial x_i} = \nabla \cdot \boldsymbol{v}. \tag{5.21}$$

Therefore, Equation (5.20) becomes

$$\boldsymbol{\sigma}^T : \boldsymbol{D} = -p\,\nabla \cdot \boldsymbol{v}. \tag{5.22}$$

Substitution of Equation (5.22) into Equation (5.19) gives

$$\rho\dot{e} + p\,(\nabla \cdot \boldsymbol{v}) - \rho\, r = 0.$$

Gathering the field equations, we have

$$\boxed{\begin{aligned} \dot{\rho} + \rho\,(\nabla \cdot \boldsymbol{v}) &= 0,\\ \rho\dot{\boldsymbol{v}} + \nabla p - \rho\,\boldsymbol{b} &= \boldsymbol{0},\\ \rho\dot{e} + p\,(\nabla \cdot \boldsymbol{v}) - \rho\, r &= 0. \end{aligned}} \tag{5.23}$$

5.6 Acoustic Waves

We have described the development of a constitutive model for an ideal gas. We will now use this constitutive model to derive and numerically solve the linear equations governing the motion of one-dimensional linear acoustic waves. Though we will use the ideal gas model, the results for real gasses or even fluids are similar. Acoustic waves are pressure waves traveling through a compressible medium. Speech and music pressure waves with an amplitude that is very small compared to ambient pressure of the gaseous medium. By definition, the waves propagate at the speed of sound through the medium. Due to the combination of small amplitude and high wave speed, the displacement of particles within the medium in which these waves are transmitted is small. In addition, these small amplitude acoustic waves do not interact significantly with one another. Because of this, your brain may interpret the pitch, amplitude, and direction of multiple acoustic waves detected by your ear. For example, your brain's interpretation of the characteristics a person's voice is independent of the background noise that may or may not be present.

For clarity, let us enumerate the assumptions for a small amplitude acoustic wave alluded to in the previous paragraph.

1. The motion, both the velocity and displacement, of gaseous medium due to the transmission of the acoustic wave is small.
2. We assume that the initial density and pressure fields are uniform and have values of ρ_o and p_o, respectively.
3. The amplitude of the pressure wave is much smaller than the magnitude of the pressure, and the amplitude of the density perturbations accompanying the wave is small.
4. The compression or rarefaction of the gas occurs adiabatically. Whereas this must be true for the ideal gas model, it is approximately true for real gases. The heat due to compression does not escape the pressure wave for low-frequency acoustic waves.

Since the pressure and density fields will vary from their initial values by small amounts, we will introduce a new variable to account for this perturbation:

$$\rho'(\boldsymbol{x},t) = \rho(\boldsymbol{x},t) - \rho_o,$$
$$p'(\boldsymbol{x},t) = p(\boldsymbol{x},t) - p_o, \tag{5.24}$$

where ρ' is the perturbation in the density field, ρ_o is the initial density, and p' is the perturbation in the pressure field, and p_o is the ambient pressure. Note that both the initial density and pressure fields are constants. Condition 4 can be stated mathematically as $\frac{\rho'}{\rho_o} \ll 1$ and $\frac{p'}{p_o} \ll 1$.

The one-dimensional conservation of mass equation can be written as

$$0 = \frac{\partial \rho}{\partial t} + \frac{\partial}{\partial x}(\rho v). \tag{5.25}$$

Substituting the density perturbation given in Equation (5.24) into Equation (5.25) gives

$$0 = \frac{\partial(\rho' + \rho_o)}{\partial t} + \frac{\partial}{\partial x}((\rho' + \rho_o)v).$$

Since $\dfrac{\partial \rho_o}{\partial t} = 0$, and $\dfrac{\partial(\rho' v)}{\partial x} \ll \rho_o \dfrac{\partial v}{\partial x}$, we can write

$$0 = \frac{\partial \rho'}{\partial t} + \rho_o \frac{\partial v}{\partial x}. \tag{5.26}$$

The one-dimensional conservation of linear momentum equation becomes

$$\rho \frac{Dv}{Dt} + \frac{\partial p}{\partial x} - \rho b = 0. \tag{5.27}$$

Substituting the density and pressure perturbations given in Equation (5.24), Equation (5.27) becomes

$$(\rho' + \rho_o)\frac{Dv}{Dt} + \frac{\partial(p' + p_o)}{\partial x} - \rho b = 0.$$

Since $\dfrac{\partial P_o}{\partial x} = 0$ and $\rho' \ll \rho_o$, this becomes

$$(\rho' + \rho_o)\frac{Dv}{Dt} + \frac{\partial(p' + p_o)}{\partial x} - \rho b = 0,$$

$$\rho_o \frac{Dv}{Dt} + \frac{\partial p'}{\partial x} - \rho b = 0.$$

Expansion of the material derivative gives

$$\rho_o \left(\frac{\partial v}{\partial t} + v\frac{\partial v}{\partial x}\right) + \frac{\partial p'}{\partial x} - \rho b = 0.$$

If we assume $v \ll 1$, $\dfrac{\partial v}{\partial t} \ll 1$, and $\dfrac{\partial v}{\partial x} \ll 1$, this equation becomes

$$\rho_o \left(\frac{\partial v}{\partial t}\right) + \frac{\partial p'}{\partial x} - \rho b = 0.$$

If we ignore gravity, the body force is equal to zero, and we can write

$$\rho_o \left(\frac{\partial v}{\partial t}\right) + \frac{\partial p'}{\partial x} = 0.$$

Taking the derivative of with respect to the spatial position gives

$$\rho_o \left(\frac{\partial^2 v}{\partial x \, \partial t}\right) + \frac{\partial^2 p'}{\partial x^2} = 0. \tag{5.28}$$

And taking the temporal derivative of Equation (5.26) gives

$$0 = \frac{\partial^2 \rho'}{\partial t^2} + \rho_o \frac{\partial^2 v}{\partial t \, \partial x}. \tag{5.29}$$

Combining Equation (5.28) and (5.29) gives

$$\frac{\partial^2 \rho'}{\partial t^2} = \frac{\partial^2 p'}{\partial x^2}. \tag{5.30}$$

The pressure may be cast as a function of the density and the entropy, s, of the system as

$$p = p(\rho, \eta).$$

Since the pressure is a state function, we may write

$$dp = \left(\frac{\partial p}{\partial \rho}\right)_\eta d\rho + \left(\frac{\partial p}{\partial s}\right)_\rho d\eta.$$

Since wave propagation through an ideal gas is an isentropic process, $d\eta = 0$, and we have

$$dp = \left(\frac{\partial p}{\partial \rho}\right)_\eta d\rho, \tag{5.31}$$

where $c^2 = \left(\frac{\partial p}{\partial \rho}\right)_\eta$ is the squared speed of sound in the gaseous medium.

By substituting Equation (5.31) into Equation (5.30), we obtain the wave equation,

$$\frac{1}{c^2}\frac{\partial^2 p'}{\partial t^2} = \frac{\partial^2 p'}{\partial x^2}. \tag{5.32}$$

5.6.1 Finite Difference Method

Equation (5.32) is a second-order partial differential equation (PDE) that can be solved analytically in the one-dimensional case. However, we will often encounter partial differential equations that cannot be solved analytically. We must resort to numerical methods such as the finite difference method or the finite element method. In this section, we will discuss the discritization process, numerical implementation, and error analysis required for the finite difference method. The first step is discritization of both the time and spatial dimensions. Imagine that we are studying a pressure wave traveling through a pipe that has length L, we cannot numerically compute the solution to Equation (5.32) at every point within this domain. Instead, we compute the solution at a select a set of nDX grid points with a uniform spacing of $dx = \frac{L}{nDX}$ as shown in Figure 5.4. The grid points, commonly known as nodes, labeled $i = 0$ and nDX are boundary nodes. In order to obtain a solution, boundary conditions must be applied to these nodes. The values of pressure on nodes 1 through $nDX - 1$ are determined via the numerical solution of the PDE.

The next step is to discritize the solution in time. We obtain a complete spatial solution for every discrete time value. The one-dimensional discritized solution can be visualized as a matrix of solution points in both time and space (Figure 5.5). We must prescribe the initial condition which is the pressure at grid points $i = 0$ to nDX for $j = 0$. This is the complete pressure distribution of the system at $t = 0$.

The choice of time step, dt, and grid spacing, dx, depend on the desired accuracy and the numerical algorithm employed. In the next sections, we will discuss the implementation of both an implicit and explicit algorithm. There are many variations of each method which may be found in a standard numerical methods textbook.

The finite difference method is based on the fact that the derivative of a function may be approximated using functional values at discrete points. The Taylor expansion for the one-dimensional pressure field about point i can be written as an infinite series:

$$p_{i+1,j} = \sum_{n=0}^{inf} \frac{f^{(n)}(a,t)}{n!} = p_{i,j} + \frac{\Delta x}{1!}\left.\frac{\partial p}{\partial x}\right|_{i,j} + \frac{\Delta x^2}{2!}\left.\frac{\partial^2 p}{\partial x^2}\right|_{i,j} + \frac{\Delta x^3}{3!}\left.\frac{\partial^3 p}{\partial x^3}\right|_{i,j} + \cdots.$$

If we assume $a = \Delta x$, we obtain

$$p_{i+1,j} = p_{i,j} + \frac{\Delta x}{1!}\left.\frac{\partial p}{\partial x}\right|_{i,j} + \frac{\Delta x^2}{2!}\left.\frac{\partial^2 p}{\partial x^2}\right|_{i,j} + \frac{\Delta x^3}{3!}\left.\frac{\partial^3 p}{\partial x^3}\right|_{i,j} + \cdots. \tag{5.33}$$

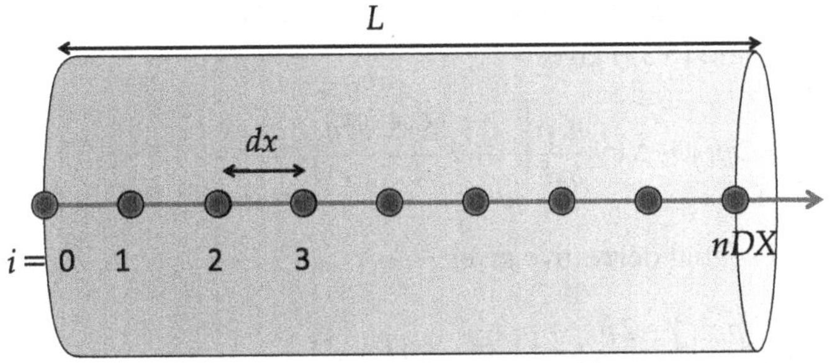

Figure 5.4. Illustration of a one-dimensional spatial grid used to model a pipe.

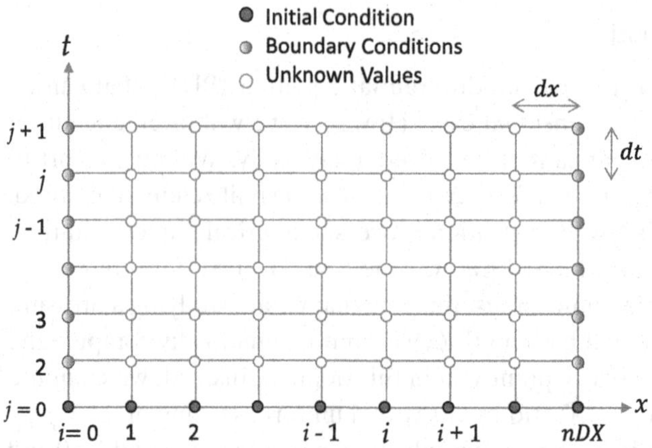

Figure 5.5. Illustration of the discretization of the solution to a differential equation obtained over time and space.

We can solve this equation for the gradient of the pressure field

$$\left.\frac{\partial p}{\partial x}\right|_{i,j} = \frac{(p_{i+1,j} - p_{i,j})}{\Delta x} - \frac{\Delta x}{2!}\left.\frac{\partial^2 p}{\partial x^2}\right|_{i,j} - \frac{\Delta x^2}{3!}\left.\frac{\partial^3 p}{\partial x^3}\right|_{i,j} - \cdots.$$

The spatial derivative can then be approximated by truncating this infinite expansion as

$$\left.\frac{\partial p}{\partial x}\right|_{i,j} = \frac{(p_{i+1,j} - p_{i,j})}{\Delta x} + O(\Delta x), \tag{5.34}$$

where the $O(\Delta x)$ is the order of the truncation error. This means that as Δx approaches zero, the truncation error approaches zero linearly. Equation (5.34) is known as the ***forward difference approximation.***

The Taylor series expansion around p_i with $a = -\Delta x$ gives

$$p_{i-1,j} = p_{i,j} - \frac{\Delta x}{1!}\left.\frac{\partial p}{\partial x}\right|_{i,j} + \frac{\Delta x^2}{2!}\left.\frac{\partial^2 p}{\partial x^2}\right|_{i,j} - \frac{\Delta x^3}{3!}\left.\frac{\partial^3 p}{\partial x^3}\right|_{i,j} + \cdots. \tag{5.35}$$

Solving for the first-order gradient and truncating the infinite solution gives

$$\left.\frac{\partial p}{\partial x}\right|_{i,j} = \frac{p_{i-1,j} - p_{i,j}}{\Delta x} + O(\Delta x). \tag{5.36}$$

Equation (5.36) is known as the ***backward difference approximation.***

Adding Equations (5.33) and (5.35) gives

$$p_{i+1,j} + p_{i-1,j} = 2p_{i,j} + \Delta x^2\left.\frac{\partial^2 p}{\partial x^2}\right|_{i,j} + 2\frac{\Delta x^4}{4!}\left.\frac{\partial^4 p}{\partial x^4}\right|_{i,j} + \cdots.$$

Solving for the second-order spatial derivative gives

$$\left.\frac{\partial^2 p}{\partial x^2}\right|_{i,j} = \frac{p_{i+1,j} - 2p_{i,j} + p_{i-1,j}}{\Delta x^2} + O\left(\Delta x^2\right). \tag{5.37}$$

A similar analysis using the Taylor series expansion in time instead of space yields the approximate second temporal derivative as

$$\left.\frac{\partial^2 p}{\partial t^2}\right|_{i,j} = \frac{p_{i,j+1} - 2p_{i,j} + p_{i,j-1}}{\Delta t^2} + O\left(\Delta x^2\right). \tag{5.38}$$

These approximations allow us to approximate the partial derivatives of continuous functions as algebraic combinations of functional values at the grid points.

5.6.2 Explicit Algorithm

The explicit finite difference formulation involves the approximation of the unknown function at a single grid point at the current time in terms of known functional values at previous times. It is explicit because the current solution at time t is explicitly written in terms of known quantities. Approximating the continuous PDE given in Equation (5.32) with second-order difference approximations for both the temporal, Equation (5.38), and spatial second derivatives, Equation (5.37), gives

$$\frac{p_{i,j+1} - 2p_{i,j} + p_{i,j-1}}{\Delta t^2} = c^2 \frac{p_{i+1,j} - 2p_{i,j} + p_{i-1,j}}{\Delta x^2}. \tag{5.39}$$

The values at the discrete time values j and $j-1$ are known; only the pressure at time $j+1$ is unknown. Therefore, we can write

$$p_{i,j+1} = \frac{c^2 \Delta t^2}{\Delta x^2}\left(p_{i+1,j} - 2p_{i,j} + p_{i-1,j}\right) + 2p_{i,j} - p_{i,j-1}. \tag{5.40}$$

The stencil for this algorithm is shown in Figure 5.6. Notice that this algorithm requires that we know the entire solution for p at two time intervals.

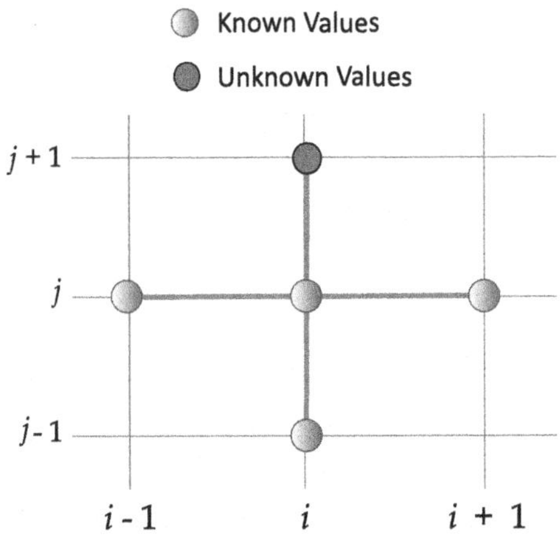

Figure 5.6. Stencil used in the explicit finite difference scheme.

5.6.3 Implicit Algorithm

If we approximate the continuous PDE given in Equation (5.32) with second-order difference approximations, Equation (5.38), but we approximate the spatial second derivative as a weighted average of the spatial derivative at times $j-1, j, j+1$, using Equation (5.37), we obtain

$$\left.\frac{\partial^2 p}{\partial x^2}\right|_{i,j} = \frac{1}{4}c^2 \left(\left.\frac{\partial^2 p}{\partial x^2}\right|_{i,j+1} + 2\left.\frac{\partial^2 p}{\partial x^2}\right|_{i,j} + \left.\frac{\partial^2 p}{\partial x^2}\right|_{i,j-1} \right).$$

Combining this with Equation (5.37) and (5.38) gives

$$\frac{p_{i,j+1}-2p_{i,j}+p_{i,j-1}}{\Delta t^2} = \frac{c^2}{4}\left(\frac{p_{i+1,j+1}-2p_{i,j+1}+p_{i-1,j+1}}{\Delta x^2} + \frac{p_{i+1,j}-2p_{i,j}+p_{i-1,j}}{\Delta x^2} + \frac{p_{i+1,j-1}-2p_{i,j-1}+p_{i-1,j-1}}{\Delta x^2}\right).$$

Solving for the unknown values at $j+1$, we obtain

$$p_{i+1,j+1} - p_{i,j+1}\left(2+\frac{4}{r^2}\right) + p_{i-1,j+1}$$
$$= \left(p_{i+1,j}-2p_{i,j}+p_{i-1,j}+p_{i+1,j-1}-2p_{i,j-1}+p_{i-1,j-1}\right) + \frac{4}{r^2}\left(-2p_{i,j}-p_{i,j-1}\right),$$

where $r^2 = \frac{c^2\Delta t^2}{\Delta x^2}$. There are three unknowns per equation, and a single equation per node. For simplicity, we define the coefficients as

$$a_{-1} = 1,$$
$$a_0 = -\left(1+2r^2\right),$$

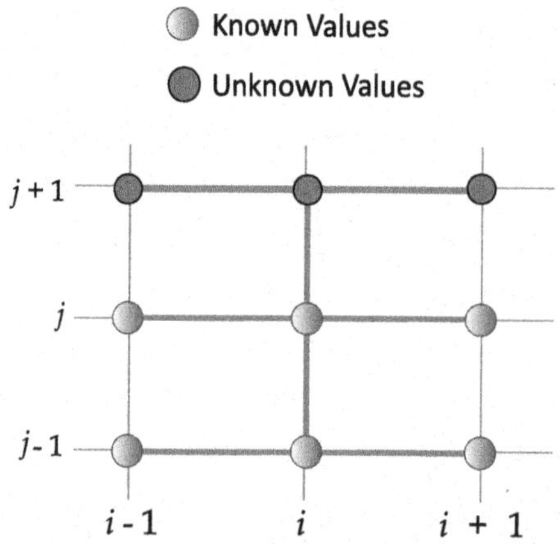

Figure 5.7. Stencil used in the implicit finite difference scheme.

$$\mathrm{a}_1 = 1, \tag{5.41}$$

$$d_i = \left(p_{i+1,j} - 2p_{i,j} + p_{i-1,j} + p_{i+1,j-1} - 2p_{i,j-1} + p_{i-1,j-1}\right) + \frac{4}{r^2}\left(2p_{i,j} - p_{i,j-1}\right).$$

Looking at the grid point that is adjacent to the left boundary condition, $i = 1$, this equation becomes

$$p_{2,j+1} - p_{1,j+1}\left(2 + \frac{4}{\mathrm{r}^2}\right) + p_{0,j+1}$$
$$= \left(p_{2,j} - 2p_{1,j} + p_{0,j} + p_{2,j-1} - 2p_{1,j-1} + p_{0,j-1}\right) + \frac{4}{r^2}\left(2p_{1,j} - p_{1,j-1}\right).$$

Since $p_{0,j+1}$ is the prescribed boundary condition at gridpoint $i = 0$, we can rewrite the equation as

$$p_{2,j+1} - p_{1,j+1}\left(2 + \frac{4}{r^2}\right) = d_1 - p_{0,j+1}.$$

We can write the system of equations in matrix form as

$$\begin{bmatrix} \mathrm{a_o} & \mathrm{a}_1 & 0 & 0 & 0 \\ \mathrm{a}_{-1} & \mathrm{a_o} & \mathrm{a}_1 & 0 & 0 \\ 0 & \ddots & \ddots & \ddots & 0 \\ 0 & 0 & \mathrm{a}_{-1} & \mathrm{a_o} & \mathrm{a}_1 \\ 0 & 0 & 0 & \mathrm{a}_{-1} & \mathrm{a_o} \end{bmatrix} \begin{bmatrix} p_1 \\ p_2 \\ \vdots \\ p_{nDX-2} \\ p_{nDX-1} \end{bmatrix} = \begin{bmatrix} d_1 - \mathrm{a}_{-1} p_o \\ d_2 \\ \vdots \\ d_{nDX-2} \\ d_{nDX-1} - \mathrm{a}_1 p_{nDX} \end{bmatrix}. \tag{5.42}$$

This represents a matrix equation $\mathbf{M} \cdot \boldsymbol{p} = \boldsymbol{d}$, where

$$\mathbf{M} = \begin{bmatrix} \mathrm{a_o} & \mathrm{a}_1 & 0 & 0 & 0 \\ \mathrm{a}_{-1} & \mathrm{a_o} & \mathrm{a}_1 & 0 & 0 \\ 0 & \ddots & \ddots & \ddots & 0 \\ 0 & 0 & \mathrm{a}_{-1} & \mathrm{a_o} & \mathrm{a}_1 \\ 0 & 0 & 0 & \mathrm{a}_{-1} & \mathrm{a_o} \end{bmatrix},$$

$$\boldsymbol{p} = \begin{bmatrix} p_1 \\ p_2 \\ \vdots \\ p_{nDX-2} \\ p_{nDX-1} \end{bmatrix}, \tag{5.43}$$

$$\boldsymbol{d} = \begin{bmatrix} d_1 - a_{-1}p_o \\ d_2 \\ \vdots \\ d_{nDX-2} \\ d_{nDX-1} - a_1 p_{nDX} \end{bmatrix}.$$

If **M** is nonsingular, then this set of equations may be solved as $\boldsymbol{p} = \boldsymbol{M}^{-1} \cdot \boldsymbol{d}$.

5.6.4 Example Problem

Assume that we have a narrow 2-cm cylindrical tube with closed ends filled with an ideal gas. The system is initially in equilibrium with a uniform pressure and temperature field. A single pressure pulse is generated at the left boundary, which has amplitude of $dP_o = 10\,\text{mPa}$ and a period of $T = 20\,\mu\text{s}$. The speed of sound within the ideal gas may be assumed to be $c = 1{,}000\,\text{m/s}$. The pressure at the right boundary condition may be assumed to remain constant.

We would like to develop a model describing the pressure wave as it travels down the length of the tube. Following the development in the previous sections, both an implicit and explicit algorithm were implemented. The numerical solution to the wave equation was found using both an implicit and explicit algorithm. The spatial dimension was discretized using 100 gridpoints. The solution was propagated for 20 μs with a 200-ns time step.

The results from both the explicit and implicit algorithm match quite closely (Figure 5.8). The pressure profile for the entire spatial domain is plotted at 5, 10, 15, and 20 μs. We can see from the pressure profile, that the wave has traveled 20 mm in 20 μs which gives a speed of 1000 m/s in agreement with the speed of sound selected for the material.

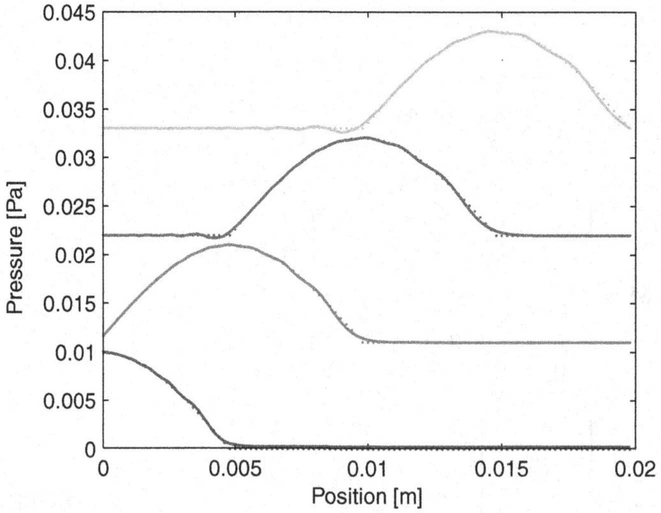

Figure 5.8. Numerical solution to the wave equation. Solution obtained using an implicit algorithm and an explicit algorithm, shown as dashed and solid lines, respectively. The curves are shifted vertically for clarity.

5.6.5 Matlab® File – Explicit Algorithm

Explicit Algorithm

1. Initialize grid spacing in both space and time

$$dX = 0.2 \text{ mm}$$
$$dT = 0.2 \text{ μs}$$

2. Initialize material properties

$$c = 1,000 \text{ m/s}$$
$$p = 1.0 \text{ kg/m}^3$$

3. Assign boundary conditions at the inlet and outlet. Note that in the code, we may assign the boundary conditions for all times, t, before solving for the pressure at the interior nodes

$$dP(1,t) = \begin{cases} dP_o \sin\left(\dfrac{2\pi t}{T}\right) & t \le \dfrac{T}{2} \\ & \\ & t > \dfrac{T}{2} \end{cases}$$

$$dP(nDx.t) = 0$$

4. Compute

$$r^2 = \frac{c^2 \, \Delta t^2}{\Delta x^2}$$

5. Calculate pressure on the interior nodes

$$dP(i, j) = r^2 \, dP(i+1, j-1) + 2(1 - r^2) \, dP(i, j-1) + r^2 dP(i-1, j-1) - dP(i, j-2)$$

Cycle over all interior nodes, i

Done

Cycle over all time increments, j

Done

6. Plot the results

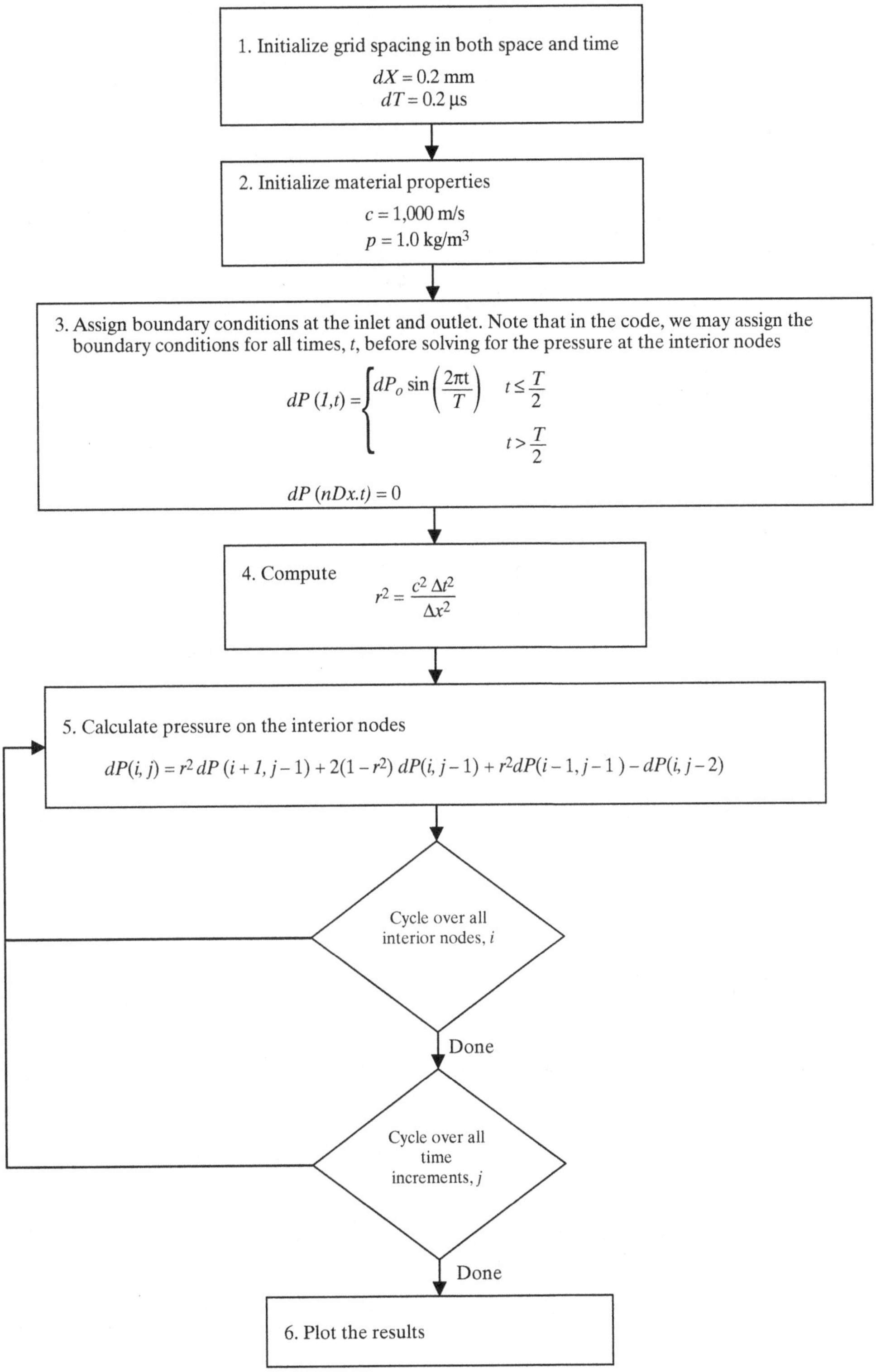

```
% ----------------------------------------------------------------------
% Numerical solution of the 1-D accoustic wave equation for an ideal gas
% using an explicit finite difference formulation. A sinusoidal pressure
% pulse is generated on the left boundary node, and the right boundary
% node is maintained at ambient pressure.
% ----------------------------------------------------------------------
%   dP(i,j) [Pa] is a matrix containing the pressure change at node
%           i for time increment j.
% ----------------------------------------------------------------------
clear;                      % clear any stored variables
% ----------------------------------------------------------------------
% 1. Initialize spatial and temporal grid
% Generate the spatial grid
dX  = 0.0002;               % [m] spacing between grid points
nDX = 100;                  % number of grid points along the X axis
X   = [0:dX:(nDX-1)*dX];    % [m] vector
 containing positions of each grid point
% ----------------------------------------------------------------------
% Generate the temporal grid
dt   = 0.0000002;           % [s] size of each time step
nDT  = 100;                 % number of time steps
time = [0:dt:nDT*dt]
;      % [s] vector containing time of each solution
% ----------------------------------------------------------------------
% 2. Initialize material parameters
c      = 1000.0;            % [m/s] speed of sound within the material
rho_o  =    1.0;            % [kg/m^3] initial density of the gas
% ----------------------------------------------------------------------
% 3. Assign boundary conditions to the inlet and outlet
dP_o            = 0.01;     % [Pa]  amplitude of the pressure wave
period          = 2*10^-5; % [s]   period of the pressure wave
for j = 1:nDT
    % The left gridpoint has a sinusoidal pressure profile.
    if (time(j) <= period/2.0)
       dP(1,j) = dP_o*sin(2*pi.*(time(j))/period);
   else
       dP(1,j) = 0;
    end

   % The right gridpoint remains at ambient pressure
   dP(nDX,j) = 0;
end

% 4. compute the constant rsqr
rsqr = (c*dt/dX)^2;

% ----------------------------------------------------------------------
% Solve the discritized wave equation
% Cycle from j = 3 to nDT in increments of 1
for j = 3:nDT
    % ------------------------------------------------------------------
    % Cycle over the interior grid points.
    for i = 2 : length(X)-1
        % 5. compute the pressure at each node point, dP(i,j)
        dP(i,j) = 2*(1-rsqr)*dP(i,j-1)    - dP(i,j-2)  ...
                     + rsqr*(dP(i+1,j-1) + dP(i-1,j-1));
    end
end
```

5.6.6 Matlab® File – Implicit Algorithm

Implicit Algorithm

1. Initialize grid spacing in both space and time

$$dX = 0.2 \text{ mm}$$
$$dT = 0.2 \text{ μm}$$

2. Initialize material properties
$$c = 1{,}000 \text{ m/s}$$
$$p = 1.0 \text{ kg/m}^3$$

3. Assign boundary conditions at the inlet and outlet. Note that we may assign the boundary conditions for all times, $t(j)$, before solving for the pressure at the interior nodes

$$dP(1,j) = \begin{cases} dP_o \sin\left(\dfrac{2\pi\, t(j)}{T}\right) & t(j) \le \dfrac{T}{2} \\ 0 & t(j) > \dfrac{T}{2} \end{cases}$$

$$dP(nDx, j) = 0$$

5. Assemble the matrix

$$\mathbf{M} = \begin{bmatrix} a_0 & a_1 & 0 & 0 & 0 \\ a_{-1} & a_0 & a_1 & 0 & 0 \\ 0 & \ddots & \ddots & \ddots & 0 \\ 0 & 0 & a_{-1} & a_0 & a_1 \\ 0 & 0 & 0 & a_{-1} & a_0 \end{bmatrix}$$

where

$$a_{-1} = a_1 = 1$$
$$a_0 = -(2 + 4/r^2)$$

Note that the matrix **M** is a constant matrix for all time steps and all nodes.

4. Compute

$$r^2 = \frac{c^2 \Delta t^2}{\Delta x^2}$$

6. Assemble the vector

$$d(i) = \left(p_{i+1,j} - 2\,p_{i,j} + p_{i-1,j} + p_{i+1,j-1} - 2\,p_{i,j-1} + p_{i-1,j-1}\right) + \frac{4}{r^2}\left(2\,p_{i,j} - p_{i,j-1}\right)$$

Cycle over all interior

Done

7. Account for the boundary conditions in the **d** vector
$$d_1 = d_1 - a_{-1}p_0$$
$$d_{nDX-1} = d_{nDX-1} - a_1 p_{nDX}$$

8. Solve for the pressure at each node using

$$dP(:,j) = \mathbf{M}^{-1} \cdot \mathbf{d}\,.$$

Cycle over all time increments, j

Done

9. Plot the results

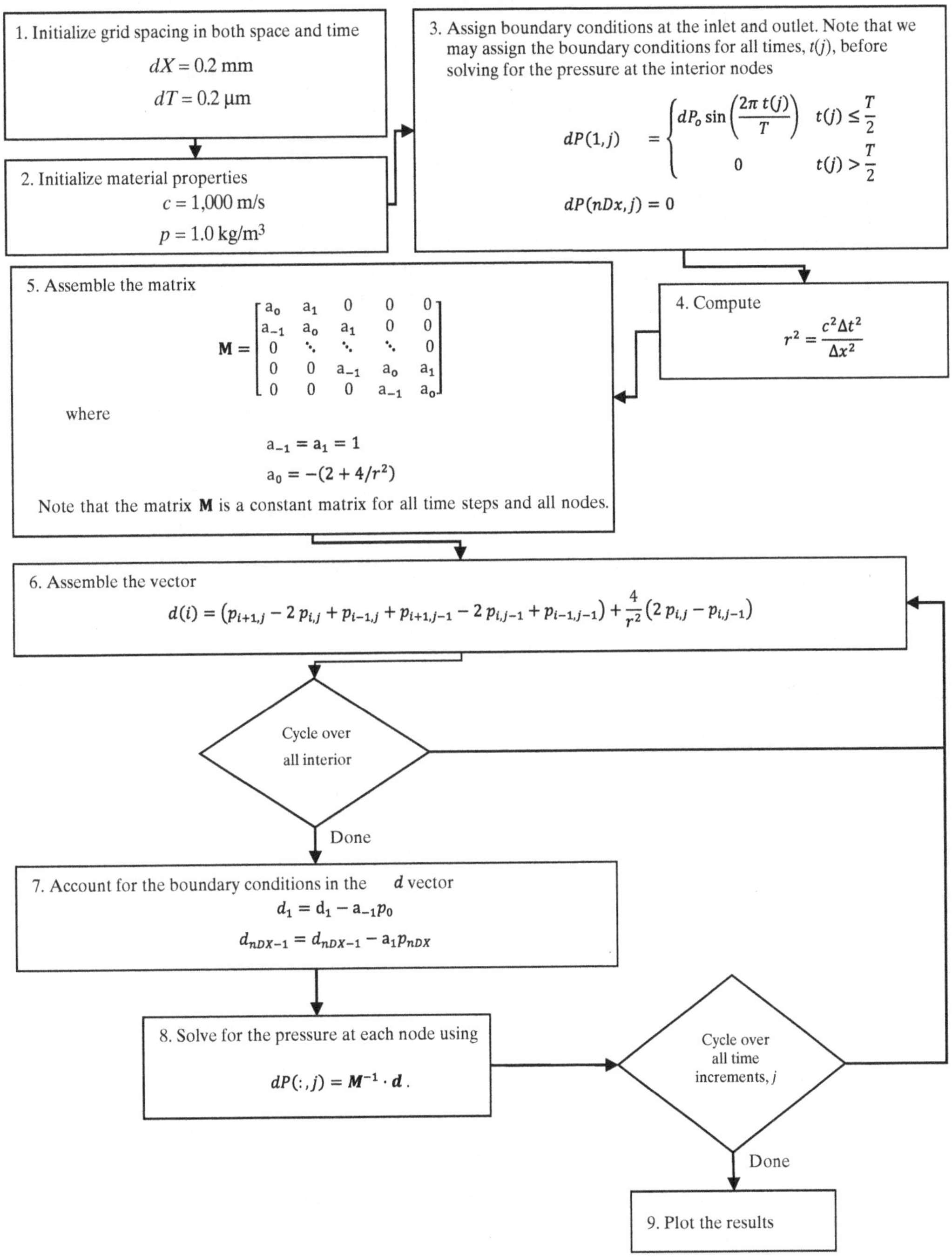

```
% -----------------------------------------------------------------------
% Numerical solution of the 1-D accoustic wave equation for an ideal gas
% using an implicit finite difference formulation.A sinusoidal pressure
% wave is generated on the left boundary node, and the right boundary
% node is maintained at ambient pressure.
% -----------------------------------------------------------------------
%   dP(i,j) [Pa] is a matrix containing the pressure change at node
%           i for time increment j.
% -----------------------------------------------------------------------
clear;                        % clear any stored variables
% -----------------------------------------------------------------------
% 1. Initialize spatial and temporal grid
% Generate the spatial grid
dX  = 0.0002;                 % [m] spacing between grid points
nDX = 100;                    % number of grid points along the X axis
X   = [0:dX:(nDX-1)*dX];      % [m] vector containing positions of each grid point
% -----------------------------------------------------------------------
% Generate the temporal grid
dt   = 0.0000002;             % [s] size of each time step
nDT  = 100;                   % number of time steps
time = [0:dt:nDT*dt];         % [s] vector containing time of each solution
% -----------------------------------------------------------------------
% 2. Initialize material parameters
c       = 1000.0;             % [m/s] speed of sound within the material
rho_o   =    1.0;             % [kg/m^3] initial density of the gas
% -----------------------------------------------------------------------
% Allocate memory for the M, b, and dP matrix to speed up calculation.
M(nDX-2,nDX-2) = 0;
b(nDX-2)       = 0;
dP(nDX,nDT)    = 0;
% -----------------------------------------------------------------------
% 3. Boundary conditions
dP_o            = 0.01;    % [Pa]  amplitude of the pressure wave
period          = 2*10^-5; % [s]   period of the pressure wave
for j = 1:nDT
    if (time(j) <= period/2.0)
       dP(1,j) = dP_o*abs(sin(2*pi.*(time(j))/period));
    else
       dP(1,j) = 0;
    end

    dP(nDX,j) = 0;
end
% 4. Compute the constant rsqr
rsqr = (c*dt/dX)^2;

% -----------------------------------------------------------------------
% 5. Assemble the M matrix and find its inverse
a_0 = 1;
a_1 = -(2.0+4.0/rsqr);
a_2 = 1;
M = a_1*diag(ones(nDX-2,1)) ...
    + a_2*diag(ones(nDX-3,1),1) ...
    + a_0*diag(ones(nDX-3,1),-1);
Minv = M^-1;
% -----------------------------------------------------------------------
% Time march
for j = 3: nDT
    % 6. Assemble the d vector for a given time step
    for i = 2:nDX-1
```

```
        b(i-1) =  4.0/rsqr*(        - 2*dP(i,j-1) + dP(i,  j-2)) ...
                - 2.0*(dP(i+1,j-1) - 2*dP(i,j-1) + dP(i-1,j-1)) ...
                -     (dP(i+1,j-2) - 2*dP(i,j-2) + dP(i-1,j-2));
    end
    % 7. subtract boundary condition from the first and last term in the d vector
    b(1) = b(1) - dP(1,j);
    % 8. Solve for the pressure
    soln = Minv*b';
    dP(2:nDX-1,j) = soln;
end
```

6 Fluids

The response of a fluid depends on the rate of deformation. In this chapter, we present the development of the constitutive law for a Newtonian fluid, the formulation of the field equations, and methods for determining the material parameters within the Newtonian fluid constitutive equations. The compressible and incompressible Navier-Stokes equation and Bernoulli's equation are derived from the constitutive equations and the balance laws for a Newtonian fluid. Finally, we include a brief discussion of non-Newtonian fluid models.

The balance between molecular interactions and thermal energy determine the state of matter. In a fluid, thermal energy is sufficient for atoms or molecules to slide relative to one another. Because of the low barrier to relative motion, the fluid cannot sustain shear stress in its equilibrium state. This leads to the familiar consequence that the fluid will flow to take the shape of the container it occupies. However, unlike a gas, the attractive interactions between atoms or molecules in a fluid are sufficient to maintain a constant density. In other words, the fluid will not expand to fill the volume of its container.

Although the fluid may not sustain an equilibrium shear stress, the molecules within a deforming fluid may interact with one another giving rise to internal friction. The ***viscosity*** of the fluid is a measure of the internal friction between molecules, which leads to transient shear stresses within the deforming fluid. Therefore, the fluid response is a function of the rate of deformation. Higher rates of deformation lead to increased internal friction and higher transient shear stresses.

For a ***Newtonian fluid***, the effective viscosity is constant. In practice, the viscosity of all fluids is somewhat dependent on the shear rate. Fluids may be modeled as Newtonian fluids if their viscosity is constant over a large range of shear rates or at a minimum over the range of shear rates encounter in the experiment or simulation of interest. ***Non-Newtonian fluids*** are fluids whose viscosity does depend on the shear rate. As the names imply, a ***shear thinning fluid*** is one whose effective viscosity drops with increasing shear rate and a ***shear thickening fluid*** is one whose effective viscosity increases as the shear rate increases.

At the end of this chapter, we will discuss the parameterization of the Newtonian fluid model using a rheometer. A rheometer is a device that can be used to determine viscosity from measurements of both the transient shear stress and the velocity gradients in a fluid. There are a variety of rheometer configurations such as capillary,

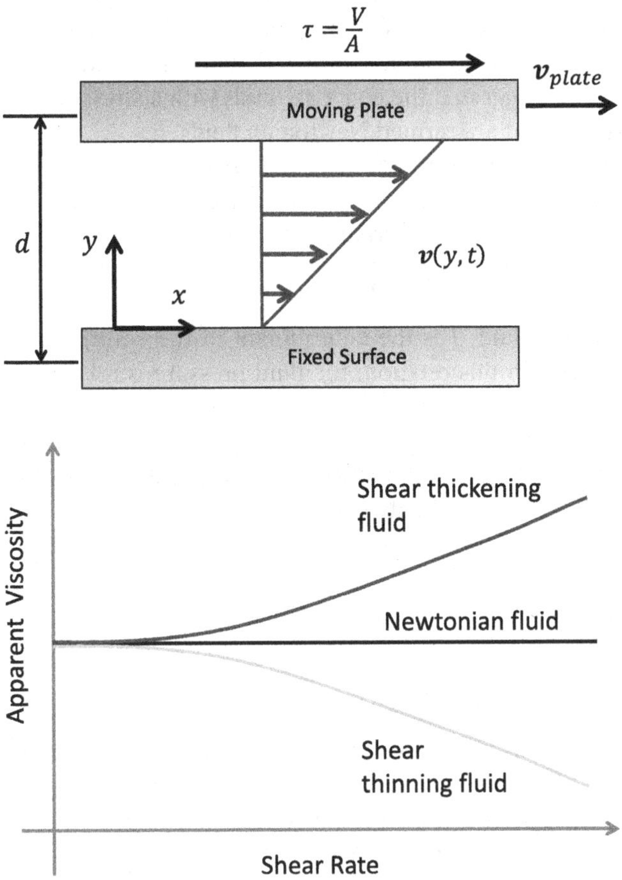

Figure 6.1. (top) Illustration of a parallel plate rheometer. (bottom) Effective viscosity, $\frac{\tau}{\dot{\gamma}}$, versus shear rate $\dot{\gamma} = \frac{v_{plate}}{d}$ for Newtonian fluids, shear thinning fluids, and shear thickening fluids.

rotational cylinder, cone and plate, and extensional rheometers. In the parallel plate rheometer (Figure 6.1), the top plate is moved at constant speed and the resultant force acting on the plate is measured. The shear rate is equal to $\dot{\gamma} = v_{plate}/d$, and the effective viscosity of the fluid is equal to the shear stress divided by the shear rate, $\tau/\dot{\gamma}$.

6.1 Historical Perspective

The study of fluids date back to the third century B.C. with Archimedes's studies and his manuscript entitled "On Floating Bodies". In it, Archimedes states the now-common principle taught in every high school physics class that the buoyancy force is equal to the weight of the displaced fluid. The concept of internal dissipation and specifically viscosity was not proposed until Newton first suggested it in his *Principia*, which was published in 1687. In it, Newton states that "the resistance which arises from the lack of slipperiness of the parts of a liquid, other things being equal, is proportional to the velocity with which the parts of the liquid are separated from one another." In terms of the familiar one-dimensional equation, $\tau \propto \frac{\partial v}{\partial x}$, the stress is proportional to the velocity gradient. The constant of proportionality in this equation is the viscosity of the fluid. Having suggested the concept, fluids with a stress which is proportional to the velocity gradient are termed Newtonian fluids.

The development of the mathematical form of the constitutive models for fluids progressed incrementally. Bernoulli made significant progress in the 18th century by proposing that any inviscid flow or fluid in equilibrium had a hydrostatic stress state

$$\boldsymbol{\sigma}(J,\theta) = -p(J,\theta)\mathbf{I},$$

where $p(J,\theta)$ is the pressure in the fluid, J is the determinant of the deformation gradient, and θ is the temperature. In this relation, the fluid pressure is related to the deformation through only the compression or expansion of the material volume element. Equivalently, one could state that the pressure depends only on the current density of the fluid, ρ, and the temperature. Although immensely useful for studying the equilibrium forces within all fluids, and the dynamic behavior of inviscid fluids, Bernoulli's constitutive model had to be augmented in order to capture the behavior of viscous fluids.

In 1821, Navier modified the Bernoulli model by adding a functional dependence on the velocity gradient such that

$$\boldsymbol{\sigma}(\rho,\theta,\boldsymbol{L}) = -p(\rho,\theta)\mathbf{I} + \boldsymbol{f}(\rho,\theta,\boldsymbol{L}),$$

where $\boldsymbol{f}(\boldsymbol{L})$ is a general tensor function with the velocity gradient, $\boldsymbol{L}$, as a parameter. Clearly, this is a generalization of what was suggested by Newton in his *Principia.* Furthermore, in 1831, Poisson proposed that for an isotropic fluid, the stress could be written as a function of the deformation rate tensor, $\boldsymbol{D}$, as

$$\boldsymbol{\sigma}(\rho,\theta,\boldsymbol{D}) = -p(\rho,\theta)\mathbf{I} + \lambda\, tr(\boldsymbol{D})\mathbf{I} + 2\mu\boldsymbol{D},$$

where λ is the dilatational viscosity and μ is the shear viscosity, and both are material parameters. We will proceed to the derivation of this equation in the next section.

6.2 Forces and Fields

The fluids discussed in this chapter are assumed to interact with the external universe via body forces, surface tractions, and heat transfer. We are neglecting other forces and fields such as electrical, magnetic, and chemical effects. Therefore, the fluid material models discussed here are specialized forms of the thermo-mechanical material model developed in Chapter 4. The material model for a thermo-mechanical fluid has a total of 22 unknown fields which must be determined at each point within the material. These fields are listed in Table 6.1. Note that the Helmholtz free energy, ψ, is not listed since it can be computed from the internal energy, temperature, and entropy.

6.3 Balance Laws

Since we have limited ourselves to thermo-mechanical fluids, the balance laws derived in Chapter 4 remain unchanged. They are reproduced in Table 6.2 for clarity. The full derivation of these balance laws can be found in Chapter 4.

Table 6.1. *Field variables required to specify the finite deformation response of the thermoelastic material model*

Variable	Name	Number of unknowns
$\boldsymbol{x}(\boldsymbol{X},t)$	Equations of motion	3
$\rho\,(\boldsymbol{x},t)$	Spatial density field	1
$\boldsymbol{v}\,(\boldsymbol{x},t)$	Spatial velocity field	3
$\boldsymbol{\sigma}\,(\boldsymbol{x},t)$	Cauchy stress tensor	9
$\boldsymbol{q}\,(\boldsymbol{x},t)$	Spatial heat flux	3
$e\,(\boldsymbol{x},t)$	Spatial internal energy density field	1
$\theta\,(\boldsymbol{x},t)$	Spatial temperature field	1
$\eta\,(\boldsymbol{x},t)$	Spatial entropy density field	1

Table 6.2. *Balance laws and kinematical relation for finite deformation of the thermo-mechanical material model*

		Number of equations
Kinematical relations	$\boldsymbol{v} = \frac{\partial \boldsymbol{x}}{\partial t}$	3
Conservation of mass	$0 = \frac{\partial \rho}{\partial t} + div\,(\rho\,\boldsymbol{v})$	1
Conservation of linear momentum	$\rho\,\dot{v} - div\,\boldsymbol{\sigma}^T - \rho\,\boldsymbol{b}_x = \boldsymbol{0}$	3
Conservation of angular momentum	$\boldsymbol{\sigma} = \boldsymbol{\sigma}^T$	3
Conservation of energy	$\rho\dot{e} - \boldsymbol{\sigma} : \boldsymbol{D} + div\,(\boldsymbol{q}_x) - \rho\,r_x = 0$	1
Entropy inequality	$-\rho\,\dot{\psi} - \rho\eta\dot{\theta} - \frac{1}{\theta}\boldsymbol{q}_x \cdot grad\,\theta + \boldsymbol{\sigma}^T : \boldsymbol{D} \geq 0$	0

6.4 Constitutive Model

Once again, the development of a closed set of equations requires 11 constitutive equations. Once again, we will choose as our dependent variables, the Helmholtz free energy, ψ the entropy, η the Cauchy stress tensor, $\boldsymbol{\sigma}$ and the heat flux vector, $\boldsymbol{q}$. When determining the functional dependence of the constitutive model, we use prior experimental observations, assumptions, and general conjecture to determine which variables should be included in the formulation.

We begin with a set of experimental observations that

1. The resistance, or shear stress, of a fluid is a function of the rate of deformation. We need not assume a linear relationship between the two at the onset of model development. Instead, we will make that assumption later.
2. The fluid pressure, and hence the stress, is a function of both the current density and temperature.
3. The fluid response depends only on the hydrostatic compression or dilation of the fluid. Therefore, we expect the response to depend on the deformation gradient through only the Jacobean.
4. The heat flux will depend on the gradient of the temperature.

Using the principle of equi-presence, we will assume that if an independent variable appears in the constitutive relation for any dependent variable, it appears in all of the constitutive equations. Taking the four experimental observations stated previously, we can make the statement that a thermo-mechanical fluid response depends on the deformation gradient, rate of deformation, temperature, and the gradient of the temperature

$$\{\psi,\eta,\boldsymbol{\sigma},\boldsymbol{q}\} = f\left(\boldsymbol{F},\dot{\boldsymbol{F}},\theta,\nabla\theta\right).$$

Note that the rate of deformation may be characterized by the velocity gradient, $\boldsymbol{L}$, or the rate of change of the deformation gradient, $\boldsymbol{F} = \dot{\boldsymbol{L}}\cdot\boldsymbol{F}$. Also, observation 3 states that the response should depend only on the hydrostatic compression or dilation. Since $J = \det\boldsymbol{F}$, and $\rho_o = \rho J$, we can include either the Jacobian or the current density as our independent variable instead of the deformation gradient. Therefore, we may write

$$\{\psi,\eta,\boldsymbol{\sigma},\boldsymbol{q}\} = f\left(\rho,\boldsymbol{L},\theta,\nabla\theta\right).$$

Similarly, we may write the velocity gradient as the sum of the spin tensor, $\boldsymbol{W}$, and the strain rate tensor, $\boldsymbol{D}$, by using the relation $\boldsymbol{L} = \boldsymbol{D} + \boldsymbol{W}$. This gives

$$\{\psi,\eta,\boldsymbol{\sigma},\boldsymbol{q}\} = f\left(\rho,\boldsymbol{D},\boldsymbol{W},\theta,\nabla\theta\right).$$

While it may seem that additional complexity has been introduced in this last step for no reason, we will later show that the constitutive model may not be a function of the spin tensor. Therefore, we will reduce the dependence of the constitutive model from the velocity gradient to the strain rate tensor.

6.4.1 Constraints

Armed with the general functional dependence of the constitutive model, let us examine the constraints imposed on these functions due to the principle of material frame indifference and from the second law of thermodynamics. Even though they may not provide the exact mathematical form of the constitutive model, these constraints will dictate the final admissible form of the constitutive model.

6.4.1.1 Material Frame Indifference

The principle of material frame indifference states that physical interpretation of any quantity must be the same regardless of the reference frame in which an observer inhabits. We will write a set of relations between the dependent variables measured in two different reference frames. Namely, we will employ the starred reference frame and the unstarred reference frame which differ by an orthogonal rotation, $\boldsymbol{Q}$, as was done in Chapter 4. First, we must have the coordinate transformation rules for all of the variables appearing in the constitutive model. Previously, we have shown that the transformation of the density, temperature, and heat flux fields are given by

$$\begin{aligned}\rho^* &= \rho,\\ \theta^* &= \theta,\\ \nabla\theta^* &= \boldsymbol{Q}\cdot\nabla\theta,\\ \boldsymbol{q}^* &= \boldsymbol{Q}\cdot\boldsymbol{q},\\ \boldsymbol{F}^* &= \boldsymbol{Q}\cdot\boldsymbol{F}.\end{aligned}$$

We may determine the coordinate transformation rule for the velocity gradient, by relating the velocity gradient to the deformation gradient

$$\boldsymbol{L} = \dot{\boldsymbol{F}}\cdot\boldsymbol{F}^{-1}.$$

The velocity gradient in the starred coordinate system can similarly be computed from the deformation gradient in the starred coordinate system such that

$$\boldsymbol{L}^* = \dot{\boldsymbol{F}}^*\cdot(\boldsymbol{F}^*)^{-1}.$$

Substituting the transformation rule of the deformation gradient, $\boldsymbol{F}^* = \boldsymbol{Q}\cdot\boldsymbol{F}$, gives

$$\begin{aligned}\boldsymbol{L}^* &= \dot{\overline{(\boldsymbol{Q}\cdot\boldsymbol{F})}}\cdot(\boldsymbol{Q}\cdot\boldsymbol{F})^{-1}\\ &= (\dot{\boldsymbol{Q}}\cdot\boldsymbol{F}+\boldsymbol{Q}\cdot\dot{\boldsymbol{F}})\left(\boldsymbol{F}^{-1}\cdot\boldsymbol{Q}^{-1}\right)\\ &= \dot{\boldsymbol{Q}}\cdot\boldsymbol{Q}^T+\boldsymbol{Q}\cdot\dot{\mathbf{F}}\cdot\mathbf{F}^{-1}\cdot\mathbf{Q}^{\mathrm{T}}\\ &= \dot{\boldsymbol{Q}}\cdot\boldsymbol{Q}^T+\boldsymbol{Q}\cdot\boldsymbol{L}\cdot\boldsymbol{Q}^T.\end{aligned}$$

The coordinate transformation rule for the strain rate tensor, $\boldsymbol{D}$, can be computed by using the relation between strain rate tensor and velocity gradient $\boldsymbol{D} = \frac{1}{2}\left(\boldsymbol{L}+\boldsymbol{L}^T\right)$ The strain rate tensor in the starred coordinate system becomes

$$\boldsymbol{D}^* = \frac{1}{2}\left(\boldsymbol{L}^*+(\boldsymbol{L}^*)^T\right).$$

Substituting the result for the velocity gradient gives

$$\begin{aligned}\boldsymbol{D}^* &= \frac{1}{2}\left(\dot{\boldsymbol{Q}}\cdot\boldsymbol{Q}^T+\boldsymbol{Q}\cdot\boldsymbol{L}\cdot\boldsymbol{Q}^T+\boldsymbol{Q}\cdot\dot{\boldsymbol{Q}}^T+\boldsymbol{Q}\cdot\boldsymbol{L}^T\cdot\boldsymbol{Q}^T\right)\\ &= \frac{1}{2}\left(\dot{\boldsymbol{Q}}\cdot\boldsymbol{Q}^T+\boldsymbol{Q}\cdot\dot{\boldsymbol{Q}}^T+2\boldsymbol{Q}\cdot\boldsymbol{D}\cdot\boldsymbol{Q}^T\right).\end{aligned}$$

We can simplify this relation further by noting that since $\boldsymbol{Q}$ is an orthogonal tensor, the transpose of $\boldsymbol{Q}$ is equal to its inverse

$$\boldsymbol{Q}\cdot\boldsymbol{Q}^T = \mathbf{I}.$$

Taking the time derivative of this equation yields

$$\dot{\boldsymbol{Q}}\cdot\boldsymbol{Q}^T+\boldsymbol{Q}\cdot\dot{\boldsymbol{Q}}^T = \dot{\mathbf{I}} = \mathbf{0}.$$

Substitution into the velocity gradient gives

$$D^* = Q \cdot D \cdot Q^T.$$

The transformation rule for the spin tensor can be determined similarly

$$\begin{aligned} W^* &= \frac{1}{2}\left(L^* - (L^*)^T\right) \\ &= \frac{1}{2}\left(\dot{Q} \cdot Q^T + Q \cdot L \cdot Q^T - Q \cdot \dot{Q}^T - Q \cdot L^T \cdot Q^T\right) \\ &= \frac{1}{2}\left(\dot{Q} \cdot Q^T - Q \cdot \dot{Q}^T + Q \cdot \left(L - L^T\right) \cdot Q^T\right). \end{aligned}$$

Once again using the fact that $\dot{Q} \cdot Q^T = -Q \cdot \dot{Q}^T$, we obtain

$$\begin{aligned} W^* &= \frac{1}{2}\left(2\dot{Q} \cdot Q^T + 2Q \cdot W \cdot Q^T\right) \\ &= \dot{Q} \cdot Q^T + Q \cdot W \cdot Q^T. \end{aligned}$$

Now that we have determined the coordinate transformation rules for all necessary variables, we may apply the principle of material frame indifference. Let us consider first the Helmholtz free energy. Since the Helmholtz free energy is a scalar-valued tensor function, the principle of material frame indifference requires that $\psi^* = \psi$. Therefore, we may write

$$\psi\left(\rho^*, D^*, W^*, \theta^*, \nabla\theta^*\right) = \psi\left(\rho, D, W, \theta, \nabla\theta\right).$$

Substitution of the coordinate transformation rules gives

$$\psi\left(\rho, Q \cdot D \cdot Q^T, \dot{Q} \cdot Q^T + Q \cdot W \cdot Q^T, \theta, Q \cdot \nabla\theta\right) = \psi\left(\rho, D, W, \theta, \nabla\theta\right).$$

Since this must be true for all possible rotations, Q, and all possible rotation rates, $\dot{Q}$, we are free to select a particular rotation and rotation rate for evaluation. Remember that the rotation and rotation rate may be selected independently because the principle of material frame indifference must apply at every instant of time. Therefore, we are free to select an instant of time that has a particular rotation and rotation rate. Even though the rotation rate will dictate what the rotation will be at any later time, at any particular instant they may be selected independently. Let us select a rotation, $Q = \mathbf{I}$, and a rotation rate, $\dot{Q} = -W$. Substitution into the principal of material frame indifference gives the condition

$$\psi\left(\rho, D, \mathbf{0}, \theta, \nabla\theta\right) = \psi\left(\rho, D, W, \theta, \nabla\theta\right).$$

Similar analysis of the remaining constitutive laws for the same rotation, $Q = I$, and rotation rate, $\dot{Q} = -W$, produce identical relations:

$$\begin{aligned} \eta\left(\rho, D, \mathbf{0}, \theta, \nabla\theta\right) &= \eta\left(\rho, D, W, \theta, \nabla\theta\right), \\ \sigma\left(\rho, D, \mathbf{0}, \theta, \nabla\theta\right) &= \sigma\left(\rho, D, W, \theta, \nabla\theta\right), \\ q\left(\rho, D, \mathbf{0}, \theta, \nabla\theta\right) &= q\left(\rho, D, W, \theta, \nabla\theta\right). \end{aligned}$$

This states for any possible value of $\boldsymbol{W}$, that the constitutive model must return the same value as when the argument has a value of $\boldsymbol{W}=0$. The only way for this to be true is for the functions to be independent of the spin tensor. Therefore, the constitutive model may not depend on the spin tensor. Therefore, we may write

$$\{\psi,\eta,\boldsymbol{\sigma},\boldsymbol{q}\}=f(\rho,\boldsymbol{D},\theta,\nabla\theta).$$

6.4.1.2 Clausius-Duhem Inequality

The second law of thermodynamics provides additional constraints on both the form of the constitutive laws as well as possible values for material parameters. Following the methodology presented in Chapter 4, we will substitute the functional form of the constitutive laws into the Clausius-Duhem inequality. For clarity, the Clausius-Duhem inequality was derived in Chapter 4 is given by

$$-\rho\dot{\psi}-\rho\eta\dot{\theta}-\frac{1}{\theta}\boldsymbol{q}_x\cdot\boldsymbol{g}+\boldsymbol{\sigma}^T:\boldsymbol{D}\geq 0.$$

We may use the chain rule to expand the material derivative of the Helmholtz free energy in terms of derivatives of the independent variables such that

$$\dot{\psi}=\frac{\partial\psi}{\partial\rho}\dot{\rho}+\frac{\partial\psi}{\partial\boldsymbol{D}}\dot{\boldsymbol{D}}+\frac{\partial\psi}{\partial\theta}\dot{\theta}+\frac{\partial\psi}{\partial\boldsymbol{g}}\dot{\boldsymbol{g}}.$$

Substitution into the CD inequality gives

$$-\rho\left(\frac{\partial\psi}{\partial\rho}\dot{\rho}+\frac{\partial\psi}{\partial\boldsymbol{D}}\dot{\boldsymbol{D}}+\frac{\partial\psi}{\partial\boldsymbol{g}}\dot{\boldsymbol{g}}\right)-\rho\left(\eta+\frac{\partial\psi}{\partial\theta}\right)\dot{\theta}-\frac{1}{\theta}\boldsymbol{q}_x\cdot\boldsymbol{g}+\boldsymbol{\sigma}^T:\boldsymbol{D}\geq 0.$$

Recall that $\dot{\theta}$ does not appear in the constitutive model; therefore, ψ, η, $\boldsymbol{q}$, $\boldsymbol{\sigma}$ have no dependence on the value of $\dot{\theta}$. Similarly, the independent variables ρ, $\boldsymbol{D}$, θ, and $\boldsymbol{g}$ can be varied independently of $\dot{\theta}$. Therefore, a thermodynamic process can be selected which has $\dot{\rho}=0$, $\dot{\boldsymbol{D}}=\boldsymbol{0}$, $\dot{\boldsymbol{g}}=\boldsymbol{0}$, $\boldsymbol{g}=\boldsymbol{0}$, $\boldsymbol{D}=\boldsymbol{0}$, and a nonzero $\dot{\theta}$. This produces the inequality

$$-\rho\left(\eta+\frac{\partial\psi}{\partial\theta}\right)\dot{\theta}\geq 0.$$

Since $\dot{\theta}$ is independent of all other quantities in the equation, we are free to select a positive or negative value of $\dot{\theta}$. The inequality may be satisfied if and only if the term in parenthesis is zero providing the condition

$$\eta=-\frac{\partial\psi}{\partial\theta}.$$

Similarly, the variable $\dot{\boldsymbol{D}}$ does not appear in the constitutive equations and can be varied independently of all other quantities in the equation. This produces the relation

$$\frac{\partial\psi}{\partial\boldsymbol{D}}=0.$$

This constraint states that the Helmholtz free energy may not depend on the variable $\boldsymbol{D}$ since the partial derivative with respect to this variable is zero. The time rate of change of the temperature gradient, $\dot{\boldsymbol{g}}$, is also independent of all other quantities producing the relation

$$\frac{\partial\psi}{\partial\boldsymbol{g}}=0.$$

These three constraints together reduce the functional dependent of the Helmholtz free energy and the entropy down to two variables such that

$$\psi = \psi(\rho, \theta),$$
$$\eta = \eta(\rho, \theta).$$

Substituting these constraints into the CD inequality leaves us with

$$-\rho\left(\frac{\partial \psi}{\partial \rho}\dot{\rho}\right) - \frac{1}{\theta}\boldsymbol{q}_x \cdot \boldsymbol{g} + \boldsymbol{\sigma}^T : \boldsymbol{D} \geq 0.$$

At this point, we recognize the fact that $\dot{\rho}$ can be written as a function of $\boldsymbol{D}$. To determine the dependence, we use the conservation of mass equation to rewrite the density in terms of the deformation rate tensor. The conservation of mass equation may be written as

$$\dot{\rho} + \rho(\nabla \cdot \boldsymbol{v}) = 0.$$

The divergence of the velocity can be written as the trace of the velocity gradient, or equivalently as the trace of strain rate tensor such that

$$\nabla \cdot \mathbf{v} = \mathbf{I} : \nabla \boldsymbol{v} = \mathbf{I} : \boldsymbol{D}.$$

Therefore, we may write the material derivative of the density as

$$\dot{\rho} = -\rho \mathbf{I} : \boldsymbol{D}.$$

Substitution into the CD inequality gives

$$\left(\rho^2 \frac{\partial \psi}{\partial \rho}\mathbf{I} + \boldsymbol{\sigma}^T\right) : \boldsymbol{D} - \frac{1}{\theta}\boldsymbol{q}_x \cdot \boldsymbol{g} \geq 0.$$

Now, we note that the strain rate tensor appears in the formulation of the constitutive law for the stress since $\boldsymbol{\sigma} = \boldsymbol{\sigma}(\rho, \boldsymbol{D}, \theta, \nabla\theta)$. Therefore, the Cauchy stress is a function of the deformation rate tensor, and $\boldsymbol{D}$ cannot be varied independently as it was during the development of the thermo-mechanical model.

To overcome this, we break up the Cauchy stress into $\boldsymbol{\sigma_o}$, which is independent of the strain rate tensor and the temperature gradient such that

$$\boldsymbol{\sigma_o} = \boldsymbol{\sigma_o}(\rho, \theta),$$

and the remainder of the stress, $\hat{\boldsymbol{\sigma}}$, such that

$$\hat{\boldsymbol{\sigma}}(\rho, \boldsymbol{D}, \theta, \nabla\theta) = \boldsymbol{\sigma}(\rho, \boldsymbol{D}, \theta, \nabla\theta) - \boldsymbol{\sigma_o}(\rho, \theta).$$

Furthermore, we may use a Taylor expansion to write $\hat{\boldsymbol{\sigma}}(\rho, \Delta\boldsymbol{D}, \theta, \mathbf{0})$, where $\|\Delta\boldsymbol{D}\| \ll 1$ in terms of the value at $\hat{\boldsymbol{\sigma}}(\rho, \mathbf{0}, \theta, \mathbf{0})$ as

$$\hat{\boldsymbol{\sigma}}(\rho, \Delta\boldsymbol{D}, \theta, \mathbf{0}) = \mathbf{0} + \frac{\partial \hat{\boldsymbol{\sigma}}}{\partial \boldsymbol{D}} : \Delta\boldsymbol{D} + \boldsymbol{O}\left(\|\Delta\boldsymbol{D}\|^2\right).$$

Note that as the norm approaches zero, $\|\Delta\boldsymbol{D}\| \to 0$, the correction term approaches zero, $\frac{\partial \hat{\boldsymbol{\sigma}}}{\partial \boldsymbol{D}} : \Delta\boldsymbol{D} \to \mathbf{0}$ and the remainder of the stress approaches zero, $\hat{\boldsymbol{\sigma}}(\rho, \Delta\boldsymbol{D}, \theta, \mathbf{0}) \to \mathbf{0}$.

The CD inequality for a thermodynamic process which has a zero temperature gradient reduces to the following form

$$\left(\rho^2 \frac{\partial \psi}{\partial \rho}\mathbf{I} + \boldsymbol{\sigma}_o\right) : \boldsymbol{D} + \frac{\partial \hat{\boldsymbol{\sigma}}}{\partial \boldsymbol{D}} : \Delta \boldsymbol{D} : \boldsymbol{D} \geq 0.$$

As $\|\Delta \boldsymbol{D}\|$ approaches zero, we obtain the condition

$$\boldsymbol{\sigma}_o = -\rho^2 \frac{\partial \psi}{\partial \rho}\mathbf{I},$$

where $-\rho^2 \frac{\partial \psi}{\partial \rho}$ is known as the thermodynamic pressure.

If the stress is a function of the temperature gradient and the heat flux depends on the strain rate tensor, we cannot proceed any further. However, if we can remove this dependence and write the constitutive relations as

$$\boldsymbol{\sigma} = \boldsymbol{\sigma}(\rho, \theta, \boldsymbol{D}), \quad \text{and} \quad \boldsymbol{q} = \boldsymbol{q}(\rho, \theta, \boldsymbol{g}),$$

the CD inequality gives us a condition on the remainder of the stress

$$\hat{\boldsymbol{\sigma}} : \boldsymbol{D} \geq 0$$

and on the heat flux vector

$$-\frac{1}{\theta}\boldsymbol{q}_x \cdot \boldsymbol{g} \geq 0.$$

Therefore, the constitutive equations must have the form

$$\begin{aligned}
\psi &= \psi(\rho, \theta), \\
\eta(\rho, \theta) &= -\frac{\partial \psi}{\partial \theta}, \\
\boldsymbol{\sigma}_o(\rho, \theta) &= -\rho^2 \frac{\partial \psi}{\partial \rho}\mathbf{I}, \\
\hat{\boldsymbol{\sigma}} &= \hat{\boldsymbol{\sigma}}(\rho, \theta, \boldsymbol{D}), \\
\boldsymbol{q} &= \boldsymbol{q}(\rho, \theta, \boldsymbol{g}), \\
\hat{\boldsymbol{\sigma}} : \boldsymbol{D} &\geq 0, \\
-\frac{1}{\theta}\boldsymbol{q}_x \cdot \boldsymbol{g} &\geq 0.
\end{aligned}$$

6.4.2 Constitutive Relations for the Newtonian Fluid

Given the constraints that arise from the principle of material frame indifference, the CD inequality, and the material symmetry for an isotropic fluid, we arrive at the constitutive relations for an isotropic Newtonian fluid

$$\boxed{\begin{aligned}
\psi &= \psi(\rho, \theta), \\
\eta(\rho, \theta) &= -\frac{\partial \psi}{\partial \theta}, \\
\boldsymbol{\sigma}(\rho, \theta, \boldsymbol{D}) &= -p\boldsymbol{I} + \lambda\, tr(\boldsymbol{D})\mathbf{I} + 2\mu \boldsymbol{D}, \\
\boldsymbol{q}(\rho, \theta, \boldsymbol{g}) &= -k\nabla\theta.
\end{aligned}} \tag{6.1}$$

This constitutive model has three material parameters, λ the dilatational viscosity, μ the shear viscosity, and k the thermal conductivity. The values for the Helmholtz free energy are determined experimentally and usually presented in tabular form for various fluids. These equations combined with the balance laws provide a closed set of equations, which can be solved for the state of any equilibrium or nonequilibrium Newtonian fluid.

Furthermore, we may use the constraints derived from the second law of thermodynamics to determine the valid range of potential values for the material parameters, λ, μ, and k in our specific constitutive model. We begin with the constraint on the remainder of the Cauchy stress derived in the previous section

$$\hat{\boldsymbol{\sigma}} : \boldsymbol{D} \geq 0.$$

Writing the deformation rate tensor in terms of a deviatoric and a hydrostatic part, we obtain

$$\hat{\boldsymbol{\sigma}} : (\mathrm{tr}\,(\boldsymbol{D})\mathbf{I} + \boldsymbol{D}_{dev}) \geq 0.$$

From the constitutive model, we have $\hat{\boldsymbol{\sigma}} = \lambda\, tr\,(\boldsymbol{D})\mathbf{I} + 2\mu\boldsymbol{D}$, which substituted into the previous inequality gives

$$(\lambda\, tr\,(\boldsymbol{D})\mathbf{I} + 2\mu\boldsymbol{D}) : (\mathrm{tr}\,(\boldsymbol{D})\mathbf{I} + \boldsymbol{D}_{dev}) \geq 0.$$

Recalling that the trace of the identity is $\mathbf{I}:\mathbf{I} = 3$, that the trace of the strain rate tensor can be written as $\mathbf{D}:\mathbf{I} = tr\,(\boldsymbol{D})$, and that the trace of a deviatoric tensor is zero $\mathbf{I}:\boldsymbol{D}_{dev} = 0$, we can simplify the inequality to obtain

$$(3\lambda + 2\mu)\,(tr\,(\boldsymbol{D}))^2 + 2\mu\boldsymbol{D}:\boldsymbol{D}_{dev} \geq 0.$$

Once again separating the hydrostatic and deviatoric parts of the deformation rate tensor gives

$$(3\lambda + 2\mu)\,(tr\,(\boldsymbol{D}))^2 + 2\mu\boldsymbol{D}:\boldsymbol{D}_{dev} \geq 0.$$

Using the fact that the deviatoric part of the deformation rate tensor can be varied independently of the hydrostatic part, and the fact that $\boldsymbol{D} : \boldsymbol{D}_{dev} = \boldsymbol{D}_{dev} : \boldsymbol{D}_{dev}$, we obtain two conditions from this inequality

$$(3\lambda + 2\mu)\,(tr\,(\boldsymbol{D}))^2 \geq 0$$

and

$$2\mu\boldsymbol{D}_{dev} : \boldsymbol{D}_{dev} \geq 0.$$

Since both $\boldsymbol{D}_{dev} : \boldsymbol{D}_{dev}$ and $(tr\,(\boldsymbol{D}))^2$ must always be positive numbers, we require that $\mu \geq 0$ and $\lambda \geq -\frac{2}{3}\mu$.

The constraint on the value of the thermal conductivity is obtain by using the inequality

$$-\frac{1}{\theta}\boldsymbol{q}_x \cdot \boldsymbol{g} \geq 0.$$

Substitution of the heat flux vector gives

$$\frac{1}{\theta}k\,(\nabla\theta \cdot \nabla\theta) \geq 0.$$

Since the temperature is always a positive number, and the dot product of the gradient of the temperature with itself must also give a positive number, this condition requires that the thermal conductivity be a greater than or equal to zero, $k \geq 0$.

6.4.3 Stokes Condition

In 1895, Stokes proposed that the thermodynamic pressure should be equal to the mechanical pressure. As mentioned earlier, the thermodynamic pressure, p, is determined from the Clasius-Duhem condition

$$\boldsymbol{\sigma}_O = -\rho^2 \frac{\partial \psi}{\partial \rho} \boldsymbol{I}$$

and is defined as

$$p = -\rho^2 \frac{\partial \psi}{\partial \rho}.$$

The mechanical pressure, $\bar{p}$, is defined as one third of the trace of the stress tensor

$$\bar{p} = \frac{1}{3} tr\,(\boldsymbol{\sigma}).$$

The difference between the thermodynamic pressure, p, and the mechanical pressure, $\bar{p}$, is given by

$$\begin{aligned} \bar{p} - p &= \frac{1}{3} tr\,(\boldsymbol{\sigma}) - p \\ &= \frac{1}{3}\left(3p + 3\lambda\, tr\,(\boldsymbol{D}) + 2\mu\, tr\,(\boldsymbol{D})\right) - p \\ &= \left(\lambda + \frac{2}{3}\mu\right)(\nabla \cdot \boldsymbol{v}). \end{aligned}$$

Notice that the thermodynamic pressure and the mechanical pressure are equal if either $\lambda = -\frac{2}{3}\mu$, which is known as Stokes' hypothesis, or if $\nabla \cdot \boldsymbol{v} = 0$ which is true for an incompressible fluid. The bulk modulus for an isotropic fluid, μ_B, is defined as

$$\mu_B = \left(\lambda + \frac{2}{3}\mu\right).$$

The hypothesis that $\mu_B = 0$ was erroneously proposed by Stokes in 1895. While it was shown that this is strictly true for monatomic gases, it is only an approximation for fluids. However, it is a good approximation for fluids at low Mach number where the density fluctuations are small and therefore $\nabla \cdot \boldsymbol{v}$ is small or when the Reynolds number is large enough that viscous effects can be neglected entirely. The bulk viscosity plays a role in the attenuation of sound waves. High-frequency measurements have shown that the bulk modulus is frequency-dependent and that at high frequency it has same order of magnitude as the shear viscosity. Therefore, it is nonnegligible in these cases. Employing the Stokes' condition, only two material parameters the viscosity, μ, and thermal conductivity, k, are needed to characterize the Newtonian fluid.

6.5 Governing Equations

By combining the constitutive equations with the balance equations, we obtain a set of governing equations specific to the Newtonian fluid. The Navier-Stokes equation and the Bernoulli equation are examples of equations that can be obtained from these governing equations using varying degrees of approximation.

6.5.1 Compressible Newtonian Fluid

In this section, we will derive the governing equations for a compressible Newtonian fluid without employing the Stokes' condition. These are known as the compressible Navier-Stokes equations. We begin with the principle of conservation of mass which cannot be simplified any further and remains

$$0=\dot{\rho}+\rho\,\nabla\cdot\boldsymbol{v}.$$

The principle of conservation of linear momentum can be simplified by combining it with the constitutive relation for the Cauchy stress. The equation for conservation of linear momentum for a thermo-mechanical material is given by

$$\rho\dot{\boldsymbol{v}}-\nabla\cdot\boldsymbol{\sigma}^T-\rho\,\boldsymbol{b}=\boldsymbol{0}. \tag{6.2}$$

Using the constitutive relation for the Cauchy stress, $\boldsymbol{\sigma}(\rho,\theta,\boldsymbol{D})=-p\boldsymbol{I}+\lambda\,tr(\boldsymbol{D})\mathbf{I}+2\mu\boldsymbol{D}$, the divergence of the stress can be written as

$$\nabla\cdot\boldsymbol{\sigma}^T=\nabla\cdot(-p\boldsymbol{I}+\lambda\,tr(\boldsymbol{D})\mathbf{I}+2\mu\boldsymbol{D})=-\nabla p+\lambda\,(\nabla\cdot(tr(\boldsymbol{D})\mathbf{I}))+2\mu\,\nabla\cdot\boldsymbol{D}.$$

Since the strain rate tensor can be written in terms of the velocity gradient, $\boldsymbol{D}=\frac{1}{2}\left(\boldsymbol{v}\overleftarrow{\nabla}+\overrightarrow{\nabla}\boldsymbol{v}\right)$, and the trace of the strain rate tensor is equal to the divergence of the velocity, $tr(\boldsymbol{D})=\nabla\cdot\boldsymbol{v}$, we can write

$$\begin{aligned}\nabla\cdot\boldsymbol{\sigma}^T&=-\nabla p+\lambda\,(\nabla\cdot((\nabla\cdot\boldsymbol{v})\mathbf{I}))+\mu\,\nabla\cdot\left(\boldsymbol{v}\overleftarrow{\nabla}+\overrightarrow{\nabla}\boldsymbol{v}\right)\\&=-\nabla p+\lambda(\nabla\cdot\nabla\boldsymbol{v})+\mu\left(\nabla\cdot\boldsymbol{v}\overleftarrow{\nabla}\right)+\mu\nabla\cdot\overrightarrow{\nabla}\boldsymbol{v}\end{aligned} \tag{6.3}$$

In addition, combining the identities

$$\nabla\cdot((\nabla\cdot\boldsymbol{v})\mathbf{I})=\frac{\partial}{\partial x_i}\left(v_{k,k}\delta_{ij}\right)\boldsymbol{e}_j=\nabla\,(\nabla\cdot\boldsymbol{v}) \tag{6.4}$$

and

$$\left(\nabla\cdot\boldsymbol{v}\overleftarrow{\nabla}\right)=\left(\frac{\partial}{\partial x_i}\left(v_{i,j}\right)\boldsymbol{e}_j\right), \tag{6.5}$$

we may write

$$\nabla\cdot((\nabla\cdot\boldsymbol{v})\mathbf{I})=\nabla\,(\nabla\cdot\boldsymbol{v})=\left(\nabla\cdot\boldsymbol{v}\overleftarrow{\nabla}\right). \tag{6.6}$$

Substituting this identity into the divergence of the stress tensor gives

$$\nabla\cdot\boldsymbol{\sigma}^T=-\nabla p+(\lambda+\mu)\,(\nabla\,(\nabla\cdot\boldsymbol{v}))+\mu\nabla\cdot\overrightarrow{\nabla}\boldsymbol{v}. \tag{6.7}$$

The equation for conservation of linear momentum can then be written as

$$\rho\dot{\boldsymbol{v}}+\nabla p-(\lambda+\mu)\,(\nabla\,(\nabla\cdot\boldsymbol{v}))-\mu\nabla\cdot\nabla\boldsymbol{v}-\rho\,\boldsymbol{b}=\boldsymbol{0}. \tag{6.8}$$

This equation is known as the compressible Navier-Stokes equation.

The principle of conservation of energy can be simplified in a similar way. We restate the principle of conservation of energy for clarity as

$$\rho\dot{e}-\boldsymbol{\sigma}^T:\boldsymbol{D}+div\,\boldsymbol{q}-\rho r=0. \tag{6.9}$$

Making use of the constitutive equations given by (6.1), we obtain

$$\begin{aligned}\boldsymbol{\sigma}^T : \boldsymbol{D} &= (-p\boldsymbol{I} + \lambda\, tr(\boldsymbol{D})\mathbf{I} + 2\mu\boldsymbol{D}) : \boldsymbol{D} \\ &= -p\,(\nabla\cdot\boldsymbol{v}) + \lambda(\nabla\cdot\boldsymbol{v})^2 + 2\mu\boldsymbol{D}:\boldsymbol{D}.\end{aligned} \tag{6.10}$$

Substitution of Equation (6.10) into Equation (6.9) gives

$$\rho\dot{e} + p\,(\nabla\cdot\boldsymbol{v}) - \lambda(\nabla\cdot\boldsymbol{v})^2 - 2\mu\boldsymbol{D}:\boldsymbol{D} - \rho r = 0.$$

Gathering these equations, we have the following governing equations for a compressible Newtonian fluid:

$$\boxed{\begin{aligned}\dot{\rho} + \rho\nabla\cdot\boldsymbol{v} &= 0, \\ \rho\dot{\boldsymbol{v}} + \nabla p - (\lambda+\mu)(\nabla(\nabla\cdot\boldsymbol{v})) - \mu\nabla\cdot\nabla\boldsymbol{v} - \rho\boldsymbol{b} &= \boldsymbol{0}, \\ \rho\dot{e} + p\,(\nabla\cdot\boldsymbol{v}) - \lambda(\nabla\cdot\boldsymbol{v})^2 - 2\mu\boldsymbol{D}:\boldsymbol{D} - \rho r &= 0.\end{aligned}} \tag{6.11}$$

6.5.2 Incompressible Newtonian Fluid

The governing equations are greatly simplified in the case of an incompressible Newtonian fluid. Specifically, if we assume that the fluid is incompressible, then the equation for conservation of mass gives

$$\nabla\cdot v = 0,$$

and the conservation of linear momentum for a compressible Newtonian fluid

$$\rho\dot{\boldsymbol{v}} + \nabla p - (\lambda+\mu)(\nabla(\nabla\cdot\boldsymbol{v})) - \mu\nabla\cdot\nabla\boldsymbol{v} - \rho\boldsymbol{b} = \boldsymbol{0} \tag{6.12}$$

can be written as

$$\rho\dot{\boldsymbol{v}} + \nabla p - \mu\,(\nabla\cdot\nabla\boldsymbol{v}) - \rho\boldsymbol{b} = \boldsymbol{0}. \tag{6.13}$$

Written in index notation, this becomes

$$\rho\dot{v}_i = -\frac{\partial p}{\partial x_i} + \mu\, v_{i,jj} + \rho b_i.$$

Similarly, the conservation of energy for a compressible Newtonian fluid

$$\rho\dot{e} + p\,(\nabla\cdot\boldsymbol{v}) - \lambda(\nabla\cdot\boldsymbol{v})^2 - 2\mu\boldsymbol{D}:\boldsymbol{D} - \rho r = 0$$

becomes

$$\rho\dot{e} - 2\mu\boldsymbol{D}:\boldsymbol{D} - \rho r = 0.$$

The governing equations for incompressible Newtonian fluid can then be written as

$$\boxed{\begin{aligned}\nabla\cdot\boldsymbol{v} &= 0, \\ \rho\dot{\boldsymbol{v}} + \nabla p - \mu\nabla\cdot\nabla\boldsymbol{v} - \rho\boldsymbol{b} &= \boldsymbol{0}, \\ \rho\dot{e} - 2\mu\boldsymbol{D}:\boldsymbol{D} - \rho r &= 0.\end{aligned}} \tag{6.14}$$

6.5.3 Irrotational Steady Flow of an Incompressible Newtonian Fluid

Assuming that we have an irrotational steady flow of an incompressible Newtonian fluid produces Bernoulli's equation. If the flow is irrotational, the curl of the velocity vector is zero such that

$$\nabla \times \boldsymbol{v} = \boldsymbol{0}.$$

The Navier-Stokes equation for an incompressible fluid is

$$\rho\dot{\boldsymbol{v}} + \nabla p - \mu\,(\nabla \cdot \nabla \boldsymbol{v}) - \rho\,\boldsymbol{b} = \boldsymbol{0}. \tag{6.15}$$

Expanding the material derivative gives

$$\rho\left(\frac{\partial \boldsymbol{v}}{\partial t} + \boldsymbol{v}\cdot\nabla\boldsymbol{v}\right) + \nabla p - \mu\,(\nabla \cdot \nabla \boldsymbol{v}) - \rho\,\boldsymbol{b} = \boldsymbol{0}.$$

If the only body force is gravity, we may write

$$\rho\boldsymbol{b} = -\nabla\left(\rho\left(\mathbf{g}\cdot\boldsymbol{h}\right)\right),$$

where $\mathbf{g}$ is the vector gravitational acceleration, and $\mathbf{h}$ is the vector distance to some reference point.

The vector identities for the curl of the velocity vector,

$$\nabla \times \nabla \times \boldsymbol{v} = \nabla\,(\nabla \cdot \boldsymbol{v}) - \nabla \cdot \nabla \boldsymbol{v},$$

simplifies for incompressible fluid, $\nabla \cdot \boldsymbol{v} = 0$, undergoing irrotational flow, $\nabla \times \boldsymbol{v} = \boldsymbol{0}$, to

$$\nabla \cdot \nabla \boldsymbol{v} = \boldsymbol{0}.$$

Therefore, the Navier-Stokes equation becomes

$$\rho\left(\frac{\partial \boldsymbol{v}}{\partial t} + \boldsymbol{v}\cdot\nabla\boldsymbol{v}\right) + \nabla p + \nabla\left(\rho\left(\mathbf{g}\cdot\boldsymbol{h}\right)\right) = \boldsymbol{0}.$$

Gathering terms gives

$$\left(\frac{\partial \boldsymbol{v}}{\partial t} + \nabla\left(\frac{\boldsymbol{v}\cdot\boldsymbol{v}}{2} + \frac{p}{\rho} + \boldsymbol{g}\cdot\boldsymbol{h}\right)\right) = \boldsymbol{0}.$$

Finally, for a steady flow, the spatial velocity field does not change with time

$$\frac{\partial \boldsymbol{v}}{\partial t} = \boldsymbol{0}.$$

This gives the Bernoulli equation for the steady irrotational flow of an incompressible fluid

$$\boxed{\left(\frac{\boldsymbol{v}\cdot\boldsymbol{v}}{2} + \frac{p}{\rho} + \boldsymbol{g}\cdot\boldsymbol{h}\right) = \text{constant.}} \tag{6.16}$$

6.6 Non-Newtonian Fluid Models

A non-Newtonian fluid is one whose viscosity changes with strain rate. We can express the functional form of the viscosity in terms of the principle invariants of the strain rate tensor such that

$$\mu = \mu\,(I_{\mathbf{D}}, II_{\mathbf{D}}, III_{\mathbf{D}}).$$

However, in the case of incompressible fluids, the first invariant is zero (due to conservation of mass), and experimental data suggest that the viscosity does not depend on the third invariant. So most models will have the reduced form

$$\mu = \mu\,(II_{\boldsymbol{D}}).$$

In this section, we will derive the governing equations for an incompressible non-Newtonian fluid. Since we have an incompressible fluid, the conservation of mass equation, $0 = \dot{\rho} + \rho\,\nabla \cdot \boldsymbol{v}$, requires that

$$\nabla \cdot \boldsymbol{v} = 0.$$

Recall that the constitutive relation for the Cauchy stress of an isotropic non-Newtonian fluid is given by

$$\boldsymbol{\sigma}\,(\rho,\theta,\boldsymbol{D}) = -p\,(\rho,\theta)\boldsymbol{I} + \lambda\,(\boldsymbol{D})\;tr\,(\boldsymbol{D})\mathbf{I} + 2\mu\,(\boldsymbol{D})\boldsymbol{D},$$

where the dependence of the viscosities, λ and μ, on the strain rate tensor, $\boldsymbol{D}$, are shown for clarity. Given the incompressibility restriction, $tr\,(\boldsymbol{D}) = 0$, we obtain

$$\boldsymbol{\sigma}\,(\rho,\theta,\boldsymbol{D}) = -p\,(\rho,\theta)\boldsymbol{I} + 2\mu\,(\boldsymbol{D})\boldsymbol{D}.$$

The divergence of the stress can be written as

$$\nabla \cdot \boldsymbol{\sigma}^T = \nabla \cdot (-p\boldsymbol{I} + 2\mu\boldsymbol{D}) = -\nabla p + 2\,(\nabla\mu)\cdot\boldsymbol{D} + 2\mu\,\nabla\cdot\boldsymbol{D}.$$

Writing the strain rate tensor in terms of the velocity gradient, $\boldsymbol{D} = \frac{1}{2}\left(\boldsymbol{v}\overleftarrow{\nabla} + \overrightarrow{\nabla}\boldsymbol{v}\right)$, and using the fact that $\left(\nabla \cdot \boldsymbol{v}\overleftarrow{\nabla}\right) = \mathbf{0}$ for an incompressible fluid gives

$$\nabla \cdot \boldsymbol{\sigma}^T = \nabla \cdot (-p\boldsymbol{I} + 2\mu\boldsymbol{D}) = -\nabla p + (\nabla\mu)\cdot\left(\boldsymbol{v}\overleftarrow{\nabla} + \overrightarrow{\nabla}\boldsymbol{v}\right) + \mu\nabla\cdot\nabla\boldsymbol{v}.$$

The general equation for conservation of linear momentum, $\rho\dot{\boldsymbol{v}} - \nabla \cdot \boldsymbol{\sigma}^T - \rho\,\boldsymbol{b} = \mathbf{0}$, may be written for an incompressible non-Newtonian fluid as

$$\rho\dot{\boldsymbol{v}} + \nabla p - (\nabla\mu)\cdot\left((\nabla\boldsymbol{v})^{\boldsymbol{T}} + \nabla\boldsymbol{v}\right) - \mu\nabla\cdot\nabla\boldsymbol{v} - \rho\,\boldsymbol{b} = \mathbf{0}. \tag{6.17}$$

The energy equation is identical to that for the incompressible Newtonian fluid.

$$\begin{aligned} \nabla \cdot \boldsymbol{v} &= 0, \\ \rho\dot{\boldsymbol{v}} + \nabla p - (\nabla\mu)\cdot\left((\nabla\boldsymbol{v})^{\boldsymbol{T}} + \nabla\boldsymbol{v}\right) - \mu\nabla\cdot\nabla\boldsymbol{v} - \rho\,\boldsymbol{b} &= 0, \\ \rho\dot{e} - 2\mu\boldsymbol{D} : \boldsymbol{D} - \rho\,r &= 0. \end{aligned} \tag{6.18}$$

6.6.1 Power Law Model

The power law is a common empirical fit to viscosity as a function of the second invariant of the strain rate tensor

$$\mu = K\,II_{\boldsymbol{D}}^{(n-1)/2},$$

where n is the power law index, and K is the consistency. When $n < 1$, the viscosity drops with higher rate of deformation known as shear thinning, while $n > 1$ the viscosity increases and the fluid is a shear thickening fluid.

6.6.2 Cross Model

The viscosity in the cross model approaches a constant at low shear rates and at high shear rates. The form

$$\mu = \mu_\infty + \frac{\mu_o - \mu_\infty}{1 + m II_{\boldsymbol{D}}^{(n-1)/2}},$$

where μ_o is the viscosity at low shear rate, μ_∞ is the viscosity measured at high shear rate, m and n are material parameters, and II_D is the second invariant of the strain rate tensor.

6.6.3 Bingham Model

A Bingham fluid is fluid-like only when the stresses rise above a threshold level. For small stresses, the fluid supports the stress as a solid would. The Bingham model has the form

$$\boldsymbol{\tau} = \left(\frac{\tau_o}{\sqrt{II_{\boldsymbol{D}}}} + 2\mu\right)\boldsymbol{D} \quad if\ \frac{1}{2}tr\left(\boldsymbol{\tau}^2\right) \geq \tau_o^2,$$

$$\boldsymbol{\tau} = 0 \quad if\ \frac{1}{2}tr\left(\boldsymbol{\tau}^2\right) < \tau_o^2.$$

In addition, the viscosity itself may also depend on the strain rate.

6.7 Couette Viscometer

The Couette viscometer is used to determine the viscosity of a fluid. Whereas there are many types of viscometers, let us focus on the rotational viscometer. A fluid is placed between two cylinders whose length is much greater than the radius. In the Couette viscometer, the cup is rotated at a constant angular frequency, ω, whereas in the Searle viscometer, the bob is rotated at a constant angular frequency. The Newtonian fluid model postulates a fluid whose viscosity is independent of the strain rate; however, in reality most fluid deviate from this assumption. By varying the angular frequency of either the bob or the cup, the fluid viscosity may be measured as a function of strain rate and the character of the fluid determined.

6.7.1 Newtonian Fluid

In this section, we will derive the equations describing the velocity profile for a Newtonian fluid within a Couette viscometer. The constitutive model for an incompressible Newtonian fluid has only a single material parameter, which is the fluid viscosity. The viscosity of the fluid, μ, may be found experimentally for a given fluid using a viscometer. For the case of a Couette or Searle viscometer, the viscosity is found by measuring the torque required to rotate the cup or bob at a given angular frequency.

We will now determine the theoretical relationship between applied torque and angular frequency by solving the governing equations given the geometry and boundary conditions found within the Couette viscometer. The solution of the governing equations for the case of the Couette or Searle viscometer is most easily achieved

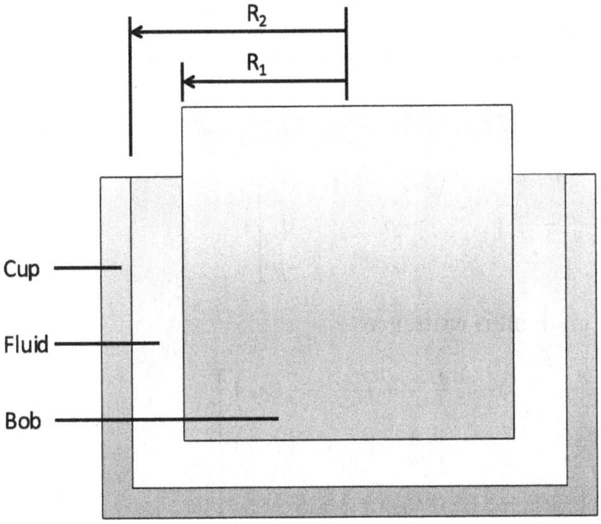

Figure 6.2. Schematic of a Couette viscometer. The cup is rotated at an angular frequency, ω, while the bob is held stationary.

by employing cylindrical coordinates. In addition, we assume that we have steady laminar flow of the fluid between the cup and bob. If we align the z axis with the axis of the bob, the components of the laminar steady-state fluid velocity vector can be written as

$$v_r = v_z = 0, \quad \text{and} \quad v_\theta = v_\theta(r).$$

The fluid must have zero velocity in the r and z directions. Only a component in the θ direction is allowed. In cylindrical coordinates, the velocity gradient may be written as

$$\boldsymbol{L} = \nabla \boldsymbol{v} = \begin{bmatrix} \frac{\partial v_r}{\partial x_r} & \frac{\partial v_\theta}{\partial x_r} & \frac{\partial v_z}{\partial x_r} \\ \left[\frac{1}{r}\frac{\partial v_r}{\partial x_\theta} - \frac{v_\theta}{r}\right] & \left[\frac{1}{r}\frac{\partial v_\theta}{\partial x_\theta} + \frac{v_r}{r}\right] & \left[\frac{1}{r}\frac{\partial v_z}{\partial x_\theta}\right] \\ \frac{\partial v_r}{\partial x_z} & \frac{\partial v_\theta}{\partial x_z} & \frac{\partial v_z}{\partial x_z} \end{bmatrix} = \begin{bmatrix} 0 & \frac{\partial v_\theta}{\partial x_r} & 0 \\ -\frac{v_\theta}{r} & 0 & 0 \\ 0 & 0 & 0 \end{bmatrix}.$$

The fluid acceleration is given by the material derivative of the velocity such that

$$\begin{aligned} \dot{\boldsymbol{v}} &= \frac{\partial \boldsymbol{v}}{\partial t} + (\nabla \boldsymbol{v}) \cdot \boldsymbol{v} \\ &= \frac{\partial \boldsymbol{v}}{\partial t} + \boldsymbol{L} \cdot \boldsymbol{v} \\ &= \begin{bmatrix} v_\theta \frac{\partial v_\theta}{\partial x_r} \\ 0 \\ 0 \end{bmatrix}. \end{aligned}$$

The strain rate tensor can be written as

$$\boldsymbol{D} = \frac{1}{2}\left(\boldsymbol{L} + \boldsymbol{L}^T\right) = \frac{1}{2}\begin{bmatrix} 0 & \left(\frac{\partial v_\theta}{\partial x_r} - \frac{v_\theta}{r}\right) & 0 \\ \left(\frac{\partial v_\theta}{\partial x_r} - \frac{v_\theta}{r}\right) & 0 & 0 \\ 0 & 0 & 0 \end{bmatrix}.$$

Employing the constitutive equations for the Cauchy stress of a Newtonian fluid, we can write

$$\boldsymbol{\sigma} = -p\boldsymbol{I} + \lambda\, tr(\boldsymbol{D})\mathbf{I} + 2\mu\boldsymbol{D}$$

$$= \begin{bmatrix} -p & \mu\left(\frac{\partial v_\theta}{\partial x_r} - \frac{v_\theta}{r}\right) & 0 \\ \mu\left(\frac{\partial v_\theta}{\partial x_r} - \frac{v_\theta}{r}\right) & -p & 0 \\ 0 & 0 & -p \end{bmatrix}.$$

The divergence of the velocity gradient is written as

$$\nabla\cdot\boldsymbol{L} = \begin{bmatrix} \left(\frac{1}{r}\frac{\partial}{\partial x_r}(rL_{rr}) + \frac{1}{r}\frac{\partial L_{\theta r}}{\partial x_\theta} + \frac{\partial L_{zr}}{\partial x_z} - \frac{1}{r}L_{\theta\theta}\right) \\ \left(\frac{1}{r}\frac{\partial L_{\theta\theta}}{\partial x_\theta} + \frac{\partial L_{z\theta}}{\partial x_z} + \frac{1}{r}\frac{\partial}{\partial x_r}(rL_{r\theta}) + \frac{1}{r}L_{\theta r}\right) \\ \left(\frac{\partial L_{zz}}{\partial x_z} + \frac{1}{r}\frac{\partial}{\partial x_r}(rL_{rz}) + \frac{1}{r}\frac{\partial L_{\theta z}}{\partial x_r}\right) \end{bmatrix}$$

$$= \begin{bmatrix} 0 \\ \frac{1}{r}\frac{\partial}{\partial x_r}(rL_{r\theta}) + \frac{1}{r}L_{\theta r} \\ 0 \end{bmatrix}$$

$$= \begin{bmatrix} 0 \\ -\frac{1}{r}\frac{\partial}{\partial x_r}\left(r\frac{\partial v_\theta}{\partial x_r}\right) + \frac{1}{r}\frac{v_\theta}{r} \\ 0 \end{bmatrix}$$

$$= \begin{bmatrix} 0 \\ -\left(\frac{\partial^2 v_\theta}{\partial x_r^2}\right) - \frac{1}{r}\frac{\partial v_\theta}{\partial x_r} + \frac{v_\theta}{r^2} \\ 0 \end{bmatrix}.$$

Neglecting gravity, the Navier-Stokes equation for an incompressible Newtonian fluid becomes

$$\rho\dot{\boldsymbol{v}} + \nabla p - \mu\nabla\cdot\nabla\boldsymbol{v} - \rho\boldsymbol{b} = \boldsymbol{0},$$

$$\rho\dot{\boldsymbol{v}} + \nabla p - \mu\nabla\cdot\boldsymbol{L} = \boldsymbol{0}.$$

Substituting the velocity gradient tensor and the acceleration vector and writing the equation in component form gives

$$\rho\begin{bmatrix} v_\theta\frac{\partial v_\theta}{\partial x_r} \\ 0 \\ 0 \end{bmatrix} + \begin{bmatrix} \frac{\partial p}{\partial x_r} \\ 0 \\ 0 \end{bmatrix} - \mu\begin{bmatrix} 0 \\ -\left(\frac{\partial^2 v_\theta}{\partial x_r^2}\right) - \frac{1}{r}\frac{\partial v_\theta}{\partial x_r} + \frac{v_\theta}{r^2} \\ 0 \end{bmatrix} = \boldsymbol{0}.$$

This represents the two equations

$$\rho v_\theta\frac{\partial v_\theta}{\partial x_r} + \frac{\partial p}{\partial x_r} = 0$$

and

$$-\left(\frac{\partial^2 v_\theta}{\partial x_r^2}\right) - \frac{1}{r}\frac{\partial v_\theta}{\partial x_r} + \frac{v_\theta}{r^2} = 0.$$

The second of these equations may be solved for the velocity profile of the fluid. Given the velocity profile, the first can be used to compute the pressure profile.

The solution of the second partial differential equation has the general polynomial form

$$v_\theta = \alpha r^n,$$

where α and n are constants. Substitution of the general solution into the partial differential equation gives

$$-\alpha n(n-1)r^{n-2} - \frac{1}{r}\alpha n r^{n-1} + \frac{1}{r^2}\alpha r^n = 0.$$

Simplifying this equation gives the condition that

$$-n(n-1) - n + 1 = 0,$$
$$-n^2 + 1 = 0.$$

The roots of this equation are $n = \pm 1$. Therefore, the solution can be written as the sum of the two particular solutions

$$v_\theta = \alpha_1 r + \frac{\alpha_2}{r}.$$

The coefficients, α_1 and α_2, are found by employing the boundary conditions. For the Couette viscometer with the rotating cup, the boundary conditions are

$$v_\theta(R_1) = 0 \text{ and } v_\theta(R_2) = \omega R_2,$$

where ω is the angular velocity of the cup. This gives the two equations

$$0 = \alpha_1 R_1 + \frac{\alpha_2}{R_1} \quad \text{and} \quad \omega R_2 = \alpha_1 R_2 + \frac{\alpha_2}{R_2}.$$

Solving these equations simultaneously form α_1 and α_2, then plugging the constants back into the velocity profile gives

$$v_\theta = \frac{\omega R_2^2}{R_2^2 - R_1^2}\left(r - \frac{R_1^2}{r}\right).$$

As mentioned earlier, the pressure profile may be determined from the velocity field as

$$\frac{\partial p}{\partial x_r} = -\rho v_\theta \frac{\partial v_\theta}{\partial x_r} = -\rho\left(\frac{\omega R_2^2}{R_2^2 - R_1^2}\right)^2\left(r - \frac{R_1^2}{r}\right)\left(r + \frac{R_1^2}{r^2}\right).$$

The total torque, T, exerted by the fluid on the cup can be written as the integral of the surface traction at the interface between the fluid and the cup such that

$$T = \int_{\partial\Omega} R_2 \boldsymbol{e}_r \times \boldsymbol{\tau}_x \, dS,$$

where $\partial\Omega$ is the surface of the cup, $R_2\boldsymbol{e}_r$ is the distance vector from the center to the surface of the cup, and $\boldsymbol{\tau}_x$ is the Cauchy traction vector acting on the surface of the cup. The traction vector can be written as

$$\boldsymbol{\tau}_x = \boldsymbol{n}_x \cdot \boldsymbol{\sigma}^T = \{1 \quad 0 \quad 0\} \cdot \begin{bmatrix} -p & \mu r\left(\frac{\partial}{\partial x_r}\left(\frac{v_\theta}{r}\right)\right) & 0 \\ \mu r\left(\frac{\partial}{\partial x_r}\left(\frac{v_\theta}{r}\right)\right) & -p & 0 \\ 0 & 0 & -p \end{bmatrix}$$

$$= \left\{-p \quad \mu r\left(\frac{\partial}{\partial x_r}\left(\frac{v_\theta}{r}\right)\right) \quad 0\right\}.$$

The total torque becomes

$$\boldsymbol{T} = \int_{\partial\Omega} R_2 \mu r\left(\frac{\partial}{\partial x_r}\left(\frac{v_\theta}{r}\right)\right) \boldsymbol{e}_z \, dS.$$

Since the integrand varies only with r, and the infinitesimal surface element in cylindrical coordinates may be written as $dS = rd\theta\, dz$, we may write

$$\boldsymbol{T} = \int_0^L \int_0^{2\pi} R_2 \mu r\left(\frac{\partial}{\partial x_r}\left(\frac{v_\theta}{r}\right)\right) \boldsymbol{e}_z \, r\, d\theta\, dz = 2\pi L \mu R_2^3 \left(\frac{\partial}{\partial x_r}\left(\frac{v_\theta}{r}\right)\right)_{r=R_2} \boldsymbol{e}_z,$$

where L is the length of the cup. Substituting the velocity profile gives

$$\left(\frac{\partial}{\partial x_r}\left(\frac{v_\theta}{r}\right)\right)_{r=R_1} = \frac{2\omega R_2^2}{R_2^2 - R_1^2}\left(\frac{R_1^2}{R_2^3}\right),$$

and the total torque becomes

$$\boldsymbol{T} = 4\pi L \mu \left(\frac{\omega R_2^2 R_1^2}{R_2^2 - R_1^2}\right) \boldsymbol{e}_z.$$

By simultaneously monitoring both the angular velocity of the cup, and the torque required to spin the cup, T_z, one may determine the viscosity of the fluid from the equation

$$\mu = \frac{(R_2^2 - R_1^2)}{4\pi L \omega R_2^2 R_1^2} T_z.$$

6.7.1.1 Finite Difference Algorithm

The field equation for a Newtonian fluid in a Couette viscometer,

$$\left(\frac{\partial^2 v_\theta}{\partial x_r^2}\right) + \frac{1}{r}\frac{\partial v_\theta}{\partial x_r} - \frac{v_\theta}{r^2} = 0,$$

is a second-order linear partial differential equation. Since the only nonzero velocity component is the component along the $\boldsymbol{e}_\theta$ direction, let us say that $v = v_\theta$. The finite difference approximation for the field equation becomes

$$\frac{(v_{i+1} - 2v_i + v_{i-1})}{\Delta x^2} + \frac{1}{r_i}\frac{(v_{i+1} - v_{i-1})}{2\,\Delta x} - \frac{v_i}{r_i^2} = 0.$$

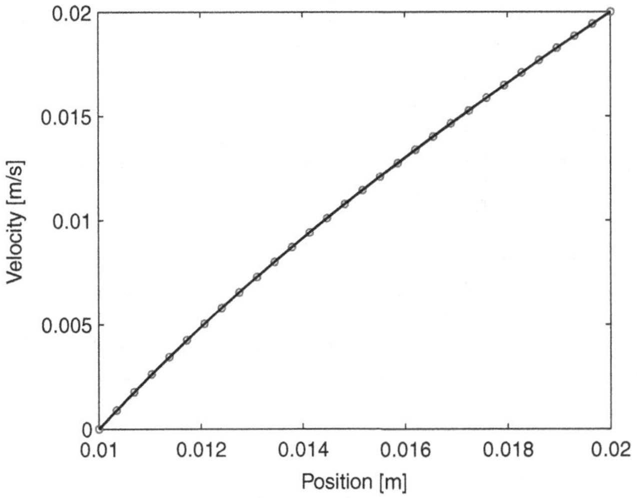

Figure 6.3. The velocity profile for a Newtonian fluid in a Couette viscometer obtained from the analytical solution and numerical finite difference approximation are shown.

This can be rewritten as

$$v_{i+1}\left(\frac{1}{\Delta x^2}+\frac{1}{2r_i\Delta x}\right)+v_i\left(-\frac{2}{\Delta x^2}-\frac{1}{r_i^2}\right)+v_{i-1}\left(\frac{1}{\Delta x^2}-\frac{1}{2r_i\Delta x}\right)=0.$$

Since we have one of these equations for each interior gridpoint, we can write the set of equations in matrix form as

$$\begin{bmatrix} \mathrm{a}_o & \mathrm{a}_1 & 0 & 0 & 0 \\ \mathrm{a}_{-1} & \mathrm{a}_o & \mathrm{a}_1 & 0 & 0 \\ 0 & \ddots & \ddots & \ddots & 0 \\ 0 & 0 & \mathrm{a}_{-1} & \mathrm{a}_o & \mathrm{a}_1 \\ 0 & 0 & 0 & \mathrm{a}_{-1} & \mathrm{a}_o \end{bmatrix}\begin{bmatrix} v_1 \\ v_2 \\ \vdots \\ v_{nDX-2} \\ v_{nDX-1} \end{bmatrix}=\begin{bmatrix} d_1-\mathrm{a}_{-1}v_o \\ d_2 \\ \vdots \\ d_{nDX-2} \\ d_{nDX-1}-\mathrm{a}_1 v_{nDX} \end{bmatrix},$$

where

$$\mathrm{a}_o=\left(-\frac{2}{\Delta x^2}-\frac{1}{r_i^2}\right),\quad \mathrm{a}_1=\left(\frac{1}{\Delta x^2}+\frac{1}{r_i\Delta x}\right),\quad \mathrm{a}_{-1}=\left(\frac{1}{\Delta x^2}-\frac{1}{r_i\Delta x}\right),\ \text{and } d_i=0.$$

6.7.1.2 Example Problem

Assume that we have a Couette viscometer as shown in Figure 6.2 with an inner and outer radius of $R_1 = 1\,\mathrm{cm}$ and $R_2 = 2\,\mathrm{cm}$. The viscometer is filled with a Newtonian fluid, which has a viscosity of $\mu = 1.0$ Pa s. We would like to determine the velocity profile within the fluid as a function of distance from the central axis if the cup is rotated at an angular frequency of $\omega_o = 1\,\mathrm{rad/s}$.

The implementation in Matlab® of the algorithm outlined in Section 6.7.1.1 can be found in Section 6.7.1.3. Thirty grid points were distributed along the $\boldsymbol{e}_r$ axis. The resulting velocity profile was shown to be in good agreement with the analytical solution obtained in Section 6.7.1 as can be seen in Figure 6.3.

6.7.1.3 Matlab® File

Newtonian Fluid

1. Initialize viscometer geometry

$$R_1 = 1\ \text{cm}$$
$$R_2 = 2\ \text{cm}$$
$$\omega_o = 1\ \text{rad/s}$$

2. Initialize grid

$$nDX = 30$$

3. Initialize the material parameters

$$\mu = 1.0\ \text{Pa}\cdot\text{s}$$

4. Allocate memory

5. Assemble the matrix

$$\mathbf{M} = \begin{bmatrix} a_o & a_1 & 0 & 0 & 0 \\ a_{-1} & a_o & a_1 & 0 & 0 \\ 0 & \ddots & \ddots & \ddots & 0 \\ 0 & 0 & a_{-1} & a_o & a_1 \\ 0 & 0 & 0 & a_{-1} & a_o \end{bmatrix}$$

where

$$a_{-1} = \left(\frac{1}{\Delta x^2} - \frac{1}{2x\Delta x}\right)$$
$$a_0 = \left(-\frac{2}{\Delta x^2} - \frac{1}{x^2}\right)$$
$$a_1 = \left(\frac{1}{\Delta x^2} + \frac{1}{2x\,\Delta x}\right)$$
$$d_i = 0$$

6. Boundary conditions

$$v(R_2) = \omega_o * R_2$$
$$v(R_1) = 0$$
$$d(R_1) = -a_{-1} * v(R_1)$$
$$d(R_2) = -a_1 * v(R_2)$$

7. Account for the boundary conditions in the *d* vector

$$d_1 = d_1 - a_{-1}p_0$$
$$d_{nDX-1} = d_{nDX-1} - a_1 p_{nDX}$$

8. Solve for the velocity at each node using

$$\boldsymbol{v} = \boldsymbol{M}^{-1}\cdot\boldsymbol{d}$$

```
% ------------------------------------------------------------------------
% Numerical solution for the steady state velocity profile of a Newtonian
% fluid in a Couette viscometer.
% ------------------------------------------------------------------------
%   v(i)  [m/s] is a vector containing the velocity at node i.
% ------------------------------------------------------------------------
clear;                        % clear any stored variables
% ------------------------------------------------------------------------
% 1. Setup viscometer geometry
R1       = 0.01;             % [m]
R2       = 0.02;             % [m]
omega_o  = 1;                % [rad/s] angular velocity of the cup
% ------------------------------------------------------------------------
% 2. Initialize the spatial grid
nDX = 30;                     % number of grid points along the X axis
dX  = (R2-R1)/(nDX-1);        % [m] spacing between grid points
X   = [R1:dX:R2]';            % [m] positions of each grid point
% ------------------------------------------------------------------------
% 3. Initialize material parameters
mu    =    1.0;               % [Pa s] material viscosity
% ------------------------------------------------------------------------
% 4. Allocate memory for the M, b, and dP matrix to speed up calculation.
M(nDX-2,nDX-2) = 0;
v(nDX,1)       = 0;
d(nDX-2,1)     = 0;
% ------------------------------------------------------------------------
% 5. Assemble the M matrix and the d vector
am1 = ( 1/dX^2 - 1./(2*X.*dX));
a0  = (-2/dX^2 - 1./(X.^2));
ap1 = ( 1/dX^2 + 1./(2*X.*dX));
for i = 2:nDX-1
    index0 = i-2;
    index1 = i-1;
    index2 = i;

    M(index1, index1) = a0(i);
    if (i ~= 2)     M(index1, index0) = am1(i); end
    if (i ~= nDX-1) M(index1, index2) = ap1(i); end

    d(index1) = 0;
end
% ------------------------------------------------------------------------
% 6. Boundary Conditions
v(1)     = 0;
d(1)     = -am1(1)*v(1);
v(nDX)   = omega_o*X(nDX);
d(nDX-2) = -ap1(nDX-1)*v(nDX);

% ------------------------------------------------------------------------
% 7.  Solve for the velocity
```

```
v(2:nDX-1) = M\d;
% --------------------------------------------------------------------
% Analytical Solution
vAnalytical = omega_o * R2^2/(R2^2-R1^2).*(X-R1^2./X);
% The error between the numerical and analytical solution
error       = sum((v -vAnalytical).^2)/sum(v.^2);
```

6.7.2 Power Law Fluid Model

In the previous section, we derived the equations relating the applied torque to the angular velocity of the viscometer when the fluid being tested was a Newtonian fluid. In addition, we solved for the velocity profile within the viscometer for a given angular velocity. These relations may also be found for non-Newtonain fluids. In fact, when presented with a novel fluid, the relationship between applied torque and angular velocity obtained from a viscometer may be used to determine the appropriate material model for the given fluid. In this section, we will derive the relationship between applied torque and angular velocity for fluid that can be modeled using a power law.

The power law model for a non-Newtonian fluid states that the viscosity is related to the second invariant of the strain rate tensor such that

$$\mu = K\, II_{\boldsymbol{D}}^{(n-1)/2},$$

where K and n are two material parameters that must be determined experimentally. In this section, we will derive the field equations for a non-Newtonian fluid that obeys the power law being sheared in the Couette viscometer. Borrowing results from the previous section, we know that the only nonzero velocity component is in the $\boldsymbol{e}_\theta$ direction,

$$v_r = v_z = 0 \quad \text{and} \quad v_\theta = v_\theta(r).$$

In cylindrical coordinates, the fluid acceleration, the velocity gradient, and the strain rate tensor may be written, respectively, as

$$\dot{\boldsymbol{v}} = \begin{bmatrix} v_\theta \frac{\partial v_\theta}{\partial x_r} \\ 0 \\ 0 \end{bmatrix}, \quad \nabla \boldsymbol{v} = \begin{bmatrix} 0 & \frac{\partial v_\theta}{\partial x_r} & 0 \\ -\frac{v_\theta}{r} & 0 & 0 \\ 0 & 0 & 0 \end{bmatrix}, \quad \text{and} \quad \boldsymbol{D} = \frac{1}{2}\begin{bmatrix} 0 & \left(\frac{\partial v_\theta}{\partial x_r} - \frac{v_\theta}{r}\right) & 0 \\ \left(\frac{\partial v_\theta}{\partial x_r} - \frac{v_\theta}{r}\right) & 0 & 0 \\ 0 & 0 & 0 \end{bmatrix}.$$

The divergence of the velocity gradient is given by

$$\nabla \cdot \boldsymbol{L} = \begin{bmatrix} 0 \\ -\left(\frac{\partial^2 v_\theta}{\partial x_r^2}\right) - \frac{1}{r}\frac{\partial v_\theta}{\partial x_r} + \frac{v_\theta}{r^2} \\ 0 \end{bmatrix}.$$

The Cauchy stress for a non-Newtonian fluid is given by

$$\begin{aligned} \boldsymbol{\sigma} &= -p\boldsymbol{I} + 2\mu\boldsymbol{D} \\ &= \begin{bmatrix} -p & \mu\left(\frac{\partial v_\theta}{\partial x_r} - \frac{v_\theta}{r}\right) & 0 \\ \mu\left(\frac{\partial v_\theta}{\partial x_r} - \frac{v_\theta}{r}\right) & -p & 0 \\ 0 & 0 & -p \end{bmatrix}. \end{aligned}$$

Neglecting gravity, the Navier-Stokes equation for an incompressible non-Newtonian fluid becomes

$$\rho\dot{\boldsymbol{v}} + \nabla p - 2(\nabla\mu)\cdot\boldsymbol{D} - \mu\nabla\cdot\nabla v = \boldsymbol{0}.$$

This is similar to the Navier-Stokes equation for an incompressible Newtonian fluid with the addition of the $2(\nabla\mu)\cdot\boldsymbol{D}$ term. Due to the symmetry of the deformation, the gradient of the viscosity becomes

$$\nabla\mu = \begin{bmatrix} \dfrac{\partial\mu}{\partial x_r} \\ \dfrac{1}{r}\dfrac{\partial\mu}{\partial x_\theta} \\ \dfrac{\partial\mu}{\partial x_z} \end{bmatrix} = \begin{bmatrix} \dfrac{\partial\mu}{\partial x_r} \\ 0 \\ 0 \end{bmatrix},$$

and the additional term in the Navier-Stokes equation becomes

$$2(\nabla\mu)\cdot\boldsymbol{D} = \begin{bmatrix} \frac{\partial\mu}{\partial x_r} & 0 & 0 \end{bmatrix} \begin{bmatrix} 0 & \left(\frac{\partial v_\theta}{\partial x_r} - \frac{v_\theta}{r}\right) & 0 \\ \left(\frac{\partial v_\theta}{\partial x_r} - \frac{v_\theta}{r}\right) & 0 & 0 \\ 0 & 0 & 0 \end{bmatrix}$$

$$= \begin{bmatrix} 0 \\ \frac{\partial\mu}{\partial x_r}\left(\frac{\partial v_\theta}{\partial x_r} - \frac{v_\theta}{r}\right) \\ 0 \end{bmatrix}.$$

Substituting the velocity gradient tensor, the acceleration vector, and the gradient of the viscosity gives the following set of equations

$$\rho\begin{bmatrix} v_\theta\frac{\partial v_\theta}{\partial x_r} \\ 0 \\ 0 \end{bmatrix} + \begin{bmatrix} \frac{\partial p}{\partial x_r} \\ 0 \\ 0 \end{bmatrix} - \mu\begin{bmatrix} 0 \\ -\left(\frac{\partial^2 v_\theta}{\partial x_r^2}\right) - \frac{1}{r}\frac{\partial v_\theta}{\partial x_r} + \frac{v_\theta}{r^2} \\ 0 \end{bmatrix} - \begin{bmatrix} 0 \\ \frac{\partial\mu}{\partial x_r}\left(\frac{\partial v_\theta}{\partial x_r} - \frac{v_\theta}{r}\right) \\ 0 \end{bmatrix} = \begin{bmatrix} 0 \\ 0 \\ 0 \end{bmatrix}.$$

This represents two equations,

$$\rho v_\theta\frac{\partial v_\theta}{\partial x_r} + \frac{\partial p}{\partial x_r} = 0$$

and

$$\mu\left(\frac{\partial^2 v_\theta}{\partial x_r^2}\right) + \frac{\mu}{r}\frac{\partial v_\theta}{\partial x_r} - \mu\frac{v_\theta}{r^2} - \frac{\partial\mu}{\partial x_r}\left(\frac{\partial v_\theta}{\partial x_r} - \frac{v_\theta}{r}\right) = 0.$$

The derivative of the viscosity can be found from

$$\frac{\partial\mu}{\partial x_r} = \frac{\partial}{\partial x_r}\left(K\,II_{\boldsymbol{D}}^{(n-1)/2}\right) = \frac{(n-1)}{2}K\,II_{\boldsymbol{D}}^{\frac{(n-3)}{2}}\frac{\partial II_{\boldsymbol{D}}}{\partial x_r} = \mu\left(\frac{(n-1)}{2}II_{\boldsymbol{D}}^{-1}\frac{\partial II_{\boldsymbol{D}}}{\partial x_r}\right).$$

Since

$$II_{\boldsymbol{D}} = \frac{1}{2}\left[(tr\,\boldsymbol{D})^2 - tr\left(\boldsymbol{D}^2\right)\right] = \frac{1}{2}\left(tr\left(\boldsymbol{D}^2\right)\right) = \frac{1}{8}\left(\frac{\partial v_\theta}{\partial x_r} - \frac{v_\theta}{r}\right)^2,$$

we can write

$$\frac{\partial II_{\boldsymbol{D}}}{\partial x_r} = \frac{1}{4}\left(\frac{\partial v_\theta}{\partial x_r} - \frac{v_\theta}{r}\right)\left(\frac{\partial^2 v_\theta}{\partial x_r^2} - \frac{1}{r}\frac{\partial v_\theta}{\partial x_r} + \frac{v_\theta}{r^2}\right).$$

The derivative of the viscosity becomes

$$\frac{\partial\mu}{\partial x_r} = \mu\frac{(n-1)}{2}2\left(\frac{\partial v_\theta}{\partial x_r} - \frac{v_\theta}{r}\right)^{-1}\left(\frac{\partial^2 v_\theta}{\partial x_r^2} - \frac{1}{r}\frac{\partial v_\theta}{\partial x_r} + \frac{v_\theta}{r^2}\right).$$

In terms of the velocities, the field equation becomes

$$\left(\frac{\partial^2 v_\theta}{\partial x_r^2}\right)+\frac{1}{r}\frac{\partial v_\theta}{\partial x_r}-\frac{v_\theta}{r^2}-\frac{(n-1)}{2}2\left(\frac{\partial v_\theta}{\partial x_r}-\frac{v_\theta}{r}\right)^{-1}$$
$$\times\left(\frac{\partial^2 v_\theta}{\partial x_r^2}-\frac{1}{r}\frac{\partial v_\theta}{\partial x_r}+\frac{v_\theta}{r^2}\right)\left(\frac{\partial v_\theta}{\partial x_r}-\frac{v_\theta}{r}\right)=0.$$

Simplifying this equation gives

$$(2-n)\left(\frac{\partial^2 v_\theta}{\partial x_r^2}\right)+n\frac{1}{r}\frac{\partial v_\theta}{\partial x_r}-n\frac{v_\theta}{r^2}=0.$$

6.7.2.1 Finite Difference Algorithm

The field equation for the power law fluid in a Couette viscometer is given by

$$(n-2)\left(\frac{\partial^2 v_\theta}{\partial x_r^2}\right)-n\frac{1}{r}\frac{\partial v_\theta}{\partial x_r}+n\frac{v_\theta}{r^2}=0.$$

Again the only nonzero component of the velocity is the component parallel to the $\boldsymbol{e}_\theta$ direction. Therefore, we set $v=v_\theta$ and use the finite difference approximations to write

$$(n-2)\frac{(v_{i+1}-2v_i+v_{i-1})}{\Delta x^2}-n\frac{1}{r_i}\frac{(v_{i+1}-v_{i-1})}{2\Delta x}+n\frac{v_i}{r_i^2}=0.$$

Gathering terms, we find that

$$v_{i+1}\left((n-2)\frac{1}{\Delta x^2}-n\frac{1}{2r_i\Delta x}\right)+v_i\left(-(n-2)\frac{2}{\Delta x^2}+n\frac{1}{r_i^2}\right)$$
$$+v_{i-1}\left((n-2)\frac{1}{\Delta x^2}+n\frac{1}{2r_i\Delta x}\right)=0.$$

Writing the set of equations in matrix form gives

$$\begin{bmatrix} \mathrm{a_o} & \mathrm{a_1} & 0 & 0 & 0 \\ \mathrm{a_{-1}} & \mathrm{a_o} & \mathrm{a_1} & 0 & 0 \\ 0 & \ddots & \ddots & \ddots & 0 \\ 0 & 0 & \mathrm{a_{-1}} & \mathrm{a_o} & \mathrm{a_1} \\ 0 & 0 & 0 & \mathrm{a_{-1}} & \mathrm{a_o} \end{bmatrix}\begin{bmatrix} v_1 \\ v_2 \\ \vdots \\ v_{nDX-2} \\ v_{nDX-1} \end{bmatrix}=\begin{bmatrix} d_1-\mathrm{a_{-1}}v_o \\ d_2 \\ \vdots \\ d_{nDX-2} \\ d_{nDX-1}-\mathrm{a_1}v_{nDX} \end{bmatrix},$$

where

$$\mathrm{a}_o=\left(-(n-2)\frac{2}{\Delta x^2}+n\frac{1}{r_i^2}\right),$$

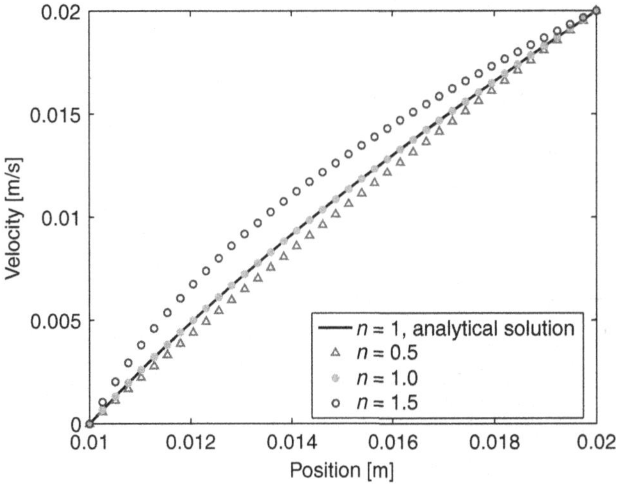

Figure 6.4. Velocity profile for a power law fluid with $n = 1.0$, $n = 1.5$, and $n = 0.5$. $K = 1.0$ $[Pa s]n = 0.5$, $n = 1.0$, and $n = 1.5$. $K = 1.0$ $[Pa s]$ in all cases.

$$a_1 = \left((n-2)\frac{1}{\Delta x^2} - n\frac{1}{2r_i\Delta x}\right),$$

$$a_{-1} = \left((n-2)\frac{1}{\Delta x^2} + n\frac{1}{2r_i\Delta x}\right),$$

$$d_i = 0.$$

6.7.2.2 Example Problem

Assume that we have a Couette viscometer as shown in Figure 6.2 with $R_1 = 1\,\mathrm{cm}$ and $R_2 = 2\,\mathrm{cm}$. The viscometer is filled with a three separate fluids. The viscosity of these three fluids may be modeled with a power law. We will refer to these fluids as the shear thinning fluid ($K = 1.0\,\mathrm{Pa\,s}, n = 0.5$), the Newtonian fluid ($K = 1.0\,\mathrm{Pa\,s}, n = 1.0$), and the shear thickening fluid ($K = 1.0\,\mathrm{Pa\,s}, n = 1.5$). We would like to determine the velocity profile if the cup is rotated at an angular velocity of $\omega_o = 1\,\mathrm{rad/s}$. We would also like to determine the torque needed to maintain a given angular velocity as a function of angular velocity.

The implementation in Matlab® of the algorithm outlined in Section 6.7.2.2 can be found in Section 6.7.2.3. Forty gridpoints were distributed along the $\boldsymbol{e}_r$ axis. The resulting velocity profile for the Newtonian fluid was shown to be in good agreement with the analytical solution obtained in Section 6.7.1 as can be seen in Figure 6.3. The torque needed to maintain an angular velocity for the Newtonian fluid is proportional to the angular velocity. However, the effective viscosity of the shear thinning fluid decreases with increasing angular velocity. Therefore, the torque required to maintain a given angular velocity grows at a slower pace than that for the Newtonian fluid. Similarly, the effective viscosity of the shear thickening fluid increases with increasing angular velocity. Therefore, the torque required to maintain a given angular velocity grows at a much faster rate than the Newtonian fluid. Most importantly, we see that the dependence of torque on angular velocity may be used to identify whether a fluid is Newtonian, shear thickening, or shear thinning.

6.7.2.3 Matlab® File

Power Law Fluid

1. Initialize viscometer geometry

$$R_1 = 1\text{ cm}$$
$$R_2 = 2\text{ cm}$$

2. Initialize grid

$$nDX = 250$$

3. Initialize the material parameters

$$K = 1.0\text{ Pa}\cdot\text{s}$$
$$n = 1.5$$

4. Allocate memory

5. Assemble the matrix

$$\mathbf{M} = \begin{bmatrix} a_o & a_1 & 0 & 0 & 0 \\ a_{-1} & a_o & a_1 & 0 & 0 \\ 0 & \ddots & \ddots & \ddots & 0 \\ 0 & 0 & a_{-1} & a_o & a_1 \\ 0 & 0 & 0 & a_{-1} & a_o \end{bmatrix}$$

where

$$a_{-1} = \left(\frac{(n-2)}{\Delta x^2} + \frac{n}{2x\Delta x}\right)$$

$$a_0 = \left(-\frac{2\,(n-2)}{\Delta x^2} + \frac{n}{x^2}\right)$$

$$a_1 = \left(\frac{(n-2)}{\Delta x^2} - \frac{n}{2x\Delta x}\right)$$

6. Boundary conditions

$$v(R_2) = \omega_o * R_2$$
$$v(R_1) = 0$$
$$d(R_1) = -a_{-1} * v(R_1)$$
$$d(R_2) = -a_1 * v(R_2)$$

7. Solve for the velocity at each node using

$$\boldsymbol{v} = \boldsymbol{M}^{-1}\cdot\boldsymbol{d}$$

```
% ---------------------------------------------------------------------
% Numerical solution for the steady state velocity profile of a Power Law
% fluid in a Couette viscometer.
% ---------------------------------------------------------------------
%   v(i)  [m/s] is a vector containing the velocity at node i.
% ---------------------------------------------------------------------
clear;                          % clear any stored variables
% ---------------------------------------------------------------------
% 1. Setup viscometer geometry
R1       = 0.01;               % [m]
R2       = 0.02;               % [m]
omega_o  = 1;                  % [rad/s] angular velocity of the cup
% ---------------------------------------------------------------------
% 2. Initialize the spatial grid
nDX = 250;                      % number of grid points along the X axis
dX  = (R2-R1)/(nDX-1);          % [m] spacing between grid points
X   = [R1:dX:R2]';              % [m] positions of each grid point
% ---------------------------------------------------------------------
% 3. Initialize material parameters
K   = 1.0;                      % [Pa s] Power law fluid viscosity
n   = 1.5;                      % [ ]    Power law exponent
% ---------------------------------------------------------------------
% 4. Allocate memory for the M, b, and dP matrix to speed up calculation.
M(nDX-2,nDX-2) = 0;
v(nDX,1)       = 0;
d(nDX-2,1)     = 0;
% ---------------------------------------------------------------------
% 5. Assemble the M matrix and the d vector
am1 = (    (n-2)/dX^2 + (n+0)./(2*X*dX));
a0  = (-2*(n-2)/dX^2 + (n+0)./(X.^2));
ap1 = (    (n-2)/dX^2 - (n+0)./(2*X*dX));
for i = 2:nDX-1
    M(i-1, i-1) = a0(i);
    if (i ~= 2)      M(i-1, i-2) = am1(i); end
    if (i ~= nDX-1) M(i-1, i)   = ap1(i); end
    d(i-1) = 0;
end
% ---------------------------------------------------------------------
% 6. Boundary Conditions
v(1)      = 0;
d(1)      = -am1(1)*v(1);
v(nDX)    = omega_o*X(nDX);
d(nDX-2) = -ap1(nDX-1)*v(nDX);
% ---------------------------------------------------------------------
% 7.  Solve for the velocity
v(2:nDX-1) = M\d;
```

6.7.3 General Non-Newtonian Fluid

In the previous section, we were able to determine the relationship between applied torque and angular velocity within a viscometer for a power law fluid. Luckily the resulting differential equation was linear. For more complex fluid material models, the governing differential equation may prove to be nonlinear. In such cases, it is necessary to use an iterative numerical method to find a solution. In this section, we will derive the general differential equation governing fluid flow within the Couette viscometer regardless of material model. For this purpose, we assume some unknown dependence of fluid viscosity on the deformation rate, $\mu = \mu(\boldsymbol{D})$. We also outline the quasi-linearization method which can be used to solve the resulting nonlinear differential equation. The implementation of the nonlinear iterative scheme is given for the case of the power law fluid.

6.7.3.1 Finite Difference Algorithm

In Section 6.7.2.1, the field equation for a non-Newtonian fluid in a Couette viscometer was found to be the nonlinear second-order partial differential equation given by

$$\mu\left(\frac{\partial^2 v_\theta}{\partial x_r^2}\right) + \frac{\mu}{r}\frac{\partial v_\theta}{\partial x_r} - \mu\frac{v_\theta}{r^2} - \frac{\partial \mu}{\partial x_r}\left(\frac{\partial v_\theta}{\partial x_r} - \frac{v_\theta}{r}\right) = 0,$$

where the general form the viscosity has not yet been specified and $\mu = \mu(v_\theta)$. In the general form, this equation cannot be solved by direct application of the finite difference approximations. Instead, we will solve the equation by using the quasi-linearization method and then iteratively solving for the velocity. Once again for simplicity, we will assume that $v = v_\theta$ and $x = x_r$. We will also use the notation

$$v'' = \frac{\partial^2 v}{\partial x^2} \quad \text{and} \quad v' = \frac{\partial v}{\partial x}.$$

Any nonlinear second-order PDE, G, can be written in the form

$$G = f\left(\frac{\partial^2 v}{\partial x^2}, \frac{\partial v}{\partial x}, v\right) = f\left(v'', v', v\right) = 0.$$

Let us select an initial guess for the solution denoted by $\left\{v''^{(k)}, v'^{(k)}, v^{(k)}\right\}$. Then the Taylor expansion of the PDE can be written in terms of the initial guess and the actual solution, $\left\{v''^{(k+1)}, v'^{(k+1)}, v^{(k+1)}\right\}$, such that

$$f\left(v''^{(k)}, v'^{(k)}, v^{(k)}\right) + \left(\frac{\partial f}{\partial v''}\right)^{(k)}\left(v''^{(k+1)} - v''^{(k)}\right) + \left(\frac{\partial f}{\partial v'}\right)^{(k)}\left(v'^{(k+1)} - v'^{(k)}\right)$$
$$+\left(\frac{\partial f}{\partial v}\right)^{(k)}\left(v^{(k+1)} - v^{(k)}\right) = 0.$$

Note that this would be exact if the actual solution was infinitesimaly close the initial guess. However, in reality this will not be the case and the derivatives are not constant between k and $k+1$. The solution at $k+1$ will simply be a better guess than the initial solution at k.

Rewriting the equations gives

$$v''^{(k+1)}\left(\frac{\partial f}{\partial v''}\right)^{(k)} + v'^{(k+1)}\left(\frac{\partial f}{\partial v'}\right)^{(k)} + v^{(k+1)}\left(\frac{\partial f}{\partial v}\right)^{(k)}$$
$$= \left(\frac{\partial f}{\partial v''}\right)^{(k)} v''^{(k)} + \left(\frac{\partial f}{\partial v'}\right)^{(k)} v'^{(k)} + \left(\frac{\partial f}{\partial v}\right)^{(k)} v^{(k)} - f\left(v''_k, v'_k, v_k\right).$$

Applying the finite difference approximation to the next iteration gives

$$\frac{\left(v_{i+1}^{k+1}-2v_i^{k+1}+v_{i-1}^{k+1}\right)}{\Delta x^2}\left(\frac{\partial f}{\partial v''}\right)^{(k)}+\frac{\left(v_{i+1}^{k+1}-v_{i-1}^{k+1}\right)}{2\,\Delta x}\left(\frac{\partial f}{\partial v'}\right)^{(k)}+v_i^{(k+1)}\left(\frac{\partial f}{\partial v}\right)^{(k)}$$
$$=\left(\frac{\partial f}{\partial v''}\right)^{(k)}v''^{(k)}+\left(\frac{\partial f}{\partial v'}\right)^{(k)}v'^{(k)}+\left(\frac{\partial f}{\partial v}\right)^{(k)}v^{(k)}-f\left(v_k'',v_k',v_k\right).$$

Rearranging this gives

$$v_{i+1}^{k+1}\mathrm{a}_{+1}+v_i^{k+1}\mathrm{a}_0+v_{i-1}^{k+1}\mathrm{a}_{-1}=d_i.$$

where the coefficients are defined by

$$\mathrm{a}_1=\left(\frac{1}{\Delta x^2}\left(\frac{\partial f}{\partial v''}\right)^{(k)}+\frac{1}{2\,\Delta x}\left(\frac{\partial f}{\partial v'}\right)^{(k)}\right),$$
$$\mathrm{a}_o=\left(-2\frac{1}{\Delta x^2}\left(\frac{\partial f}{\partial v''}\right)^{(k)}+\left(\frac{\partial f}{\partial v}\right)^{(k)}\right),$$
$$\mathrm{a}_{-1}=\left(\frac{1}{\Delta x^2}\left(\frac{\partial f}{\partial v''}\right)^{(k)}-\frac{1}{2\Delta x}\left(\frac{\partial f}{\partial v'}\right)^{(k)}\right),$$
$$d_i=\left(\frac{\partial f}{\partial v''}\right)^{(k)}v''^{(k)}+\left(\frac{\partial f}{\partial v'}\right)^{(k)}v'^{(k)}+\left(\frac{\partial f}{\partial v}\right)^{(k)}v^{(k)}-f\left(v_k'',v_k',v_k\right).$$

where the partial differential equation is defined as

$$f\left(v'',v',v\right)=\mu\left(\frac{\partial^2 v_\theta}{\partial x_r^2}\right)+\frac{\mu}{r}\frac{\partial v_\theta}{\partial x_r}-\mu\frac{v_\theta}{r^2}-\frac{\partial\mu}{\partial x_r}\left(\frac{\partial v_\theta}{\partial x_r}-\frac{v_\theta}{r}\right)$$

and

$$\frac{\partial f}{\partial v''}=\frac{\partial\mu}{\partial v''}\left(\frac{\partial^2 v_\theta}{\partial x_r^2}+\frac{1}{r}\frac{\partial v_\theta}{\partial x_r}-\frac{v_\theta}{r^2}\right)-\frac{\partial^2\mu}{\partial v''\partial x_r}\left(\frac{\partial v_\theta}{\partial x_r}-\frac{v_\theta}{r}\right)+\mu,$$
$$\frac{\partial f}{\partial v'}=\frac{\partial\mu}{\partial v'}\left(\frac{\partial^2 v_\theta}{\partial x_r^2}+\frac{1}{r}\frac{\partial v_\theta}{\partial x_r}-\frac{v_\theta}{r^2}\right)-\frac{\partial^2\mu}{\partial v'\partial x_r}\left(\frac{\partial v_\theta}{\partial x_r}-\frac{v_\theta}{r}\right)+\frac{\mu}{r}-\frac{\partial\mu}{\partial x_r},$$
$$\frac{\partial f}{\partial v}=\frac{\partial\mu}{\partial v}\left(\frac{\partial^2 v_\theta}{\partial x_r^2}+\frac{1}{r}\frac{\partial v_\theta}{\partial x_r}-\frac{v_\theta}{r^2}\right)-\frac{\partial^2\mu}{\partial v\partial x_r}\left(\frac{\partial v_\theta}{\partial x_r}-\frac{v_\theta}{r}\right)-\frac{\mu}{r^2}+\frac{1}{r}\frac{\partial\mu}{\partial x_r}.$$

Once again, we have one equation per interior gridpoint. The set of equations may be written in matrix form as

$$\begin{bmatrix} a_o & a_1 & 0 & 0 & 0 \\ a_{-1} & a_o & a_1 & 0 & 0 \\ 0 & \ddots & \ddots & \ddots & 0 \\ 0 & 0 & a_{-1} & a_o & a_1 \\ 0 & 0 & 0 & a_{-1} & a_o \end{bmatrix} \begin{bmatrix} v_1^{(k)} \\ v_2^{(k)} \\ \vdots \\ v_{nDX-2}^{(k)} \\ v_{nDX-1}^{(k)} \end{bmatrix} = \begin{bmatrix} d_1 - a_{-1} v_o \\ d_2 \\ \vdots \\ d_{nDX-2} \\ d_{nDX-1} - a_1 v_{nDX} \end{bmatrix}. \quad (6.19)$$

We now have a procedure for iteratively improving the solution for the velocity profile. Notice the coefficients a_o, a_1, a_{-1}, and d_1 to d_{nDX} each depend on the previous solution. The basic procedure consists of (1) assume an initial solution, (2) use the initial solution to compute the coefficients in Equation 6.19 (3) solve Equation 6.19 to obtain an improved solution, (4) compute new coefficients using the improved solution and return to step 3 while monitoring the solution for convergence.

6.7.3.2 Example Problem

Assume that we have a Couette viscometer as shown in Figure 6.2 with $R_1 = 1\,\text{cm}$ and $R_2 = 2\,\text{cm}$. The viscometer is filled a non-Newtonain fluid. While the methods outlined in Section 6.7.3.1 may be applied to any general fluid model, we will assume that the non-Newtonian fluid may be modeled as a power law fluid. This allows us to compare the quasi-linear algorithm to the finite difference algorithm obtained for the power law fluid in Section 6.7.2.1. The parameters for the power law fluid are given as $K = 1.0$ Pa s and $n = 1.5$.

The implementation in Matlab® of the quasi-linear algorithm outlined in Section 6.7.3.1 can be found in Section 6.7.3.3. Forty gridpoints were distributed along the $\boldsymbol{e}_r$ axis. As expected, the velocity profile matches that found using the direct implementation of the finite difference method for the power law fluid. Even though there is no advantage to using the quasi-linear method for the power law fluid, the quasi-linear method may be used for more detailed models of viscosity including models for which a direct implementation is not possible.

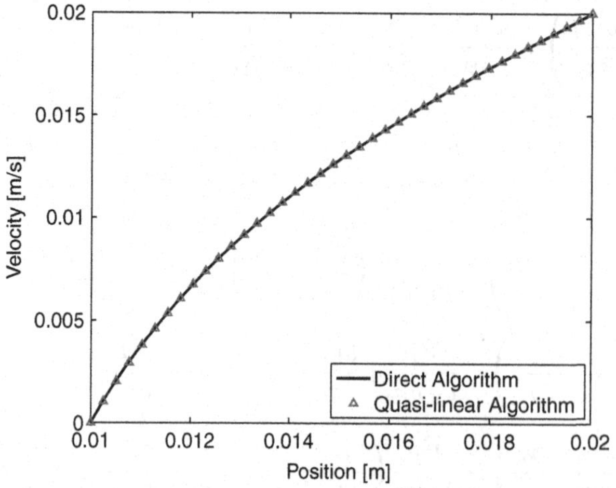

Figure 6.5. Velocity profile found for a power law fluid using a general form of the quasilinear method for non-Newtonian fluids and a direct implementation of the finite difference method to the power law fluid.

6.7.3.3 Matlab® File

Quasi-linear Finite Difference Solution

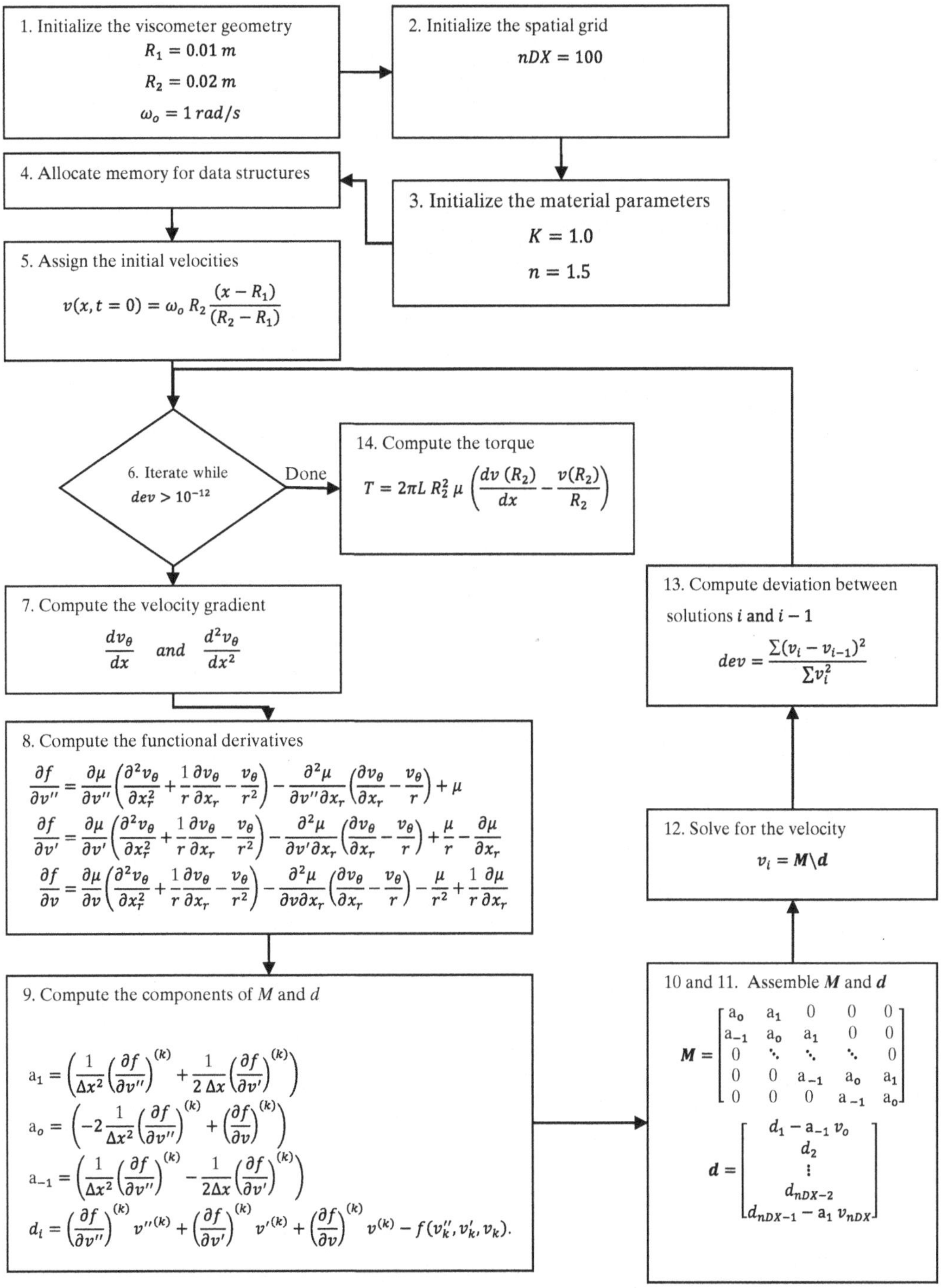

```
% ------------------------------------------------------------------------
% Numerical solution for the steady state velocity profile of a power law
% fluid modeled in a Couette viscometer obtained by using the quasilinear
% approximation of the governing PDE.
% ------------------------------------------------------------------------
%   v(i)  [m/s] is a vector containing the velocity at node i.
% ------------------------------------------------------------------------
clear;                          % clear any stored variables
clf;
% ------------------------------------------------------------------------
% 1. Setup viscometer geometry
R1        = 0.01;             % [m]
R2        = 0.02;             % [m]
omega_o   = 1;                % [rad/s] angular velocity of the cup
Length    = 0.06;             % [m]     length of the bob
% ------------------------------------------------------------------------
% 2. Initialize the spatial grid
nDX = 100;                      % number of grid points along the X axis
dX  = (R2-R1)/(nDX-1);        % [m] spacing between grid points
X   = [R1:dX:R2]';            % [m] positions of each grid point
% ------------------------------------------------------------------------
% 3. Initialize material parameters
K   = 1.0;                      % [Pa s] Power law fluid viscosity
n   = 1.5;                      % [ ]    Power law exponent
% ------------------------------------------------------------------------
% 4. Allocate memory for the M, d, and dP matrix to speed up calculation.
M(nDX-2,nDX-2) = 0;
vn(nDX,1)      = 0;
d(nDX-2,1)     = 0;
% ------------------------------------------------------------------------
% 5. use a linear velocity profile for an initial guess
v              = omega_o * R2 * (X-R1)./(R2-R1);
% ------------------------------------------------------------------------
% 6. iterate to find the true solution
deviation(1) = 1;
iteration    = 1;
while (deviation(iteration) > 10^-12)
    % --------------------------------------------------------------------
    % 7. compute the derivatives of the velocity
    dvdx   = gradient(v,    dX);
    d2vdx2 = gradient(dvdx, dX);
    % --------------------------------------------------------------------
    % 8. compute the functional derivatives
    IID        = 1/8*(dvdx-v./X).^2;
    dIIDdx     = 1/8*2*(dvdx-v./X).*(d2vdx2-1./X.*dvdx+v./X.^2);
    dIIDdv     = 1/8*2*(dvdx-v./X).*(-1./X);
    dIIDdv1    = 1/8*2*(dvdx-v./X).*(1);
    d2IIDdxdv1 = 1/8*2*(d2vdx2-1./X.*dvdx+v./X.^2);
    d2IIDdxdv  = 1/8*((-2./X.*(d2vdx2-1./X.*dvdx+v./X.^2) ...
                      +2*(dvdx-v./X)./X.^2));
```

```
mu         = K * IID.^((n-1)/2);
dmudIID    = K * (n-1)/2 * IID.^((n-3)/2);
d2mudIID2  = K * (n-1)/2 * (n-3)/2 * (IID.^((n-5)/2));
% -------------------------------------------------------------------
dmudx      = dmudIID.*dIIDdx;
dmudv      = dmudIID.*dIIDdv;
dmudv1     = dmudIID.*dIIDdv1;
dmudv1dx   = d2mudIID2.*dIIDdx.*dIIDdv1 + dmudIID.*d2IIDdxdv1;
dmudvdx    = d2mudIID2.*dIIDdx.*dIIDdv  + dmudIID.*d2IIDdxdv;
% -------------------------------------------------------------------
const1 = d2vdx2 + 1./X.*dvdx - v./X.^2;
const2 =  dvdx  - v./X;
dGdv   = -mu./X.^2+ 1./X.*dmudx + dmudv .*const1 - dmudvdx .*const2;
dGdv1  =  mu./X    - dmudx      + dmudv1.*const1 - dmudv1dx.*const2;
dGdv2  =  mu;
% -------------------------------------------------------------------
% 9. Compute the components of the M matrix and the d vector
ap1 = (   dGdv2/dX^2 + dGdv1/(2*dX));
a0  = (-2*dGdv2/dX^2 + dGdv);
am1 = (   dGdv2/dX^2 - dGdv1/(2*dX));
do  = (dGdv2.*d2vdx2 + dGdv1.*dvdx + dGdv.*v) ...
          - (mu.*(d2vdx2+dvdx./X-v./X.^2)-dmudx.*(dvdx-v./X));
% -------------------------------------------------------------------
% 10. Assemble the M matrix and the d vector
for i = 2:nDX-1
    index0 = i-2;
    index1 = i-1;
    index2 = i;

    M(index1, index1) = a0(i);
    if (i ~= 2)     M(index1, index0) = am1(i); end
    if (i ~= nDX-1) M(index1, index2) = ap1(i); end

    d(index1) = do(i);
end
% -------------------------------------------------------------------
% 11. Assign boundary conditions
vnew(1,1)       = 0;
vnew(nDX,1)     = omega_o*X(nDX);
d(1)     = d(1)     - am1(1)*vnew(1,1);
d(nDX-2) = d(nDX-2) - ap1(nDX-1)*vnew(nDX,1);
% -------------------------------------------------------------------
% 12.  Solve for the velocity
vnew(2:nDX-1)    = M\d;
% -------------------------------------------------------------------
% 13. Check for convergence
deviation(iteration+1) = sum((vnew-v).^2)/sum(v.^2);
v                      = vnew;
fprintf(1,'Iteration %d, Deviation %e. \n', iteration, deviation);
iteration = iteration + 1;
```

```
end
% -----------------------------------------------------------------------
% 14. Compute the torque exerted by the fluid on the cup
dvdx        = gradient(v,    dX);
viscosity   = K*(abs(dvdx(nDX)-v(nDX)/R2))^(n-1);
Torque      = 2*pi*Length* R2^2 * viscosity *(dvdx(nDX)-v(nDX)/R2);
```

7 Elastic Material Models

In this chapter, we discuss constitutive model development for elastic materials. A material is considered ***purely elastic*** if it lacks the means to internally dissipate energy during deformation. Thus, a purely elastic material will not exhibit hysteresis or rate dependence. In addition, the deformation of an elastic material is reversible. Removal of load will allow the material to regain its original shape. A thermoelastic material model was used to introduce the concepts of constitutive modeling in Chapter 4. While many of the results from the analysis in Chapter 4 will be restated here, the reader may refer back to Chapter 4 for the detailed derivations.

While first developing the general framework for finite thermoelasticity, we will also discuss specialization of the model for ***isothermal finite elasticity***, which assumes constant uniform temperature fields, ***hyperelastic materials*** for which there exists a strain energy function, and ***linear thermoelastic materials*** for which stress and strain are proportional. While the linear dependence between stress and strain works well for some materials, many engineering materials such as natural and synthetic polymers exhibit strong nonlinearity of the stress-strain curve even before yielding. This leads to a ***material nonlinearity*** in the constitutive response functions. In addition, subjecting any material model to large strains leads to a geometric nonlinearity. Large strain deformation is referred to as ***finite deformation***. Interestingly, some of these polymers may be deformed to very large strains and recover their original shape when released. In effect, they act elastically even at large deformations. At the end of the chapter, we will discuss the determination of material parameters using tensile testing methods.

7.1 Historical Perspective

Understanding the deformation of beams, trusses, and columns was critical to construction even in early times. The concepts of load and weight have been understood for quite some time, but the quantification of material properties and the mathematical formulation of constitutive models is young in comparison. Early contributors include the likes of Leonardo da Vinci (1452–1519) who is credited with developing a simple tensile testing machine for determining the strength of wires and conducting rudimentary analysis of beams and columns and Galileo Galilei (1564–1642) who conducted studies on the strength of materials determining that the strength and

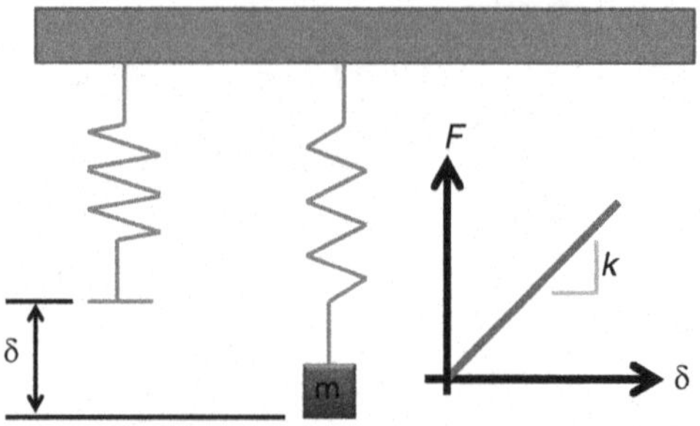

Figure 7.1. Robert Hooke found that the displacement, δ, of a spring is linearly related to the applied force, $F = mg$.

cross-sectional area of a solid are proportional. Later, in 1676, Robert Hooke published the results from a series of experiments on a set of springs showing that the applied force was proportional to the spring's elongation, $F \propto \delta x$. Whereas this early work focused on springs, later investigators found that the relationship between force and displacement in solid material structures was also linear for small displacements. In 1727 Leonhard Euler published his theory of deformation related to columns and deformable materials and developed the concept of strain energy. Later in 1807, Thomas Young, for whom the Young's modulus is named, published a work in which he demonstrated the linear relationship between stress and strain, $\sigma = E\epsilon$, where E is the Young's modulus. Augustin Louis Cauchy (1789–1857) later generalized this concept to a fully three-dimensional stress state with the stress-strain relation given by $\boldsymbol{\sigma} = \mathbb{C} \cdot \boldsymbol{\epsilon}$, where stress and strain are both second-order tensors and $\mathbb{C}$ is the fourth-order stiffness tensor.

7.2 Finite Thermoelastic Material Model

In this section, we restate many of the important developments made in Chapter 4 for the thermoelastic material model. A thermoelastic material has no mechanism for the internal dissipation of energy. As the material is deformed, energy is stored internally in the form of stretched bonds or internal molecular interactions. When the load is removed, the internal energy is released and the material returns to its initial shape. There is no hysteresis meaning that the stress-strain curve for a cyclically loaded material does not vary with the number of cycles. The response of a thermoelastic material depends both on temperature and deformation gradient. We expect internal energy and entropy changes as well as expansion or contraction in response to temperature changes, and the full material response to be dependent on material strain through a constitutive law. The model developed in this section is applicable for finite deformation. This may be a useful framework for some polymers or biological materials; however, many materials undergo plastic deformation or failure at very small strains. For these materials, a linear elastic model is more appropriate.

7.2.1 Forces and Fields

Given that we are considering only thermal effects and mechanical forces, the thermo-mechanical material is assumed to interact with the external universe via

Table 7.1. *Field variables required to specify the finite deformation response of the thermoelastic material model*

Variable	Name	Number of unknowns
$\boldsymbol{x}(\boldsymbol{X},t)$	Equations of motion	3
$\rho\,(\boldsymbol{x},t)$	Spatial density field	1
$\boldsymbol{v}\,(\boldsymbol{x},t)$	Spatial velocity field	3
$\boldsymbol{\sigma}\,(\boldsymbol{x},t)$	Cauchy stress tensor	9
$\boldsymbol{q}\,(\boldsymbol{x},t)$	Spatial heat flux	3
$e\,(\boldsymbol{x},t)$	Spatial internal energy density field	1
$\theta\,(\boldsymbol{x},t)$	Spatial temperature field	1
$\eta\,(\boldsymbol{x},t)$	Spatial entropy density field	1

Table 7.2. *Balance laws and kinematical relation for finite deformation of the thermo-mechanical material model*

		Number of equations
Kinematical relations	$\boldsymbol{v} = \dfrac{\partial \boldsymbol{x}}{\partial t}$	3
Conservation of mass	$0 = \dfrac{\partial \rho}{\partial t} + div\,(\rho\,\boldsymbol{v})$	1
Conservation of linear momentum	$\rho\,\dot{v} - div\,\boldsymbol{\sigma}^T - \rho\,\boldsymbol{b}_x = \boldsymbol{0}$	3
Conservation of angular momentum	$\boldsymbol{\sigma} = \boldsymbol{\sigma}^T$	3
Conservation of energy	$\rho\dot{e} - \boldsymbol{\sigma} : \boldsymbol{D} + div\,(\boldsymbol{q}_x) - \rho\,r_x = 0$	1
Entropy inequality	$-\rho\,\dot{\psi} - \rho\eta\dot{\theta} - \dfrac{1}{\theta}\boldsymbol{q}_x \cdot grad\,\theta + \boldsymbol{\sigma}^T : \boldsymbol{D} \geq 0$	0

body forces, surface tractions, and heat transfer. We are neglecting other forces and fields such as electrical, magnetic, and chemical effects. Thermoelastic material models are a subset of thermomechanical models.

Thermoelastic material models have a total of 22 unknown fields, which must be determined at each point within the material. These fields are listed below in Table 7.1.

7.2.2 Balance Laws

In Chapter 4, we derived the local form of the principles of conservation of mass, linear momentum, angular momentum, and energy and the entropy inequality for a general thermo-mechanical material model. Since the thermoelastic material model is a specialization of the general thermo-mechancial material model, the balance laws remain unchanged. This analysis produced a total of 11 equations which are reproduced for clarity in Table 7.2. For a full derivation of the balance laws refer back to Chapter 4.

7.2.3 Constitutive Model

For the thermo-mechanical material model subject to finite deformation, we have a total of 22 unknown field variables and 11 equations. Therefore, we require 11 constitutive equations to form a closed set of equations and to solve for these unknown fields. Given the lack of dissipative mechanisms, and the fact that the entire material response may be determined from the deformation gradient and the thermal variables, the general thermoelastic constitutive model will have the form

$$\{\psi, \boldsymbol{\sigma}, \eta, \boldsymbol{q}\} = f(\boldsymbol{F}, \theta, \nabla\theta).$$

This is the identical to the form of the constitutive model assumed in Chapter 4.

7.2.4 Constraints Due to Material Frame Indifference

For the sake of clarity, we reproduce the results obtained by applying the principle of material frame indifference to the thermo-mechanical material model constitutive equations. In Chapter, 4, we had determined that the principal of material frame indifference required that the constitutive law must obey the following relations

$$\begin{aligned}
\psi(\boldsymbol{Q}\cdot\boldsymbol{F},\theta,\boldsymbol{Q}\cdot\boldsymbol{g}) &= \psi(\boldsymbol{F},\theta,\boldsymbol{g}),\\
\eta(\boldsymbol{Q}\cdot\boldsymbol{F},\theta,\boldsymbol{Q}\cdot\boldsymbol{g}) &= \eta(\boldsymbol{F},\theta,\boldsymbol{g}),\\
\boldsymbol{\sigma}(\boldsymbol{Q}\cdot\boldsymbol{F},\theta,\boldsymbol{Q}\cdot\boldsymbol{g}) &= \boldsymbol{Q}\cdot\boldsymbol{\sigma}(\boldsymbol{F},\theta)\cdot\boldsymbol{Q}^T,\\
\boldsymbol{q}(\boldsymbol{Q}\cdot\boldsymbol{F},\theta,\boldsymbol{Q}\cdot\boldsymbol{g}) &= \boldsymbol{Q}\cdot\boldsymbol{q}(\boldsymbol{F},\theta,\boldsymbol{g}).
\end{aligned}$$

Since these constraints must be satisfied for each and every possible choice of $\boldsymbol{Q}$, we were able to select a particular rotation, $\boldsymbol{Q} = \boldsymbol{R}^T$, and write the constraint as

$$\begin{aligned}
\psi(\boldsymbol{F},\theta,\boldsymbol{g}) &= \psi\left(\boldsymbol{U},\theta,\boldsymbol{R}^T\right),\\
\eta(\boldsymbol{F},\theta,\boldsymbol{g}) &= \eta\left(\boldsymbol{U},\theta,\boldsymbol{R}^T\cdot\boldsymbol{g}\right),\\
\boldsymbol{\sigma}(\boldsymbol{F},\theta,\boldsymbol{g}) &= \boldsymbol{R}^T\cdot\boldsymbol{\sigma}\left(\boldsymbol{U},\theta,\boldsymbol{R}^T\cdot\boldsymbol{g}\right)\cdot\boldsymbol{R},\\
\boldsymbol{q}(\boldsymbol{F},\theta,\boldsymbol{g}) &= \boldsymbol{R}^T\cdot\boldsymbol{q}\left(\boldsymbol{U},\theta,\boldsymbol{R}^T\cdot\boldsymbol{g}\right),
\end{aligned}$$

where we had used the fact that $\boldsymbol{R}^T\cdot\boldsymbol{F} = \boldsymbol{U}$. These constraints require that the form of the constitutive model be reducible to a function of the right stretch tensor. The reduced form of the constitutive model can then be written as

$$\{\psi, \boldsymbol{\sigma}, \eta, \boldsymbol{q}\} = f(\boldsymbol{U}, \theta, \nabla\theta).$$

We also noted that since $\boldsymbol{U}^2 = \boldsymbol{C}$, we could equivalently write

$$\{\psi, \boldsymbol{\sigma}, \eta, \boldsymbol{q}\} = f(\boldsymbol{U}, \theta, \nabla\theta).$$

7.2.5 Constraints Due to the Second Law of Thermodynamics

In Chapter 4, we used the principle of dissipation to eliminate the temperature gradient from the Helmholtz free energy, and to show that the stress and entropy may both be determined from the Helmholtz free energy such that

$$\begin{aligned}
\psi &= \psi\left(\boldsymbol{C},\theta\right), \\
\boldsymbol{\sigma} &= \rho \frac{\partial \psi}{\partial \boldsymbol{F}} \boldsymbol{F}^T, \\
\eta &= \frac{\partial \psi}{\partial \theta}, \\
\boldsymbol{q} &= \boldsymbol{q}\left(\boldsymbol{C},\theta,\boldsymbol{g}\right).
\end{aligned}$$

Since the Helmholtz free energy is written in terms of the right Cauchy deformation tensor, $\boldsymbol{C}$, we would like to convert the derivative of the Helmholtz free energy with respect to the deformation gradient into a derivative with respect to $\boldsymbol{C}$. This is accomplished by using the chain rule such that

$$\boldsymbol{\sigma} = \rho \frac{\partial \psi}{\partial \boldsymbol{F}} \boldsymbol{F}^T = \rho \frac{\partial \psi}{\partial \boldsymbol{C}} : \frac{\partial \boldsymbol{C}}{\partial \boldsymbol{F}} \cdot \boldsymbol{F}^T.$$

Noting that $\boldsymbol{C} = \boldsymbol{F}^T \cdot \boldsymbol{F}$, we can write the forth-order tensor, $\frac{\partial \boldsymbol{C}}{\partial \boldsymbol{F}}$, as

$$\frac{\partial \boldsymbol{C}}{\partial \boldsymbol{F}} = \frac{\partial}{\partial \boldsymbol{F}} \left(\boldsymbol{F}^T \cdot \boldsymbol{F}\right) = \frac{\partial \boldsymbol{F}^T}{\partial \boldsymbol{F}} \cdot \boldsymbol{F} + \boldsymbol{F}^T \cdot \frac{\partial \boldsymbol{F}}{\partial \boldsymbol{F}}.$$

In index notation, this becomes

$$\begin{aligned}
\frac{\partial C_{qr}}{\partial F_{ik}} &= \frac{\partial F_{oq}}{\partial F_{ik}} F_{or} + F_{oq} \frac{\partial F_{or}}{\partial F_{ik}} \\
&= \delta_{oi}\delta_{qk} F_{or} + F_{oq}\delta_{oi}\delta_{rk} \\
&= \delta_{qk} F_{ir} + F_{iq}\delta_{rk}.
\end{aligned}$$

Substitution into the stress equation gives

$$\begin{aligned}
\sigma_{ij} &= \rho \frac{\partial \psi}{\partial C_{qr}} \frac{\partial C_{qr}}{\partial F_{ik}} F_{jk} \\
&= \rho \frac{\partial \psi}{\partial C_{qr}} \left(\delta_{qk} F_{ir} + F_{iq}\delta_{rk}\right) F_{jk} \\
&= \rho \frac{\partial \psi}{\partial C_{qr}} F_{ir} F_{jq} + \rho \frac{\partial \psi}{\partial C_{qr}} F_{iq} F_{jr}.
\end{aligned}$$

Since the right Cauchy deformation tensor is symmetric, we can write

$$\frac{\partial \psi}{\partial C_{qr}} = \frac{\partial \psi}{\partial C_{rq}}.$$

The Cauchy stress becomes

$$\sigma_{ij} = 2\rho F_{ir} \frac{\partial \psi}{\partial C_{qr}} F_{jq}.$$

Finally, the constitutive model for the finite deformation of a thermo-mechanical material may be written in the form

$$\boxed{\begin{aligned}&\psi = \psi(\boldsymbol{C},\theta),\\ &\boldsymbol{\sigma} = 2\rho\boldsymbol{F}\cdot\tfrac{\partial\psi}{\partial\boldsymbol{C}}\cdot\boldsymbol{F}^T,\\ &\eta = \tfrac{\partial\psi}{\partial\theta},\\ &\boldsymbol{q} = \boldsymbol{q}(\boldsymbol{C},\theta,\boldsymbol{g}).\end{aligned}}$$

7.3 Hyperelastic Material Model

If we take the model for finite deformation of a thermoelastic material and apply it to the case of isothermal deformation, where the temperature is uniform and constant at all points in the system, we may drop the temperature dependence within the Helmholtz free energy and eliminate all thermal variables from the equations. A hyperelastic material is one for which there exists a ***strain energy function***, $\mathcal{W}(\boldsymbol{F})$, which is defined as the Helmholtz free energy per unit reference volume such that

$$\mathcal{W}(\boldsymbol{F}) = \rho_o\psi(\boldsymbol{F}),$$

where ρ_o is the reference density of the material. Recall that the Helmholtz free energy has units of energy per unit mass.

Since the temperature field is prescribed, and we assume there is no heat flux or heat generation, the list of unknown quantities for the hyperelastic material model is significantly shorter than that for the thermoelastic material model. Listed in Table 7.3 are the 17 unknown quantities in the hyperelastic material model.

7.3.1 Balance Laws

The balance laws for the hyperelastic model are identical to those for the finite thermoelastic material model except that they may be simplified for the case of constant thermal fields. The conservation of energy equation is not needed, and the heat flux is removed from the CD inequality to give the equation found in Table 7.4.

7.3.2 Constitutive Model

For the isothermal hyperelastic model, we have 17 unknown variables and 10 equations. Therefore, we need only 7 constitutive equations. It is sufficient to specify the strain energy density and the stress to form a closed set of equations. However, we will later see that the stress may be determined directly from the strain energy density. Therefore, only a single equation for the strain energy density is required to specify the response of a hyperelastic material. Let us begin with a constitutive model, which has the form

$$\{\mathcal{W},\boldsymbol{\sigma}\} = f(\boldsymbol{F}).$$

Table 7.3. *Field variables required to specify the finite deformation response of the hyperelastic material model*

Variable	Name	Number of unknowns
$\boldsymbol{x}(\boldsymbol{X},t)$	Equations of motion	3
$\rho\,(\boldsymbol{x},t)$	Spatial density field	1
$\boldsymbol{v}\,(\boldsymbol{x},t)$	Spatial velocity field	3
$\boldsymbol{\sigma}\,(\boldsymbol{x},t)$	Cauchy stress tensor	9
$\mathcal{W}\,(\boldsymbol{x},t)$	Strain energy density	1

Table 7.4. *Balance laws and kinematical relation for finite deformation of the hyperelastic material model*

		Number of equations
Kinematical relations	$\boldsymbol{v}=\dfrac{\partial \boldsymbol{x}}{\partial t}$	3
Conservation of mass	$0=\dfrac{\partial \rho}{\partial t}+div\,(\rho\,\boldsymbol{v})$	1
Conservation of linear momentum	$\rho\,\dot{\boldsymbol{v}}-div\,\boldsymbol{\sigma}^T-\rho\,\boldsymbol{b}_x=\boldsymbol{0}$	3
Conservation of angular momentum	$\boldsymbol{\sigma}=\boldsymbol{\sigma}^T$	3
Entropy inequality	$-\rho\,\dot{\psi}+\boldsymbol{\sigma}^T:\boldsymbol{D}\geq 0$	0

7.3.3 Constraints Due to Material Frame Indifference

The principle of material frame indifference applied to the isothermal hyperelastic constitutive model produces the constraints that

$$\mathcal{W}\,(\boldsymbol{Q}\cdot\boldsymbol{F})=\mathcal{W}\,(\boldsymbol{F}),$$
$$\boldsymbol{\sigma}\,(\boldsymbol{Q}\cdot\boldsymbol{F})=\boldsymbol{Q}\cdot\boldsymbol{\sigma}\,(\boldsymbol{F})\cdot\boldsymbol{Q}^T.$$

Since this must hold for all possible values of the rotation, $\boldsymbol{Q}$, we are free to select a particular value $\boldsymbol{Q}=\boldsymbol{R}$. This particular choice gives the constrain that

$$\mathcal{W}\,(\boldsymbol{F})=\mathcal{W}\,(\boldsymbol{U}),$$
$$\boldsymbol{\sigma}\,(\boldsymbol{F})=\boldsymbol{R}^T\cdot\boldsymbol{\sigma}\,(\boldsymbol{U})\cdot\boldsymbol{R}.$$

Therefore, we must be able to write the strain energy density as a function of $\boldsymbol{U}$. Equivalently, the strain energy may be written as a function of $\boldsymbol{C}=\boldsymbol{U}^2$ such that

$$\{\mathcal{W},\boldsymbol{\sigma}\}=f\,(\boldsymbol{C}).$$

7.3.4 Clausius-Duhem Inequality

The Clausius-Duhem inequality may be used to further specify the form of the constitutive relations and to eliminate any inadmissible parameters in the constitutive equations. For the case of the hyperelastic material model, the temperature field has a uniform constant value. Since the heat flux and heat generation are zero, the CD inequality can then be written as

$$-\rho\dot{\psi}+\boldsymbol{\sigma}^T:\boldsymbol{D}\geq 0.$$

Substituting $\mathcal{W}=\rho_o\,\psi$, we can write this equation in terms of the strain energy density

$$-J^{-1}\dot{\mathcal{W}}+\boldsymbol{\sigma}^T:\boldsymbol{D}\geq 0.$$

Since $\dot{\mathcal{W}}(\boldsymbol{F})=\frac{\partial\mathcal{W}}{\partial\boldsymbol{F}}:\frac{\partial\boldsymbol{F}}{\partial t}$, we may write

$$\left(-J^{-1}\frac{\partial\mathcal{W}}{\partial\boldsymbol{F}}\boldsymbol{F}^T+\boldsymbol{\sigma}^T\right):\boldsymbol{D}\geq 0.$$

Because this equation must hold for all thermodynamically admissible variations of $\boldsymbol{D}$, we obtain the constraint

$$\boldsymbol{\sigma}=J^{-1}\frac{\partial\mathcal{W}}{\partial\boldsymbol{F}}\cdot\boldsymbol{F}^T.$$

If we choose to write the constitutive equations in terms of $\boldsymbol{C}$, this produces the constraint that

$$\boldsymbol{\sigma}=\frac{1}{J}\frac{\partial\mathcal{W}(\boldsymbol{C})}{\partial\boldsymbol{C}}:\frac{\partial\boldsymbol{C}}{\partial\boldsymbol{F}}\cdot\boldsymbol{F}^T=2J^{-1}\boldsymbol{F}\cdot\frac{\partial\mathcal{W}}{\partial\boldsymbol{C}}\cdot\boldsymbol{F}^T.$$

Therefore, for the hyperelastic constitutive model, specifying the strain energy density is all that is required to fully specify the material response:

$$\boxed{\begin{aligned}&\mathcal{W}=\mathcal{W}(\boldsymbol{C}),\\&\psi=\tfrac{1}{\rho_o}\mathcal{W},\\&\boldsymbol{\sigma}=2J^{-1}\boldsymbol{F}\cdot\tfrac{\partial\mathcal{W}}{\partial\boldsymbol{C}}\cdot\boldsymbol{F}^T.\end{aligned}}$$

7.3.5 Material Symmetry

The form of the admissible constitutive equations for a hyperelastic model may be further simplified if we consider material symmetry. In this section, we will consider the admissible form for isotropic and transversely isotropic hyperelastic material models.

7.3.6 Isotropic Materials

The material properties of an isotropic material are uniform in all directions. Therefore, the material response must be invariant to rotation of the reference configuration. In the case of the hyperelastic model that has only one independent equation for the constitutive model, we may write

$$\mathcal{W}(\boldsymbol{C})=\mathcal{W}\left(\boldsymbol{Q}\cdot\boldsymbol{C}\cdot\boldsymbol{Q}^T\right).$$

Therefore, the strain energy density for the isotropic hyperelastic material is a scalar-valued isotropic function. As we proved in Section 4.6.1, given this constraint, the only admissible forms of the strain energy are those which can be written in terms of the invariants of the right Cauchy deformation tensor such that

$$\mathcal{W}(\boldsymbol{C})=\mathcal{W}(I_C,II_C,III_C),$$

where invariants of the right Cauchy deformation tensor are given by

$$I_C = tr\boldsymbol{C}, \quad II_C = \frac{1}{2}\left(tr(\boldsymbol{C})^2 - tr\left(\boldsymbol{C}^2\right)\right), \quad \text{and} \quad III_C = \det \boldsymbol{C}.$$

Furthermore, we may simplify the equation for obtaining the stress tensor from this strain energy density. By using the chain rule, the Cauchy stress tensor may be written as

$$\boldsymbol{\sigma} = 2J^{-1}\boldsymbol{F}\cdot\left(\frac{\partial \mathcal{W}}{\partial I_C}\frac{\partial I_C}{\partial \boldsymbol{C}} + \frac{\partial \mathcal{W}}{\partial II_C}\frac{\partial II_C}{\partial \boldsymbol{C}} + \frac{\partial \mathcal{W}}{\partial III_C}\frac{\partial III_C}{\partial \boldsymbol{C}}\right)\cdot \boldsymbol{F}^T.$$

The derivative of the first invariant with respect to the right Cauchy deformation tensor is given by

$$\frac{\partial I_C}{\partial \boldsymbol{C}} = \frac{\partial (tr\,\boldsymbol{C})}{\partial \boldsymbol{C}}.$$

Written in index notation, this gives

$$\frac{\partial I_C}{\partial C_{pq}} = \frac{\partial C_{ii}}{\partial C_{pq}} = \delta_{ip}\delta_{iq} = \delta_{qp}$$

or, equivalently,

$$\frac{\partial I_C}{\partial \boldsymbol{C}} = \boldsymbol{I}.$$

The derivative of the second invariant gives

$$\frac{\partial II_C}{\partial \boldsymbol{C}} = \frac{\partial\left\{\frac{1}{2}\left(tr(\boldsymbol{C})^2 - tr\left(\boldsymbol{C}^2\right)\right)\right\}}{\partial \boldsymbol{C}}.$$

Writing this in index notation, we have

$$\begin{aligned}
\frac{\partial II_C}{\partial C_{pq}} &= \frac{1}{2}\frac{\partial\left(C_{ii}C_{jj} - C_{ji}C_{ij}\right)}{\partial C_{pq}} \\
&= \frac{1}{2}\left(C_{ii}\frac{\partial C_{jj}}{\partial C_{pq}} + \frac{\partial C_{ii}}{\partial C_{pq}}C_{jj} - \frac{\partial C_{ji}}{\partial C_{pq}}C_{ij} - C_{ji}\frac{\partial C_{ij}}{\partial C_{pq}}\right) \\
&= \frac{1}{2}\left(C_{ii}\delta_{pq} + \delta_{pq}C_{jj} - \delta_{jp}\delta_{iq}C_{ij} - C_{ji}\delta_{ip}\delta_{jq}\right) \\
&= \left(C_{ii}\delta_{pq} - C_{qp}\right)
\end{aligned}$$

or, equivalently,

$$\frac{\partial II_C}{\partial \boldsymbol{C}} = tr(\boldsymbol{C})\,\boldsymbol{I} - \boldsymbol{C}^T.$$

The derivative of the third invariant becomes

$$\frac{\partial III_C}{\partial \boldsymbol{C}} = \frac{\partial (\det \boldsymbol{C})}{\partial \boldsymbol{C}}.$$

We can evaluate this derivative by first using the Cayley-Hamilton theorem to rewrite the determinant in terms of powers of the right Cauchy deformation tensor. The Cayley-Hamilton theorem states that a tensor satisfies its own characteristic equation such that

$$-\boldsymbol{C}^3 + I_C\boldsymbol{C}^2 - II_C\boldsymbol{C} + III_C\mathbf{I} = 0.$$

Taking the trace of the equation gives

$$tr\left(-\mathbf{C}^3 + I_C\mathbf{C}^2 - II_C\mathbf{C} + III_C\mathbf{I}\right) = 0,$$

$$-tr\left(\mathbf{C}^3\right) + tr\left(I_C\mathbf{C}^2\right) - tr\left(II_C\mathbf{C}\right) + tr\left(III_C\mathbf{I}\right) = 0,$$

$$-tr\left(\mathbf{C}^3\right) + I_C tr\left(\mathbf{C}^2\right) - II_C tr\left(\mathbf{C}\right) + 3III_C = 0.$$

Solving this equation for the third invariant gives

$$III_C = \frac{1}{3}\left(tr(\mathbf{C}^3) - I_C tr(\mathbf{C}^2) + II_C tr(\mathbf{C})\right).$$

Substitution into the derivative of the third invariant gives

$$\frac{\partial III_C}{\partial \mathbf{C}} = \frac{1}{3}\frac{\partial}{\partial \mathbf{C}}\left(tr\left(\mathbf{C}^3\right) - I_C\, tr\left(\mathbf{C}^2\right) + II_C\, I_C\right).$$

Expanding the derivative gives

$$\frac{\partial III_C}{\partial \mathbf{C}} = \frac{1}{3}\left\{\frac{\partial\left(tr\left(\mathbf{C}^3\right)\right)}{\partial \mathbf{C}} - I_C\frac{\partial\left(tr\left(\mathbf{C}^2\right)\right)}{\partial \mathbf{C}} + \left\{II_C - tr\left(\mathbf{C}^2\right)\right\}\frac{\partial I_C}{\partial \mathbf{C}} + I_C\frac{\partial II_C}{\partial \mathbf{C}}\right\}.$$

The derivative in the first term on the right is given by

$$\begin{aligned}
\frac{\partial\left(tr\left(\mathbf{C}^3\right)\right)}{\partial C_{ij}} &= \frac{\partial\left(C_{mn}C_{no}C_{om}\right)}{\partial C_{ij}} \\
&= \delta_{mi}\delta_{nj}C_{no}C_{om} + C_{mn}\delta_{ni}\delta_{oj}C_{om} + C_{mn}C_{no}\delta_{oi}\delta_{mj} \\
&= C_{jo}C_{oi} + C_{mi}C_{jm} + C_{jn}C_{ni} \\
&= 3C_{jo}C_{oi}.
\end{aligned}$$

In tensor notation, this becomes

$$\frac{\partial\left(tr\left(\mathbf{C}^3\right)\right)}{\partial C_{ij}} = 3\left(\mathbf{C}\cdot\mathbf{C}\right)^T.$$

The derivative in the second term is given by

$$\frac{\partial\left(tr\left(\mathbf{C}^2\right)\right)}{\partial C_{ij}} = \frac{\partial\left(C_{mn}C_{nm}\right)}{\partial C_{ij}} = 2C_{ji}.$$

The derivative in the third and forth terms have already been found. Therefore, the derivative of the third invariant can then be written as

$$\frac{\partial III_C}{\partial \mathbf{C}} = \frac{1}{3}\left\{3\left(\mathbf{C}\cdot\mathbf{C}\right)^T - I_C\left(2\mathbf{C}^T\right) + \left\{II_C - tr\left(\mathbf{C}^2\right)\right\}\mathbf{I} + \left(tr\left(\mathbf{C}\right)\right)^2\mathbf{I} - I_C\mathbf{C}^T\right\}.$$

Noting that $\left(\left(tr\left(\mathbf{C}\right)\right)^2 - tr\left(\mathbf{C}^2\right)\right) = 2II_C$, we have

$$\begin{aligned}
\frac{\partial III_C}{\partial \mathbf{C}} &= \frac{1}{3}(3\left(\mathbf{C}\cdot\mathbf{C}\right)^T - 3I_C\mathbf{C}^T + 3II_C\mathbf{I} \\
&= \left(\mathbf{C}\cdot\mathbf{C}\right)^T - I_C\mathbf{C}^T + II_C\mathbf{I}.
\end{aligned}$$

Substituting the Cayley-Hamilton theorem for the inverse of the right Cauchy deformation tensor and using the fact that the right Cauchy deformation tensor is symmetric gives

$$\frac{\partial III_C}{\partial \boldsymbol{C}} = III_C \boldsymbol{C}^{-1}.$$

Therefore, the Cauchy stress tensor may be written in the form

$$\sigma = 2J^{-1}\boldsymbol{F}\cdot\left(\frac{\partial \mathcal{W}}{\partial I_C}\boldsymbol{I} + \frac{\partial \mathcal{W}}{\partial II_C}(I_C\boldsymbol{I} - \boldsymbol{C}) + \frac{\partial \mathcal{W}}{\partial III_C} III_C \boldsymbol{C}^{-1}\right)\cdot\boldsymbol{F}^T.$$

We may equivalently write this equation in terms of the left Cauchy deformation tensor, $\boldsymbol{B} = \boldsymbol{F}\cdot\boldsymbol{F}^T$, by substitution of

$$\boldsymbol{F}\cdot\boldsymbol{C}\cdot\boldsymbol{F}^T = \boldsymbol{F}\cdot\boldsymbol{F}^T\cdot\boldsymbol{F}\cdot\boldsymbol{F}^T = \boldsymbol{B}^2$$

and

$$\boldsymbol{F}\cdot\boldsymbol{C}^{-1}\cdot\boldsymbol{F}^T = \boldsymbol{F}\cdot\left(\boldsymbol{F}^T\cdot\boldsymbol{F}\right)^{-1}\cdot\boldsymbol{F}^T = \boldsymbol{F}\cdot\boldsymbol{F}^{-1}\cdot\boldsymbol{F}^{-T}\cdot\boldsymbol{F}^T = \mathbf{I}$$

into the Cauchy stress tensor. The result is

$$\sigma = 2J^{-1}\left(\left(\frac{\partial \mathcal{W}}{\partial I_C} + I_C\frac{\partial \mathcal{W}}{\partial II_C}\right)\boldsymbol{B} - \frac{\partial \mathcal{W}}{\partial II_C}\boldsymbol{B}^2 + III_C\frac{\partial \mathcal{W}}{\partial III_C}\mathbf{I}\right).$$

Furthermore, we may represent the strain energy density as a power series in terms of the invariants. Assuming the strain energy density is zero in the reference state, $\mathcal{W}(I_C = 3, II_C = 3, III_C = 1) = 0$, we can write

$$\mathcal{W}(I_C, II_C, III_C) = \sum_{\mathrm{i,j,k}=0}^{\infty} \alpha_{ijk}(I_1-3)^i(I_2-3)^j(I_3-1)^k.$$

Therefore, all admissible isotropic hyperelastic material models will have the basic form given by the following equations:

$$\boxed{\begin{aligned}&\mathcal{W} = \mathcal{W}(I_C, II_C, III_C),\\ &\psi = \tfrac{1}{\rho_o}\mathcal{W},\\ &\sigma = 2J^{-1}\left(\left(\tfrac{\partial \mathcal{W}}{\partial I_C} + I_C\tfrac{\partial \mathcal{W}}{\partial II_C}\right)\boldsymbol{B} - \tfrac{\partial \mathcal{W}}{\partial II_C}\boldsymbol{B}^2 + III_C\tfrac{\partial \mathcal{W}}{\partial III_C}\mathbf{I}\right).\end{aligned}}$$

7.3.7 Transversely Isotropic Materials

The material properties of a transversely isotropic material are symmetric about a single axis. We may define a constitutive model that captures the transverse isotropy of the material by introducing an orientation vector. In this derivation, the orientation of the axis of symmetry in the reference configuration will be specified by the unit vector $\hat{\boldsymbol{a}}_o(\boldsymbol{X})$. In order to find the orientation of the axis of symmetry in the current configuration, $\hat{\boldsymbol{a}}(\boldsymbol{x},t)$, we can transform using the deformation gradient such that

$$\hat{\boldsymbol{a}} = \left|\boldsymbol{F}\cdot\hat{\boldsymbol{a}}_o\right|.$$

Although the most direct way to incorporate the axis of symmetry would be to formulate a constitutive model as

$$\mathcal{W} = \mathcal{W}\left(\boldsymbol{U}, \hat{\boldsymbol{a}}_o(\boldsymbol{X})\right),$$

We will take a slightly more complicated route because it is more easily extended to multiple axes of symmetry. Specifically, we introduce the orientation tensor, $\boldsymbol{M}$, given by

$$\boldsymbol{M} = \hat{\boldsymbol{a}}_o \otimes \hat{\boldsymbol{a}}_o.$$

The orientation tensor captures all of the symmetry information required for this derivation. The strain energy function can then be expressed as a function of both the right Cauchy deformation tensor and the orientation tensor such that

$$\mathcal{W} = \mathcal{W}(\boldsymbol{C}, \boldsymbol{M}).$$

Material symmetry for a transversely isotropic material with fiber orientation $\hat{\boldsymbol{a}}_o$ requires that the constitutive response be invariant to any rotation about the axis defined by $\hat{\boldsymbol{a}}_o$. The set of permissible rotations, $\left\{\boldsymbol{Q}^{(i)}\right\}$, has the property $\boldsymbol{Q} \cdot \hat{\boldsymbol{a}}_o = \hat{\boldsymbol{a}}_o$, and the strain density must satisfy

$$\mathcal{W}(\boldsymbol{C}, \boldsymbol{M}) = \mathcal{W}\left(\boldsymbol{Q} \cdot \boldsymbol{C} \cdot \boldsymbol{Q}^T, \boldsymbol{Q} \cdot \boldsymbol{M} \cdot \boldsymbol{Q}^T\right).$$

Here the strain energy density is shown to be a scalar transversely isotropic function with two tensor parameters. The derivation is not provided here; however, it can be shown that the only admissible functions which conform to this constraint are those that can be written in the form

$$\mathcal{W} = \mathcal{W}(I_{\boldsymbol{C}}, II_{\boldsymbol{C}}, III_{\boldsymbol{C}}, I_4, I_5),$$

where we have introduced two new invariants that are functions of both $\boldsymbol{C}$ and $\boldsymbol{M}$

$$I_4\left(\boldsymbol{C}, \hat{\boldsymbol{a}}_o\right) = \hat{\boldsymbol{a}}_o \cdot \boldsymbol{C} \cdot \hat{\boldsymbol{a}}_o,$$
$$I_5\left(\boldsymbol{C}, \hat{\boldsymbol{a}}_o\right) = \hat{\boldsymbol{a}}_o \cdot \boldsymbol{C}^2 \cdot \hat{\boldsymbol{a}}_o.$$

The derivative of the fourth invariant with respect to $\boldsymbol{C}$ is

$$\begin{aligned}\frac{\partial I_4}{\partial C_{ij}} &= \frac{\partial}{\partial C_{ij}}\left\{\left(\hat{a}_o\right)_m C_{mn}\left(\hat{a}_o\right)_n\right\}\\ &= \left(\hat{a}_o\right)_m \delta_{mi}\delta_{nj}\left(\hat{a}_o\right)_n\\ &= \left(\hat{a}_o\right)_i\left(\hat{a}_o\right)_j\end{aligned}$$

or, equivalently,

$$\frac{\partial I_4}{\partial \boldsymbol{C}} = \hat{\boldsymbol{a}}_o \otimes \hat{\boldsymbol{a}}_o.$$

The derivative of the fifth invariant with respect to $\boldsymbol{C}$ is

$$\begin{aligned}\frac{\partial I_5}{\partial C_{ij}} &= \frac{\partial}{\partial C_{ij}}\left\{\left(\hat{a}_o\right)_k C_{km}C_{mn}\left(\hat{a}_o\right)_n\right\}\\ &= \left(\hat{a}_o\right)_k \delta_{ki}\delta_{mj} C_{mn}\left(\hat{a}_o\right)_n + \left(\hat{a}_o\right)_k C_{km}\delta_{mi}\delta_{nj}\left(\hat{a}_o\right)_n\\ &= \left(\hat{a}_o\right)_i C_{jn}\left(\hat{a}_o\right)_n + \left(\hat{a}_o\right)_k C_{ki}\left(\hat{a}_o\right)_j\end{aligned}$$

or, equivalently,

$$\frac{\partial I_5}{\partial \boldsymbol{C}} = \hat{\boldsymbol{a}}_o \otimes \boldsymbol{C} \cdot \hat{\boldsymbol{a}}_o + \hat{\boldsymbol{a}}_o \cdot \boldsymbol{C} \otimes \hat{\boldsymbol{a}}_o.$$

The Cauchy stress for the transversely isotropic material can be determined directly from the strain energy density as

$$\boldsymbol{\sigma} = 2J^{-1}\boldsymbol{F}\cdot\left(\frac{\partial \mathcal{W}}{\partial I_C}\frac{\partial I_C}{\partial \boldsymbol{C}} + \frac{\partial \mathcal{W}}{\partial II_C}\frac{\partial II_C}{\partial \boldsymbol{C}} + \frac{\partial \mathcal{W}}{\partial III_C}\frac{\partial III_C}{\partial \boldsymbol{C}} + \frac{\partial \mathcal{W}}{\partial I_4}\frac{\partial I_4}{\partial \boldsymbol{C}} + \frac{\partial \mathcal{W}}{\partial I_5}\frac{\partial I_5}{\partial \boldsymbol{C}}\right)\cdot\boldsymbol{F}^T.$$

Before substituting the derivatives of the invariants with respect to the right Cauchy deformation tensor, let us rewrite these derivatives in terms of the symmetry axis in the current configuration, $\hat{\boldsymbol{a}}$. Noting that $\lambda\hat{\boldsymbol{a}} = \boldsymbol{F}\cdot\hat{\boldsymbol{a}}_o$, we can write

$$\boldsymbol{F}\cdot\frac{\partial I_4}{\partial \boldsymbol{C}}\cdot\boldsymbol{F}^T = \boldsymbol{F}\cdot\hat{\boldsymbol{a}}_o\otimes\hat{\boldsymbol{a}}_o\cdot\boldsymbol{F}^T = \lambda^2\,\hat{\boldsymbol{a}}\otimes\hat{\boldsymbol{a}}$$

and

$$\begin{aligned}\boldsymbol{F}\cdot\frac{\partial I_5}{\partial \boldsymbol{C}}\cdot\boldsymbol{F}^T &= \boldsymbol{F}\cdot\left\{\hat{\boldsymbol{a}}_o\otimes\boldsymbol{C}\cdot\hat{\boldsymbol{a}}_o + \hat{\boldsymbol{a}}_o\cdot\boldsymbol{C}\otimes\hat{\boldsymbol{a}}_o\right\}\cdot\boldsymbol{F}^T\\ &= \lambda^2\left\{\hat{\boldsymbol{a}}\otimes\boldsymbol{C}\cdot\hat{\boldsymbol{a}} + \hat{\boldsymbol{a}}\cdot\boldsymbol{C}\otimes\hat{\boldsymbol{a}}\right\}.\end{aligned}$$

Substituting the derivatives of the invariants into this equation and writing in terms of the left Cauchy deformation tensor gives

$$\begin{aligned}\boldsymbol{\sigma} = 2J^{-1}\Bigg(&\left(\frac{\partial \mathcal{W}}{\partial I_C} + I_C\frac{\partial \mathcal{W}}{\partial II_C}\right)\boldsymbol{B} - \frac{\partial \mathcal{W}}{\partial II_C}\boldsymbol{B}^2 + III_C\frac{\partial \mathcal{W}}{\partial III_C}\mathbf{I} + I_4\frac{\partial \mathcal{W}}{\partial I_4}\hat{\boldsymbol{a}}\otimes\hat{\boldsymbol{a}}\\ &+ I_4\frac{\partial \mathcal{W}}{\partial I_5}\left\{\hat{\boldsymbol{a}}\otimes\boldsymbol{C}\cdot\hat{\boldsymbol{a}} + \hat{\boldsymbol{a}}\cdot\boldsymbol{C}\otimes\hat{\boldsymbol{a}}\right\}\Bigg).\end{aligned}$$

Therefore, all admissible material models for a transversely isotropic hyperelastic material must have the form

$$\begin{aligned}&\mathcal{W} = \mathcal{W}(I_C, II_C, III_C, I_4, I_5),\\ &\psi = \tfrac{1}{\rho_o}\mathcal{W},\\ &\boldsymbol{\sigma} = 2J^{-1}\left(\left(\tfrac{\partial \mathcal{W}}{\partial I_C} + I_C\tfrac{\partial \mathcal{W}}{\partial II_C}\right)\boldsymbol{B} - \tfrac{\partial \mathcal{W}}{\partial II_C}\boldsymbol{B}^2 + III_C\tfrac{\partial \mathcal{W}}{\partial III_C}\mathbf{I} + I_4\tfrac{\partial \mathcal{W}}{\partial I_4}\hat{\boldsymbol{a}}\otimes\hat{\boldsymbol{a}}\right.\\ &\qquad\left. + I_4\tfrac{\partial \mathcal{W}}{\partial I_5}\left\{\hat{\boldsymbol{a}}\otimes\boldsymbol{C}\cdot\hat{\boldsymbol{a}} + \hat{\boldsymbol{a}}\cdot\boldsymbol{C}\otimes\hat{\boldsymbol{a}}\right\}\right).\end{aligned}$$

7.3.8 Incompressible Materials

An incompressible material is one whose volume does not change during deformation. This approximation is often made with some polymers and fluids. In the case of polymers, the bulk modulus tends to be many orders of magnitude larger than the shear modulus making this approximation quite reasonable. In an incompressible material, the stress field is no longer explicitly determined by the strain field. The boundary conditions must be considered to fully define the stress within the material. Let us conduct a simple thought experiment. Let us take a small sphere of an incompressible material and subject it to a hydrostatic pressure, P_1. By definition, the volume of the incompressible sphere does not change. Similarly, if we increase the hydrostatic pressure by a factor of 2, then the volume of the sphere still does not change. In fact, in both cases, there is no deformation field associated with the applied hydrostatic pressure. We cannot determine the internal stress state from the deformation of the object. Yet we know intuitively that the internal stress in the second case must be higher than that of the first case.

The hydrostatic pressure does not do any work on the incompressible material, and so it is not included within the strain energy function. Instead, we must amend the stress formulation in order to account for incompressibility. We do this by breaking the stress into two parts:

$$\boldsymbol{\sigma} = \hat{\boldsymbol{\sigma}} + \tilde{\boldsymbol{\sigma}}.$$

We will say that $\hat{\boldsymbol{\sigma}}$ is due to the hydrostatic pressure, and $\tilde{\boldsymbol{\sigma}}$ is the part of the stress we can determine using the strain energy function. We can write this stress function as

$$\boldsymbol{\sigma} = -p\mathbf{I} + 2\rho\left(\frac{\partial\psi}{\partial \boldsymbol{B}} \cdot \boldsymbol{B}^T\right).$$

We have introduced the additional pressure term within the stress that is due to the incompressibility constraint. In the section on experimental determination of constitutive parameters, we will demonstrate how the pressure is determined by using the boundary conditions within the system.

7.3.9 Common Hyperelastic Constitutive Models

Due to the ease of use, and the wide range of problems to which they are applicable, the list of hyperelastic material models is very long. Even though we have formulated the constraints on the constitutive equations using the invariants I_C, II_C, and III_C, many common constitutive models are formulated in terms of the modified invariants

$$\overline{I}_C = \frac{I_C}{(III_C)^{2/3}}, \quad \overline{II}_C = \frac{II_C}{(III_C)^{4/3}}, \quad \text{and} \quad J = \sqrt{III_C} = \det \boldsymbol{F}.$$

The purpose for this is to separate the volumetric expansion or rarefaction from the distortion of the material. The first two modified invariants, $\overline{I}_C$ and $\overline{II}_C$, capture the distortion, whereas the third, J, captures volumetric expansion or contraction. Using the modified invariants, the generalized form of the strain energy density can be written as the power series

$$\mathcal{W}(I_C, II_C, III_C) = \sum_{i,j}^{\infty} \alpha_{ij} \left(\overline{I}_C - 3\right)^i \left(\overline{II}_C - 3\right)^j + \sum_{i}^{\infty} \beta_i (J-1)^{2i}.$$

7.3.10 Freely Jointed Chain

In this section, we will derive a simple approximation for the strain energy function of a polymer based on the microscopic details of the system. There are many more involved models for polymer chain interactions, but this serves as an illustration of the methodology and the resulting connection between molecular structure and macroscopic constitutive models. A polymer chain consists of a series of repeating monomers. Polyethylene, for example, consists of repeating ethylene monomers, C_2H_4. The polymer chain is flexible, and if the length is large compared to the monomer size, the polymer chain resembles a flexible rope. The macroscopic stress-strain behavior of a polymer can be modeled by considering the behavior of a large

Table 7.5. *Table of common hyperelastic strain energy functions*

Incompressible neo-Hookean model	$\mathcal{W} = \alpha_1 (I_C - 3)$
Generalized neo-Hookean model	$\mathcal{W} = \alpha_1 \left(\bar{I}_C - 3\right) + \alpha_2 (J-1)^2$
Generalized Mooney-Rivlin model	$\mathcal{W} = \alpha_1 \left(\bar{I}_C - 3\right) + \alpha_2 \left(\bar{II}_C - 3\right) + \alpha_3 (J-1)^2$
Incompressible Yeoh model	$\mathcal{W} = \alpha_1 (I_C - 3) + \alpha_2 (I_C - 3)^2 + \alpha_3 (I_C - 3)^3$
Generalized Yeoh model	$\mathcal{W} = \sum_i \alpha_i \left(\bar{I}_C - 3\right)^i + \sum_i \beta_i (J-1)^{2i}$
Arruda-Boyce 8 chain model	$\mathcal{W} = \alpha_1 \left\{ \frac{1}{2}\left(\bar{I}_C - 3\right) + \frac{1}{20\alpha_2^2}\left(\bar{I}_C^2 - 9\right) + \frac{11}{1050\alpha_2^4}\left(\bar{I}_C^3 - 27\right) + \cdots \right\} + \alpha_3 (J-1)^2$
Incompressible Gent model	$\mathcal{W} = -\alpha_1 \ln\left(1 - \frac{(I_C - 3)}{\alpha_2}\right)$

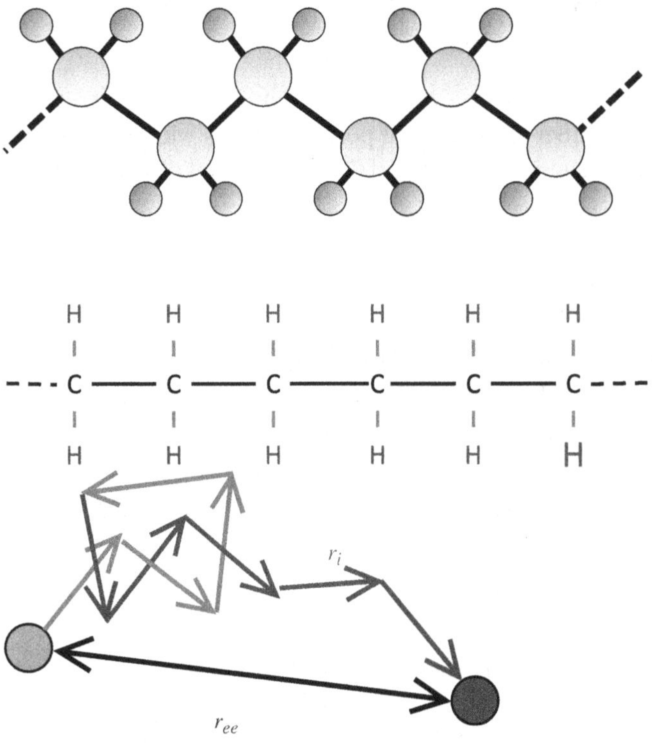

Figure 7.2. (top) Illustration of the molecular structure of polyethylene. (bottom) Illustration of the random walk used to model the polymer chain.

number of individual polymer chains. The behavior of a single chain can be modeled with varying degrees of approximation. In this section, we will consider the ***freely jointed chain model***. This model assumes that the chain can be modeled as a random walk. That is, each chain is made of individual rigid links whose orientation does not depend on the orientation of the preceding or proceeding links. Each link is randomly oriented with respect to any other link.

The bonds do not occupy any real space and can pass through one another. The vector distance between the two ends of the polymer chain, $\boldsymbol{r}_{ee}$, can then be written as the sum of the bond vectors, $\boldsymbol{r}_i$ over the length of the chain

$$\boldsymbol{r}_{ee} = \sum_{i=1}^{n} \boldsymbol{r}_i,$$

where n is the number of bonds within a single polymer chain and the length of an individual bond is given by $|\boldsymbol{r}_i| = \ell$.

The magnitude squared of the end-to-end distance is given by

$$|\boldsymbol{r}_{ee}|^2 = \boldsymbol{r}_{ee} \cdot \boldsymbol{r}_{ee} = \left(\sum_{i=1}^{n} \boldsymbol{r}_i\right) \cdot \left(\sum_{j=1}^{n} \boldsymbol{r}_j\right).$$

We can break this summation into two parts, the first are the terms where $i = j$ and the second term includes all terms where $i \neq j$ such that

$$|\boldsymbol{r}_{ee}|^2 = \left(\sum_{i=1}^{n} (\boldsymbol{r}_i \cdot \boldsymbol{r}_i)\right) + 2\sum_{i=1}^{n} \sum_{j=i+1}^{n} (\boldsymbol{r}_i \cdot \boldsymbol{r}_j).$$

Here we have used the fact that $\boldsymbol{r}_i \cdot \boldsymbol{r}_j = \boldsymbol{r}_j \cdot \boldsymbol{r}_i$ and $\sum_{i=1}^{n} \sum_{j \neq i}^{n} (\boldsymbol{r}_i \cdot \boldsymbol{r}_j) = 2\sum_{i=1}^{n} \sum_{j=i+1}^{n} (\boldsymbol{r}_i \cdot \boldsymbol{r}_j)$. The number of individual polymer chains within a polymer melt is large. If we average the squared end-to-end distance over all of the molecules within the melt, we obtain

$$\langle \boldsymbol{r}_{ee} \cdot \boldsymbol{r}_{ee} \rangle = \frac{1}{N} \sum_{\mathbf{1}}^{N} \left[\left(\sum_{i=1}^{n} (\boldsymbol{r}_i \cdot \boldsymbol{r}_i)\right) + 2\sum_{i=1}^{n} \sum_{j=i+1}^{n} (\boldsymbol{r}_i \cdot \boldsymbol{r}_j) \right],$$

where N is the number of molecules within the polymer melt. Since the chain is freely jointed, the direction of each bond vector is independent of any other bond vector. Therefore, the average value of the projection of one bond onto another is zero. That means the second term $\frac{2}{N} \sum_{\mathbf{1}}^{N} \left[\sum_{i=1}^{n} \sum_{j=i+1}^{n} (\boldsymbol{r}_i \cdot \boldsymbol{r}_j)\right]$ goes to zero as N grows larger. Recall, that the first term is the projection of the bond onto itself such that $\boldsymbol{r}_i \cdot \boldsymbol{r}_i = \ell^2$ and

$$\langle \boldsymbol{r}_{ee} \cdot \boldsymbol{r}_{ee} \rangle = \frac{1}{N} \sum_{\mathbf{1}}^{N} \left[\left(\sum_{i=1}^{n} (\boldsymbol{r}_i \cdot \boldsymbol{r}_i)\right) \right] = \frac{1}{N} N n \ell^2.$$

The probability of finding a chain with a particular end-to-end distance vector, $p(\boldsymbol{r}_{ee})$, is given by

$$p(\boldsymbol{r}_{ee}) = \int p(\boldsymbol{r}_1, \boldsymbol{r}_2, \cdots, \boldsymbol{r}_n)\, \delta \left(\boldsymbol{r}_{ee} - \sum_{i=1}^{n} \boldsymbol{r}_i \right) d\boldsymbol{r}_1 d\boldsymbol{r}_2 \cdots d\boldsymbol{r}_n,$$

where $p(\boldsymbol{r}_1, \boldsymbol{r}_2, \cdots, \boldsymbol{r}_n)$ is the joint probability that the individual bond vectors take on the values of $\boldsymbol{r}_1 \boldsymbol{r}_2, \cdots, \boldsymbol{r}_n$, and $\delta(\boldsymbol{x})$ is the delta function which has the value of 1 if $\boldsymbol{x} = 0$ and a value of 0 otherwise. Since the bond vectors are independent, the joint probability can be written as the multiple of the individual probability functions

$$p(\boldsymbol{r}_1, \boldsymbol{r}_2, \cdots, \boldsymbol{r}_n) = p(\boldsymbol{r}_1)\, p(\boldsymbol{r}_2) \cdots p(\boldsymbol{r}_n) = \prod_{j=1}^{n} p(\boldsymbol{r}_j).$$

Combining these results, we have

$$p(\boldsymbol{r}_{ee}) = \int \left[\prod_{j=1}^{n} p(\boldsymbol{r}_j)\right]\left[\delta\left(\boldsymbol{r}_{ee} - \sum_{i=1}^{n}\boldsymbol{r}_i\right)\right] d\boldsymbol{r}_1 d\boldsymbol{r}_2 \cdots d\boldsymbol{r}_n.$$

The probability of finding a single bond vector in a particular orientation is given by

$$p(\boldsymbol{r}_j) = \frac{1}{4\pi\ell^2}\delta\left(|\boldsymbol{r}_j| - \ell\right).$$

Since this is a normalized probability distribution, it has the property

$$\int p(\boldsymbol{r}_j)\, d\boldsymbol{r}_j = 1.$$

This gives

$$p(\boldsymbol{r}_{ee}) = \int \left[\prod_{j=1}^{n} \frac{1}{4\pi\ell^2}\delta\left(|\boldsymbol{r}_j| - \ell\right)\right]\left[\delta\left(\boldsymbol{r}_{ee} - \sum_{i=1}^{n}\boldsymbol{r}_i\right)\right] d\boldsymbol{r}_1 d\boldsymbol{r}_2 \cdots d\boldsymbol{r}_n.$$

The delta function may be represented with the complex integral

$$\delta(\boldsymbol{x}) = \frac{1}{(2\pi)^3}\int \exp\left(-i\boldsymbol{k}\cdot\boldsymbol{x}\right) d\boldsymbol{k}.$$

This gives

$$p(\boldsymbol{r}_{ee}) = \frac{1}{(2\pi)^3}\int\int \left[\prod_{j=1}^{n} \frac{1}{4\pi\ell^2}\delta\left(|\boldsymbol{r}_j| - \ell\right)\right] \times \left[exp\left\{-i\boldsymbol{k}\cdot\left(\boldsymbol{r}_{ee} - \sum_{i=1}^{n}\boldsymbol{r}_i\right)\right\}\right] d\boldsymbol{r}_1 d\boldsymbol{r}_2 \cdots d\boldsymbol{r}_n d\boldsymbol{k}.$$

Rewriting the exponent and using the fact that $\exp\left\{\left(\sum_{i=1}^{n}\boldsymbol{x}_i\right)\right\} = \prod_{i=1}^{n}\exp\{\boldsymbol{x}_i\}$ gives

$$p(\boldsymbol{r}_{ee}) = \frac{1}{(2\pi)^3}\int exp(-\mathrm{ik}\cdot\boldsymbol{r}_{ee})\int\left[\prod_{j=1}^{n} \frac{1}{4\pi\ell^2}\delta\left(|\boldsymbol{r}_j| - \ell\right)\right] \times \left[\prod_{i=1}^{n} exp\left\{i\boldsymbol{k}\cdot(\boldsymbol{r}_i)\right\}\right] d\boldsymbol{r}_1 d\boldsymbol{r}_2 \cdots d\boldsymbol{r}_n d\boldsymbol{k}.$$

Isolating the jth bond vector in each multiplication allows us to write

$$p(\boldsymbol{r}_{ee}) = \frac{1}{(2\pi)^3}\int \exp(-i\boldsymbol{k}\cdot\boldsymbol{r}_{ee})\left[\prod_{j=1}^{n}\int \frac{1}{4\pi\ell^2}\delta\left(|\boldsymbol{r}_j| - \ell\right)\exp\left\{i\boldsymbol{k}\cdot(\boldsymbol{r}_j)\right\} d\boldsymbol{r}_j\right] d\boldsymbol{k}.$$

It can be shown that

$$\int \frac{1}{4\pi\ell^2}\delta\left(|\boldsymbol{r}_j| - \ell\right)\exp\left\{i\boldsymbol{k}\cdot(\boldsymbol{r}_j)\right\} d\boldsymbol{r}_j = \frac{\sin|\boldsymbol{k}|\,l}{|\boldsymbol{k}|\,l}.$$

Completing the integration, we find that the probability of finding a chain with a particular end-to-end distance, r_{ee}, is described by the function

$$p(r_{ee}) = \left(\frac{3}{2\pi \ell^2 n}\right)^{\frac{3}{2}} \exp\left(-\frac{3r_{ee}^2}{2\ell^2 n}\right) = \left(\frac{b}{\pi}\right)^{\frac{3}{2}} \exp\left(-b r_{ee}^2\right),$$

where $b = \frac{3}{2\ell^2 n}$ is a constant.

The entropy within a system is related to the number of available conformations. In fact, Boltzmann's equation relates the probability of finding a system within a given configuration to the entropy:

$$\eta = \eta^* + k_b lnp(r),$$

where k_b is the Boltzmann's constant, and η^* is the entropy of the reference state. Substitution of the Gaussian distribution gives

$$\eta = \eta^\dagger - k_b b^2 r^2,$$

where $\eta^\dagger = \eta^* + k_b \ln\left(\frac{b}{\pi}\right)^{\frac{3}{2}}$.

If we consider affine deformation of an isotropic and incompressible polymer, the equations of motion can be written as

$$x_i = \lambda_i X_i,$$

with the constraint that $\lambda_1 \lambda_2 \lambda_3 = 1$.

The individual molecules within a material are either compressed or stretched when the sample is deformed. The change in entropy due to the stretch in a single chain can be calculated using Boltzmann's equation:

$$\begin{aligned} \Delta\eta_j &= -k_b b^2 \left(\sum_i^3 \left(\lambda_i X_i^2\right) - \sum_i^3 X_i^2\right) \\ &= -k_b b^2 \sum_i^3 \left(\left(\lambda_i^2 - 1\right) X_i^2\right). \end{aligned}$$

The entropy change for all of the chains in the system is simply the sum of entropy changes for each of the chains:

$$\begin{aligned} \Delta\eta &= \sum_1^N \Delta\eta_j = \sum_1^N -k_b b^2 \left(\left(\lambda_1^2 - 1\right) X_1^2 + \left(\lambda_2^2 - 1\right) X_2^2 + \left(\lambda_3^2 - 1\right) X_3^2\right) \\ &= -k_b b^2 \left[\left(\lambda_1^2 - 1\right) \sum_1^N \left(X_1^2\right) + \left(\lambda_2^2 - 1\right) \sum_1^N \left(X_2^2\right) + \left(\lambda_3^2 - 1\right) \sum_1^N \left(X_3^2\right)\right]. \end{aligned}$$

If we assume that the polymer chains are randomly oriented in the reference configuration, then

$$\sum_1^N \left(X_1^2\right) = \sum_1^N \left(X_2^2\right) = \sum_1^N \left(X_3^2\right) = c,$$

where c is a constant. Therefore, the total entropy change can be written as

$$\Delta\eta = -k_b b^2 c\left(\lambda_1^2+\lambda_2^2+\lambda_3^2-3\right).$$

The Helmholtz free energy is related to the entropy via

$$\psi = e - \theta\eta.$$

If we consider an isothermal process, the change in free energy is given by

$$\Delta\psi = \Delta e - \theta\Delta\eta.$$

The internal energy of the polymer does not change giving us

$$\Delta\psi = -\theta\Delta\eta = c^*\theta\left(\lambda_1^2+\lambda_2^2+\lambda_3^2-3\right),$$

where the constant $c^* = k_b b^2 c$. Taking the undeformed material as the reference state gives

$$\psi = c^*\theta\left(\lambda_1^2+\lambda_2^2+\lambda_3^2-3\right).$$

7.4 Linear Thermoelastic Material Model

If one is attempting to model the behavior of a material at very small strains, a linear thermoelastic material model may be sufficient to obtain the material response. For example, metal components of many machines and structures are designed to have very small strains during normal operating conditions. The added complexity of a finite thermoelastic material model might not be necessary in these cases.

In this section, we eliminate geometric nonlinearity by assuming the modeled material will be subject to infinitesimal deformation, and we eliminate material nonlinearity by adopting a linear thermoelastic constitute equation for the material response. Specifically, we assume infinitesimal deformation, $\|\nabla \boldsymbol{u}\| \ll 1$, small temperature changes, $|\theta-\theta_o| \ll 1$, and a linear elastic material model, $\boldsymbol{S} = \boldsymbol{S}_o + \rho_o\left[\mathrm{C}:\boldsymbol{E}+\boldsymbol{\beta}\left(\hat{\theta}-\theta_o\right)\right]$. These will be discussed in more detail in the following text. We start by postulating a constitutive model that is a function of the Green strain tensor, temperature, and temperature gradient. We then determine the admissible forms for the constitutive model by applying the Clausius-Duhem inequality. Next, we adopt a linear constitutive relation for the stress, strain, temperature equation. Finally, we combine these with the balance laws to derive the governing equations for linear thermoelastic materials.

There are 22 unknown field variables for the linear thermoelastic material model. These are listed in Table 7.6. They are identical to those presented for the finite thermoelastic material model.

7.4.1 Balance Laws

The balance laws are also identical to those presented for the finite deformation thermoelastic material model. They are reproduced in Table 7.7 for clarity.

Table 7.6. *Field variables required to specify the finite deformation response of the thermelastic material model*

Variable	Name	Number of unknowns
$\boldsymbol{x}(\boldsymbol{X},t)$	Equations of motion	3
$\rho\,(\boldsymbol{x},t)$	Spatial density field	1
$\boldsymbol{v}\,(\boldsymbol{x},t)$	Spatial velocity field	3
$\boldsymbol{\sigma}\,(\boldsymbol{x},t)$	Cauchy stress tensor	9
$\boldsymbol{q}\,(\boldsymbol{x},t)$	Spatial heat flux	3
$e\,(\boldsymbol{x},t)$	Spatial internal energy density field	1
$\theta\,(\boldsymbol{x},t)$	Spatial temperature field	1
$\eta\,(\boldsymbol{x},t)$	Spatial entropy density field	1

Table 7.7. *Balance laws and kinematical relation for finite deformation of a thermo-mechanical material model*

		Number of equations
Kinematical relations	$\boldsymbol{v}=\dfrac{\partial \boldsymbol{x}}{\partial t}$	3
Conservation of mass	$0=\dfrac{\partial \rho}{\partial t}+div\,(\rho\,\boldsymbol{v})$	1
Conservation of linear momentum	$\rho\,\dot{v}-div\,\boldsymbol{\sigma}^T-\rho\,\boldsymbol{b}_x=\boldsymbol{0}$	3
Conservation of angular momentum	$\boldsymbol{\sigma}=\boldsymbol{\sigma}^T$	3
Conservation of energy	$\rho\dot{e}-\boldsymbol{\sigma}:\boldsymbol{D}+div\,(\boldsymbol{q}_x)-\rho\,r_x=0$	1
Entropy inequality	$-\rho\,\dot{\psi}-\rho\eta\dot{\theta}-\dfrac{1}{\theta}\boldsymbol{q}_x\cdot grad\,\theta+\boldsymbol{\sigma}^T:\boldsymbol{D}\geq 0$	0

7.4.2 Constitutive Model

Where we diverge slightly from the finite thermoelastic model development, is in the formulation of the constitutive model. Instead of formulating the constitutive model as a function of the general deformation gradient, we will select the Green strain, $\boldsymbol{E}=\frac{1}{2}\left(\boldsymbol{U}^2-1\right)$, temperature, and the temperature gradient as the independent variables. The constitutive model then has the form

$$\left\{\psi,\eta,\boldsymbol{\sigma},\boldsymbol{q}\right\}=f\left\{\boldsymbol{E},\theta,\nabla\theta\right\}.$$

We will later make use of the fact that the Green strain is approximately equal to the infinitesimal strain $\boldsymbol{E}\cong\boldsymbol{\varepsilon}$ for infinitesimal deformation.

7.4.3 Clausius-Duhem Inequality

The constraints on the constitutive law due to the second law of thermodynamics can easily be found using the CD inequality. The CD inequality states that

$$-\rho\dot{\psi}-\rho\eta\dot{\theta}-\frac{1}{\theta}\boldsymbol{q}\cdot grad\,\theta+\boldsymbol{\sigma}^T:\boldsymbol{D}\geq 0.$$

The linear thermoelastic constitutive model has been formulated in terms of the Green strain. Therefore, the material derivative of the Helmholtz free energy can be written as

$$\dot{\psi}\left(\boldsymbol{E},\theta,\nabla\theta\right)=\frac{\partial\psi}{\partial\boldsymbol{E}}:\frac{D\boldsymbol{E}}{Dt}+\frac{\partial\psi}{\partial\theta}\frac{D\theta}{Dt}+\frac{\partial\psi}{\partial\boldsymbol{g}}\cdot\frac{D\boldsymbol{g}}{Dt}.$$

Substitution into the CD inequality gives

$$-\rho\left(\frac{\partial\psi}{\partial\boldsymbol{E}}:\frac{D\boldsymbol{E}}{Dt}+\frac{\partial\psi}{\partial\theta}\frac{\partial\theta}{\partial t}+\frac{\partial\psi}{\partial\boldsymbol{g}}\cdot\frac{D\boldsymbol{g}}{Dt}\right)-\rho\eta\dot{\theta}-\frac{1}{\theta}\boldsymbol{q}\cdot\nabla\theta+\boldsymbol{\sigma}^T:\boldsymbol{D}\geq 0.$$

Using the relation $\dot{F}=\boldsymbol{L}\cdot\boldsymbol{F}$ and $\boldsymbol{D}=\frac{1}{2}\left(\boldsymbol{L}+\boldsymbol{L}^T\right)$, the material time derivative of strain tensor can be written as

$$\begin{aligned}\frac{D\boldsymbol{E}}{Dt}&=\frac{D}{Dt}\left(\frac{1}{2}\left(\boldsymbol{F}^T\cdot\boldsymbol{F}-\boldsymbol{I}\right)\right)\\&=\frac{1}{2}\left(\dot{F}^T\cdot\boldsymbol{F}+\boldsymbol{F}^T\cdot\dot{F}\right)\\&=\frac{1}{2}\left((\boldsymbol{L}\cdot\boldsymbol{F})^T\cdot\boldsymbol{F}+\boldsymbol{F}^T\cdot(\boldsymbol{L}\cdot\boldsymbol{F})\right)\\&=\frac{1}{2}\left(\boldsymbol{F}^T\cdot\boldsymbol{L}^T\cdot\boldsymbol{F}+\boldsymbol{F}^T\cdot(\boldsymbol{L}\cdot\boldsymbol{F})\right)\\&=\boldsymbol{F}^T\cdot\boldsymbol{D}\cdot\boldsymbol{F}.\end{aligned}$$

Rearranging the CD inequality, we obtain

$$-\rho\frac{\partial\psi}{\partial\boldsymbol{E}}:\boldsymbol{F}^T\cdot\boldsymbol{D}\cdot\boldsymbol{F}+\boldsymbol{\sigma}^T:\boldsymbol{D}-\rho\left(\frac{\partial\psi}{\partial\theta}-\eta\right)\dot{\theta}-\rho\frac{\partial\psi}{\partial\boldsymbol{g}}\cdot\dot{g}-\frac{1}{\theta}\boldsymbol{q}\cdot\nabla\theta\geq 0,$$

$$\left(-\rho\boldsymbol{F}\cdot\frac{\partial\psi}{\partial\boldsymbol{E}}\cdot\boldsymbol{F}^T+\boldsymbol{\sigma}^T\right):\boldsymbol{D}-\rho\left(\frac{\partial\psi}{\partial\theta}-\eta\right)\dot{\theta}-\rho\frac{\partial\psi}{\partial\boldsymbol{g}}\cdot\dot{g}-\frac{1}{\theta}\boldsymbol{q}\cdot\nabla\theta\geq 0.$$

Since this must be true for all admissible thermodynamic processes, we obtain the following set of constraints on the constitutive equations

$$\begin{aligned}\psi&=\psi\left(\boldsymbol{E},\theta\right),\\\boldsymbol{\sigma}^T&=\rho\boldsymbol{F}\cdot\frac{\partial\psi}{\partial\boldsymbol{E}}\cdot\boldsymbol{F}^T,\\\eta&=\frac{\partial\psi}{\partial\theta},\\-\boldsymbol{q}\cdot\nabla\theta&\geq 0.\end{aligned}$$

Since the deformation is small, the reference and current configurations are similar and the Piola-Kirchhoff stress is sometimes used in the formulation of the model. Making use of the relation $\boldsymbol{S}=J\,\boldsymbol{F}^{-1}\cdot\boldsymbol{\sigma}\cdot\boldsymbol{F}^{-T}$, we may equivalently write the admissible form of the constitutive model in terms of both the Piola-Kirchhoff and Cauchy stress tensor as

$$\boxed{\begin{aligned}&\psi=\psi\left(\boldsymbol{E},\theta\right),\\&\boldsymbol{S}=\rho_0\tfrac{\partial\psi}{\partial\boldsymbol{E}}\ or\ \boldsymbol{\sigma}^T=\rho\boldsymbol{F}\cdot\tfrac{\partial\psi}{\partial\boldsymbol{E}}\cdot\boldsymbol{F}^T,\\&\eta=\tfrac{\partial\psi}{\partial\theta},\\&\boldsymbol{q}=\boldsymbol{q}\left(\boldsymbol{E},\theta,\nabla\theta\right).\end{aligned}}$$

7.4.4 Linear Thermoelastic Constitutive Relation

We will derive the general form of the linear thermoelastic constitutive model by taking the Taylor expansion of the Piola-Kirchhoff stress tensor for small displacements and temperature changes and keeping only the linear terms. However, before we proceed, let us examine the implications of the small strain assumption on the kinematic variables. The deformation gradient may be written in terms of the displacement field as

$$\boldsymbol{F} = \frac{\partial}{\partial \boldsymbol{X}}(\boldsymbol{X} - \boldsymbol{u}) = \mathbf{I} + Grad\ \boldsymbol{u}.$$

Let us define a new tensor, $\boldsymbol{H}$, to represent the gradient of the displacement vector field

$$\boldsymbol{H} = Grad\ \boldsymbol{u}.$$

The assumption that the gradient of the displacement field is small allows us to write

$$\|\boldsymbol{H}\| \ll 1.$$

The right Cauchy deformation tensor may be written as

$$\begin{aligned} \boldsymbol{C} = \boldsymbol{F}^T\boldsymbol{F} &= \left(1 + \boldsymbol{H}^T\right)(1 + \boldsymbol{H}) \\ &= 1 + \boldsymbol{H} + \boldsymbol{H}^T + \boldsymbol{H}^T \cdot \boldsymbol{H} \\ &= 1 + 2\boldsymbol{E} + \boldsymbol{H}^T \cdot \boldsymbol{H}. \end{aligned}$$

The Green strain tensor may be written as

$$\boldsymbol{E} = \frac{1}{2}(\mathbf{C} - \boldsymbol{I}) = \frac{1}{2}\left(\boldsymbol{H} + \boldsymbol{H}^T\right) + O\left(\|\boldsymbol{H}\|^2\right).$$

Because, the norm of $\boldsymbol{H}$ is very small, we may neglect the second-order terms in the Green strain tensor. Therefore, to a first-order approximation, the Green strain tensor is equal to the infinitesimal strain tensor, $\boldsymbol{\epsilon}$, and we have

$$\boldsymbol{E} \cong \boldsymbol{\epsilon} = \frac{1}{2}\left(\boldsymbol{H} + \boldsymbol{H}^T\right).$$

The temperature field may be written as

$$\theta(\boldsymbol{X},t) = \theta_o(\boldsymbol{X}) + \hat{\theta}(\boldsymbol{X},t),$$

where $\theta_o(\boldsymbol{X})$ is the reference temperature field, and $\hat{\theta}(\boldsymbol{X},t)$ is the deviation from the reference temperature. Since the temperature change is assumed to be very small, we have the condition that

$$\left|\hat{\theta}(\boldsymbol{X},t)\right| \ll 1.$$

If we perform a Taylor expansion of the second Piola-Kirchhoff stress where $\boldsymbol{S} = \boldsymbol{S}(\boldsymbol{E},\theta)$ about the reference configuration given by $\boldsymbol{E} = 0$ and $\theta = \theta_o(\boldsymbol{X})$, we obtain

$$\begin{aligned} \boldsymbol{S} = \rho_o \frac{\partial \psi}{\partial \mathbf{E}}\bigg|_{\substack{\mathbf{E}=\mathbf{0} \\ \theta=\theta_o}} &+ \rho_o \left[\frac{\partial}{\partial \mathbf{E}}\frac{\partial \psi}{\partial \mathbf{E}}\bigg|_{\substack{\mathbf{E}=\mathbf{0} \\ \theta=\theta_o}} (\mathbf{E} - \mathbf{0}) + \rho_o \frac{\partial}{\partial \theta}\frac{\partial \psi}{\partial \mathbf{E}}\bigg|_{\substack{\mathbf{E}=\mathbf{0} \\ \theta=\theta_o}} \left(\hat{\theta} - \theta_o\right)\right] \\ &+ O\left(\hat{\theta}^2\right) + O\left(\boldsymbol{E}^2\right), \end{aligned}$$

where $\boldsymbol{S}_o = \rho_o \left.\frac{\partial \psi}{\partial \boldsymbol{E}}\right|_{\substack{\boldsymbol{E}=\boldsymbol{0}\\ \theta=\theta_o}}$ is the residual stress in the material, $C = \rho_o \left.\frac{\partial^2 \psi}{\partial \boldsymbol{E}^2}\right|_{\substack{\boldsymbol{E}=\boldsymbol{0}\\ \theta=\theta_o}}$ is the fourth-order tensor representing the material stiffness, and $\boldsymbol{\beta} = \rho_o \frac{\partial}{\partial \theta} \left.\frac{\partial \psi}{\partial \boldsymbol{E}}\right|_{\substack{\boldsymbol{E}=\boldsymbol{0}\\ \theta=\theta_o}}$ is related to the thermal expansion of the material. At this point, we may linearize this equation by neglecting higher order terms in both temperature change and strain

$$\boldsymbol{S} = \boldsymbol{S}_o + \rho_o\left[\mathrm{C} : \boldsymbol{E} + \boldsymbol{\beta}\left(\hat{\boldsymbol{\theta}} - \theta_o\right)\right].$$

Note that because we are keeping only first-order terms in the Taylor expansion, we should keep only first-order terms in stress and strain. Therefore, we may use the fact that the first-order approximations of each of the stress tensors are equal, $\boldsymbol{\sigma} = \boldsymbol{S} = \boldsymbol{P}$, and the first-order approximations of the strain tensors are also approximately equal, $\boldsymbol{E} = \boldsymbol{\epsilon}$. The linearized equation may also be written as

$$\boxed{\boldsymbol{\sigma} = \boldsymbol{\sigma}_o + \rho_o\left[\mathrm{C} : \boldsymbol{\epsilon} + \boldsymbol{\beta}\left(\hat{\boldsymbol{\theta}} - \theta_o\right)\right].}$$

Written in index notation, we have

$$\sigma_{ij} = (\sigma_o)_{ij} + \rho_o\left[C_{ijkl}\epsilon_{kl} + \beta_{ij}\left(\hat{\boldsymbol{\theta}} - \theta_o\right)\right].$$

As written here, the stiffness tensor has 81 components, and the $\boldsymbol{\beta}$ tensor has 9 components. However, the symmetry of both the stress and strain tensors impose requirements on the stiffness tensors before we even account for any material symmetry. Specifically, we have the condition that both the Cauchy stress and the infinitesimal strain be symmetric giving

$$\sigma_{ij} = \sigma_{ji} \quad and \quad \epsilon_{ij} = \epsilon_{ji}.$$

From the symmetry in the stress, we have the condition

$$(\sigma_o)_{ij} + \rho_o\left[C_{ijkl}\epsilon_{kl} + \beta_{ij}\left(\hat{\boldsymbol{\theta}} - \theta_o\right)\right] = (\sigma_o)_{ji} + \rho_o\left[C_{jikl}\epsilon_{kl} + \beta_{ji}\left(\hat{\boldsymbol{\theta}} - \theta_o\right)\right].$$

Since this must be true for all possible temperature changes, strains, and residual stresses, we have three independent conditions

$$C_{ijkl} = C_{jikl},$$
$$(\sigma_o)_{ij} = (\sigma_o)_{ji},$$
$$\beta_{ij} = \beta_{ji}.$$

From the symmetry in the strain tensor, we have the condition

$$(\sigma_o)_{ij} + \rho_o\left[C_{ijkl}\epsilon_{kl} + \beta_{ij}\left(\hat{\boldsymbol{\theta}} - \theta_o\right)\right] = (\sigma_o)_{ij} + \rho_o\left[C_{ijkl}\epsilon_{kl} + \beta_{ij}\left(\hat{\boldsymbol{\theta}} - \theta_o\right)\right],$$

which requires

$$C_{ijkl} = C_{ijlk}.$$

Given these conditions, the number of independent constants in the stiffness tensor reduces to 36 and the residual stress and $\boldsymbol{\beta}$ are each symmetric tensors with 6 independent constants.

If for a moment we assume that there are no temperature changes or residual stresses, we may write the relations between stress and strain as

$$\sigma_{ij} = \rho_o \left(C_{ij11}\epsilon_{11} + C_{ij22}\epsilon_{22} + C_{ij33}\epsilon_{33} + 2C_{ij12}\epsilon_{12} + 2C_{ij13}\epsilon_{13} + 2C_{ij23}\epsilon_{23}\right).$$

We can write this equation in a simplified matrix expression as

$$\begin{bmatrix} \sigma_{11} \\ \sigma_{22} \\ \sigma_{33} \\ \sigma_{23} \\ \sigma_{31} \\ \sigma_{12} \end{bmatrix} = \begin{bmatrix} c_{1111} & c_{1122} & c_{1133} & c_{1123} & c_{1113} & c_{1112} \\ c_{1122} & c_{2222} & c_{2233} & c_{2223} & c_{2213} & c_{2212} \\ c_{1133} & c_{2233} & c_{3333} & c_{3323} & c_{3313} & c_{3312} \\ c_{1123} & c_{2223} & c_{3323} & c_{2323} & c_{2313} & c_{2312} \\ c_{1113} & c_{2213} & c_{3313} & c_{2313} & c_{1313} & c_{1312} \\ c_{1112} & c_{2212} & c_{3312} & c_{2312} & c_{1312} & c_{1212} \end{bmatrix} \begin{bmatrix} \epsilon_{11} \\ \epsilon_{22} \\ \epsilon_{33} \\ 2\epsilon_{23} \\ 2\epsilon_{31} \\ 2\epsilon_{12} \end{bmatrix}.$$

This manner of writing the stress and strain tensors as one-dimensional vectors is known as the Voigt notation. Reintroducing the possible temperature change and the residual stress, we may write

$$\begin{bmatrix} \sigma_{11} \\ \sigma_{22} \\ \sigma_{33} \\ \sigma_{23} \\ \sigma_{31} \\ \sigma_{12} \end{bmatrix} = \begin{bmatrix} (\sigma_o)_{11} \\ (\sigma_o)_{22} \\ (\sigma_o)_{33} \\ (\sigma_o)_{23} \\ (\sigma_o)_{31} \\ (\sigma_o)_{12} \end{bmatrix} + \begin{bmatrix} c_{1111} & c_{1122} & c_{1133} & c_{1123} & c_{1113} & c_{1112} \\ c_{1122} & c_{2222} & c_{2233} & c_{2223} & c_{2213} & c_{2212} \\ c_{1133} & c_{2233} & c_{3333} & c_{3323} & c_{3313} & c_{3312} \\ c_{1123} & c_{2223} & c_{3323} & c_{2323} & c_{2313} & c_{2312} \\ c_{1113} & c_{2213} & c_{3313} & c_{2313} & c_{1313} & c_{1312} \\ c_{1112} & c_{2212} & c_{3312} & c_{2312} & c_{1312} & c_{1212} \end{bmatrix} \begin{bmatrix} \epsilon_{11} \\ \epsilon_{22} \\ \epsilon_{33} \\ 2\epsilon_{23} \\ 2\epsilon_{31} \\ 2\epsilon_{12} \end{bmatrix}$$

$$+ \left(\hat{\theta} - \theta_o\right) \begin{bmatrix} \beta_{11} \\ \beta_{22} \\ \beta_{33} \\ \beta_{23} \\ \beta_{31} \\ \beta_{12} \end{bmatrix}.$$

This will become useful later when we implement this model using finite element analysis.

7.4.5 Material Symmetry

For an isotropic material, the material response must be invariant under all possible rotations of the reference coordinate system. This places additional constraints on the stiffness tensor allowing for only two independent constants. The isotropic stiffness tensor is given by

$$[C] = \begin{bmatrix} \lambda+2\mu & \lambda & \lambda & 0 & 0 & 0 \\ \lambda & \lambda+2\mu & \lambda & 0 & 0 & 0 \\ \lambda & \lambda & \lambda+2\mu & 0 & 0 & 0 \\ 0 & 0 & 0 & \mu & 0 & 0 \\ 0 & 0 & 0 & 0 & \mu & 0 \\ 0 & 0 & 0 & 0 & 0 & \mu \end{bmatrix},$$

where λ and μ are the Lame coefficients. The Lame coefficients, Young's modulus, Poisson's ratio are related via

$$\mathrm{E} = \frac{\mu\,(3\lambda + 2\mu)}{(\lambda + \mu)} \quad \text{and} \quad \nu = \frac{\lambda}{2\,(\lambda + \mu)}.$$

The isotropic material also has the temperature coefficients given by

$$[\boldsymbol{\beta}] = \begin{bmatrix} \beta_{11} & 0 & 0 \\ 0 & \beta_{11} & 0 \\ 0 & 0 & \beta_{11} \end{bmatrix}.$$

In tensor notation, Hooke's law for the isothermal isotropic material can be written as

$$\boldsymbol{\sigma} = \lambda\, tr\,(\boldsymbol{\epsilon})\,\mathbf{I} + 2\mu\,\boldsymbol{\epsilon}.$$

The linear isotropic thermoelastic constitutive law can be written as

$$\boxed{\boldsymbol{\sigma} = \frac{E}{(1+\nu)}\left[\boldsymbol{\epsilon} + \frac{\nu}{(1-2\nu)}\, tr\,(\boldsymbol{\epsilon})\,\mathbf{I}\right] + \frac{\mathrm{E}\alpha\Delta T}{1-2\nu}\,\mathbf{I}.}$$

Only three material parameters are needed to determine the relationship between stress, strain, and temperature change for a linear thermoelastic isotropic material.

7.4.6 Governing Equations for the Isotropic Linear Elastic Material

Since we have assumed that the deformation is small, and we are keeping only first-order terms, the Cauchy stress is equal to the second Piola-Kirchhoff stress, and we may write the conservation of linear momentum as

$$\rho_o \frac{\partial^2 \boldsymbol{u}}{\partial t^2} - \nabla_X \cdot (\boldsymbol{\sigma}) - \rho_o\,\boldsymbol{b} = 0.$$

Substitution of the constitutive relation for an isotropic linear elastic material gives

$$\rho_o \frac{\partial^2 \boldsymbol{u}}{\partial t^2} - \nabla_X \cdot\, (\lambda\, tr\,(\boldsymbol{\epsilon})\,\boldsymbol{I} + 2\mu\,\boldsymbol{\epsilon}) - \rho_o\,\boldsymbol{b} = 0,$$

$$\rho_o \frac{\partial^2 \boldsymbol{u}}{\partial t^2} - \lambda\,(\nabla_X \cdot (tr\,(\boldsymbol{\epsilon})\,\boldsymbol{I})) - 2\mu\,(\nabla_X \cdot \boldsymbol{\epsilon}) - \rho_o\,\boldsymbol{b} = 0.$$

Written in index notation, this becomes

$$\rho_o \frac{\partial^2 u_i}{\partial t^2} - \lambda \left(\frac{\partial}{\partial X_i}\left(\frac{\partial u_k}{\partial X_k}\delta_{ij}\right)\right) - 2\mu\left(\frac{\partial}{\partial X_i}\frac{1}{2}\left(\frac{\partial u_i}{\partial X_j} + \frac{\partial u_j}{\partial X_i}\right)\right) - \rho_o \boldsymbol{b} = 0,$$

$$\rho_o \frac{\partial^2 u_i}{\partial t^2} - \lambda\,\left(u_{k,kj}\right) - \mu\,\left(u_{i,ji} + u_{j,ii}\right) - \rho_o \boldsymbol{b} = 0.$$

The energy equation becomes

$$\rho_o \dot{e}_o - \boldsymbol{P} : \dot{F} - \rho_o\, r_o + \nabla_X \cdot (\boldsymbol{q}_o) = 0,$$

$$\rho_o \dot{e}_o - \lambda\, u_{k,k}\, \dot{F}_{jj} - \mu\,\left(u_{i,j} + u_{j,i}\right) \dot{F}_{ij} - \rho_o\, r_o + (q_o)_{i,i} = 0.$$

Since $\dot{F} = \frac{\partial \dot{u}}{\partial X}$, we may write

$$\rho_o \dot{e}_o - \lambda u_{k,k} \dot{u}_{j,j} - \mu \left(u_{i,j} + u_{j,i}\right) \dot{u}_{i,j} - \rho_o r_o + (q_o)_{i,i} = 0.$$

The linearized governing equations are given by

$$\boxed{\begin{aligned} \rho_o \ddot{u} \, (\lambda + \mu) \, \nabla (\nabla \cdot \boldsymbol{u}) - \mu \nabla^2 \boldsymbol{u} - \rho_o \boldsymbol{b}_o &= 0, \\ \rho_o \dot{e}_o - \lambda \, (\nabla \cdot \boldsymbol{u}) \nabla \dot{u} - \mu \left(\nabla \boldsymbol{u} + \boldsymbol{u} \overleftarrow{\nabla}\right) : \nabla \dot{u} - \rho_o r_o + \nabla \cdot \boldsymbol{q}_o &= 0. \end{aligned}}$$

In index notation, this can be written as

$$\boxed{\begin{aligned} \rho_o \ddot{u}_j - \lambda \left(u_{k,kj}\right) - \mu \left(u_{i,ji} + u_{j,ii}\right) - \rho_o \, (b_o)_j &= 0, \\ \rho_o \dot{e}_o - \lambda \, u_{k,k} \, \dot{u}_{i,j} - \mu \left(u_{i,j} + u_{j,i}\right) \dot{u}_{i,j} - \rho_o r_o + (q_o)_{i,i} &= 0. \end{aligned}}$$

7.5 Uniaxial Tension Test

Every constitutive model has a set of material parameters that are specific to the exact material being modeled. For example, the isotropic linear thermoelastic material model has three parameters, which are the Young's modulus, E, the Poisson's ratio, ν, and the thermal conductivity, α. In comparison, the incompressible neo-Hookean model has only one material parameter, α_1. These material parameters vary from material to material. For example, even though the isotropic thermoelastic material model can be used to describe the small strain behavior of aluminum and carbon steel, the values of the material parameters for each material differs.

In this section, we will develop the field equations for the uniaxial tension test, which can be used to determine the parameters for elastic materials. We will examine closely the field equations that result for the linear thermoelastic material subject to a uniaxial tension test and an incompressible hyperelastic material subject to a uniaxial tension test.

In order to conduct a uniaxial tension test, a dogbone sample is cut out of the material to be tested. The sample is placed between the grips of a testing machine and secured in place. The grips of the testing machine are displaced at a constant rate, and both the displacement of the sample and the load applied to the sample are measured and recorded. Displacement of the sample is often measured by attaching an extensometer directly to the sample as shown in Figure 7.3. The change in lateral dimension may also be monitored using a second extensometer. The recorded force-displacement data can be used to reconstruct a stress-strain curve for the material since the stress in the narrow section of the dogbone is very nearly uniform and axial. The single nonzero component of the Cauchy stress tensor is given by $\sigma_{11} = F_a/A$, where F_a is the instantaneous force applied to the sample, and A is the current cross-sectional area of the sample. Note that as the sample deforms, the cross-sectional area changes. The material model is then fit to the stress-strain curve, and the desired material parameters are extracted from the fitting procedure.

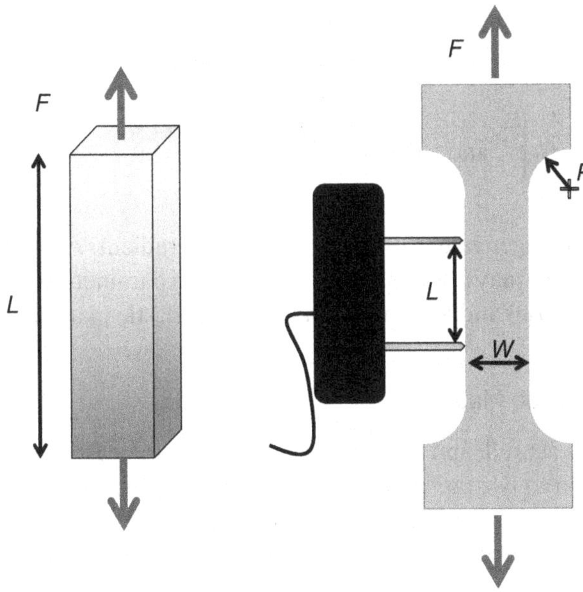

Figure 7.3. Illustration of a tension test used to determine the Young's modulus of a metallic material.

7.5.1 Kinematics

The equations of motion for a homogenous isotropic sample undergoing uniaxial deformation are easily obtained. If we assume that the axis of the specimen is parallel to the $\boldsymbol{e}_1$ basis vector as shown in Figure 7.3, then the equations of motion for the simple uniaxial tension test of an *isotropic* elastic material can be written as

$$x_1 = \lambda_1 X_1,$$
$$x_2 = \lambda_2 X_2,$$
$$x_3 = \lambda_2 X_3.$$

Notice that because there is no difference in the material properties for an isotropic material in the $\boldsymbol{e}_2$ and $\boldsymbol{e}_3$ directions, the stretch ratio, λ_2, will be the same. The same equations of motion are obtained for a transversely isotropic material with the axis of symmetry aligned along the $\boldsymbol{e}_1$ direction. Taking the gradient of the equations of motion, we find that deformation gradient is given by

$$\boldsymbol{F} = \begin{bmatrix} \lambda_1 & 0 & 0 \\ 0 & \lambda_2 & 0 \\ 0 & 0 & \lambda_2 \end{bmatrix}.$$

The infinitesimal strain tensor is given by

$$\epsilon = \frac{1}{2}\left(\boldsymbol{u}\overleftarrow{\nabla}_{\boldsymbol{X}} + \overrightarrow{\nabla}_{\boldsymbol{X}}\boldsymbol{u}\right) = \begin{bmatrix} \lambda_1 - 1 & 0 & 0 \\ 0 & \lambda_2 - 1 & 0 \\ 0 & 0 & \lambda_2 - 1 \end{bmatrix}.$$

The left and right Cauchy deformation tensor may be determined directly from the deformation gradient such that

$$\boldsymbol{B} = \boldsymbol{F} \cdot \boldsymbol{F}^T = \begin{bmatrix} \lambda_1^2 & 0 & 0 \\ 0 & \lambda_2^2 & 0 \\ 0 & 0 & \lambda_2^2 \end{bmatrix} \quad \text{and} \quad \boldsymbol{C} = \boldsymbol{F}^T \cdot \boldsymbol{F} = \begin{bmatrix} \lambda_1^2 & 0 & 0 \\ 0 & \lambda_2^2 & 0 \\ 0 & 0 & \lambda_2^2 \end{bmatrix}.$$

Remember that λ_1 and λ_2 are the eigenvalues of the deformation gradient. Armed with these constitutive equations, we may now determine the material parameters for both the linear thermoelastic material and the neo-Hookean hyperelastic material.

7.5.2 Isotropic Linear Thermoelastic Material

Since the displacement field is measured, and the material is assumed to be isotropic and homogenous, we do not need to use the balance laws to solve for displacement fields. Instead, we can use the constitutive relation directly to relate the measured stress and strain. The constitutive relation for a linear thermoelastic material is given by

$$\sigma = \frac{\mathrm{E}}{(1+\nu)} \left[\epsilon + \frac{\nu}{(1-2\nu)} tr(\epsilon)\, \mathbf{I} \right] + \frac{\mathrm{E}\alpha \Delta T}{1-2\nu} \mathbf{I}.$$

The temperature does not change during the uniaxial test, so $\Delta T = 0$. For this model, we will use only small strain data to calibrate the modulus and Poisson's ratio. As mentioned earlier, the stress state in the material during the uniaxial tension test can be written as

$$\sigma = \begin{bmatrix} F_a/\mathrm{A} & 0 & 0 \\ 0 & 0 & 0 \\ 0 & 0 & 0 \end{bmatrix}.$$

The infinitesimal strain for the uniaxial deformation state is given in the previous section. Combining the stress state, and the equations of motion, the constitutive relation for the axial stress can be written as

$$\sigma_{11} = \frac{\mathrm{E}}{1+\nu} \left[\epsilon_{11} + \frac{\nu}{1-2\nu} (\epsilon_{11} + 2\epsilon_{22}) \right].$$

Since $\epsilon_{22} = \epsilon_{33}$, the normal stress in the $\boldsymbol{e}_2$ and $\boldsymbol{e}_3$ dimensions is given by

$$\sigma_{22} = \sigma_{33} = 0 = \frac{\mathrm{E}}{1+\nu} \left[\left(1 + \frac{2\nu}{1-2\nu} \right) \epsilon_{22} + \frac{\nu}{1-2\nu} \epsilon_{11} \right].$$

Rearranging gives the original definition of the Poisson's ratio

$$\epsilon_{22} = -\nu \epsilon_{11}.$$

The axial stress becomes

$$\sigma_{11} = \mathrm{E} \epsilon_{11}. \tag{7.1}$$

Substituting the stress relation and strain gives

$$\frac{F}{A} = \mathrm{E} \left(\frac{L - L_o}{L_o} \right),$$

where F is the applied load, A is the cross sectional area of the sample, E is the modulus, L is the current length, and L_o is the reference length of the sample. Experimentally, we determine the force and length as a function of time. Determining the Cauchy stress is complicated by the fact that the cross-sectional area changes slightly with time although for small strains this change can be neglected. An additional extensometer may be used to measure the lateral strain and compute the cross-sectional area as a function of time. The modulus can be found by determining the slope of the stress versus strain curve. The Poisson's ratio can be found from the negative of the slope of the lateral strain versus the axial strain curve.

7.5.3 Incompressible Isotropic Neo-Hookean Model

Once again, the kinematics of the uniaxial deformation test are determined from experimental data, and the fact that the material is isotropic and homogenous means that the internal stress can be determined from the axial force. We do not need to use the balance equations to solve for either the displacement or stress fields. Instead, we use the known stress and displacement field to solve for the unknown material parameters. The incompressible neo-Hookean model for a hyperelastic material is given by

$$w = w(I_C) = \alpha_1 (I_C - 3),$$

$$\sigma = -p\mathbf{I} + \frac{2\alpha_1}{\mathrm{J}}\boldsymbol{B}.$$

Since the material is incompressible, the Jacobean must be equal to one throughout the deformation process. Therefore, we have the additional condition that $\det \boldsymbol{F} = 1$. Written in terms of the principal values of the deformation gradient, this gives the condition $\lambda_1\lambda_2\lambda_3 = 1$. Since $\lambda_2 = \lambda_3$, we may write

$$\lambda_2^2 = \frac{1}{\lambda_1}.$$

The deformation gradient may then be written as

$$\boldsymbol{F} = \begin{bmatrix} \lambda_1 & 0 & 0 \\ 0 & \sqrt{\lambda_1}^{-1} & 0 \\ 0 & 0 & \sqrt{\lambda_1}^{-1} \end{bmatrix}.$$

The left Cauchy deformation tensor $\boldsymbol{B} = \boldsymbol{F} \cdot \boldsymbol{F}^T$ becomes

$$\boldsymbol{B} = \begin{bmatrix} \lambda_1^2 & 0 & 0 \\ 0 & \lambda_1^{-1} & 0 \\ 0 & 0 & \lambda_1^{-1} \end{bmatrix}.$$

Substituting the left Cachy deformation tensor into the stress equation gives

$$\sigma = -p \begin{bmatrix} 1 & 0 & 0 \\ 0 & 1 & 0 \\ 0 & 0 & 1 \end{bmatrix} + \frac{2\alpha_1}{\mathrm{J}} \begin{bmatrix} \lambda^2 & 0 & 0 \\ 0 & \lambda^{-1} & 0 \\ 0 & 0 & \lambda^{-1} \end{bmatrix}.$$

Practically speaking, the load cell in a uniaxial test is mounted in series with the sample along the $\boldsymbol{e}_1$ axis. This allows us to measure the axial force and, therefore, the axial stress, $\boldsymbol{\sigma}_{11}$, as a function of time. Simultaneously, strain gauges are used to measure the axial stretch, λ_1. The stress on the lateral faces is zero due to the traction free boundary condition giving $\sigma_{22} = 0$ and $\sigma_{33} = 0$. This gives us the three equations:

$$\sigma_{11} = -p + \frac{2\alpha_1}{J}\lambda_1^2,$$

$$\sigma_{22} = -p + \frac{2\alpha_1}{J}\lambda_1^{-1},$$

$$\sigma_{33} = -p + \frac{2\alpha_1}{J}\lambda_1^{-1}.$$

There are two unknowns parameters, namely, the hydrostatic pressure, p, and the shear modulus, α_1. The equations for the stress components, σ_{22} and σ_{33}, are identical. Therefore, we have two independent equations. Substituting the zero traction condition into the σ_{22} component gives

$$\sigma_{22} = 0 = -p + \frac{2\alpha_1}{J}\lambda_1^{-1}.$$

Solving this equation for the hydrostratic pressure, we find

$$p = \frac{2\alpha_1}{J}\lambda_1^{-1}.$$

Eliminating the hydrostatic pressure from the equation for the axial stress, σ_{11}, gives

$$\sigma_{11} = -\frac{2\alpha_1}{J}\lambda_1^{-1} + \frac{2\alpha_1}{J}\lambda_1^2.$$

Since the material is incompressible, the Jacobian is equal to one, and we have the final equation

$$\sigma_{11} = 2\alpha_1\left(\lambda_1^2 - \lambda_1^{-1}\right).$$

We can now relate the stress that we measure experimentally, σ_{11}, to the experimentally determined stretch $\lambda_1 = \frac{L}{L_o}$.

We have found an equation relating the Cauchy stress, which is the stress per unit current area to the stretch. Experimentally, we measure the force and then convert the force into a stress. Dividing the force by the current area gives us the Cauchy stress while dividing by the reference area gives us the Piola-Kirchhoff stress. For the case of an incompressible material, it is easy to convert between Cauchy and Piola-Kirchhoff stress measurements since the cross-sectional area can be computed as a function of time. However, you could also rewrite the governing equation in terms of the Piola-Kirchhoff stress using the relation

$$\boldsymbol{S} = J\boldsymbol{\sigma}\boldsymbol{F}^{-T}.$$

For the uniaxial tension test, we obtain

$$\boldsymbol{S} = \begin{bmatrix} 2\alpha_1\left(\lambda_1^2 - \frac{1}{\lambda_1}\right) & 0 & 0 \\ 0 & 0 & 0 \\ 0 & 0 & 0 \end{bmatrix} \begin{bmatrix} \lambda_1 & 0 & 0 \\ 0 & \sqrt{\lambda_1}^{-1} & 0 \\ 0 & 0 & \sqrt{\lambda_1}^{-1} \end{bmatrix}.$$

As expected, the only nonzero component of this tensor is the axial stress component

$$S_{11} = 2\alpha_1 \left(\lambda_1 - \lambda_1^{-2}\right).$$

It is also important to note that uniaxial tests for parameterizing hyperelastic models tend to involve large strains. Therefore, the difference between the Cauchy stress and Piola-Kirchhoff stress may be large.

8 Continuum Mixture Theory

When discussing the development of constitutive models for the ideal gas, fluids, and elastic solids, we had restricted the discussion to materials that consist of a single phase. However, there are many applications where a single material model may be used to represent the behavior of multiple interacting phases. For example, consider biological tissue, which we may think of as being made up of a solid component that consists of the extracellular matrix, a fluid component that contains significant quantities of water and smaller amounts of charged particles, and living cells that secrete substances and react to chemical and mechanical stimuli. In addition, cells undergo growth or death and cause remodeling of the extracellular matrix in response to external stimuli. It is the interaction among all of these phenomena over time that gives biological materials such a diverse and interesting response.

Using continuum mixture theory, we can model the extracellular matrix and the permeating fluid with a single model. This model consists of two continuously distributed phases that interact with one another through the transfer of momentum and energy. In reality, the solid phase and the fluid phase cannot occupy the same region of space. The true microstructure of the material may consist of an extracellular network with channels through which fluid is transported, but the model treats both phases as coexisting at the same spatial points (Figure 8.1). The continuum multiphase model performs well at capturing the aggregate behavior for many materials when the alternative, which is to model the detailed microscopic structure and phase interactions directly, is computationally intractable.

In this chapter, we will demonstrate the application of continuum mixture theory to a biphasic material consisting of two thermo-mechanical phases. Specifically, we will look at a material that consists of a thermo-mechanical solid infused with a thermo-mechanical fluid. Keep in mind that the theory is general and could be applied to more than two phases or to phases with more complex behavior.

8.1 Forces and Fields

As stated in the previous section, we have limited ourselves to a biphasic material that consists of a thermo-mechanical solid and a thermo-mechanical fluid phase. Each of these phases interacts with its surroundings via mechanical and thermal forces. Therefore, we must include in the formulation all of the variables discussed

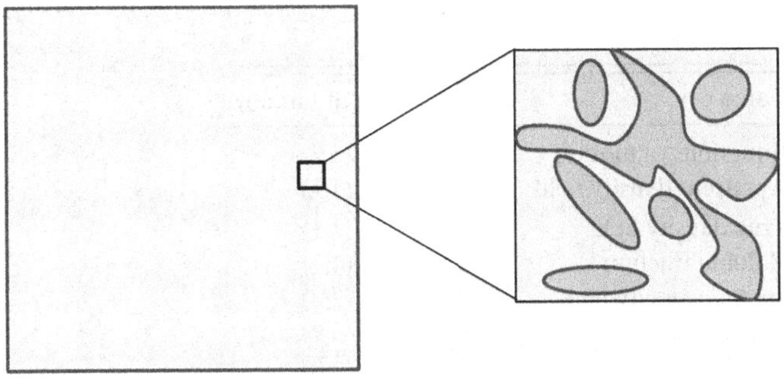

Figure 8.1. Illustration showing the actual distribution of solid and fluid phases in a porous material. At any given point in space is occupied by either the solid of fluid phase.

in the development of the thermo-mechancial fluid and thermo-mechancial solid models. In addition, the multiphase models include variables specifying the amount of each phase located at each point within the material and variables that account for transfer of momentum and a transfer of energy between the phases.

Let us assume that we have a material that consists of n phases labeled as $\alpha = 1 \ldots n$. Our continuum model consists of a set of overlapping continuous fields. Therefore, at any given location, $\boldsymbol{x}$, within the material, both fluid and solid are present. If we take some small infinitesimal volume of the material, dV_x, it is occupied by some amount of each phase. In order to quantify the distribution of material, we need to know how much mass of each phase is located within the infinitesimal volume, and what fraction of the infinitesimal volume each phase occupies. Therefore, we introduce the volume fraction of each phase, ϕ^α, which is given by

$$\phi^\alpha(\boldsymbol{x},t) \equiv \frac{dV_x^\alpha}{dV_x},$$

where dV_x^α is the part of the infinitesimal volume occupied by the α phase, and dV_x is the total infinitesimal volume. Notice that since the sum of the volumes occupied by each phase must equal the total infinitesimal volume, $\sum_\alpha dV_x^\alpha = dV_x$, the sum of the volume fractions at any given point within a material must be equal to one,

$$\sum_\alpha \phi^\alpha = 1. \tag{8.1}$$

Next, we use the variable dm^α to represent the incremental mass of the α phase within an infinitesimal volume of material, dV_x. We are now free to define density fields in the material. There are two density fields that can be defined for each phase in the mixture. We have the *apparent density*, ρ^α, at each point in the material, which is the mass of each phase divided by the total infinitesimal volume,

$$\rho^\alpha(\boldsymbol{x},t) = \frac{dm^\alpha}{dV_x}.$$

The *true density* of each phase is equal to the incremental mass divided by the volume occupied by the given phase such that

$$\rho_T^\alpha \equiv \frac{dm^\alpha}{dV_x^\alpha} = \frac{\rho^\alpha}{\phi^\alpha}. \tag{8.2}$$

Table 8.1. *List of variables for a two-phase mixture*

Fluid phase	Solid phase	Name	Number of unknowns
	$\boldsymbol{x}^s(\boldsymbol{X},t)$	Equations of motion	(0,3)
$\rho^f(\boldsymbol{x},t)$	$\rho^s(\boldsymbol{x},t)$	Apparent density field	(1,1)
$\rho^f_T(\boldsymbol{x},t)$	$\rho^s_T(\boldsymbol{x},t)$	True density field	(1,1)
$\phi^f(\boldsymbol{x},t)$	$\phi^s(\boldsymbol{x},t)$	Volume fraction	(1,1)
$\boldsymbol{v}^f(\boldsymbol{x},t)$	$\boldsymbol{v}^s(\boldsymbol{x},t)$	Spatial velocity field	(3,3)
$\boldsymbol{\sigma}^f(\boldsymbol{x},t)$	$\boldsymbol{\sigma}^s(\boldsymbol{x},t)$	Cauchy stress tensor	(9,9)
$\boldsymbol{b}^f(\boldsymbol{x},t)$	$\boldsymbol{b}^s(\boldsymbol{x},t)$	Spatial body force field	(0,0)
$\boldsymbol{\pi}^{fs}(\boldsymbol{x},t)$	$\boldsymbol{\pi}^{sf}(\boldsymbol{x},t)$	Linear momentum transfer	(3,3)
$\Pi^{fs}(\boldsymbol{x},t)$	$\Pi^{sf}(\boldsymbol{x},t)$	Energy transfer	(1,1)
$r^f(\boldsymbol{x},t)$	$r^s(\boldsymbol{x},t)$	Heat generation	(0,0)
$\boldsymbol{q}^f(\boldsymbol{x},t)$	$\boldsymbol{q}^s(\boldsymbol{x},t)$	Heat flux	(3,3)
$e^f(\boldsymbol{x},t)$	$e^s(\boldsymbol{x},t)$	Internal energy	(1,1)
$\theta^f(\boldsymbol{x},t)$	$\theta^s(\boldsymbol{x},t)$	Temperature	(1,1)
$\eta^f(\boldsymbol{x},t)$	$\eta^s(\boldsymbol{x},t)$	Entropy	(1,1)

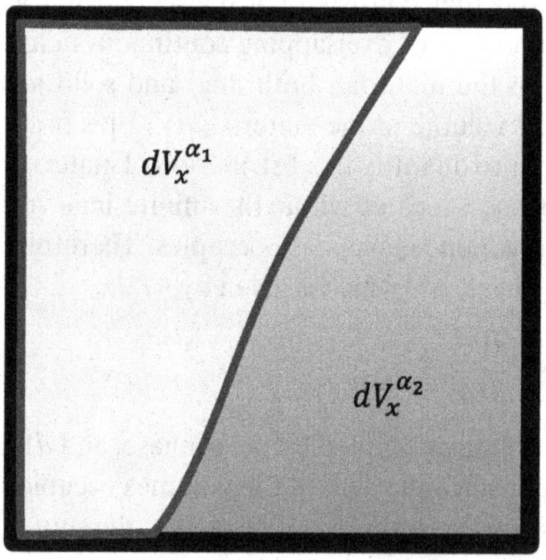

Figure 8.2. Illustration of the distribution of material within an infinitesimal volume $dV_x = dV_x^{\alpha_1} + dV_x^z$, where $dV_x^{\alpha_1}$ is the volume occupied by phase α_1, and $dV_x^{\alpha_z}$ is the volume occupied by α_z.

To illustrate the difference between these, let us imagine that we have a two-phase mixture where one phase is a gas. In this mixture, the ideal gas law could be used to relate the gas pressure to its true density, ρ^{α}_T, not the apparent density.

A full list of variables for a thermal biphasic model consisting of a single fluid phase and a single solid phase are listed in Table 8.1.

8.2 Balance Laws

Continuum mixture theory involves the assumption that we have multiple interacting phases that can each be modeled as a continuum field. The balance laws must be derived for each field and the mixture. In this section, we will derive each of the balance laws as they apply to continuum mixture theory.

8.2.1 Conservation of Mass

Now that we have a definition of both mass and density, we may now write down the conservation of mass equation for our mixture. Remember that conservation of mass may be applied to individual components and to the mixture as a whole. For each component, we obtain the integral form of the conservation of mass as

$$\frac{D}{Dt}\left[m^{\alpha}(t)\right] = \frac{D}{Dt}\int_{\Omega_t} \rho^{\alpha}(\boldsymbol{x},t)\, dV_x,$$

where m^{α} and ρ^{α} is the mass and the apparent density of phase α. Notice that we have used the apparent density because the incremental mass is given by $dm^{\alpha} = \rho^{\alpha} dV_x$. If we assume that the system contains neither a source nor a sink for either material phase, then the rate of change of the mass for each phase is zero and

$$\frac{D}{Dt}\left[m^{\alpha}(t)\right] = 0.$$

Using the Reynolds transport theorem in the form given in Equation (2.54), the differential form of the conservation of mass becomes

$$\frac{D\rho^{\alpha}}{Dt} + \rho^{\alpha}\, div\left(\boldsymbol{v}^{\alpha}\right) = 0,$$

where $\boldsymbol{v}^{\alpha}$ is the velocity of the α phase. Expanding the material derivative, we may write the equivalent equation

$$\frac{\partial\rho^{\alpha}}{\partial t} + div\left(\rho^{\alpha}\boldsymbol{v}^{\alpha}\right) = 0.$$

We may also write a statement of conservation of mass for the entire mixture. In this case, the rate of change of the total mass of the system, m, is given by

$$\frac{D}{Dt}\left[m(t)\right] = \sum_{\alpha}\frac{D}{Dt}\int_{\Omega_t} \rho^{\alpha}(\boldsymbol{x},t)\, dV_x. \tag{8.3}$$

Instead, imagine we have an external observer who has no knowledge of the internal structure or phase makeup of the material. The observer sees only an aggregate density of the total mixture, $\rho(\boldsymbol{x},t)$. This external observer could write the conservation of mass for the system as

$$\frac{D}{Dt}\left[m(t)\right] = \frac{D}{Dt}\int_{\Omega_t} \rho(\boldsymbol{x},t)\, dV_x. \tag{8.4}$$

Equations (8.3) and (8.4) are both valid; therefore, we may define the density of the mixture, $\rho(\boldsymbol{x},t)$, as the sum of the apparent densities for each phase

$$\rho(\boldsymbol{x},t) = \sum_{\alpha}\rho^{\alpha}(\boldsymbol{x},t).$$

8.2.2 Conservation of Momentum

Conservation of momentum requires that for each phase, the sum of the change in linear momentum is equal to the sum of the forces acting on the phase. However, when considering the conservation of linear momentum for each individual phase, we must account for the transfer of linear momentum between phases. Let us consider the example of a porous solid filled with a fluid. As the solid material is compressed, the pores containing the fluid are constricted, increasing the fluid pressure and driving a flow field through the pores. Also, the fluid passing through the small channels in the material exerts both a normal force due to the fluid pressure and a shear force due to the fluid viscosity on the solid walls of the channel. In this manner, the fluid and solid transfer momentum at a microscopic level.

We will denote the linear momentum transferred from phase β into the α phase as $\boldsymbol{\pi}^{\alpha\beta}$. We begin with the general statement of conservation of linear momentum for each individual phase,

$$\frac{D}{Dt}\left(m^{\alpha}\boldsymbol{v}^{\alpha}\right)=\sum \boldsymbol{f}^{\alpha}.$$

The external forces applied to a phase include the surface traction, $\boldsymbol{t}^{\alpha}$, the external body force, $\boldsymbol{b}^{\alpha}$, and the transfer of linear momentum, $\boldsymbol{\pi}^{\alpha\beta}$, such that the conservation of linear momentum for that phase becomes

$$\frac{D}{Dt}\int_{\Omega_t}\rho^{\alpha}\boldsymbol{v}^{\alpha}\,dV_x=\int_{\partial\Omega_t}\boldsymbol{t}^{\alpha}\,dS_x+\int_{\Omega_t}\rho^{\alpha}\boldsymbol{b}^{\alpha}\,dV_x+\sum_{\beta}\int_{\Omega_t}\boldsymbol{\pi}^{\alpha\beta}\,dV_x.$$

From this equation, we obtain the local differential form as

$$\rho^{\alpha}\dot{\boldsymbol{v}}^{\alpha}-div\left(\boldsymbol{\sigma}^{\alpha}\right)-\rho^{\alpha}\boldsymbol{b}^{\alpha}-\sum_{\beta}\boldsymbol{\pi}^{\alpha\beta}=\boldsymbol{0}.$$

We may also write the statement of conservation of linear momentum for the mixture by adding the equations for each component such that

$$\sum_{\alpha}\frac{D}{Dt}\int_{\Omega_t}\rho\boldsymbol{v}\,dV_x=\sum_{\alpha}\int_{\partial\Omega_t}\boldsymbol{t}\,dS_x+\sum_{\alpha}\int_{\Omega_t}\rho\boldsymbol{b}\,dV_x+\sum_{\alpha}\sum_{\beta}\int_{\Omega_t}\boldsymbol{\pi}^{\alpha\beta}\,dV_x,$$

with the local differential form becoming

$$\sum_{\alpha}\left(\rho^{\alpha}\dot{\boldsymbol{v}}^{\alpha}-div\left(\boldsymbol{\sigma}^{\alpha}\right)-\rho^{\alpha}\boldsymbol{b}^{\alpha}-\sum_{\beta}\boldsymbol{\pi}^{\alpha\beta}\right)=\boldsymbol{0}. \tag{8.5}$$

Again, we imagine that an external observer with no knowledge of the microstructure is writing a statement of conservation of linear momentum for the mixture. This observer would have no knowledge of the interactions between phases and would write the conservation of linear momentum for the mixture as

$$\frac{D}{Dt}\int_{\Omega_t}\rho\boldsymbol{v}\,dV_x=\int_{\partial\Omega_t}\boldsymbol{t}\,dS_x+\int_{\Omega_t}\rho\boldsymbol{b}\,dV_x.$$

The local differential form becomes

$$\rho\dot{\boldsymbol{v}}-div\left(\boldsymbol{\sigma}\right)-\rho\boldsymbol{b}=\boldsymbol{0}. \tag{8.6}$$

where ρ is the average density of the mixture, $\boldsymbol{v}$ is the average velocity, $\boldsymbol{t}$ is the average surface traction, and $\boldsymbol{b}$ is an averaged body force. These average values are the quantities measured by the external observer. In order to reconcile Equations (8.5) and (8.6), we define the stress, $\boldsymbol{\sigma}$, the body force, $\boldsymbol{b}$, and the time rate of change of the linear momentum of the mixture, $\rho\dot{\boldsymbol{v}}$, as

$$\rho\dot{\boldsymbol{v}} = \sum_{\alpha} \left(\rho^{\alpha}\dot{\boldsymbol{v}}^{\alpha}\right),$$

$$\boldsymbol{\sigma} = \sum_{\alpha} \boldsymbol{\sigma}^{\alpha},$$

$$\rho\boldsymbol{b} = \sum_{\alpha} \rho^{\alpha}\boldsymbol{b}^{\alpha}.$$

In addition, we have the requirement that the transfer of linear momentum from one phase to another must be matched by an equal and opposite transfer that requires the following constraint on the transfer of linear momentum

$$\sum_{\alpha}\sum_{\beta} \boldsymbol{\pi}^{\alpha\beta} = 0. \tag{8.7}$$

8.2.3 Conservation of Angular Momentum

Conservation of angular momentum applied to each phase in the material states that the time rate of change of the angular momentum for each phase is equal to the sum of the moments, $\mathbf{m}^{\alpha}$ applied to that phase such that

$$\frac{D}{Dt}\int_{\Omega_t} \left(\boldsymbol{r} \times \rho^{\alpha}\boldsymbol{v}^{\alpha}\right) dV_x = \mathbf{m}^{\alpha}.$$

In addition to the applied moment due to the surface traction, body forces, and transfer of linear momentum, there may be a direct transfer of angular momentum between the two phases we will not consider in this chapter. The statement of conservation of angular momentum for the each phase within the mixture become

$$\frac{D}{Dt}\int_{\Omega_t} \left(r \times \rho^{\alpha}\boldsymbol{v}^{\alpha}\right) dV_x = \int_{\partial\Omega_t} \left(r \times \boldsymbol{t}^{\alpha}\right) dS_x + \int_{\Omega_t} \left(r \times \rho^{\alpha}\boldsymbol{b}^{\alpha}\right) dV_x$$
$$+\sum_{\beta}\left(\int_{\Omega_t} \left(r \times \boldsymbol{\pi}^{\alpha\beta}\right) dV_x + \int_{\Omega_t} \boldsymbol{m}^{\alpha\beta}\, dV_x\right).$$

By using the Reynolds transport theorem, we can write the conservation of angular momentum for the α phase as

$$\int_{\Omega_t} \left(r \times \rho^{\alpha}\dot{\boldsymbol{v}}^{\alpha}\right) dV_x = \int_{\partial\Omega_t} \left(r \times \boldsymbol{t}^{\alpha}\right) dS_x + \int_{\Omega_t} \left(r \times \rho^{\alpha}\boldsymbol{b}^{\alpha}\right) dV_x + \sum_{\beta}\left(\int_{\Omega_t} \left(r \times \boldsymbol{\pi}^{\alpha\beta}\right) dV_x\right).$$

We next write this in index notation and use the divergence theorem to convert the surface integral into a volume integral

$$0 = \int_{\Omega_t} \left(\varepsilon_{ijk} r_j \left(\rho^{\alpha} b_k^{\alpha} + \sum_{\beta} \pi_k^{\alpha\beta} + \rho^{\alpha}\dot{v}_k^{\alpha} + \sigma_{lj,l}^{\alpha}\right) + \varepsilon_{ijk}\sigma_{jk}^{\alpha}\right) dV_x.$$

The statement of conservation of linear momentum, Equation (8.5), may be used to simplify this equation giving the final result

$$0 = \varepsilon_{ijk}\sigma^{\alpha}_{jk}. \tag{8.8}$$

If we expand this equation, we find

$$0 = \sigma^{\alpha}_{23} - \sigma^{\alpha}_{32},$$
$$0 = \sigma^{\alpha}_{31} - \sigma^{\alpha}_{13},$$
$$0 = \sigma^{\alpha}_{12} - \sigma^{\alpha}_{21}.$$

Equivalently, we may write

$$\boldsymbol{\sigma}^{\alpha} = \left(\boldsymbol{\sigma}^{\alpha}\right)^{T}.$$

Therefore, the stress tensor will be symmetric for each phase in the mixture provided there is no transfer of angular momentum between phases. The statement of conservation of angular momentum for the entire mixture can be found by summing Equation (8.8) for each phase. This gives

$$0 = \sum_{\alpha} \varepsilon_{ijk}\sigma^{\alpha}_{jk},$$
$$0 = \varepsilon_{ijk}\sum_{\alpha} \sigma^{\alpha}_{jk}. \tag{8.9}$$

Note that an external observer who has no knowledge of the microstructure and the transfer of angular momentum between phases, would write the statement for conservation of angular momentum of the mixture as

$$\frac{D}{Dt}\int_{\Omega_t} (r \times \rho \boldsymbol{v})\, dV_x = \int_{\partial\Omega_t} (r \times \boldsymbol{t})\, dS_x + \int_{\Omega_t} \left(r \times \rho \boldsymbol{b}\right) dV_x.$$

and the local form would be

$$0 = \varepsilon_{ijk}\sigma_{jk}. \tag{8.10}$$

Therefore, we learn that in order for Equation (8.9) and Equation (8.10) to agree, we require only that $\boldsymbol{\sigma} = \sum_{\alpha} \boldsymbol{\sigma}^{\alpha}$. This is identical to the requirement obtained from the conservation of momentum.

8.2.4 Conservation of Energy

When formulating the conservation of energy equation for each phase, we must account for the energy transferred between phases. The conservation of energy for each phase requires that

$$\frac{D}{Dt}\int_{\Omega_t} \left(\rho^{\alpha} e^{\alpha} + \frac{1}{2}\rho^{\alpha} \boldsymbol{v}^{\alpha} \cdot \boldsymbol{v}^{\alpha}\right) dV_x = \int_{\partial\Omega_t} (\boldsymbol{t}^{\alpha} \cdot \boldsymbol{v}^{\alpha} - \boldsymbol{q}^{\alpha} \cdot \boldsymbol{n})\, dS_x$$
$$+ \int_{\Omega_t} \left(\rho^{\alpha} \boldsymbol{b}^{\alpha} \cdot \boldsymbol{v}^{\alpha} + \rho^{\alpha} r^{\alpha} + \sum_{\beta} \Pi^{\alpha\beta}\right) dV_x,$$

where $\Pi^{\alpha\beta}$ is the energy transferred from the β phase to the α phase by means other than the work done by the transfer of linear momentum. Using the Reynolds transport theorem and the divergence theorem, $div\left(\boldsymbol{\sigma}^T \cdot \boldsymbol{v}\right) = div\,\boldsymbol{\sigma}^T \cdot \boldsymbol{v} + \boldsymbol{\sigma} : \left(\boldsymbol{v}\overleftarrow{\nabla}\right)$, and we can write this as

$$\int_{\Omega_t} \left(\rho^\alpha \dot{e}^\alpha + \boldsymbol{v}^\alpha \cdot \left(\rho^\alpha \dot{\boldsymbol{v}}^\alpha - div\left(\boldsymbol{\sigma}^\alpha\right) - \rho^\alpha \boldsymbol{b}^\alpha\right) - \boldsymbol{\sigma}^\alpha : \boldsymbol{L}^\alpha - \rho^\alpha r^\alpha + div\,\boldsymbol{q}^\alpha - \sum_\beta \Pi^{\alpha\beta} \right) dV_x = 0.$$

Substitution of the conservation of linear momentum, Equation (8.5), gives

$$\rho^\alpha \dot{e}^\alpha - \boldsymbol{\sigma}^\alpha : \boldsymbol{L}^\alpha - \rho^\alpha r^\alpha + div\,\boldsymbol{q}^\alpha + \sum_\beta \left(\boldsymbol{\pi}^{\alpha\beta} \cdot \boldsymbol{v}^\alpha\right) - \sum_\beta \Pi^{\alpha\beta} = 0.$$

Conservation of energy for the entire mixture can be found by summing this equation over all of the phases in the system such that

$$\sum_\alpha \left(\rho^\alpha \dot{e}^\alpha - \boldsymbol{\sigma}^\alpha : \boldsymbol{L}^\alpha - \rho^\alpha r^\alpha + div\,\boldsymbol{q}^\alpha + \sum_\beta \left(\boldsymbol{\pi}^{\alpha\beta} \cdot \boldsymbol{v}^\alpha\right) - \sum_\beta \Pi^{\alpha\beta} \right) = 0. \tag{8.11}$$

An external observer who lacks the ability to observe individual phases would write the conservation of energy for the mixture as

$$\rho \dot{e} - \boldsymbol{\sigma} : \boldsymbol{L} - \rho r + div\,\boldsymbol{q} = 0. \tag{8.12}$$

In order to reconcile Equation (8.11) and (8.12), the average quantities must be related to the microscopic quantities through the transformation

$$\rho \dot{e} = \sum_\alpha \rho^\alpha \dot{e}^\alpha,$$

$$\rho r = \sum_\alpha \rho^\alpha r^\alpha,$$

$$\boldsymbol{q} = \sum_\alpha \boldsymbol{q}^\alpha,$$

$$\boldsymbol{\sigma} : \boldsymbol{L} = \sum_\alpha \boldsymbol{\sigma}^\alpha : \boldsymbol{L}^\alpha.$$

In addition, we have the constraint that the sum of the energy transferred between all of the phases and work done by the internal transfer forces must be zero

$$\sum_{\alpha,\beta} \left(\boldsymbol{\pi}^{\alpha\beta} \cdot \boldsymbol{v}^\alpha - \Pi^{\alpha\beta}\right) = 0.$$

8.2.5 Second Law of Thermodynamics

The second law of thermodynamics must be satisfied for the entire mixture and not for the individual phases

$$\sum_\alpha \left(\rho^\alpha \theta^\alpha \frac{D\eta^\alpha}{Dt} + div\,\boldsymbol{q}^\alpha - \frac{\boldsymbol{q}^\alpha \cdot grad\,\theta^\alpha}{\theta^\alpha} - \rho^\alpha r^\alpha \right) \geq 0.$$

Table 8.2. *List of variables for a two-phase mixture*

Fluid phase	Solid phase	Name	Number of unknowns
	$\boldsymbol{x}^s(\boldsymbol{X},t)$	Equations of motion	(0,3)
$\rho^f(\boldsymbol{x},t)$	$\rho^s(\boldsymbol{x},t)$	Apparent density field	(1,1)
$\rho_T^f(\boldsymbol{x},t)$	$\rho_T^s(\boldsymbol{x},t)$	True density field	(1,1)
$\phi^f(\boldsymbol{x},t)$	$\phi^s(\boldsymbol{x},t)$	Volume fraction	(1,1)
$\boldsymbol{v}^f(\boldsymbol{x},t)$	$\boldsymbol{v}^s(\boldsymbol{x},t)$	Spatial velocity field	(3,3)
$\boldsymbol{\sigma}^f(\boldsymbol{x},t)$	$\boldsymbol{\sigma}^s(\boldsymbol{x},t)$	Cauchy stress tensor	(9,9)
$\boldsymbol{b}^f(\boldsymbol{x},t)$	$\boldsymbol{b}^s(\boldsymbol{x},t)$	Spatial body force field	(0,0)
$\boldsymbol{\pi}^{fs}(\boldsymbol{x},t)$	$\boldsymbol{\pi}^{sf}(\boldsymbol{x},t)$	Linear momentum transfer	(3,3)
$\Pi^{fs}(\boldsymbol{x},t)$	$\Pi^{sf}(\boldsymbol{x},t)$	Energy transfer	(1,1)
$r^f(\boldsymbol{x},t)$	$r^s(\boldsymbol{x},t)$	Heat generation	(0,0)
$\boldsymbol{q}^f(\boldsymbol{x},t)$	$\boldsymbol{q}^s(\boldsymbol{x},t)$	Heat flux	(3,3)
$e^f(\boldsymbol{x},t)$	$e^s(\boldsymbol{x},t)$	Internal energy	(1,1)
$\theta^f(\boldsymbol{x},t)$	$\theta^s(\boldsymbol{x},t)$	Temperature	(1,1)
$\eta^f(\boldsymbol{x},t)$	$\eta^s(\boldsymbol{x},t)$	Entropy	(1,1)

Substitution of the conservation of energy gives

$$\sum_\alpha \left(\rho^\alpha \theta^\alpha \dot{\eta}^\alpha - \frac{\boldsymbol{q}^\alpha \cdot grad\, \theta^\alpha}{\theta^\alpha} - \rho^\alpha \dot{e}^\alpha + \boldsymbol{\sigma}^\alpha : \boldsymbol{L}^\alpha - \sum_\beta (\Pi^{\alpha\beta}) \right) \geq 0.$$

Substituting the condition on energy transfer

$$\sum_\alpha \left(\rho^\alpha \theta^\alpha \dot{\eta}^\alpha - \frac{\boldsymbol{q}^\alpha \cdot grad\, \theta^\alpha}{\theta^\alpha} - \rho^\alpha \dot{e}^\alpha + \boldsymbol{\sigma}^\alpha : \boldsymbol{L}^\alpha + \sum_\beta (\boldsymbol{\pi}^{\alpha\beta} \cdot \boldsymbol{v}^\alpha) \right) \geq 0.$$

Written in terms of the Helmholtz free energy we have the relation

$$\sum_\alpha \left(-\rho^\alpha \dot{\psi}^\alpha - \rho^\alpha \dot{\theta}^\alpha \eta^\alpha - \frac{\boldsymbol{q}^\alpha \cdot grad\, \theta^\alpha}{\theta^\alpha} + \boldsymbol{\sigma}^\alpha : \boldsymbol{L}^\alpha + \sum_\beta (\boldsymbol{\pi}^{\alpha\beta} \cdot \boldsymbol{v}^\alpha) \right) \geq 0.$$

8.3 Biphasic Model

Applying continuum mixture theory to a two-phase material consisting of a porous solid phase saturated with liquid phase, we obtain a total of 53 unknowns (Table 8.2), and a system of 27 equations (Table 8.3). This requires the addition of 26 constitutive equation.

8.4 Isothermal Biphasic Model

By limiting ourselves to an isothermal model two-phase model, we greatly simplify the system of governing equations. We obtain 39 variables (Table 8.4), with 23 equations (Table 8.5). Therefore, we must specify 16 constitutive equations.

Table 8.3. *List of the spatial forms of the balance equations and other relevant equations for a two-phase mixture*

Equation	Fluid phase	Solid phase	Independent equations (material, spatial)
Density–volume fraction relation	$\rho^f = \phi^f \rho_T^f$	$\rho^s = \phi^s \rho_T^s$	(1,1)
Kinematical relations		$\boldsymbol{v}^s = \frac{\partial \boldsymbol{x}^s}{\partial t}$	(0,3)
Conservation of mass	$\frac{\partial \rho^f}{\partial t} + div\left(\rho^f \boldsymbol{v}^f\right) = 0$	$\frac{\partial \rho^s}{\partial t} + div\left(\rho^s \boldsymbol{v}^s\right) = 0$	(1,1)
Conservation of linear momentum	$\rho^f \dot{\boldsymbol{v}}^f - div\left(\boldsymbol{\sigma}^f\right) - \rho^f \boldsymbol{b}^f - \boldsymbol{\pi}^{fs} = 0$	$\rho^s \dot{\boldsymbol{v}}^s - div\left(\boldsymbol{\sigma}^s\right) - \rho^s \boldsymbol{b}^s - \boldsymbol{\pi}^{sf} = 0$	(3,3)
Conservation of angular momentum	$\boldsymbol{\sigma}^f = \left(\boldsymbol{\sigma}^f\right)^T$	$\boldsymbol{\sigma}^s = \left(\boldsymbol{\sigma}^s\right)^T$	(3,3)
Conservation of energy	$\rho^f \dot{e}^f - \boldsymbol{\sigma}^f : \boldsymbol{L}^f - \rho^f r^f + div\, \boldsymbol{q}^f - \Pi^{fs} = 0$	$\rho^s \dot{e}^s - \boldsymbol{\sigma}^s : \boldsymbol{L}^s - \rho^s r^s + div\, \boldsymbol{q}^s - \Pi^{sf} = 0$	(1,1)
Constraint on temperature	$\theta^f = \theta^s$		(1)
Constraint on transfer of linear momentum	$\boldsymbol{\pi}^{sf} + \boldsymbol{\pi}^{fs} = \boldsymbol{0}$		(3)
Constraint on transfer of energy	$\Pi^{fs} - \Pi^{sf} = 0$		(1)
Constraint on volume fraction	$\phi^s + \phi^f = 1$		(1)
Second law	$\sum_{\alpha=s,f}\left(-\rho^\alpha \dot{\psi}^\alpha - \rho^\alpha \dot{\theta}^\alpha \eta^\alpha - \frac{\boldsymbol{q}^\alpha \cdot grad\, \theta^\alpha}{\theta^\alpha} + \boldsymbol{\sigma}^\alpha : \boldsymbol{L}^\alpha + \sum_\beta \left(\boldsymbol{\pi}^{\alpha\beta} \cdot \boldsymbol{v}^\alpha\right)\right) \geq 0$		(0)

Table 8.4. *List of variables for a two-phase isothermal continuum model*

Fluid phase	Solid phase	Name	Number of unknowns
	$\boldsymbol{x}^s(\boldsymbol{X},t)$	Equations of motion	(0,3)
$\rho^f(\boldsymbol{x},t)$	$\rho^s(\boldsymbol{x},t)$	Apparent density field	(1,1)
$\rho_T^f(\boldsymbol{x},t)$	$\rho_T^s(\boldsymbol{x},t)$	True density field	(1,1)
$\phi^f(\boldsymbol{x},t)$	$\phi^s(\boldsymbol{x},t)$	Volume fraction	(1,1)
$\boldsymbol{v}^f(\boldsymbol{x},t)$	$\boldsymbol{v}^s(\boldsymbol{x},t)$	Spatial velocity field	(3,3)
$\boldsymbol{\sigma}^f(\boldsymbol{x},t)$	$\boldsymbol{\sigma}^s(\boldsymbol{x},t)$	Cauchy stress tensor	(9,9)
$\boldsymbol{b}^f(\boldsymbol{x},t)$	$\boldsymbol{b}^s(\boldsymbol{x},t)$	Spatial body force field	(0,0)
$\boldsymbol{\pi}^{fs}(\boldsymbol{x},t)$	$\boldsymbol{\pi}^{sf}(\boldsymbol{x},t)$	Linear momentum transfer	(3,3)

Table 8.5. *List of the spatial forms of the balance equations and other relevant equations for an isothermal two-phase model*

Equation	Fluid phase	Solid phase	Independent equations (material, spatial)
Density – volume fraction relation	$\rho^f = \phi^f \rho_T^f$	$\rho^s = \phi^s \rho_T^s$	(1,1)
Kinematical relations		$\boldsymbol{v}^s = \frac{\partial \boldsymbol{x}^s}{\partial t}$	(0,3)
Conservation of mass	$\frac{\partial \rho^f}{\partial t} + div\,(\rho^f \boldsymbol{v}^f) = 0$	$\frac{\partial \rho^s}{\partial t} + div\,(\rho^s \boldsymbol{v}^s) = 0$	(1,1)
Conservation of linear momentum	$\rho^f \dot{\boldsymbol{v}}^f - div\,(\boldsymbol{\sigma}^f) - \rho^f \boldsymbol{b}^f - \boldsymbol{\pi}^{fs} = 0$	$\rho^s \dot{\boldsymbol{v}}^s - div\,(\boldsymbol{\sigma}^s) - \rho^s \boldsymbol{b}^s - \boldsymbol{\pi}^{sf} = 0$	(3,3)
Conservation of angular momentum	$\boldsymbol{\sigma}^f = (\boldsymbol{\sigma}^f)^T$	$\boldsymbol{\sigma}^s = (\boldsymbol{\sigma}^s)^T$	(3,3)
Constraint on transfer of linear momentum	$\boldsymbol{\pi}^{sf} + \boldsymbol{\pi}^{fs} = \boldsymbol{0}$		(3)
Constraint on volume fraction	$\phi^s + \phi^f = 1$		(1)
Second law	$\sum_{\alpha=s,f} \left(-\rho^\alpha \dot{\psi}^\alpha + \boldsymbol{\sigma}^\alpha : \boldsymbol{L}^\alpha + \sum_\beta (\boldsymbol{\pi}^{\alpha\beta} \cdot \boldsymbol{v}^\alpha) \right) \geq 0$		(0)

8.5 Application to Soft Tissue

Let us explore the application of the isothermal biphasic model to the study of biological tissue. Assuming the tissue can be modeled as a solid and fluid phase, we obtain the governing equations outlined in the previous section. The constitutive model for the isothermal biphasic model requires an equation describing the stress in the fluid, an equation for the stress in the solid, and an equation for the transfer of linear momentum between the two phases. The stress in the solid component will depend on the deformation gradient of the solid, $\boldsymbol{F}^s$, the stress in the fluid phase will be some function of the strain rate tensor for the fluid, $\boldsymbol{D}^f$, and the transfer of linear momentum between the solid and the fluid will depend on the relative velocities between the two components which requires the addition of $\boldsymbol{v}^f$ and $\boldsymbol{v}^s$. Therefore, we propose a constitutive model, which may be written in generic form as

$$\left\{ \psi^s, \psi^f, \boldsymbol{\sigma}^s, \boldsymbol{\sigma}^f, \boldsymbol{\pi}^s \right\} = f\left(\boldsymbol{F}^s, \boldsymbol{D}^f, \boldsymbol{v}^s - \boldsymbol{v}^f \right).$$

We will make a series of additional assumptions in order to simplify the analysis. Specifically, we will make the following assumptions.

1. The transfer of linear momentum between the fluid and solid phases is proportional to the relative velocity between solid and fluid such that

$$\boldsymbol{\pi}^s = -\boldsymbol{\pi}^f = K\left(\boldsymbol{v}^f - \boldsymbol{v}^s\right) + p\,\nabla\phi^s, \tag{8.13}$$

where $K = \frac{(\phi^f)^2}{k}$ is the coefficient of diffusive resistance, and k is the coefficient of permeability.
2. Both the solid and fluid phases are intrinsically incompressible. Therefore, we require that the true fluid and solid densities be a constant such that

$$\frac{\partial \rho_T^s}{\partial t} = 0 \quad \text{and} \quad \frac{\partial \rho_T^f}{\partial t} = 0.$$

3. The solid and fluid are each homogenous in terms of material properties and density. Therefore, the true density does not vary with position, and we may write

$$\frac{\partial \rho_T^s}{\partial \boldsymbol{x}} = \mathbf{0} \quad \text{and} \quad \frac{\partial \rho_T^f}{\partial \boldsymbol{x}} = \mathbf{0}.$$

4. The solid and fluid phases are each isotropic and homogenous.
5. The solid phase is a linearly isotropic incompressible material such that the constitutive equation for the stress is given by

$$\boldsymbol{\sigma}^s = -\phi^s p\mathbf{I} + \lambda\,\mathrm{tr}(\boldsymbol{E})\,\mathbf{I} + 2\mu\boldsymbol{E}, \tag{8.14}$$

where λ and μ are material parameters.
6. The fluid is an incompressible inviscid fluid such that the constitutive model may be given by

$$\boldsymbol{\sigma}^f = -\phi^f p\mathbf{I}. \tag{8.15}$$

8.5.1 Confined Compression Experiment

A confined compression test may be used to determine the material parameters for the isothermal biphasic constitutive model outlined in the previous section. In this section, we will outline the kinematics of the confined compression test and then derive the governing equations in terms of the specified constitutive model.

During a confined compression test, a sample is compressed while the lateral dimensions are constrained, Figure 8.3. A sample is placed within an impermeable fixture made of a material which is significantly stiffer than the sample. A cylindrical porous indenter made of a stiff material with higher permeability than the test sample is used to apply an axial compressive force. As the indenter moves downward, the fluid pressure increases driving a flow of fluid out of the material and through the porous indenter.

Before we begin the analysis, let us outline several important approximations made in the derivation of the governing equations. These approximations follow:

1. The solid and fluid phases are modeled according to the assumptions listed in Section 8.5.

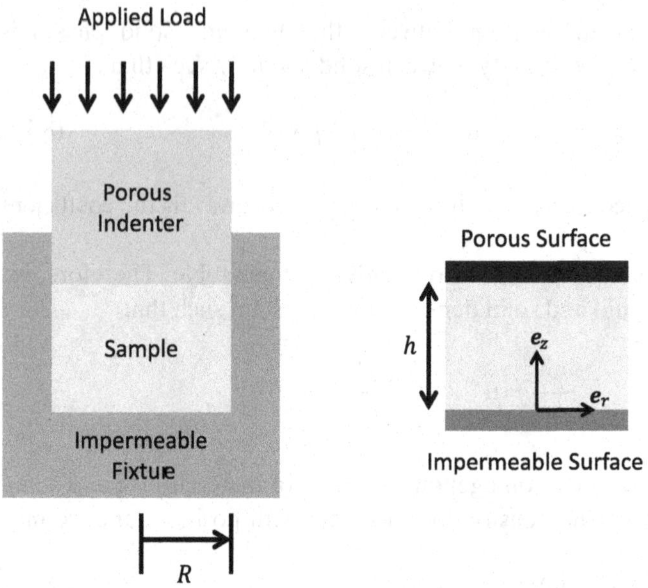

Figure 8.3. Illustration of the confined compression experimental setup.

2. The inertial forces may be neglected such that $\dot{\boldsymbol{v}}^s \cong \mathbf{0}$.
3. Gravitational forces may be neglected such that $\boldsymbol{b}^s \cong \mathbf{0}$.
4. Due to the symmetry of configuration and the fact that the phases are isotropic and homogenous, both lateral fluid velocity and lateral solid velocity must be zero such that

$$v_r^f = v_r^f = 0, \quad v_\theta^s = v_\theta^s = 0, \quad \pi_r^f = \pi_r^f = 0, \quad \text{and} \quad \pi_\theta^s = \pi_\theta^s = 0.$$

8.5.1.1 Kinematics

Given the assumptions listed in the previous section, the equations of motion for the ***solid phase*** can be written as

$$x_r^s = X_r^s,$$
$$x_\theta^s = X_\theta^s,$$
$$x_z^s = \lambda_z \left(X^s, t\right) X_z^s,$$

where X_r^s, X_θ^s, and X_z^s are the reference positions of the solid phase, $\lambda_z \left(X^s, t\right)$ is the axial stretch of the solid, which in its general form may vary with both time and location within the sample.

The displacement field can be found directly from the equations of motion such that

$$\boldsymbol{u}_z^s \left(X^s, t\right) = \boldsymbol{x}_z^s \left(X^s, t\right) - \boldsymbol{X}_z^s = (\lambda_z - 1) X_z^s.$$

The velocity field for the solid is then given by the time derivative of the current positions,

$$\boldsymbol{v}_z^s \left(\boldsymbol{X}^s, t\right) = \frac{\partial \boldsymbol{x}_z^s}{\partial t} = \frac{\partial \boldsymbol{u}_z^s}{\partial t} = \left(\frac{\partial \lambda_3}{\partial t}\right) X_3^s.$$

The deformation gradient becomes

$$F^s\left(X^s,t\right)=\begin{bmatrix}1 & 0 & 0\\ 0 & 1 & 0\\ 0 & 0 & \lambda_z\end{bmatrix}.$$

The left Cauchy deformation tensor is given by

$$B^s\left(X^s,t\right)=F\cdot F^T=\begin{bmatrix}1 & 0 & 0\\ 0 & 1 & 0\\ 0 & 0 & \lambda_z{}^2\end{bmatrix}.$$

The invariants of the left Cauchy deformation tensor are

$$\begin{aligned} I_B &= trB = 2+\lambda_z{}^2,\\ II_B &= \lambda_r\lambda_\theta+\lambda_r\lambda_z+\lambda_\theta\lambda_z = 1+2\lambda_z{}^2,\\ III_B &= \det B = \lambda_z{}^2. \end{aligned}$$

The Green strain tensor for this deformation is given by

$$E=\frac{1}{2}\left(F^TF-\mathbf{I}\right)=\frac{1}{2}\begin{bmatrix}0 & 0 & 0\\ 0 & 0 & 0\\ 0 & 0 & \lambda_z{}^2-1\end{bmatrix}. \tag{8.16}$$

8.5.1.2 Intrinsic Incompressibility Requirement

Intrinsic incompressibility means that the true density of each phase may not change. However, the apparent density of each phase may change in response to loading because the volume fraction of material may change as fluid is forced into or out of the solid. This constraint allows us to relate the solid phase and fluid phase velocities. Substituting the true density into the statement of conservation of mass for the fluid and solid phases gives

$$\frac{\partial\left(\phi^f\rho_T^f\right)}{\partial t}+div\left(\phi^f\rho_T^f v^f\right)=0 \quad \text{and} \quad \frac{\partial\left(\phi^s\rho_T^s\right)}{\partial t}+div\left(\phi^s\rho_T^s v^s\right)=0.$$

The true densities, ρ_T^f, and ρ_T^s, have constant values since both the fluid and solid are intrinsically incompressible. Therefore, they drop out of the equation giving

$$\frac{\partial\left(\phi^f\right)}{\partial t}+div\left(\phi^f v^f\right)=0 \quad \text{and} \quad \frac{\partial\left(\phi^s\right)}{\partial t}+div\left(\phi^s v^s\right)=0.$$

Adding the two equations gives

$$\frac{\partial\left(\phi^s+\phi^f\right)}{\partial t}+div\left(\phi^s v^s+\phi^f v^f\right)=0.$$

Since $\phi^s+\phi^f=1$, we have

$$div\left(\phi^s v^s+\phi^f v^f\right)=0.$$

For the confined compression test, only the component of the velocity along the $\boldsymbol{e}_z$ direction is nonzero, so we can write

$$\frac{\partial}{\partial x_z}\left(\phi^s v_z^s + \phi^f v_z^f\right) = 0.$$

Since the derivative is zero, the term in the brackets must be a constant giving us

$$\left(\phi^s v_z^s + \phi^f v_z^f\right) = v_o,$$

where v_o is not a function of x_z. Therefore, we may use the boundary conditions to solve for the value of the constant. Looking at the bottom of the sample ($x_z = 0$), we have $v_z^s(t) = v_z^f(t) = 0$ for all times. Therefore, $v_o(t) = 0$, and we may write

$$\left(\phi^s v_z^s + \phi^f v_z^f\right) = 0.$$

Rearranging this equation, we obtain an equation relating fluid phase velocity to solid phase velocity

$$v_z^f = \frac{-\phi^s}{\phi^f} v_z^s.$$

We will also make use of the fact that

$$v_z^f - v_z^s = \frac{-\phi^s}{\phi^f} v_z^s - v_z^s = -\frac{\phi^s + \phi^f}{\phi^f} v_z^s = -\frac{v_z^s}{\phi^f}. \tag{8.17}$$

8.5.1.3 Governing Equation

Neglecting gravity and inertial effects, the equation for conservation of linear momentum for the solid and fluid phases can be written as

$$div\left(\boldsymbol{\sigma}^s\right) + \boldsymbol{\pi}^s = 0,$$
$$div\left(\boldsymbol{\sigma}^f\right) + \boldsymbol{\pi}^f = 0.$$

Given an isotropic linear elastic solid phase and an isotropic inviscid fluid phase, we find that the shear stress terms are zero. The $\boldsymbol{e}_z$ component for the conservation of linear momentum for solid and fluid phases becomes

$$\frac{\partial \sigma_{zz}^s}{\partial x_z} = -\pi_z^s \quad \text{and} \quad \frac{\partial \sigma_{zz}^f}{\partial x_z} = -\pi_z^f.$$

Substitution of our constitutive law for the permeability, Equation (8.13), into the conservation of linear momentum gives

$$\frac{\partial \sigma_{zz}^s}{\partial x_z} = -K\left(v_z^f - v_z^s\right) + p\frac{\partial \phi^s}{\partial x_z} \quad \text{and} \quad \frac{\partial \sigma_{zz}^f}{\partial x_z} = K\left(v_z^f - v_z^s\right) - p\frac{\partial \phi^s}{\partial x_z}.$$

Next, we use the constitutive models to eliminate the pressure from these equations. Specifically, we have an isotropic linear elastic solid with the following constitutive equation

$$\boldsymbol{\sigma}^s = -\phi^s p\mathbf{I} + \lambda\,\mathrm{tr}(\boldsymbol{E})\mathbf{I} + 2\mu\boldsymbol{E}.$$

Substituting the Green strain tensor for the confined compression test, Equation (8.16), gives

$$\sigma^s_{zz} = -\phi^s p + \frac{1}{2}(\lambda + 2\mu)\left(\lambda_z^2 - 1\right).$$

Taking the derivative with respect to the stretch gives

$$\frac{\partial \sigma^s_{zz}}{\partial x_z} = -p\frac{\partial \phi^s}{\partial x_z} - \phi^s\frac{\partial p}{\partial x_z} + (\lambda + 2\mu)\lambda_z\frac{\partial \lambda_z}{\partial x_z}.$$

For the fluid, we have the constitutive relation

$$\boldsymbol{\sigma}^f = -\phi^f p\mathbf{I}.$$

Taking the derivative with respect to x_z gives

$$\frac{\partial \sigma^f_{zz}}{\partial x_z} = -p\frac{\partial \phi^f}{\partial x_z} - \phi^f\frac{\partial p}{\partial x_z}.$$

The conservation of momentum for the fluid becomes

$$-p\frac{\partial \phi^f}{\partial x_z} - \phi^f\frac{\partial p}{\partial x_z} = K\left(v^f_z - v^s_z\right) - p\frac{\partial \phi^s}{\partial x_z}.$$

Noting that $\frac{\partial \phi^f}{\partial x_z} = -\frac{\partial \phi^s}{\partial x_z}$, we can write

$$-\phi^f\frac{\partial p}{\partial x_z} = K\left(v^f_z - v^s_z\right). \tag{8.18}$$

Conservation of momentum for the solid becomes

$$-p\frac{\partial \phi^s}{\partial x_z} - \phi^s\frac{\partial p}{\partial x_z} + \frac{\partial}{\partial x_z}\left(\lambda_z(\lambda + 2\mu)\right) = -K\left(v^f_z - v^s_z\right) + p\frac{\partial \phi^s}{\partial x_z}$$

$$-\phi^s\frac{\partial p}{\partial x_z} + (\lambda + 2\mu)\lambda_z\frac{\partial \lambda_z}{\partial x_z} = -K\left(v^f_z - v^s_z\right). \tag{8.19}$$

Adding Equations (8.18) and (8.19) gives

$$-\phi^f\frac{\partial p}{\partial x_z} - \phi^s\frac{\partial p}{\partial x_z} + (\lambda + 2\mu)\lambda_z\frac{\partial \lambda_z}{\partial x_z} = 0.$$

Since $\phi^f + \phi^s = 1$, we can write

$$\frac{\partial p}{\partial x_z} = (\lambda + 2\mu)\lambda_z\frac{\partial \lambda_z}{\partial x_z}.$$

Substitution back into Equation (8.19) gives

$$\left(1 - \phi^s\right)(\lambda + 2\mu)\lambda_z\frac{\partial \lambda_z}{\partial x_z} = -K\left(v^f_z - v^s_z\right). \tag{8.20}$$

For the case of confined compression, we can convert the derivative of the stress with respect to the current configuration into the derivative of the stress with respect to the reference configuration as follows:

$$\frac{\partial \lambda_z}{\partial x_z} = \frac{\partial \lambda_z}{\partial X_z}\frac{\partial X_z}{\partial x_z} = \frac{\partial \lambda_z}{\partial X_z}\lambda_z^{-1}.$$

Noting that $(1-\phi^s)=\phi^f$ and converting to a derivative with respect to the reference coordinate, Equation (8.20) becomes

$$\phi^f(\lambda+2\mu)\frac{\partial\lambda_z}{\partial X_z}=-K\left(v_z^f-v_z^s\right).$$

Given that $\frac{\partial u_z^s}{\partial X_z}=\lambda_z-1$ and $\frac{\partial^2 u_z^s}{\partial X_z^2}=\frac{\partial\lambda_z}{\partial X_z}$ and substituting $v_z^f-v_z^s=-\frac{1}{\phi^f}\frac{\partial u_z^s}{\partial t}$, we can write this as

$$\phi^f(\lambda+2\mu)\frac{\partial^2 u_z^s}{\partial X_z^2}=K\frac{1}{\phi^f}\frac{\partial u_z^s}{\partial t}.$$

Rearranging the equation and substituting $K=\frac{\left(\phi^f\right)^2}{k}$ gives the governing equation

$$\boxed{\frac{\partial^2 u_z^s}{\partial X_z^2}=\frac{1}{k(\lambda+2\mu)}\frac{\partial u_z^s}{\partial t}.} \tag{8.21}$$

We have derived a single governing equation for the confined compression of a material modeled with an isothermal biphasic material. Solving this equation gives us the response of the material as a function of both reference position and time. Next we will implement the numerical solution to this partial differential equation.

8.5.1.4 Finite Difference Algorithm

For the confined compression of a material modeled with an isotropic isothermal biphasic model, the governing equation is a parabolic partial differential equation, (8.21). The PDE is given by

$$\frac{\partial^2 u_z^s}{\partial X_z^2}=\frac{1}{k(\lambda+2\mu)}\frac{\partial u_z^s}{\partial t}.$$

Introducing a new constant, $c=k(\lambda+2\mu)$, and for simplicity of notation, setting $u=u_z^s$, we have

$$c\frac{\partial^2 u}{\partial Z^2}=\frac{\partial u}{\partial t}.$$

We use a forward difference to discretize the time variable

$$\frac{\partial u_{i,j}}{\partial t}=\frac{u_{i,j+1}-u_{i,j}}{\Delta t}.$$

We use a central difference formula to discretize the spatial dimension

$$\frac{\partial^2 u_{i,j}}{\partial Z^2}=\frac{u_{i+1,j}-2u_{i,j}+u_{i-1,j}}{\Delta x^2}.$$

Substitution into the partial differential equations, we have

$$\frac{u_{i,j+1}-u_{i,j}}{\Delta t}=c\left(\frac{u_{i+1,j}-2u_{i,j}+u_{i-1,j}}{\Delta x^2}\right).$$

Rearranging this equation and solving for $u_{i,j+1}$, we obtain

$$u_{i,j+1} = \frac{c\,\Delta t}{\Delta x^2}\left(u_{i+1,j} - 2\,u_{i,j} + u_{i-1,j}\right) + u_{i,j}.$$

Introducing a new constant, $r = \frac{c\,\Delta t}{\Delta x^2}$, we have

$$u_{i,j+1} = r\left(u_{i+1,j} + u_{i-1,j}\right) + (1-2r)\,u_{i,j}. \tag{8.22}$$

This gives an explicit routine for determining the solution or our differential equation. We solve for the current displacement, $u_{i,j+1}$, by using the values of the displacements at the previous time step, $u_{i,j}$, $u_{i+1,j}$, and $u_{i-1,j}$. The initial conditions must be specified giving us the solution at $t = 0$ for all gridpoints. We can then iterate to find the solution at all future times.

8.5.1.5 Example Problem

Assume that we have a confined compression testing apparatus with a radius $R = 5\,\text{cm}$ and a height $h = 2\,\text{cm}$. We place within a material that may be modeled as a biphasic material with an incompressible linear elastic solid filled with an incompressible inviscid fluid. The aggregate modulus, permeability, and initial solid content of the material are given by $H_a = \lambda + 2\mu = 1 \times 10^6$ [MPa], $k = 1 \times 10^{-11}\left[\frac{\text{kg}}{\text{m}^2\,\text{s}}\right]$, and $\phi(\boldsymbol{X}, t = 0\,\text{s}) = 0.5$, respectively. The displacement of the top surface is prescribed as a boundary condition, $u_z^s(X_3 = 0, t)$, is shown in Figure 8.4. This is equivalent to controlling the displacement of the top piston during the course of the experiment. Find the stress exerted on the top piston as a function of time.

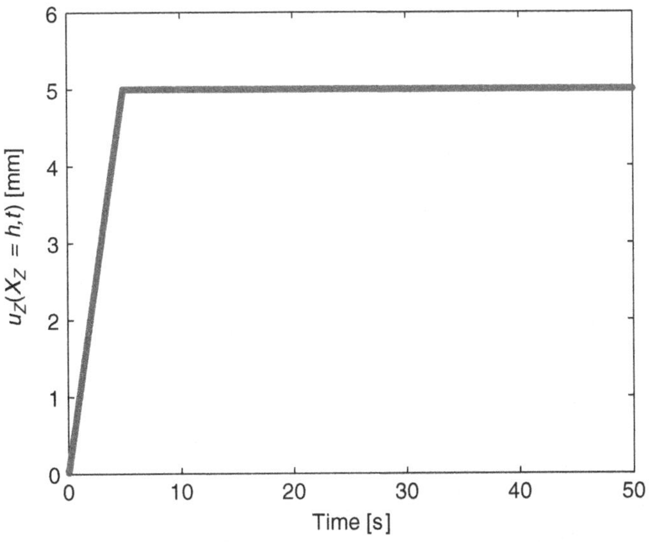

Figure 8.4. Downward surface displacement versus time for poroelastic material in a confined compression stress relaxation experiment.

The governing equations for this experimental setup were given in Section 8.5.1.3. The finite difference solution for these governing equations was outlined in Section 8.5.1.4. The implementation in Matlab® is given in Section 8.5.1.6. Notice that we are solving for the displacement of the solid component of the material. The stress, fluid pressure, and fluid velocities may all be determined directly from the solid displacements combined with the boundary conditions. The stress on the top surface of the material is given by $\sigma^s_{zz}(X^s = h, t)$. The force applied to the piston would be the surface stress on the solid multiplied by the cross-sectional area of the sample, $F = \sigma^s_{zz} \times \pi R^2$. The surface stress is shown in Figure 8.5. As we can see, after the initial ramp in displacement, the sample is held fixed, and the surface stress decays away. This occurs as fluid rushes out of the sample, and the fluid pressure drops. Eventually, an equilibrium stress is obtained, it represents the state where the stress is maintained solely by the elastic solid.

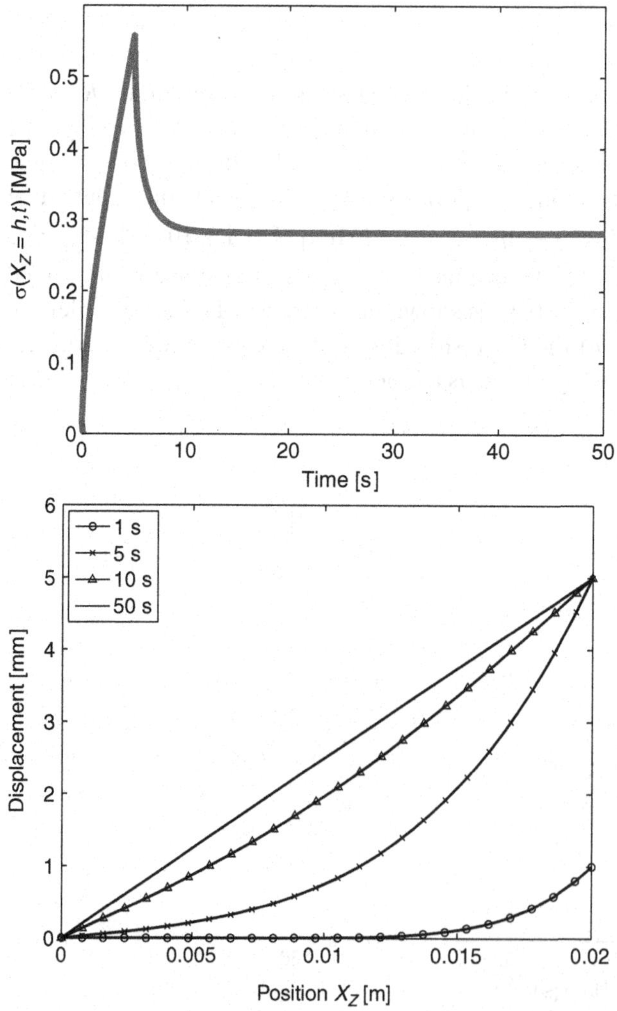

Figure 8.5. (top) Compressive surface stress versus time for poroelastic material in a confined compression stress relaxation experiment. (bottom) Downward displacement versus position within the sample as a function of time.

8.5.1.6 Matlab® File

Explicit Algorithm

1. Initialize sample geometry $R = 5$ cm and $h = 2$ cm.
   Setup spatial and temporal grids

   $$ndZ = 100$$

   $$dT = 1 \text{ ms}$$

2. Initialize material properties

   $$H_a = 1 \times 10^6 \text{ Pa}$$

   $$k_o = 1 \times 10^{-11}$$

   $$\phi_o^s = 0.5$$

3. Allocate memory and initial conditions

4. Assign boundary conditions at the top and bottom surfaces

   $$\dot{u}(1, t) = \begin{cases} dispRate & totalDisp \leq dispRate \times t \\ 0 & totalDisp > dispRate \times t \end{cases}$$

   $$\dot{u}(nDx, t) = 0$$

5. Compute

   $$r^2 = \frac{k_o H_a \Delta t}{\Delta Z^2}$$

6. Calculate pressure on the interior nodes

   $$u(i, j) = r\, u(i + 1, j - 1) + u(i - 1, j - 1) + (1 - 2r)\, u(i, j - 1)$$

   Cycle over all interior nodes, i

   Done

   Cycle over all time increments, j

   Done

7. Plot the results

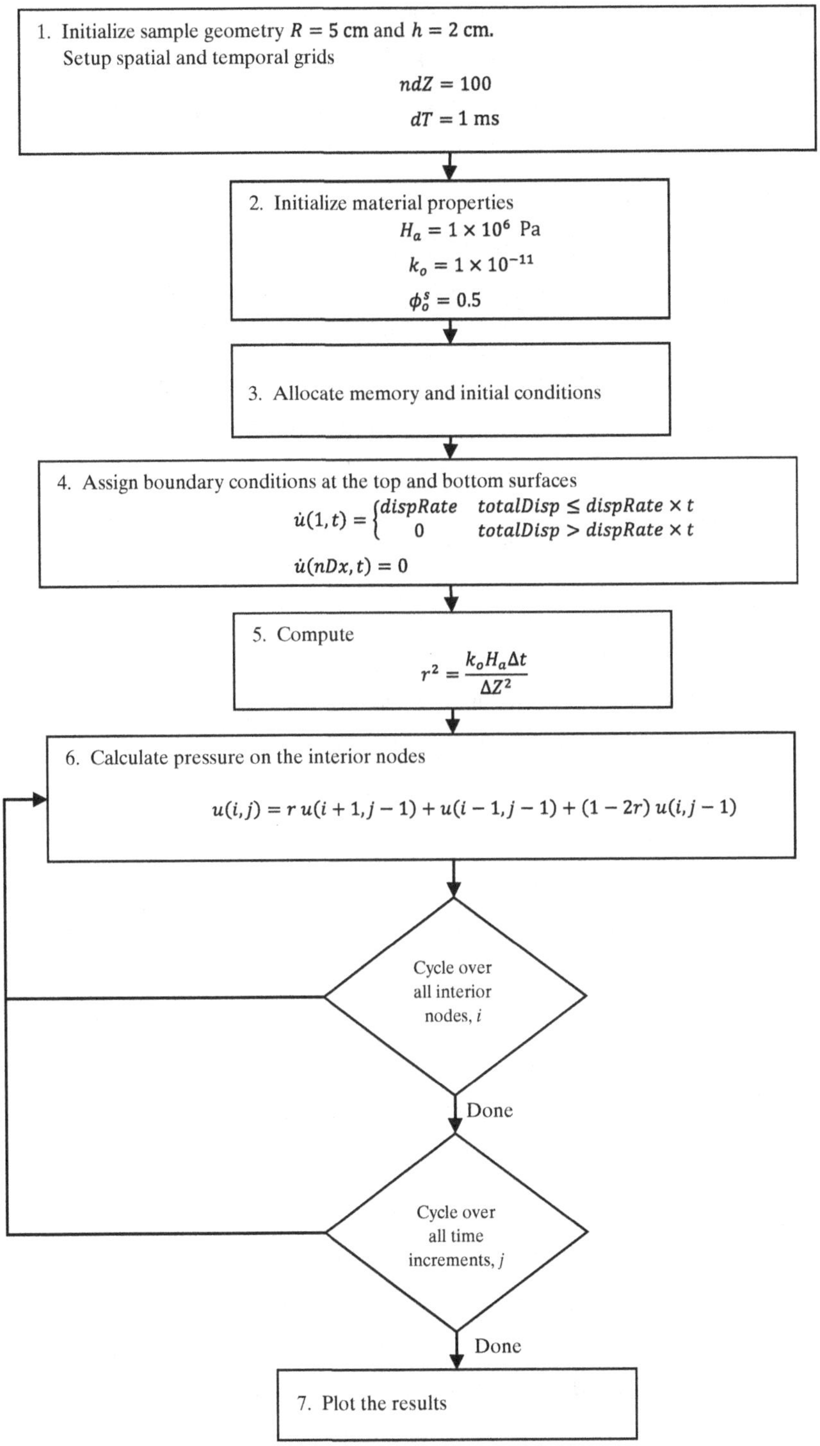

```
%---------------------------------------------------------------------------
% Numerical solution of the confined compression test of the isothermal
% biphasic model with an isotropic homogeneous linear elastic solid and
% an isotropic homogeneous inviscid fluid. The displacement of the top
% surface is controlled and the bottom surface is fixed.
%---------------------------------------------------------------------------
% u(i,j) [m] is a matrix containing the displacement at each node i
%           for the time increment j.
%---------------------------------------------------------------------------
clear;
%---------------------------------------------------------------------------
% 1. Initialize the spatial and temporal grid
% Setup sample geometry
w             = 0.05;  % [m] sample width
h             = 0.02;  % [m] sample height
%---------------------------------------------------------------------------
% Initialize test parameters
dispRate      = 0.002; % [m/s] Displacement rate
totalDisp     = 0.005; % [m]   Total displacement of sample
simTime       = 25.0;  % [s]   Total simulation time
%---------------------------------------------------------------------------
% Setup spatial grid
nrZ   = 10;                  % number of gridpoints along z axis
dZ    = h/(nrZ-1);           % spacing between gridpoints
z     = [0:dZ:h];            % z position of each gridpoint
%---------------------------------------------------------------------------
% Setup the temporal grid
dT    = 0.01;                % [s] time steps for explicit algorithm
nrT   = round(simTime/dT); % number of time steps in simulation
%---------------------------------------------------------------------------
% 2. Initialize the material properties
HAO           = 1e6;   % [Pa]        aggregate modulus
k_o           = 1e-11; % [kg/m^2/s] permeability
solidContent = 0.5;    % volume fraction of solid
%---------------------------------------------------------------------------
% 3. Allocate memory and setup initial condition
u             = zeros(nrZ, nrT); % aso sets up initial condition u(:,0) = 0
currentDisp   = 0;
%---------------------------------------------------------------------------
% 4. Assign the boundary conditions to the top and bottom surfaces
for j = 1:nrT
    u(1, j)   = 0;             % bottom surface is fixed
    if (currentDisp < totalDisp)
        currentDisp = currentDisp + dispRate*dT;
    end
    u(nrZ, j) = currentDisp; % top surface is displaced
end
%---------------------------------------------------------------------------
% 5. Compute the constant r
r          = k_o*HAO*dT/dZ^2;
%---------------------------------------------------------------------------
% 6. Solve the discritized governing equation
for j = 2:nrT
    for i = 2:nrZ-1
        u(i,j) = r*(u(i+1,j-1) + u(i-1,j-1)) + (1-2*r)*u(i,j-1);
    end
end
```

8.5.2 Unconfined Compression

Using only a confined compression test, one may determine the aggregate modulus of a material. However, by performing an unconfined compression test in conjunction with a confined compression test, one may determine the individual Lame coefficients for a linear elastic material. In an unconfined compression experiment, a sample is placed between two impermeable platens, Figure 8.6. The sample is allowed to freely expand in the radial direction as it is compressed along the z axis.

As we did for the case of confined compression, we will make the following assumptions:

1. The solid and fluid phases are modeled according to the assumptions listed in Section 8.5.
2. Inertial and gravitational forces may be neglected.
3. Due to the symmetry of the configuration and the fact that the phases are isotropic and homogenous, material moves radially but does not rotate as the sample is compressed.
4. The axial stretch is uniform across the cross section such that $\lambda_z = \lambda_z(t)$.
5. There is no relative motion between the fluid and solid in the z direction, $v_z^f(\boldsymbol{x},t) = v_z^s(\boldsymbol{x},t)$.
6. The pressure is assumed to vary in only the radial direction, $p = p(x_r,t)$.
7. The contact surface between the sample and the platen is frictionless such that the radial stretch may be written as $\lambda_r = \lambda_r(x_r,t)$.
8. We will limit ourselves to infinitesimal deformation.

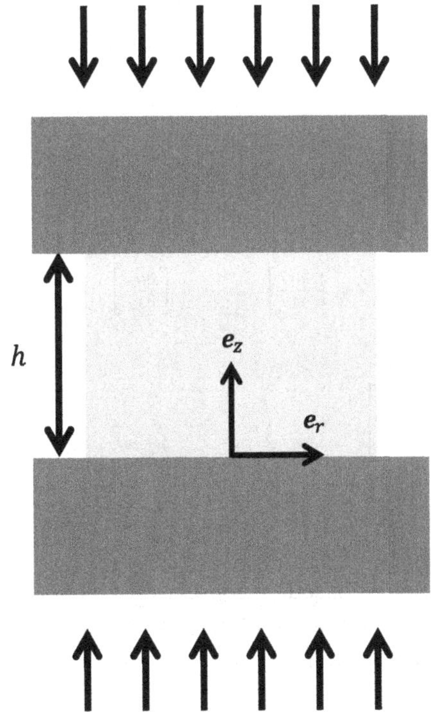

Figure 8.6. Experimental setup for uniaxial compression test.

9. Boundary conditions are as follows. The radial displacement is zero at the center of the sample due to symmetry considerations $u_r(x_r = 0, t) = 0$. There is a traction free boundary condition on the outer surface of the solid such that $\sigma_{rr} + \phi^s p = 0$ at $x_r = R$.

8.5.2.1 Kinematics

In the case of uniaxial compression, it is preferable to write the equations of motion in cylindrical coordinates. In the most general form, we can write the equations of motion for the solid, which conform to the assumptions listed in the previous section as

$$\begin{aligned} x_r^s &= \lambda_r(x_r,t)\, X_r^s, \\ x_\theta^s &= X_\theta^s, \\ x_z^s &= \lambda_z(t)\, X_z^s. \end{aligned}$$

The displacement field becomes

$$\begin{aligned} u_r^s &= (\lambda_r - 1)\, X_r^s, \\ u_\theta^s &= X_\theta^s, \\ u_z^s &= (\lambda_z - 1)\, X_z^s. \end{aligned}$$

The deformation gradient in cylindrical coordinates is given by

$$\boldsymbol{F} = \begin{bmatrix} 1+\frac{\partial u_r}{\partial x_r} & \left[\frac{1}{r}\frac{\partial u_r}{\partial x_\theta} - \frac{u_\theta}{r}\right] & \frac{\partial u_r}{\partial x_z} \\ \frac{\partial u_\theta}{\partial x_r} & 1+\left[\frac{1}{r}\frac{\partial u_\theta}{\partial x_\theta} + \frac{u_r}{r}\right] & \frac{\partial u_\theta}{\partial x_z} \\ \frac{\partial u_z}{\partial x_r} & \left[\frac{1}{r}\frac{\partial u_z}{\partial x_\theta}\right] & 1+\frac{\partial u_z}{\partial x_z} \end{bmatrix} = \begin{bmatrix} 1+\frac{\partial u_r}{\partial x_r} & 0 & 0 \\ 0 & \frac{u_r}{r} & 0 \\ 0 & 0 & 1+\frac{\partial u_z}{\partial x_z} \end{bmatrix}.$$

The infinitesimal strain tensor is given by

$$\boldsymbol{\varepsilon} = \frac{1}{2}\left(\nabla \boldsymbol{u} + \boldsymbol{u}\overleftarrow{\nabla}\right) = \begin{bmatrix} \left[\frac{\partial u_r}{\partial x_r}\right] & \left[\frac{1}{r}\frac{\partial u_r}{\partial x_\theta} - \frac{u_\theta}{r} + \frac{\partial u_\theta}{\partial x_r}\right] & \left[\frac{\partial u_r}{\partial x_z} + \frac{\partial u_z}{\partial x_r}\right] \\ \left[\frac{1}{r}\frac{\partial u_r}{\partial x_\theta} - \frac{u_\theta}{r} + \frac{\partial u_\theta}{\partial x_r}\right] & \left[\frac{1}{r}\frac{\partial u_\theta}{\partial x_\theta} + \frac{u_r}{r}\right] & \left[\frac{\partial u_\theta}{\partial x_z} + \frac{1}{r}\frac{\partial u_z}{\partial x_\theta}\right] \\ \left[\frac{\partial u_r}{\partial x_z} + \frac{\partial u_z}{\partial x_r}\right] & \left[\frac{\partial u_\theta}{\partial x_z} + \frac{1}{r}\frac{\partial u_z}{\partial x_\theta}\right] & \left[\frac{\partial u_z}{\partial x_z}\right] \end{bmatrix}.$$

Substituting the displacement field gives

$$\boldsymbol{\varepsilon} = \begin{bmatrix} \frac{\partial u_r}{\partial x_r} & 0 & 0 \\ 0 & \frac{u_r}{r} & 0 \\ 0 & 0 & \frac{\partial u_z}{\partial x_z} \end{bmatrix}.$$

8.5.2.2 Intrinsic Incompressibility

In this section, we will derive the constraint equation based on conservation of mass, which applies to our mixture. The statement of conservation of mass in terms of the true densities of the fluid and solid phases can be written as

$$\frac{\partial\left(\phi^f \rho_T^f\right)}{\partial t} + div\left(\phi^f \rho_T^f \boldsymbol{v}^f\right) = 0 \quad \text{and} \quad \frac{\partial\left(\phi^s \rho_T^s\right)}{\partial t} + div\left(\phi^s \rho_T^s \boldsymbol{v}^s\right) = 0.$$

The true densities cancel, and using the equation, $\phi^s + \phi^f = 1$, we may sum the two equations to obtain

$$div\left(\phi^s \mathbf{v}^s + \phi^f \mathbf{v}^f\right) = 0.$$

Expanding this equation in cylindrical coordinates gives

$$\left(\frac{1}{r}\frac{\partial\left(r\phi^s \mathbf{v}_r^s\right)}{\partial x_r} + \frac{1}{r}\frac{\partial\left(\phi^s \mathbf{v}_\theta^s\right)}{\partial x_\theta} + \frac{\partial\left(\phi^s \mathbf{v}_z^s\right)}{\partial x_z}\right)$$
$$+\left(\frac{1}{r}\frac{\partial\left(r\phi^f \mathbf{v}_r^f\right)}{\partial x_r} + \frac{1}{r}\frac{\partial\left(\phi^f \mathbf{v}_\theta^f\right)}{\partial x_\theta} + \frac{\partial\left(\phi^f \mathbf{v}_z^f\right)}{\partial x_z}\right) = 0.$$

Let us review the approximations made in this problem. Due to axial symmetry, we know that $v_\theta^s = 0$ and $v_\theta^f = 0$. We have assumed that there is no variation of the radial velocity in the axial direction so that $v_r^f = v_r^f(x_r,t)$ and $v_r^s = v_r^s(x_r,t)$. We have assumed that the axial stretch is constant throughout the sample $\lambda_z = \lambda_z(t)$:

$$\phi^s = \phi^s(x_r),$$
$$\phi^s = \phi^s(x_r).$$

Therefore, we obtain

$$\left(\frac{1}{r}\frac{\partial\left(r\phi^s \mathbf{v}_r^s\right)}{\partial x_r} + \phi^s\frac{\partial\left(\mathbf{v}_z^s\right)}{\partial x_z}\right) + \left(\frac{1}{r}\frac{\partial\left(r\phi^f \mathbf{v}_r^f\right)}{\partial x_r} + \phi^f\frac{\partial\left(\mathbf{v}_z^f\right)}{\partial x_z}\right) = 0.$$

There is no relative motion between the fluid and the solid in the z direction so that $v_z^s = v_z^f$, which gives

$$\frac{1}{r}\frac{\partial\left(r\phi^s \mathbf{v}_r^s\right)}{\partial x_r} + \frac{1}{r}\frac{\partial\left(r\phi^f \mathbf{v}_r^f\right)}{\partial x_r} + \left(\phi^s + \phi^f\right)\frac{\partial \mathbf{v}_z^s}{\partial x_z} = 0,$$
$$\frac{1}{r}\frac{\partial\left(r\phi^s \mathbf{v}_r^s\right)}{\partial x_r} + \frac{1}{r}\frac{\partial\left(r\phi^f \mathbf{v}_r^f\right)}{\partial x_r} + \frac{\partial \mathbf{v}_z^s}{\partial x_z} = 0.$$

Rearranging the equation, we have

$$\frac{\partial\left(r\phi^s \mathbf{v}_r^s + r\phi^f \mathbf{v}_r^f\right)}{\partial x_r} = -r\frac{\partial \mathbf{v}_z^s}{\partial x_z}.$$

Integrating the equation with respect to r, we have

$$\left(r\phi^s \mathbf{v}_r^s + r\phi^f \mathbf{v}_r^f\right)\Big|_0^r = \int_0^r -r\frac{\partial \mathbf{v}_z^s}{\partial x_z}dr.$$

Since the solid velocity is a function of x_z and t, $v_z^s = v_z^s(x_z,t)$, we have

$$\left(r\phi^s \mathbf{v}_r^s + r\phi^f \mathbf{v}_r^f\right)\Big|_0^r = -\frac{r^2}{2}\frac{\partial \mathbf{v}_z^s}{\partial x_z}\Big|_0^r.$$

Evaluating this function at the limits, we use the fact that due to the symmetry of the problem, at $x_r = 0$ there is no radial flow of either solid or fluid. This gives us

$$r\phi^s \mathbf{v}_r^s + r\phi^f \mathbf{v}_r^f = -\frac{r^2}{2}\frac{\partial \mathbf{v}_z^s}{\partial x_z}.$$

Rearranging the equation, we have

$$\phi^s \mathbf{v}_r^s + \phi^f \mathbf{v}_r^f = -\frac{r}{2}\frac{\partial \mathbf{v}_z^s}{\partial x_z}.$$

Rearranging the equation, we have

$$\mathbf{v}_r^f = -\frac{1}{\phi^f}\left(\frac{r}{2}\frac{\partial \mathbf{v}_z^s}{\partial x_z} + \phi^s \mathbf{v}_r^s\right).$$

Therefore, we may write

$$v_r^f - v_r^s = -\frac{r}{2\phi^f}\frac{\partial \mathbf{v}_z^s}{\partial x_z} - \left(\frac{\phi^s}{\phi^f} + 1\right)\mathbf{v}_r^s = -\frac{1}{\phi^f}\left(\frac{r}{2}\frac{\partial \mathbf{v}_z^s}{\partial x_z} + v_r^s\right). \tag{8.23}$$

8.5.2.3 Governing Equations

Neglecting gravity and inertial effects, we may write the equations of conservation of linear momentum for the solid and fluid phases as

$$div\left(\boldsymbol{\sigma}^s\right) + \boldsymbol{\pi}^s = 0,$$
$$div\left(\boldsymbol{\sigma}^f\right) + \boldsymbol{\pi}^f = 0.$$

Let us examine the radial components of the momentum equations. We have

$$\frac{1}{r}\frac{\partial}{\partial x_r}\left(r\sigma_{rr}^s\right) + \frac{1}{r}\frac{\partial \sigma_{\theta r}^s}{\partial x_\theta} + \frac{\partial \sigma_{zr}^s}{\partial x_z} - \frac{1}{r}\sigma_{\theta\theta}^s + \pi_r^s = 0$$

and

$$\frac{1}{r}\frac{\partial}{\partial x_r}\left(r\sigma_{rr}^f\right) + \frac{1}{r}\frac{\partial \sigma_{\theta r}^f}{\partial x_\theta} + \frac{\partial \sigma_{zr}^f}{\partial x_z} - \frac{1}{r}\sigma_{\theta\theta}^f + \pi_r^f = 0.$$

For the sake of clarity, let us restate the material model. The solid is an intrinsically incompressible homogenous linear isotropic elastic material such that

$$\boldsymbol{\sigma}^s = -\phi^s p\mathbf{I} + \lambda\,\mathrm{tr}\,(\boldsymbol{\varepsilon})\,\mathbf{I} + 2\mu\boldsymbol{\varepsilon}.$$

Substituting the infinitesimal strain, we find the only non-zero terms in the stress equation are given by

$$\sigma_{rr}^s = -\phi^s p + \lambda\left(\frac{\partial u_r^s}{\partial x_r} + \frac{u_r^s}{r} + \frac{\partial u_z^s}{\partial x_z}\right) + 2\mu\frac{\partial u_r^s}{\partial x_r},$$

$$\sigma_{\theta\theta}^s = -\phi^s p + \lambda\left(\frac{\partial u_r^s}{\partial x_r} + \frac{u_r^s}{r} + \frac{\partial u_z^s}{\partial x_z}\right) + 2\mu\frac{u_r^s}{r},$$

$$\sigma_{zz}^s = -\phi^s p + \lambda\left(\frac{\partial u_r^s}{\partial x_r} + \frac{u_r^s}{r} + \frac{\partial u_z^s}{\partial x_z}\right) + 2\mu\frac{\partial u_z^s}{\partial x_z}.$$

The fluid is an intrinsically incompressible homogenous inviscid fluid such that

$$\boldsymbol{\sigma}^f = -\phi^f p\mathbf{I}.$$

The transfer of momentum between the fluid and solid is governed by the permeability such that

$$\boldsymbol{\pi}^s = -\boldsymbol{\pi}^f = K\left(\mathbf{v}^f - \mathbf{v}^s\right) + p\nabla\phi^s.$$

Noting that the shear stress terms are zero, we have

$$\frac{\partial \sigma_{rr}^s}{\partial x_r} - \frac{1}{r}\left(\sigma_{\theta\theta}^s - \sigma_{rr}^s\right) + \pi_r^s = 0.$$

We next substitute the constitutive model into the preceding equation. Noting that $\sigma_{\theta\theta}^s - \sigma_{rr}^s = 2\mu\left(\frac{u_r^s}{r} - \frac{\partial u_r^s}{\partial x_r}\right)$, and that the axial displacement does not vary with radial position, $\frac{\partial^2 u_z}{\partial x_r \partial x_z} = 0$, we have

$$\frac{\partial}{\partial x_r}\left(-\phi^s p + \lambda\left(\frac{\partial u_r^s}{\partial x_r} + \frac{u_r^s}{r} + \frac{\partial u_z^s}{\partial x_z}\right) + 2\mu\frac{\partial u_r^s}{\partial x_r}\right)$$
$$-\frac{2\mu}{r}\left(\frac{u_r^s}{r} - \frac{\partial u_r^s}{\partial x_r}\right) + K\left(v_r^f - v_r^s\right) + p\frac{\partial \phi^s}{\partial x_r} = 0$$
$$-\phi^s\frac{\partial p}{\partial x_r} + (\lambda + 2\mu)\left(\frac{\partial^2 u_r^s}{\partial x_r^2} - \frac{u_r^s}{r^2} + \frac{1}{r}\frac{\partial u_r^s}{\partial x_r}\right) + K\left(v_r^f - v_r^s\right) = 0.$$

Since $\sigma_{\theta\theta}^f - \sigma_{rr}^f = 0$, the equation for conservation of linear momentum for the fluid phase becomes

$$\frac{\partial \sigma_{rr}^f}{\partial x_r} - \pi_r^s = 0.$$

Substituting the constitutive model into the preceding equation and substituting $\frac{\partial \phi^s}{\partial x_r} = -\frac{\partial \phi^f}{\partial x_r}$ gives

$$\frac{\partial\left(-\phi^f p\right)}{\partial x_r} - K\left(v_r^f - v_r^s\right) - p\frac{\partial \phi^s}{\partial x_r} = 0,$$
$$-\phi^f\frac{\partial p}{\partial x_r} - K\left(v_r^f - v_r^s\right) = 0.$$

Combining the conservation of linear momentum for the solid and fluid, we may eliminate the pressure term giving

$$K\left(1 + \frac{\phi^s}{\phi^f}\right)\left(v_r^f - v_r^s\right) + (\lambda + 2\mu)\left(\frac{\partial^2 u_r^s}{\partial x_r^2} - \frac{u_r^s}{r^2} + \frac{1}{r}\frac{\partial u_r^s}{\partial x_r}\right) = 0.$$

Noting that $\left(1 + \frac{\phi^s}{\phi^f}\right) = \frac{1}{\phi^f}$, and eliminating the fluid velocity term by using Equation (8.23), we obtain

$$-\frac{K}{\left(\phi^f\right)^2}\left(\frac{r}{2}\frac{\partial \mathbf{v}_z^s}{\partial x_z} + v_r^s\right) + (\lambda + 2\mu)\left(\frac{\partial^2 u_r^s}{\partial x_r^2} - \frac{u_r^s}{r^2} + \frac{1}{r}\frac{\partial u_r^s}{\partial x_r}\right) = 0.$$

The solid velocity may be written as $v_r^s = \frac{\partial u_r^s}{\partial t}$, and $\frac{\partial v_z^s}{\partial x_z} = \frac{\partial}{\partial x_z}\frac{\partial u_z^s}{\partial t} = \frac{\partial}{\partial t}\frac{\partial u_z^s}{\partial x_z} = \frac{\partial \lambda_z}{\partial t}$. Noting that $\frac{K}{(\phi^f)^2} = \frac{1}{k}$, the governing equation becomes

$$\boxed{\frac{\partial^2 u_r^s}{\partial x_r^2} - \frac{u_r^s}{r^2} + \frac{1}{r}\frac{\partial u_r^s}{\partial x_r} - \frac{1}{k(\lambda+2\mu)}\left(\frac{r}{2}\frac{\partial \lambda_z}{\partial t} + \frac{\partial u_r^s}{\partial t}\right) = 0.}$$

8.5.2.4 Finite Difference Algorithm

We now have the governing equation for unconfined compression of a biphasic material with an isotropic homogenous linear elastic solid phase and an isotropic homogenous inviscid fluid phase. Simplifying the notation by setting $u = u_r^s$, we can write

$$\frac{\partial^2 u}{\partial x_r^2} - \frac{u}{r^2} + \frac{1}{r}\frac{\partial u}{\partial x_r} - \frac{1}{k(\lambda+2\mu)}\left(\frac{r}{2}\frac{\partial \lambda_z}{\partial t} + \frac{\partial u}{\partial t}\right) = 0.$$

We will use a forward difference algorithm to discretize time

$$\frac{\partial u_{i,j}}{\partial t} = \frac{u_{i,j+1} - u_{i,j}}{\Delta t}.$$

We use a central difference formula to discretize the second spatial derivative

$$\frac{\partial^2 u_{i,j}}{\partial x_r^2} = \frac{u_{i+1,j} - 2u_{i,j} + u_{i-1,j}}{\Delta x_r^2}.$$

The first-order spatial derivative is given by

$$\frac{\partial u_{i,j}}{\partial x_r} = \frac{u_{i+1,j} - u_{i-1,j}}{2\Delta x_r}.$$

Substituting into the governing equation gives

$$\frac{u_{i+1,j} - 2u_{i,j} + u_{i-1,j}}{\Delta x_r^2} - \frac{u_{i,j}}{r_i^2} + \frac{1}{r_i}\frac{u_{i+1,j} - u_{i-1,j}}{2\Delta x_r} - \frac{1}{k(\lambda+2\mu)}\left(\frac{r_i}{2}\frac{\partial \lambda_z}{\partial t} + \frac{u_{i,j+1} - u_{i,j}}{\Delta t}\right) = 0.$$

Rearranging the equation and solving for $u_{i,j+1}$, gives

$$u_{i,j+1} = u_{i,j} + \Delta t\, k(\lambda+2\mu)\left[\left(\frac{1}{\Delta x_r^2} + \frac{1}{2r_i\Delta x_r}\right)u_{i+1,j} - \left(\frac{2}{\Delta x_r^2} + \frac{1}{r_i^2}\right)u_{i,j} + \left(\frac{1}{\Delta x_r^2} - \frac{1}{2r_i\Delta x_r}\right)u_{i-1,j}\right] - \Delta t\frac{r_i}{2}\frac{\partial \lambda_z}{\partial t}.$$

The traction free boundary condition may be enforced by using the constitutive equation for the radial stress in the solid. This gives

$$0 = \sigma_{rr}^s = -\phi^s p + (2\mu+\lambda)\frac{\partial u_r^s}{\partial x_r} + \lambda\left(\frac{u_r^s}{r} + \frac{\partial u_z^s}{\partial x_z}\right).$$

Discretizing the boundary conditions gives

$$0 = (2\mu+\lambda)\left(\frac{u_{n,j} - u_{n-1,j}}{\Delta x_r}\right) + \lambda\left(\frac{u_{n,j}}{r} + (\lambda_z - 1)\right),$$

where n is the index of the gridpoint at R. Rearranging the equation gives the boundary condition

$$u_{n,j} = \frac{\left((2\mu+\lambda)\,u_{n-1,j} - \Delta x_r\,\lambda\,(\lambda_z - 1)\right)}{\left((2\mu+\lambda) + \frac{\Delta x_r \lambda}{r_n}\right)}.$$

8.5.2.5 Example Problem

We would like to simulate the uniaxial unconfined compression of a sample with a radius $R = 5\,\text{cm}$ and a height $h = 2\,\text{cm}$. The sample will be modeled as a biphasic material with an incompressible linear elastic solid filled with an incompressible inviscid fluid. The Lame coefficients, permeability, and the initial solid content of the material are given by $\lambda = 1$ [MPa], $\mu = 0.1$ [MPa] $k = 1\times 10^{-11}\left[\frac{\text{kg}}{\text{m}^2\,\text{s}}\right]$, and $\phi\,(\boldsymbol{X}, t = 0\,\text{s}) = 0.5$,

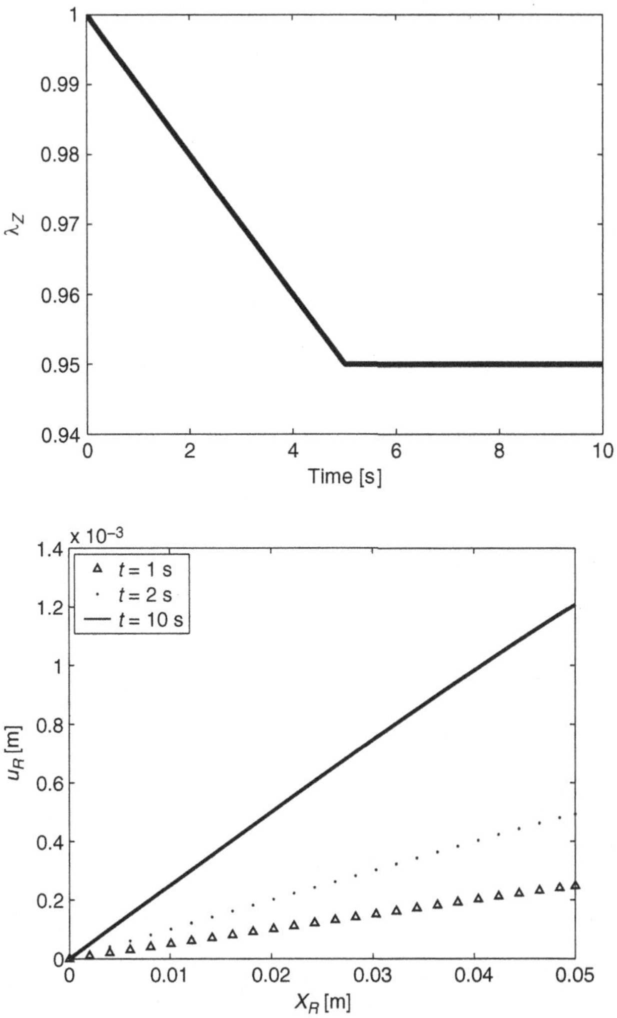

Figure 8.7. (top) Axial stretch versus time. (bottom) Radial displacement versus radial position at $t = 1\,\text{s}$, $2\,\text{s}$, and $10\,\text{s}$. $t = 1\,\text{s}$, $2\,\text{s}$, $10\,\text{s}$.

respectively. The axial stretch, $\lambda_z^s(t)$, is prescribed as a boundary condition and is shown in Figure 8.7. This is equivalent to controlling the displacement of the top piston during the course of the experiment. We would like to find the radial displacement as a function of radial position within the sample at various times.

The governing equations for this experimental setup were given in Section 8.5.2.3. The finite difference solution for these governing equations was outlined in Section 8.5.2.4. The implementation in Matlab® is given in Section 8.5.2.6. The radial displacements of the solid component of the material are shown in Figure 8.7.

8.5.2.6 Matlab® File

```
%------------------------------------------------------------------------
% Numerical solution of the unconfined compression test of the isothermal
% biphasic model with an isotropic homogeneous linear elastic solid and
% an isotropic homogeneous inviscid fluid. The displacement of the top
% surface is controlled and the bottom surface is fixed.
%------------------------------------------------------------------------
% u(i,j) [m] is a matrix containing the radial displacement at each node i
%            for the time increment j.
%------------------------------------------------------------------------
clear;
%------------------------------------------------------------------------
% 1. Initialize the spatial and temporal grid
% Setup sample geometry
sampleRadius    = 0.05;  % [m] sample radius
h               = 0.02;  % [m] sample height
%------------------------------------------------------------------------
% Initialize test parameters
stretchRate     = -0.01;  % [stretch/s] axial stretch rate DLambdaDt
finalStretch    = 0.95;   % Maximum axial stretch of sample
simTime         = 10;     % [s]   Total simulation time
%------------------------------------------------------------------------
% Setup spatial grid
nrR   = 100;                     % number of gridpoints along z axis
dR    = sampleRadius/(nrR-1);    % spacing between gridpoints
r     = [0:dR:sampleRadius];     % z position of each gridpoint
%------------------------------------------------------------------------
% Setup the temporal grid
dT    = 0.001;                % [s] time steps for explicit algorithm
nrT   = round(simTime/dT);    % number of time steps in simulation
%------------------------------------------------------------------------
% 2. Initialize the material properties
mu           = 1e5;    % [Pa]
lambda       = 1e6;    % [Pa]
k_o          = 1e-11;  % [kg/m^2/s] permeability
solidContent = 0.5;    % volume fraction of solid
%------------------------------------------------------------------------
% 3. Allocate memory and setup initial condition
u            = zeros(nrR, nrT); % aso sets up initial condition u(:,0) = 0
currentDisp  = 0;
%------------------------------------------------------------------------
```

```
% 4. Assign the axial stretch conditions
lambdaZ = ones(nrT,1);
for j = 2:nrT
    if (lambdaZ(j-1) > finalStretch)
        DlambdaDt(j) = stretchRate;
    else
        lambdaZ(j)   = lambdaZ(j-1);
        DlambdaDt(j) = 0;
    end
    lambdaZ(j)   = lambdaZ(j-1) + dT*DlambdaDt(j);
end
%-----------------------------------------------------------------------
% 5. Solve the discritized governing equation
for j = 2:nrT
    u(1,j) = 0;     % boundary condition u(r=0,t) = 0
    for i = 2:nrR-1
        u(i,j) = u(i,j-1) + dT*k_o*(lambda+2*mu)*(   ...
                  (1/dR^2+1/(2*r(i)*dR))*u(i+1,j-1)  ...
                 -(2/dR^2+1/(r(i)^2))   *u(i,  j-1)  ...
                 +(1/dR^2-1/(2*r(i)*dR))*u(i-1,j-1)) ...
                 - dT * r(i)/2.0*DlambdaDt(j);
    end
    % boundary condition sigma_rr(r=R,t) = 0
    coef = lambda*dR/(2*mu+lambda);
    u(nrR,j) = (u(nrR-1,j) - coef*(lambdaZ(j)-1))/(1+coef/r(nrR));
end
```

9 Growth Models

In this chapter, we introduce the use of internal variables. Specifically, we will consider a thermo-mechanical solid, which grows in response to the stress state within the material. This could be used as a simple model for growing or remodeling biological tissue. Internal variables can be used to capture a wide range of phenomena such as material damage, plasticity, and crystallinity. This chapter is meant to illustrate the use of an internal variable and the numerical methods used to implement such a model. The internal variable in the model outlined in this chapter is used to account for the amount of growth within the material. A simple evolution equation that couples the internal variable representing growth to the stress state of the material is written. The formulation of this model is significantly different from the previous models in that mass may be introduced or removed from the system. As mass is removed at any given point within the system, the momentum and energy associated with that mass is also removed. In effect, the mass instantaneously disappears. When mass is added at a given point, it is introduced with the same velocity, temperature, and energy of the mass that currently occupies the point.

9.1 Forces and Fields

In this chapter, we consider a thermo-mechanical solid model, which includes an internal variable accounting for the growth of the solid. Growth is captured in this model by introducing the rate of mass change per unit volume, $\varphi_g(\boldsymbol{x},t)$. Therefore, the list of fields that must be determined in the system includes all of the variables introduced in Chapter 4 as well as the rate of mass change per unit volume. The full list of variables within this growth model are given in Table 9.1.

9.2 Balance Laws

Although we are considering a thermo-mechanical solid, the balance laws will differ significantly from those derived in Chapter 4 due to the presence of mass sources and sinks. The balance laws for the growth model must account for the flow of mass into or out of the system. As mentioned earlier, the flow of mass into or out of the system is accompanied by a flow of momentum and energy. Any mass taken out of

Table 9.1. *List of unknown variables within the thermo-mechanical model*

Variable	Name
$\boldsymbol{x}(\boldsymbol{X},t)$	Equations of motion
$\rho\,(\boldsymbol{x},t)$	Spatial density field
$\boldsymbol{v}\,(\boldsymbol{x},t)$	Spatial velocity field
$\boldsymbol{\sigma}\,(\boldsymbol{x},t)$	Cauchy stress tensor
$\boldsymbol{q}\,(\boldsymbol{x},t)$	Spatial heat flux
$e\,(\boldsymbol{x},t)$	Spatial internal energy density field
$\theta\,(\boldsymbol{x},t)$	Spatial temperature field
$\eta\,(\boldsymbol{x},t)$	Spatial entropy density field
$\boldsymbol{b}\,(\boldsymbol{x},t)$	Spatial body force field
$\varphi_g\,(\boldsymbol{x},t)$	Rate of mass change per unit volume

the system is removed instantaneously and has an associated momentum and energy that is removed at the same time.

9.2.1 Conservation of Mass

The rate of change of the mass of a system, $\dot{m}$, must be equal to the material derivative of the integral of the density field, $\rho\,(\boldsymbol{x},t)$, over the current configuration, $\Omega\,(t)$, such that

$$\dot{m} = \frac{D}{Dt}\int_{\Omega(t)} \rho\,(\boldsymbol{x},t)\,dV_x.$$

In the case of growth, mass may be added or subtracted based on some growth criteria. Let us assume that that the rate of mass change per unit current volume is given by φ_g such that the total mass added in the current volume element, dV_x, is given by

$$dm = \varphi_g\,dV_x.$$

The conservation of mass can then be written as

$$\int_{\Omega(t)} \varphi_g\,dV_x = \frac{D}{Dt}\int_{\Omega(t)} \rho\,(\boldsymbol{x},t)\,dV_x.$$

Converting this to an integral over the reference configuration gives

$$\int_{\Omega_o} \varphi_g J\,dV_X = \frac{D}{Dt}\int_{\Omega_o} \rho J\,dV_X.$$

Commuting the integral and derivative and then grouping terms gives

$$0 = \int_{\Omega_o} \frac{D}{Dt}(\rho J) - \varphi_g J\,dV_X.$$

Finally, substituting $\dot{J} = J\,div\,(\boldsymbol{v})$ gives

$$0 = \int_{\Omega_o} \left(\dot{\rho} + div\,(\boldsymbol{v}) - \varphi_g\right) J\,dV_X.$$

Converting back to an integral over the current configuration gives

$$0 = \int_{\Omega(\mathrm{t})} \left(\dot{\rho} + div\,(\boldsymbol{v}) - \varphi_g\right) dV_x.$$

Since this must hold every subsection of the volume, the integrand must vanish at all points within the material. The differential form of the conservation of mass becomes

$$0 = \frac{D\rho}{Dt} + \rho\, div(\boldsymbol{v}) - \varphi_g.$$

9.2.2 Reynolds' Transport Theorem

When deriving the conservation of mass equation, we purposefully did not use the Reynolds' transport theorem as it was derived in Section 2.16. In this previous derivation, we made use of the fact that there were no sink or source terms and the equation of conservation of mass was used to write multiple forms of the Reynolds' transport theorem. Several of these can no longer be used for a material with growth. Therefore, in this section, we will derive the proper forms to be used with growth. We assume that we have a spatial scalar field, $\phi = \phi(\boldsymbol{x},t)$ which is multiplied by the density field, $\rho(\boldsymbol{x},t)$. The integral of this scalar field over an enclosed volume in the current configuration is given by

$$\zeta(t) = \int_{\Omega(t)} \rho(\boldsymbol{x},t)\,\phi(\boldsymbol{x},t)\, dV_x.$$

The material derivative of the integral may be written as

$$\dot{\zeta}(t) = \frac{D}{Dt}\int_{\Omega(t)} \rho\phi\, dV_x.$$

Converting the integral over the current configuration to an integral over the reference configuration, we obtain

$$\dot{\zeta}(t) = \frac{D}{Dt}\int_{\Omega_o} \rho\phi J\, dV_X.$$

Commuting the derivative, we may write

$$\dot{\zeta}(t) = \frac{D}{Dt}\int_{\Omega_o} \left(\left(\dot{\rho}\phi + \rho\dot{\phi}\right)J + (\rho\phi)\dot{J}\right) dV_X.$$

Substituting $\dot{J} = J\, div(\boldsymbol{v})$, we find

$$\dot{\zeta}(t) = \int_{\Omega_o} \left(\dot{\rho}\phi + \rho\dot{\phi} + \rho\phi\, div(\boldsymbol{v})\right) J\, dV_X.$$

Converting back to the current configuration gives

$$\dot{\zeta}(t) = \int_{\Omega(t)} \left(\dot{\rho}\phi + \rho\dot{\phi} + \rho\phi\, div(\boldsymbol{v})\right) dV_x.$$

Using the equation for conservation of mass, we may rewrite this equation in terms of the growth rate such that

$$\dot{\zeta}(t) = \int_{\Omega(t)} \left(-\rho\phi\, div(\boldsymbol{v}) + \phi\varphi_g + \rho\dot{\phi} + \rho\phi\, div(\boldsymbol{v})\right) dV_x.$$

Simplifying the equation gives

$$\dot{\zeta}(t) = \int_{\Omega(t)} \left(\rho\dot{\phi} + \varphi_g\phi\right) dV_x.$$

9.2.3 Conservation of Momentum

The principle of conservation of linear momentum states that the rate of change of the linear momentum of the system is equal to the sum of forces acting on the system. Due to the introduction or depletion of mass within the system, we must add an additional term. When mass is created at some point in the system, at the instance of creation, it has the same velocity as the mass that was already at that point. Therefore, we are adding additional momentum to the system due to growth equal to

$$\int_{\Omega(t)} \varphi_g \boldsymbol{v}\, dV_x.$$

The conservation of linear momentum can then be written as

$$\frac{D}{Dt}\int_{\Omega(\mathrm{t})} \rho\, \boldsymbol{v}\, dV_x = \int_{\partial\Omega(\mathrm{t})} \boldsymbol{t}_x\, dS_x + \int_{\Omega(\mathrm{t})} \rho\, \boldsymbol{b}_x\, dV_x + \int_{\Omega(t)} \varphi_g \boldsymbol{v}\, dV_x. \tag{9.1}$$

Using the Reynolds' transport theorem, we obtain

$$\int_{\Omega(\mathrm{t})} \left(\rho\, \dot{\boldsymbol{v}} + \varphi_g \boldsymbol{v}\right) dV_x = \int_{\partial\Omega(\mathrm{t})} \boldsymbol{t}_x\, dS_x + \int_{\Omega(\mathrm{t})} \rho\, \boldsymbol{b}_x\, dV_x + \int_{\Omega(t)} \varphi_g \boldsymbol{v}\, dV_x. \tag{9.2}$$

As we can see, the additional growth terms cancel out giving us the local differential form

$$\rho\, \dot{v} - div\, \boldsymbol{\sigma}^T - \rho\, \boldsymbol{b}_x = \boldsymbol{0}.$$

9.2.4 Conservation of Angular Momentum

The principle of conservation of angular momentum states that the rate of change of the angular momentum in a system is equal to the sum of the couples acting on the system. Due to the creation of mass, we have an additional term that accounts for the addition of angular momentum. When mass is added to the system, it is added with the same properties as the material that is already there. Therefore, the additional angular momentum added to the system when mass is created is given by

$$\int_{\Omega(t)} \left(\boldsymbol{r} \times \varphi_g \boldsymbol{v}\right) dV_x.$$

The statement of conservation of angular momentum becomes

$$\begin{aligned}\frac{D}{Dt}\int_{\Omega(\mathrm{t})} (\boldsymbol{r} \times \rho\, \boldsymbol{v})\, dV_x = &\int_{\partial\Omega(\mathrm{t})} \left(\boldsymbol{r} \times \left(\boldsymbol{\sigma}^T \cdot \boldsymbol{n}_x\right)\right) dS_x + \int_{\Omega(\mathrm{t})} \left(\boldsymbol{r} \times \rho\, \boldsymbol{b}_x\right) dV_x \\ &+ \int_{\Omega(t)} \left(\boldsymbol{r} \times \varphi_g \boldsymbol{v}\right) dV_x.\end{aligned}$$

Using the Reynolds' transport theorem and the conservation of mass, we may rewrite this as

$$\int_{\Omega(t)} (\boldsymbol{r}\times(\rho\dot{\boldsymbol{v}}+\varphi_g\boldsymbol{v}))\, dV_x = \int_{\partial\Omega(t)} \left(\boldsymbol{r}\times\left(\boldsymbol{\sigma}^T\cdot\boldsymbol{n_x}\right)\right) dS_x + \int_{\Omega(t)} (\boldsymbol{r}\times\rho\,\boldsymbol{b}_x)\, dV_x + \int_{\Omega(t)} (\boldsymbol{r}\times\varphi_g\boldsymbol{v})\, dV_x.$$

Once again the growth terms cancel, and we are left with a statement that the Cauchy stress tensor is symmetric such that

$$\boldsymbol{\sigma} = \boldsymbol{\sigma}^T.$$

9.2.5 Conservation of Energy

Again, when mass is introduced into the system, it is introduced with the properties of the material that is already there. Therefore, the addition of mass brings with it additional kinetic and internal stored energy equal to

$$\xi_{create} = \int_{\Omega(t)} \varphi_g\left(e+\frac{1}{2}\boldsymbol{v}\cdot\boldsymbol{v}\right) dV_x.$$

The conservation of energy equation becomes

$$\frac{D}{Dt}\int_{\Omega(t)}\left(\rho e+\frac{1}{2}\rho\,(\boldsymbol{v}\cdot\boldsymbol{v})\right) dV_x = \int_{\partial\Omega(t)} (\boldsymbol{t}_x\cdot\boldsymbol{v}-\boldsymbol{q}_x\cdot\boldsymbol{n}_x)\, dS_x + \int_{\Omega(t)} (\rho\boldsymbol{b}_x\cdot\boldsymbol{v}+\rho\, r_x)\, dV_x + \int_{\Omega(t)} \varphi_g\left(e+\frac{1}{2}\boldsymbol{v}\cdot\boldsymbol{v}\right) dV_x.$$

Applying the Reynolds' transport theorem, we have

$$\int_{\Omega}^{(t)}\left(\rho\dot{e}+\rho\,(\boldsymbol{v}\cdot\dot{\boldsymbol{v}})+\varphi_g\left(e+\frac{1}{2}\boldsymbol{v}\cdot\boldsymbol{v}\right)\right) dV_x$$
$$= \int_{\partial\Omega(t)} (\boldsymbol{t}_x\cdot\boldsymbol{v}-\boldsymbol{q}_x\cdot\boldsymbol{n}_x)\, dS_x + \int_{\Omega(t)} (\rho\boldsymbol{b}_x\cdot\boldsymbol{v}+\rho\, r_x)\, dV_x + \int_{\Omega(t)} \varphi_g\left(e+\frac{1}{2}\boldsymbol{v}\cdot\boldsymbol{v}\right) dV_x.$$

The growth terms once again cancel, and we are left with

$$\rho\dot{e}-\boldsymbol{\sigma}:\boldsymbol{D}+div\,(\boldsymbol{q}_x)-\rho\, r_x=0.$$

9.3 Decomposition of the Deformation Gradient

Imagine we have a stress free segment of biological material, both material growth and the application of forces will cause changes in the shape and size of the material sample. We can decompose our deformation gradient into two parts. The first accounts, $\boldsymbol{F}^g$, for the growth of the tissue and the second, $\boldsymbol{F}^e$, accounts for the change in shape due to the application of stress. This approach is similar to the decomposition often used in constitutive models for plasticity. In that case, the deformation gradient is broken up into an elastic and a plastic part.

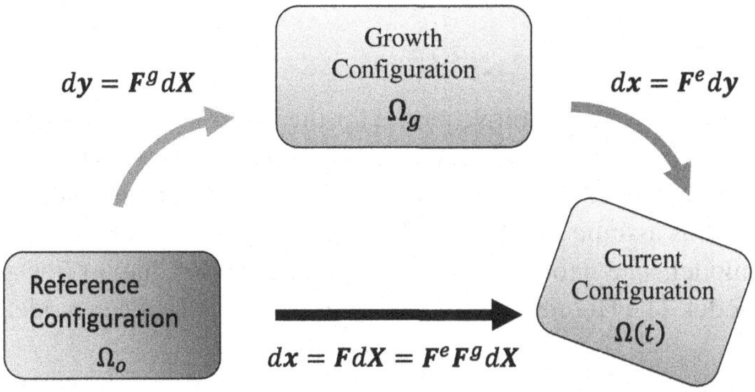

Figure 9.1. The deformation gradient, $\boldsymbol{F}$, is decomposed into the deformation due to growth, $\boldsymbol{F}^g$, and the deformation due to the application of stress, $\boldsymbol{F}^e$.

9.4 Summary of the Field Equations

Table 9.2. *List of variables within the thermo-mechanical model*

Spatial form		Unknowns
Variable	Name	(material)
$\boldsymbol{x}(\boldsymbol{X},t)$	Equations of motion	3
$\rho(\boldsymbol{x},t)$	Spatial density field	1
$\boldsymbol{v}(\boldsymbol{x},t)$	Spatial velocity field	3
$\boldsymbol{\sigma}(\boldsymbol{x},t)$	Cauchy stress tensor	9
$\boldsymbol{q}(\boldsymbol{x},t)$	Spatial heat flux	3
$e(\boldsymbol{x},t)$	Spatial internal energy density field	1
$\theta(\boldsymbol{x},t)$	Spatial temperature field	1
$\eta(\boldsymbol{x},t)$	Spatial entropy density field	1
$\boldsymbol{b}(\boldsymbol{x},t)$	Spatial body force field	0
$\varphi_g(\boldsymbol{x},t)$	Rate of mass change per unit volume	1

Table 9.3. *List of the spatial and material forms of the balance equations*

Balance law	Spatial form	Independent equations (material)
Kinematical relations	$\boldsymbol{v} = \frac{\partial \boldsymbol{x}}{\partial t}$	3
Conservation of mass	$0 = \frac{\partial \rho}{\partial t} + div(\rho\boldsymbol{v}) - \varphi_g$	1
Conservation of linear momentum	$\rho\dot{v} - div\,\boldsymbol{\sigma}^T - \rho\boldsymbol{b}_x = \boldsymbol{0}$	3
Conservation of angular momentum	$\boldsymbol{\sigma} = \boldsymbol{\sigma}^T$	3
Conservation of energy	$\rho\dot{e} - \boldsymbol{\sigma} : \boldsymbol{D} + div\left(\boldsymbol{q}_x\right) - \rho r_x = 0$	1
Entropy inequality	$-\rho\dot{\psi} - \rho\eta\dot{\theta} - \frac{1}{\theta}\boldsymbol{q}_x \cdot grad\,\theta + \boldsymbol{\sigma}^T : \boldsymbol{D} \geq 0$	0

9.5 Constitutive Model

At this point, we must formulate a constitutive model for the material. This material model is typically based on experimental observation. For example, one might perform a series of experiments on dead tissue to ascertain behavior in the absence of growth. For simplicity, let us assume that the dead tissue behavior can be modeled using a neo-Hookean model. The neo-Hookean model discussed in Chapter 7 is a hyperelastic material model. The Helmholtz free energy for the neo-Hookean model is given by

$$\psi = \frac{G}{\rho_o}\left(I_{\boldsymbol{B}^e} - 3\right),$$

where $I_{\boldsymbol{B}^e}$ is the first invariant of the tensor $\boldsymbol{B}^e = \boldsymbol{F}^e \cdot \left(\boldsymbol{F}^e\right)^T$. Notice that the material response depends on the elastic part of the deformation tensor.

Additional experiments might reveal that growth depends on the current stress state in the material. We would then need an equation relating deformation rate due to growth to the current stress state in the material such that

$$\dot{\boldsymbol{F}}^g = f\left(\boldsymbol{\sigma}\right).$$

In actuality, there are many different forms of this equation found within the literature. For example, some believe this function depends on cyclical stressing, others believe a constant stress may also induce growth. Logically, this should vary from tissue type to tissue type. For the purposes of this discussion, let us make a few additional assumptions. Let us assume that the growth occurs isotropically. Since there is no preferred growth orientation, we can write the deformation gradient in terms of a single stretch

$$\boldsymbol{F}^g = \lambda^g\ \mathbf{I}.$$

Now we seek a scalar evolution equation for this stretch rate

$$\dot{\lambda}^g = f\left(\boldsymbol{\sigma}\right).$$

As an aside, more complex models would most likely account for the presence of various biochemical components such that

$$\dot{\lambda}^g = f\left(\boldsymbol{\sigma}, c_1, c_2, \ldots, c_n\right),$$

where c_1, c_2, ..., c_n are the concentrations of various bioactive compounds within the tissue.

The form of the evolution equation would be based on experimental observation. Here, we assume a simple form for illustration. Let us assume that both growth and resorption of material occur in proportion to the hydrostatic stress within the material. The evolution equation can then be written as

$$\dot{\lambda}^g = a_1\left(tr\left(\boldsymbol{\sigma}\right) - \sigma_o\right), \tag{9.3}$$

where a_1 and σ_o are material parameters. The complete material model then consists of the two equations

$$\psi = \frac{G}{\rho_o}\left(I_{B^e} - 3\right),$$
$$\dot{\lambda}^g = a_1\left(tr\left(\boldsymbol{\sigma}\right) - \sigma_o\right). \tag{9.4}$$

9.6 Uniaxial Loading

Now that we have the balance laws and the constitutive model for this material, we may use the model to predict material behavior. In this section, we will assume that the material is subjected to a uniaxial load and predict the material response. We begin by deriving the governing differential equation for this configuration then use the finite difference method to obtain a numerical solution to the differential equation. For a more detailed discussion of the uniaxial loading configuration, refer back to Section 7.5.

9.6.1 Kinematics

Assume that a specimen of length L, is subject to a unaixal loading state as shown in Figure 9.2. The uniaxial load leads to a single internal stress parallel to the $\boldsymbol{e}_1$ direction.

The equations of motion for this configuration are given by

$$x_1 = \lambda_1 X_1,$$
$$x_2 = \lambda_2 X_2,$$
$$x_3 = \lambda_3 X_3.$$

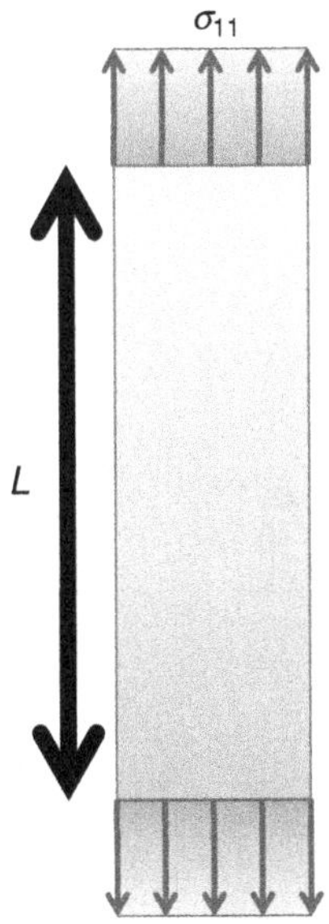

Figure 9.2. Uniaxial deformation sample geometry and loading configuration.

Due to the symmetry of the geometry and the loading, as well as the isotropic nature of the material, the stretch in the lateral dimensions will be equal, $\lambda_2 = \lambda_3$. The deformation gradient can be written as

$$\boldsymbol{F} = \begin{bmatrix} \lambda_1 & 0 & 0 \\ 0 & \lambda_2 & 0 \\ 0 & 0 & \lambda_2 \end{bmatrix}.$$

We can decompose the deformation gradient into an elastic part and a growth part such that

$$\boldsymbol{F} = \boldsymbol{F}^e \boldsymbol{F}^g.$$

We have assumed that the deformation due to growth is isotropic and may be written as

$$\boldsymbol{F}^g = \lambda^g \mathbf{I}.$$

Therefore, the total deformation gradient may be written as

$$\boldsymbol{F} = \boldsymbol{F}^e \lambda^g \mathbf{I}.$$

The elastic part of the deformation gradient may be written as

$$\boldsymbol{F}^e = \frac{1}{\lambda^g} \boldsymbol{F}.$$

The Jacobian can also be broken down into a growth part and an elastic part such that

$$J = \det \boldsymbol{F} = \det \boldsymbol{F}^e \det \boldsymbol{F}^g = J^e J^g.$$

Assuming we have an elastically incompressible material, we have the condition that

$$J^e = 1.$$

This requires that

$$\frac{\lambda_1 \lambda_1^2}{(\lambda^g)^3} = 1$$

and

$$J = J^g = \left(\lambda^g\right)^3$$

and

$$\lambda_2 = \sqrt{\frac{(\lambda^g)^3}{\lambda_1}}.$$

The deformation gradient is

$$\boldsymbol{F} = \begin{bmatrix} \lambda_1 & 0 & 0 \\ 0 & \sqrt{\frac{(\lambda^g)^3}{\lambda_1}} & 0 \\ 0 & 0 & \sqrt{\frac{(\lambda^g)^3}{\lambda_1}} \end{bmatrix},$$

and the elastic deformation gradient is

$$\boldsymbol{F}^e = \frac{1}{\lambda^g} \begin{bmatrix} \lambda_1 & 0 & 0 \\ 0 & \sqrt{\frac{(\lambda^g)^3}{\lambda_1}} & 0 \\ 0 & 0 & \sqrt{\frac{(\lambda^g)^3}{\lambda_1}} \end{bmatrix}.$$

The elastic part of the left Cauchy deformation tensor is given by

$$\begin{aligned} \boldsymbol{B}^e &= \boldsymbol{F} \cdot \boldsymbol{F}^T \\ &= \begin{bmatrix} \left(\frac{\lambda_1}{\lambda^g}\right)^2 & 0 & 0 \\ 0 & \frac{\lambda^g}{\lambda_1} & 0 \\ 0 & 0 & \frac{\lambda^g}{\lambda_1} \end{bmatrix}. \end{aligned}$$

9.6.2 Governing Equation

We can combine the geometry, loading configuration, constitutive equations, evolution equation, and conservation equations to determine the governing equation for this experimental setup. The stress strain relation for the neo-Hookean model may be written as

$$\boldsymbol{\sigma} = -p\mathbf{I} + \frac{2\alpha_1}{J}\boldsymbol{B}^e.$$

Substituting the elastic left Cauchy deformation gradient gives

$$\boldsymbol{\sigma} = \begin{bmatrix} \frac{2\alpha_1}{J}\left(\frac{\lambda_1}{\lambda^g}\right)^2 - p & 0 & 0 \\ 0 & \frac{2\alpha_1}{J}\frac{\lambda_1}{\lambda^g} - p & 0 \\ 0 & 0 & \frac{2\alpha_1}{J}\frac{\lambda_1}{\lambda^g} - p \end{bmatrix}.$$

The boundary conditions from the uniaxial loading require that the lateral dimensions be stress free

$$0 = \frac{2\alpha_1}{J}\frac{\lambda_1}{\lambda^g} - p.$$

The axial load may be determined from the applied force

$$\sigma_{11} = \frac{2\alpha_1}{J}\left(\frac{\lambda_1}{\lambda^g}\right)^2 - p.$$

This simplifies our stress equation in the axial direction

$$\sigma_{11} = \frac{2\alpha_1}{J}\left(\left(\frac{\lambda_1}{\lambda^g}\right)^2 - \frac{\lambda_1}{\lambda^g}\right).$$

Ignoring gravity, assuming that inertial effects are negligible, and using the conservation of momentum equation, we obtain

$$div\,(\boldsymbol{\sigma}) = 0.$$

Due to the symmetry of the loading and lack of shear stress, we have

$$\begin{aligned} \frac{\partial\sigma_{11}}{\partial x_1} + \frac{\partial\sigma_{12}}{\partial x_2} + \frac{\partial\sigma_{13}}{\partial x_3} &= 0, \\ \frac{\partial\sigma_{11}}{\partial x_1} &= 0. \end{aligned}$$

Therefore, the stress must be constant along the axial direction and the system of governing equations can be written as

$$\boxed{\begin{aligned} \sigma_{11} &= \frac{2\alpha_1}{J}\left(\left(\frac{\lambda_1}{\lambda^g}\right)^2 - \frac{\lambda_1}{\lambda^g}\right), \\ \dot{\lambda}^g &= a_1\left(tr\left(\boldsymbol{\sigma}\right) - \sigma_o\right). \end{aligned}} \tag{9.5}$$

9.6.3 Finite Difference Algorithm

In the previous section, we have written a set of differential equations that can be used to predict the response of the material system to uniaxial loading. In addition, we have shown that the both the stress and stretch are uniform along the axial direction as long as the material is homogenous. Therefore, we do not need to discretize the solution along the spatial axial direction, and we remove the dependence on $\boldsymbol{x}$ from both the stress and the stretch variables. In order to solve for the stress as a function of time, we will numerically solve the governing equation. First, we expand the growth rate using a forward difference formula as follows:

$$\frac{\lambda^g_{i+1} - \lambda^g_i}{\Delta t} = a_1\left(tr\left(\boldsymbol{\sigma}_i\right) - \sigma_o\right),$$

where λ^g_i is the current stretch due to growth, λ^g_{i+1} is the future stretch due to growth, and $\boldsymbol{\sigma}_i$ is the current stress within the material. Rearranging the equation, we obtain the evolution equation

$$\lambda^g_{i+1} = \lambda^g_i + \Delta t\,\alpha_1\left(tr\left(\boldsymbol{\sigma}_i\right) - \sigma_o\right).$$

The axial stress can be written in terms of the current stretch such that

$$\left(\sigma_{11}\right)_i = \frac{2\alpha}{\left(\lambda^g_i\right)^3}\left(\frac{\lambda_i}{\lambda^g_i} - \frac{\lambda^g_i}{\lambda_i}\right).$$

In this case, we solve for the stress at time t, and then determine the stretch at $t+\Delta t$. We assume that initially no growth has taken place such that

$$\lambda^g_0 = 1.$$

We can then iteratively solve for the stress subject to the displacement boundary condition at the top surface

$$u_3\left(X_3 = l_o, t\right) = U_o\left(t\right).$$

This boundary condition can be recast in terms of a stretch boundary condition

$$\lambda\left(t\right) = \frac{l_o + U_o\left(t\right)}{l_o}.$$

9.6.4 Example Problem

Let us consider the response of this material model to uniaxial load. Let us assume that the equilibrium stress for the material is given by $\sigma_o = 0$ MPa and $a_1 = 10$ MPa. Let us assume that the deformation of the sample is given when the material is stretched at a rate of 0.01/s for a total of 5 s and then is held at constant stretch for another 5 s. The prescribed stretch as a function of time is shown in Figure 9.3. The resulting growth and internal stress are shown in Figure 9.4 and Figure 9.5, respectively. The internal stress increases initially while the material is being stretched. This induces growth, and the material's equilibrium length increases. As the material grows, the internal stress decays away, and the growth rate slows.

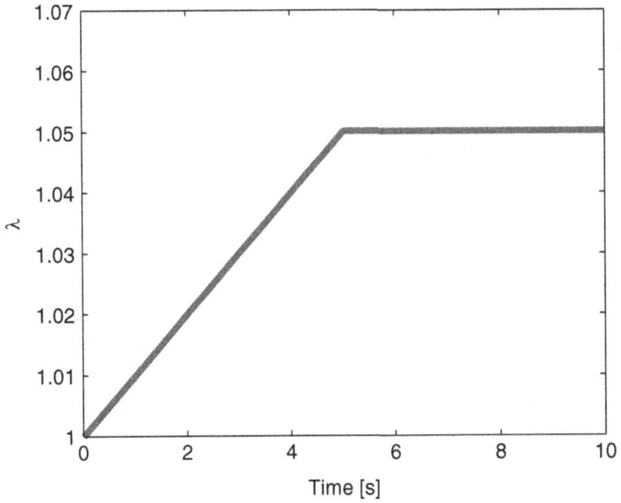

Figure 9.3. Stretch boundary condition as a function of time. Initially the stretch is increased at a constant rate for 2 seconds and then held constant for 8 seconds.

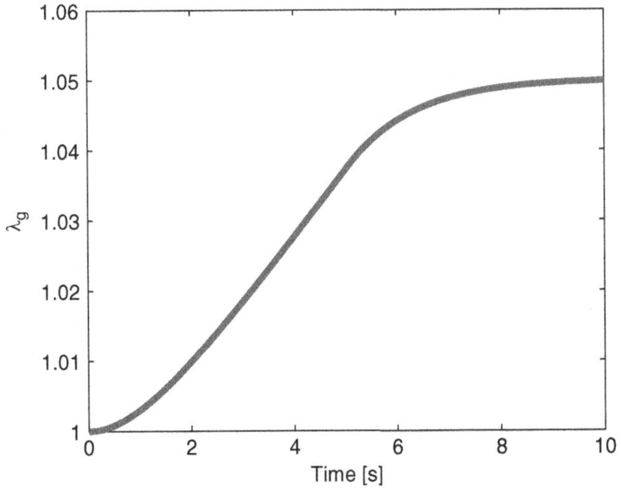

Figure 9.4. Growth response of the tissue as a function of time.

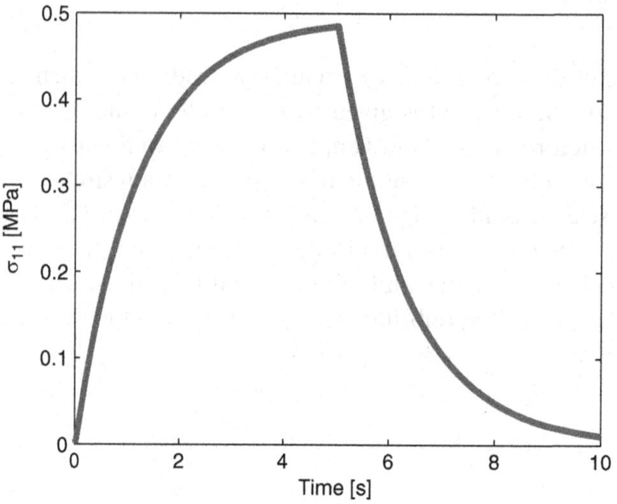

Figure 9.5. Axial stress as a function of time.

9.6.5 Matlab® File

```
%------------------------------------------------------------------------
% Numerical solution of uniaxial tension test of a material which grows in
% response to stress stimuli.
%------------------------------------------------------------------------
%  sigma_11(j) is the internal stress in the sample at time increment j
%------------------------------------------------------------------------
clear;
%------------------------------------------------------------------------
% 1. Initialize the spatial and temporal grid
% Initialize test parameters
stretchRate  = 0.01;       % [stretch/s] Stretch rate
totalStretch = 1.05;       % Total stretch of sample
simTime      = 10;         % [s]   Total simulation time
%------------------------------------------------------------------------
% Setup the temporal grid
dT    = 0.01;              % [s] time steps for explicit algorithm
nrT   = round(simTime/dT); % number of time steps in simulation
%------------------------------------------------------------------------
% 2. Initialize the material properties
a            = 0.02;       % [stretch/MPa] growth rate
sigma_o      = 0;          % [MPa] equilibrium stress
alpha_1      = 10;         % [MPa]
%------------------------------------------------------------------------
% 3. Allocate memory and setup initial condition
lambda        = ones(nrT,1);
lambda_g      = ones(nrT,1);
sigma_11      = zeros(nrT,1);
currentDisp   = 0;
%------------------------------------------------------------------------
```

```
% 4. Assign the boundary conditions
for j = 2:nrT
    if (lambda(j-1) < totalStretch)
        lambda(j) = lambda(j-1) + stretchRate*dT;
    else
        lambda(j) = lambda(j-1);
    end
end
%-----------------------------------------------------------------------
% 5. Solve the discritized governing equations
for j = 2:nrT
    lam_ratio   = lambda(j-1)/lambda_g(j-1);
    sigma_11(j) = 2*alpha_1*(lam_ratio - lam_ratio^-1);
    lambda_g(j) = lambda_g(j-1)+dT*a*(sigma_11(j-1)-sigma_o);
end
```

10 Parameter Estimation and Curve Fitting

As we have seen in this textbook, constitutive models have many material parameters that must be determined experimentally. These parameters are often found by fitting the model's predicted behavior to the experimentally observed behavior. It is critical that the uncertainty in these material parameters be reported along with the values of the parameters. In this chapter, we introduce the methodology for providing uncertainty estimates for experimental measurements and for parameters obtained from curve fitting.

10.1 Propagation of Error

Experimental measurements suffer from both systematic and random error. For example, measurements from a force transducer used to measure load have systematic and random error due to the physical sensor and the data acquisition system used to acquire data. The error in the force measurements due to the sensor is a combination of systematic uncertainty due to nonlinearity and hysteresis of the sensor and random uncertainties due to thermal-stability error and repeatability error. The error from the data acquisition system is a combination of systematic uncertainty due to nonlinearity and gain error and random uncertainty due to quantization and noise. Often these errors are well documented by the producers of the measurement equipment. However, one must often take experimentally determined values and combine or manipulate them to report calculated quantities. For example, one might report a stress that was computed using a force measurement and measurement of the cross-sectional area of a specimen. In order to compute the error for a computed quantity that is a function of the distance or force measurement, we will need to propagate these errors through the equations used to compute the desired quantities.

Let us consider a function, Ξ, of several measured variables, ξ_n, such that

$$\Xi = f(\xi_1, \xi_2, \ldots, \xi_n).$$

Since each parameter ξ_n has uncertainty $\delta\xi_n$, we would like to determine the change in the functional value $\delta\Xi$ for small changes in the function parameters $\delta\xi_n$. If we assume that the error in each parameter is small but finite and that the error in each parameter is independent of the error in any other parameter, we can approximate this by examining the Taylor expansion of the function

$$\Xi = \Xi_o + \delta\Xi = \Xi_o + \sum_{i=1}^{n} \frac{\partial \Xi}{\partial \xi_i} \delta\xi_i,$$

which gives

$$\delta\Xi = \sum_{i=1}^{n} \frac{\partial \Xi}{\partial \xi_i} \delta\xi_i.$$

However, the quantity $\delta\Xi$ does not serve as a good estimate of the error in the function Ξ because the partial derivative $\frac{\partial \Xi}{\partial \xi_i}$ may be positive or negative allowing for cancellation of error between parameters. Therefore, a better estimate of the maximum error would be the root of the sum of the squares (RSS) in the function

$$\delta\Xi^2 = \sum_{i=1}^{n} \left(\frac{\partial \Xi}{\partial \xi_i} \delta\xi_i \right)^2. \tag{10.1}$$

This function is commonly used to estimate the error in a computed quantity. The following example illustrates the application of this concept to measurements taken during uniaxial tension experiment.

EXAMPLE 10.1. *During a uniaxial tension test of an isotropic material, we determine axial, δ_a, and lateral, δ_l, displacements using an extensometer, which has a measurement error of z_e, and the axial force, F_a, using a force transducer which has a measurement error of z_f. In addition, let us assume that these errors are constant over the range of our data measurements. Measurements of specimen length, thickness, and width were taken using a micrometer, which has measurement error of z_m. Find the error in the computed engineering strain, original and final cross-sectional area, and the engineering stress.*

Solution:
The axial strain, ϵ_{11}, is given by the measured axial displacement, δ_a, divided by the original sample length, l_o, as $\epsilon_{11} = \frac{\delta_a}{l_o}$. Using Equation (10.1), the RSS estimate of the error in the axial strain measurement gives

$$\delta\epsilon_{11}^2 = \left(\frac{\partial \epsilon_{11}}{\partial l_o} \delta l_o \right)^2 + \left(\frac{\partial \epsilon_{11}}{\partial \delta_a} \delta\delta_{\mathrm{a}} \right)^2.$$

An estimate of the error in length measurement is given by the error in the micrometer measurements, $\delta l_o = z_e$, while the error in the axial displacement measurement is given by the error in the extensometer used to measure it. Taking the partial derivatives of the strain with respect to the measured variables gives $\frac{\partial \epsilon_{11}}{\partial l_o} = \frac{\delta_a}{l_o^2}$, and $\frac{\partial \epsilon_{11}}{\partial \delta_a} = \frac{1}{l_o}$. Combining these equations with the RSS estimate gives

$$\delta\epsilon_{11} = \sqrt{\left(\frac{\delta_a}{l_o^2} z_m \right)^2 + \left(\frac{1}{l_o} z_e \right)^2}.$$

Similarly, the RSS estimate of the error in the lateral strain, $\epsilon_{22} = \frac{\delta_l}{w_o}$, is given by

$$\delta\epsilon_{22}^2 = \left(\frac{\partial \epsilon_{22}}{\partial w_o} \delta w_o \right)^2 + \left(\frac{\partial \epsilon_{11}}{\partial l} \delta\delta_{\mathrm{l}} \right)^2.$$

Substituting the partial derivatives and error estimates for width and lateral extension measurements gives

$$\delta\epsilon_{22} = \sqrt{\left(\frac{1}{w_o} z_e\right)^2 + \left(\frac{\delta_l}{w_o^2} z_m\right)^2}.$$

The estimated error in the area of the reference cross section, $A_o = w_o \cdot t_o$, is

$$\delta A_o = \sqrt{(w_o \cdot z_m)^2 + (t_o \cdot z_m)^2}.$$

The estimated error in the current area measurements where $A = A_o\,(1+\epsilon_{22})^2$ is given by

$$\delta A = \sqrt{\left((1+\epsilon_{22})^2 \cdot \delta A_o\right)^2 + (2A_o\,(1+\epsilon_{22}) \cdot \delta\epsilon_{22})^2}.$$

Notice that we have used the fact that we know the error estimates for both the original cross-sectional area, δA_o, and the lateral strain, $\delta\epsilon_{22}$, from previous calculations.

Finally, the estimated error in the axial component of the engineering stress, $\sigma_{11} = F_a/A_o$, is given by

$$\delta\sigma_{11} = \sqrt{\left(\frac{1}{A_o} \cdot z_f\right)^2 + \left(\frac{F_a}{A_o^2} \cdot \delta A_o\right)^2}.$$

10.2 Least Squares Fit

Often material parameters are not the result of direct computation but are found by fitting model predictions to experimental observation. In these cases, curve fits are used to obtain the best approximation of the true material parameters. As a simple example, we will consider the determination of the material parameters for the linear elastic material model from stress-strain data.

For a linear elastic material, the axial stress is linearly related to the axial strain in the uniaxial tension experiment such that $\sigma_{11} = \mathrm{E}\epsilon_{11}$, and the lateral and axial strains are related linearly such that $\epsilon_{22} = -\nu\epsilon_{11}$. The constants of proportionality, E and ν, are the Young's modulus the Poisson's ratio, respectively. In this section, we will estimate the value of these material parameters by finding the slope of the axial stress versus axial strain plot and the lateral versus axial strain plot. We will also compute the confidence interval for these value values. *For simplicity, in the following derivations, we will assume that there is no random error in the x measurements.* We will correct this assumption at the end by adjusting the error in the y data to account for error in the x data.

The least squares method is commonly used to determine the equation of the line that best fits the experimental data. Assume that we have a set of n experimental data points consisting of pairs of values (x_i, y_i), which, in our case, correspond to measurements of stress and strain. Our model consists of two linear equations. Therefore, in each case, we would like to fit the equation of a line

$$\tilde{y} = m \cdot x + b$$

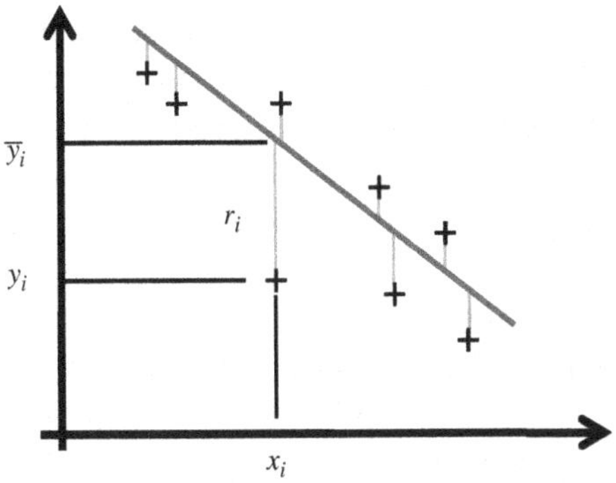

Figure 10.1. Illustration of best-fit line passing through several data points. The residual, r_i, is the difference between the actual value, y_i, and the value predicted by the best-fit line, Y_i.

to the data set where m is the slope of the line, and b is the y intercept. For any given experimental point (x_i,y_i), we have the actual experimental value y_i and the predicted value $\tilde{y}_i$, which is computed as $\tilde{y}_i = m \cdot x_i + b$. The best fit line is the one that minimizes the difference between actual and predicted values. In the following analysis, we will use an overbar to denote the mean of a quantity. Therefore, the mean values of x and y would be $\bar{x} = \frac{1}{n}\sum_{i=1}^{n} x_i$ and $\bar{y} = \frac{1}{n}\sum_{i=1}^{n} y_i$.

Whenever, we fit a line to experimental data, there will be some discrepancy between the predicted and actual values due to either error in measurements or nonlinearity of the observed response. For each point, the residual, r_i, can be computed as $r_i = (\tilde{y}_i - y_i)$ as shown in the Figure 10.1. We can express the total deviation between actual and predicted values along the entire curve as the sum of the square of the residuals (SSR) for each point such that

$$SSR = \sum_{i=1}^{n} (\tilde{y}_i - y_i)^2 = \sum_{i=1}^{n} (m \cdot x_i + b - y_i)^2 .$$

Ultimately, the best fit line will be the one that minimizes the SSR. We must now determine the values of the parameters m and b, which minimize the SSR. This is accomplished by taking the partial derivatives with respect to the slope and intercept and setting them equal to zero to obtain

$$\frac{\partial}{\partial m}(SSR) = 0 = \sum_{i=1}^{n} 2x_i\,(m \cdot x_i + b - y_i),$$

$$\frac{\partial}{\partial b}(SSR) = 0 = \sum_{i=1}^{n} 2\,(m \cdot x_i + b - y_i).$$

Solving these two equations for m and b gives

$$m = \frac{\sum_{i=1}^{n} x_i y_i - n\bar{x}\bar{y}}{\sum_{i=1}^{n} x_i^2 - n(\bar{x})^2} \tag{10.2}$$

and

$$b = \frac{\left(\bar{y}\sum_{i=1}^{n} x_i^2 - \bar{x}\left(\sum_{i=1}^{n} x_i y_i\right)\right)}{\sum_{i=1}^{n} x_i^2 - n(\bar{x})^2}. \tag{10.3}$$

You may have noticed that the form of the stress-strain relation has no y intercept. We could fit an equation with the form $\tilde{y} = m \cdot x$, which forces the line to pass through the origin. However, practically speaking, determining the zero strain/zero stress point from the force displacement measurements is approximate. Forcing the equation of the line to pass through the origin unnecessarily emphasizes the importance of this approximate quantity.

It is important to realize that the slope and y intercept computed using the least squares method is an estimate of the true slope and y intercept. Even though we can determine the values of the slope and y intercept which best approximate our experimental data, we must also provide a reasonable estimate of the uncertainty in these parameters. Without an estimate of error, the parameters are useless for engineering applications. For example, if for a given material we determined that the Young's modulus is 1 ± 200 GPa, the material could be as compliant as a rubber band or as stiff as steel. There is far too much uncertainty to make this a useful design parameter. In contrast, an experimentally determined Young's modulus of 1.00 ± 0.01 GPa is far more useful.

Ultimately, we need to determine the confidence interval for the estimated parameter. The procedure consists of finding an estimate of error variance for each parameter and using the Student's T-test to determine the confidence interval based on this estimate of error variance.

The ***estimate of error variance***, S_y, is the average of the standard deviation of the differences between the actual data points and those predicted by the best fit line such that

$$S_y = S(y_i) = \sqrt{\frac{\sum_{i=1}^{n} (y_i - \tilde{y}_i)^2}{n-2}}.$$

Notice that this is the square root of the sum of the residuals divided by the number of degrees of freedom. We use $n-2$ since we have n data points and 2 estimated parameters.

In order to determine the estimate of error variance for each parameter, we make use of an important statistical relation that the square of the estimate of error variance of a random variable, y, times a constant, c, is given by the constant squared times the estimate of error of the random variable such that

$$S^2(cy) = c^2 S^2(y). \tag{10.4}$$

The estimate of error variance for the slope of the best fit line can be found by rewriting Equation (10.3) in the equivalent form

$$m = \frac{\sum_{i=1}^{n} (x_i - \bar{x})(y_i)}{\sum_{i=1}^{n} (x_i - \bar{x})^2}.$$

Let us note once again that the x measurements are assumed to contain no random error. Only the y measurements have variance. Therefore, the denominator of the slope has no variance and becomes a constant multiplier. Let us define a new variable for this denominator $S_x = \sqrt{\frac{1}{N} \sum_{i=1}^{n} (x_i - \bar{x})^2}$. The square of the estimate of error variance in the slope, S_m, is given by

$$\begin{aligned}
S_m^2 = S^2(m) &= S^2\left(\frac{\sum_{i=1}^{n}(x_i-\bar{x})(y_i)}{\sum_{i=1}^{n}(x_i-\bar{x})^2}\right)\\
&= \left(\frac{1}{nS_x^2}\right)^2\left(\sum_{i=1}^{n}(x_i-\bar{x})^2 S^2(y_i)\right)\\
&= \left(\frac{1}{nS_x^2}\right)^2\left(\sum_{i=1}^{n}(x_i-\bar{x})^2 S_y^2\right)\\
&= \left(\frac{1}{nS_x^2}\right)^2\left(nS_x^2 S_y^2\right)\\
&= \frac{S_y^2}{nS_x^2}.
\end{aligned}$$

The estimate of error variance for the y intercept, S_b, can be found by rewriting Equation (10.2) in the equivalent form

$$b = \sum_{i=1}^{n}\left(\frac{1}{n} - \frac{\bar{x}(x_i-\bar{x})}{\sum_{i=1}^{n}(x_i-\bar{x})^2}\right)y_i.$$

The estimate of error variance becomes

$$\begin{aligned}
S_b^2 = S^2(b) &= S^2\left(\sum_{i=1}^{n}\left(\frac{1}{n} - \frac{\bar{x}(x_i-\bar{x})}{\sum_{i=1}^{n}(x_i-\bar{x})^2}\right)y_i\right)\\
&= \sum_{i=1}^{n}\left(\frac{1}{n} - \frac{\bar{x}(x_i-\bar{x})}{nS_x^2}\right)^2 S^2(y_i)\\
&= \sum_{i=1}^{n}\left(\frac{1}{n^2} - 2\frac{\bar{x}(x_i-\bar{x})}{nS_x^2} + \frac{\bar{x}^2(x_i-\bar{x})^2}{\left(nS_x^2\right)^2}\right)S^2(y_i).
\end{aligned}$$

Since $\sum \bar{x}(x_i-\bar{x}) = 0$ and $\sum \bar{x}^2(x_i-\bar{x})^2 = \bar{x}^2 nS_x^2$, we have the final result

$$S_b^2 = \left(\frac{1}{n} + \frac{\bar{x}^2}{nS_x^2}\right)S_y^2.$$

Once we have computed the estimate of error variance for each material parameter, we can determine the confidence interval.

Using a Student's T-test, we can determine the ***confidence interval*** for a given ***confidence level***. The confidence level is the level of certainty that the value of a parameter is within the confidence interval. For example, if we determine that there is a 95% probability that the value of the Young's modulus is $E = 1.0 \pm 0.01$ GPa, then the confidence level is 95% and the confidence interval is ± 0.01 GPa.

Ultimately, we select the confidence level required for your application. An aerospace application may require a greater confidence level than a consumer product. Once we have selected a confidence level, we determine the number of degrees of freedom within the fit. This is the total number of data points, n, minus the two determined parameters m and b. Therefore we have $n-2$ independent degrees of freedom. We then use the Student's T-test to find the t-statistic, $t_{cl,n}$, given the number of degrees of freedom and the confidence level. Note that the Student's T-test is actually a test of the null hypothesis. Therefore, for the fitting procedure outlined here, we use the Student's T-test to test the null hypothesis that the data are randomly distributed and do not fit the line with the given slope. The parameters for the T-test are actually $\alpha = 1 - \textit{confidence level}$. The estimate of the slope can then be written as

$$Slope = m \pm t_{\alpha,n} \cdot S_m.$$

The estimate of the y intercept can also be written as

$$y - intercept = b \pm t_{\alpha,n} \cdot S_b.$$

If the number of degrees of freedom is large (larger than 1,000), then the values of the t-statistic are relatively insensitive to further increases in n. In this case, the t-statistic is effectively a function of only the confidence level. For very large n, we have

$$Slope = m \pm 1.65\, S_m \quad \text{for a confidence level of } 90\%,$$

$$Slope = m \pm 1.96\, S_m \quad \text{for a confidence level of } 95\%,$$

$$Slope = m \pm 2.58\, S_m \quad \text{for a confidence level of } 99\%.$$

If n is smaller than 1,000, use a Student's T-test lookup table or computer program to determine the value of the t-statistic.

In order to minimize the uncertainty in the slope and y intercept, we should ensure that S_x is large. This requires that the x_i values are spread as widely as possible relative to their mean. In addition, increasing the number of data points will also decrease uncertainty. However, as the number of data points increases, the estimate of the error decreases only as a function of $1/\sqrt{n}$. Therefore, large increases in n may give only modestly better confidence intervals.

If both the x and y data measurements include random error, the analysis becomes significantly more complicated. However, we can simplify the problem if we convert the uncertainty in the x data to uncertainty in the y data. Specifically, we can say that the equivalent estimate of error variance in the y data is equal to

$$S_{y-eqiuv} = \sqrt{(S_y)^2 + (m\,\delta x)^2},$$

where δx is the uncertainty in the x data measurements. The estimates for the slope and y intercept are then written in terms of the equivalent estimate of error variance.

EXAMPLE 10.2. Assume that we perform a uniaxial tension test on a dogbone sample cut from a metallic specimen. The $\boldsymbol{e}_1$ axis is aligned parallel to the axis of the test specimen. The width of the central portion of the dog bone is $w_o = 2$ cm and the thickness is $t_o = 2$ mm measured using a micrometer with an estimated error of $\pm 5\,\mu$m. The axial force, F_a, is determined from load cell measurements with an estimated error of ± 100 N. The axial displacement, δ_a, and lateral displacements, δ_l, are measured using two distinct extensometers, which each have an estimated error of $\pm 0.5\,\mu$m. The original displacement between the pins of both the axial and lateral extensometers is $l_o = 2$ cm. The force and extension data are found in the following table.

Experimental data		
$F_a\,(N)$	$\delta_a\,(m)$	$\delta_l\,(m)$
9.30E+01	3.94E−07	−9.58E−08
2.58E+02	1.09E−06	−2.47E−07
5.27E+02	2.24E−06	−4.97E−07
8.29E+02	3.04E−06	−6.44E−07
1.06E+03	4.15E−06	−9.71E−07
1.31E+03	5.17E−06	−1.10E−06
1.56E+03	6.28E−06	−1.39E−06
1.81E+03	7.22E−06	−1.54E−06
2.03E+03	8.13E−06	−1.76E−06
2.25E+03	9.01E−06	−1.91E−06
2.53E+03	1.02E−05	−2.19E−06
2.78E+03	1.14E−05	−2.50E−06

Find the Young's modulus and the Poisson's ratio for the material.

Solution:
The basic procedure consists of the following three steps:

1. Convert the force-displacement data obtained experimentally into stress-strain data.
2. Determine the uncertainty in the stress-strain data.
3. Use least squares regression to estimate the value of the Young's modulus, Poisson's ratio, and the uncertainty in these parameters.

Step 1.
We can see from the experimental values of the displacement and measured length that the uniaxial tension test was performed for very small strains. For this reason, we will use the engineering stress-strain curve to determine the Young's modulus.

The original cross-sectional area of the sample is given by $A_o = 4 \times 10^{-5}$ m^2. The axial strain is computed as $\epsilon_{11} = \delta_a / l_o$ where l_o is the distance between the axial extensometer pins in the reference configuration. Since we have an isotropic material, the strains in the two lateral dimensions, $\boldsymbol{e}_2$ and $\boldsymbol{e}_3$, are equal. The axial component of the engineering stress are given by $\sigma_{11} = F_a / A_o$.

Step 2.
The error axial strain was previously found to be

$$\delta\epsilon_{11} = \sqrt{\left(\frac{\delta_a}{l_o^2} z_m\right)^2 + \left(\frac{1}{l_o} z_e\right)^2},$$

where the estimated error in the extensometer is $z_e = \pm 0.5\,\mu$m and estimated error in the micrometer is $z_m = \pm 5\,\mu$m. The error in the lateral strain is given by

$$\delta\epsilon_{22} = \sqrt{\left(\frac{1}{w_o} z_e\right)^2 + \left(\frac{\delta_l}{w_o^2} z_m\right)^2}.$$

Finally, the error in the engineering stress is given by

$$\delta\sigma_{11} = \sqrt{\left(\frac{1}{A_o}\cdot z_f\right)^2 + \left(\frac{F_a}{A_o^2}\cdot\delta A_o\right)^2},$$

where the estimated error in the force transducer measurement is $z_f = \pm 100$ N.

$\epsilon_{11}[\mu m/m]$	$\epsilon_{22}[\mu m/m]$	$\sigma_{11}[MPa]$
19.70 ± 25.00	−4.79 ± 25.00	2.3 ± 2.50
54.46 ± 25.01	−12.34 ± 25.00	6.4 ± 2.50
112.17 ± 25.04	−24.84 ± 25.00	13.2 ± 2.50
152.08 ± 25.07	−32.19 ± 25.00	20.7 ± 2.50
207.52 ± 25.13	−48.56 ± 25.01	26.6 ± 2.50
258.57 ± 25.21	−55.21 ± 25.01	32.7 ± 2.50
313.89 ± 25.31	−69.50 ± 25.02	38.9 ± 2.50
360.89 ± 25.40	−76.95 ± 25.02	45.3 ± 2.50
406.64 ± 25.51	−87.99 ± 25.02	50.7 ± 2.50
450.62 ± 25.63	−95.34 ± 25.03	56.3 ± 2.50
508.54 ± 25.80	−109.69 ± 25.04	63.3 ± 2.51
570.76 ± 26.00	−124.77 ± 25.05	69.5 ± 2.51

Step 3.
The slope of the stress strain curve is used to find the Young's modulus. The x data in this case is the engineering axial strain while the y data is the axial engineering stress. We will use Equation (7.1) to find the slope. The number of data points, n, is equal to 10. The average of the x data, $\bar{x}$, is equal to 273 μm. The average of the y data, $\bar{y}$, is equal to 0.34 MPa. Similarly, $\sum x_i y_i = 1560$ N/m and $\sum x_i^2 = 1.25\times 10^{-6}$ m^2. Plugging these values into Equation (7.1) gives the Young's modulus, $E = 1.28$ GPa.
The error in the Young's modulus is found using

$$\text{Confidence Interval} = t_{\alpha,n}\cdot S_m = t_{\alpha,n}\cdot\frac{S_y^2}{nS_x^2}.$$

The value of the t-parameter is obtained by selecting a confidence level, $CL =$ 95%, determining the α parameter where $\alpha = 1 - CL = 0.05$, and determining the degrees of freedom in the system, $dof = n-2 = 10$. Using a student's T-table gives the value of $t_{0.05,10} = 2.23$.

We have previously defined $S_x = \sqrt{\frac{1}{N}\sum_{i=1}^{n}(x_i-\bar{x})^2}$ and S_y is the variance in the y data. As discussed, the estimate of the data variance

$$S_y = \sqrt{\frac{\sum_{i=1}^{n}(y_i-\tilde{y}_i)^2}{n-2}}$$

is an estimated based on the closeness of the fit. This gives use a value of $S_y = 33.9$ kPa and a corresponding confidence interval of 0.13 GPa. However, we have access to the actual error estimates for each data point in both the x and y data. Using only the error in the y data gives $S_{y-act} = 25$ kPa and cooresponding

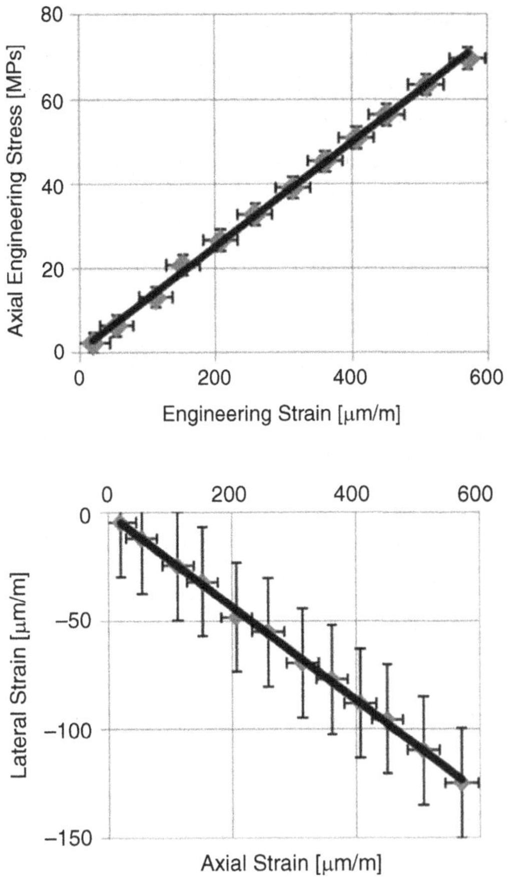

Figure 10.2. Determination of the elastic modulus and Poisson's ratio using experimental data with both x and y error bars shown.

confidence interval of 0.094 GPa. Even though this is lower than the estimate of the data variance, it does not account for the significant error in the x data. We can use the modified error variance to account for the x data variance which gives

$$S_{y-eqiuv} = \sqrt{(S_y)^2 + (m\delta x)^2} = 40.5\,\text{kPa}.$$

The corresponding confidence interval is 0.153 GPa.

Using the modified error variance, which accounts for the known error in both the x and y data, is probably best in this case. Had we not known the variance for each data point, we could simply use the estimate of error variance obtaining quite similar results. Given the analysis, we can state that the Young's modulus is equal to 1.28 ± 0.153 GPa.

The Poisson's ratio is determined from the slope the lateral strain versus axial strain curve. Here the x data is the axial strain, and the y data is the lateral strain. Analysis of the experimental data gives, $n = 12, \bar{x} = 273\,\mu\text{m}, \bar{y} = -57.4\,\mu\text{m}$, $\sum x_i y_i = 2.62 \times 10^{-7}\text{m}^2$, and $\sum x_i^2 = 1.25 \times 10^{-6}\text{m}^2$. Plugging these values into Equation (7.1) gives a slope of 0.21.

The t-parameter for the 95% confidence level is the same as that used for the Young's modulus, $t = 2.23$. The variance for the Poisson's ratio was computed using the actual experimental variance in both the x and y data gives $S_{y-equiv} = 25.5\,\mu\text{m}$, a variance in the slope of $S_m = 0.042$ and a corresponding confidence interval of $CI = 0.094$. Therefore, we can say that the Poisson's ratio for this material is 0.21 ± 0.094.

11 Finite Element Method

r, s	Coordinates in the natural coordinate system
xy	Coordinates in the global coordinate system
$N_i(r,s)$	Shape function
$G_i(r,s)$	Geometric interpolation functions
ϵ	Infinitesimal strain tensor
$\boldsymbol{E}$	Green strain tensor
σ	Cauchy stress tensor
$\boldsymbol{P}$	First Piola-Kirchhoff stress tensor
$\boldsymbol{S}$	Second Piola-Kirchhoff stress tensor
$\mathbb{C}$	Stiffness tensor
W_i	Gaussian quadrature weights
$\{\boldsymbol{u}_n\}$	Nodal displacement vector
$\{\boldsymbol{X}_n\}$	Nodal position vector

11.1 Introduction

The finite element method is a numerical technique for solving systems of partial differential equations. We will begin by describing the mechanics of the finite element method. Namely, we will discuss the discretization process, geometric interpolation functions and shape functions for quadrilateral elements, integration and differentiation of field variables, and numerical integration using Gaussian quadrature. We will then apply the method to plane stress, plane strain, and axisymmetric deformation of elastic solids before progressing to more complex material models. In this brief description of the method and its implementation, we will limit ourselves to the implementation of two-dimensional isoparametric quadrilateral elements.

The formulation of a finite element problem begins with the construction of the governing equations. We will find later that these equations contain surface and volume integrals that are difficult to evaluate analytically for complex geometries and loading configurations. The basic concept of the finite element method is to discretize space so that the volume and surface integrals in the governing equations can be written as a sum of integrals taken over individual elements. For example, instead of integrating over the entire volume of an object, $\Omega(t)$, we can write

$$\int_{\Omega(t)} \delta\epsilon : \mathbb{C} : \epsilon \, dV \cong \sum_{i=1}^{n_e} \int_{\Omega_i} \delta\epsilon : \mathbb{C} : \epsilon \, dV,$$

where Ω_i is the volume of element i, and n_e is the total number of elements.

Within each element, we will represent the field variables as continuous quantities that can be interpolated from the values at the nodal points. After rewriting the governing equation as the summation of volume or surface integrals over the elements, we will use Gaussian quadrature to numerically evaluate the integrals and solve the governing equations for discrete number of nodal values of the field variables.

11.1.1 Element Types

Finite element analysis begins with the discretization of an objects volume. Referred to as meshing the object, we create a set of connected elements, which creates some approximation of the object geometry.

First, we must select an element geometry that is capable of representing the object geometry. For example, several common two-dimensional element types include triangular, rectangular, and quadrilateral elements. Triangular elements have three sides, whereas rectangular and quadrilateral elements have four sides. As the name suggests, the internal angles of rectangular elements are restricted to 90°, whereas the quadrilateral elements are not. Meshing a two-dimensional object with curved edges in Figure 11–1 with rectangular elements may not be optimal. By

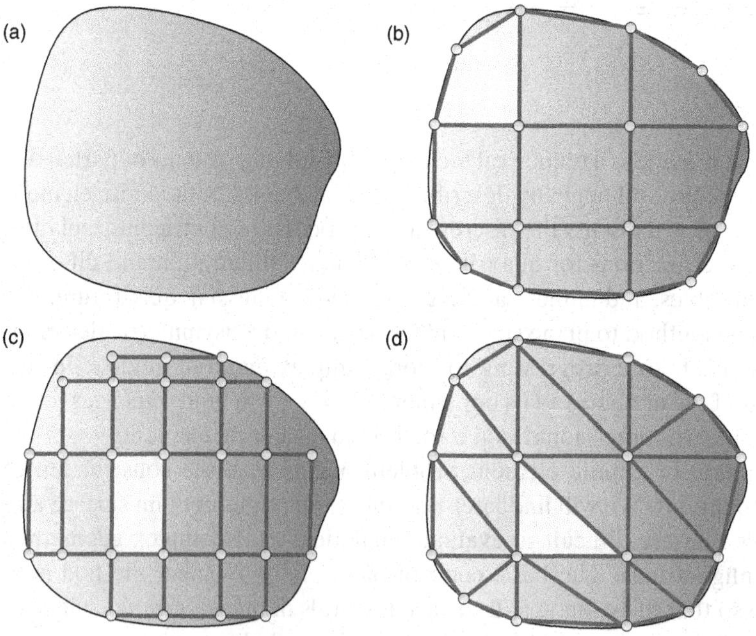

Figure 11.1. (A) An illustration of a continuum body. (B) Object meshed using quadrilateral elements. (C) Object meshed using rectangular elements. (D) Object meshed using triangular elements.

using triangular or quadrilateral elements the shape of the boundary may be more closely approximated.

Also, we shall see that fields within elements are approximated from nodal values using continuous interpolation functions. Even though linear interpolation functions may be satisfactory for some problems, higher order interpolation is sometimes needed for accurate results. Higher order interpolation functions require additional nodal points within the element. For example, a four-node quadrilateral element is sufficient for a linear interpolation function, but an eight-node element is required for quadratic interpolation of field variables.

In this text, we will focus on the formulation and implementation of the finite element method and will limit ourselves to four-node quadrilateral elements.

11.1.2 Natural Versus Global Coordinates for a Quadrilateral Element

When we mesh an object, we create a set of elements and assign (x,y) positions to the nodes such that the geometry of the object is approximated. However, we will see that integration over an element's (x,y) positions is computationally expensive.

Instead of computing integrals in the ***global coordinate system*** (x,y) space, we create a geometric mapping that relates the coordinates of a quadrilateral element to the coordinates of a square element defined in a new coordinate system, (r,s). The new r-s coordinate system is referred to as the ***natural coordinate system***. A geometric mapping function is used to convert between the global coordinate system, (x,y) and the natural coordinate system, (r,s).

The rectangular element defined in the natural coordinate system is referred to as the ***parent element***. In the natural coordinate system, the positions of the nodes are defined according to the following table.

Node	r_i	s_i	x_i	y_i
1	-1	-1	x_1	y_1
2	1	-1	x_2	y_2
3	1	1	x_3	y_3
4	-1	1	x_4	y_4

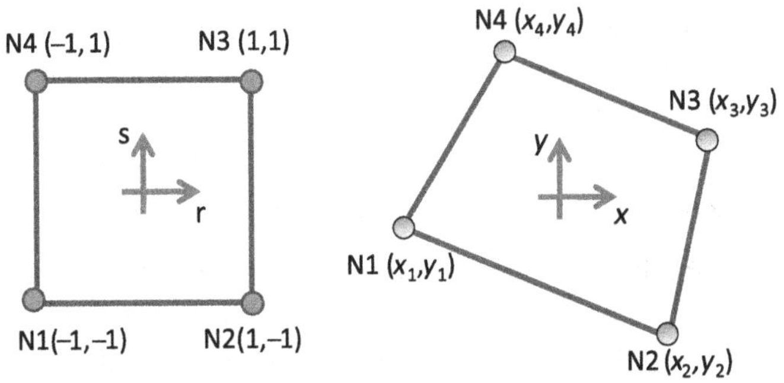

Figure 11.2. Illustration of mapping between (left) natural coordinate system and (right) Cartesian coordinate system.

A ***geometric interpolation function***, $G_i(r,s)$, is used to transform between coordinates of a point within the natural coordinate system, (r,s), and the corresponding point within the global coordinate system, (x,y), such that

$$x = \sum_{i=1}^{4} G_i(r,s)x_i,$$

$$y = \sum_{i=1}^{4} G_i(r,s)y_i,$$

where x_i and y_i are the x and y positions of the nodes of the quadrilateral. The transformation must have the properties that the nodal points map onto one another such that $(r_i,s_i) \to (x_i,y_i)$ where i is the node number. For example, we have the condition that

$$x_j = \sum_{i=1}^{4} G_i\left(r_j,s_j\right)x_i \quad \text{and} \quad y_j = \sum_{i=1}^{4} G_i\left(r_j,s_j\right)y_i.$$

As an example, the geometric interpolation for the first node must give

$$x_1 = \sum_{i=1}^{4} G_i(-1,-1)x_i \quad \text{and} \quad y_1 = \sum_{i=1}^{4} G_i(-1,-1)y_i.$$

Given these conditions, the geometric interpolation function has the properties that $G_i\left(r_j,s_j\right) = \delta_{ij}$.

For our four-node quadrilateral element, we will use a linear geometric interpolation function such that

$$G_1(r,s) = \frac{1}{4}(1-r)(1-s),$$

$$G_2(r,s) = \frac{1}{4}(1+r)(1-s),$$

$$G_3(r,s) = \frac{1}{4}(1+r)(1+s),$$

$$G_4(r,s) = \frac{1}{4}(1-r)(1+s).$$

You can easily verify that these interpolation functions satisfy the required condition. Note also, that by using a linear geometric interpolation function the sides of the quadrilateral element must remain linear. If we use an element with additional nodes and a higher order interpolation function, the edges of the element could be curved.

11.1.3 Field Variable Representation Within an Element

In the displacement-based formulation, the nodal displacements are treated as unknowns. This finite set of displacements is determined by solving the governing equations. However, the entire displacement field at all interior points within the object is required for the determination of a solution. The displacement field for a point within an element is approximated by interpolating between the nodal values.

Note that we are approximating the actual displacement field with a continuous interpolated function between discrete nodal displacement values.

The value of the field variable at a given location in the natural coordinate system, $\phi(r,s)$, is determined from the values of the field variable at the nodal positions, ϕ_i, and an interpolation function such that

$$\phi(r,s) = \sum_{i=1}^{4} N_i(r,s)\phi_i.$$

The interpolation function, also known as a ***shape function***, $N_i(r,s)$, has the same properties as the geometric interpolation function. Namely, the interpolation must return the field variable at the nodal points is the nodal positions are input. This requires that

$$\phi_j = \sum_{i=1}^{4} N_i(r_j,s_j)\phi_i.$$

Therefore, the shape function must then have the property that

$$N_i(r_j,s_j) = \delta_{ij}.$$

Keep in mind that the value of the field variable at a point in the natural coordinate system is equal to the value of the field variable at the corresponding point in the global coordinate system. For the nodal values, this requirement takes the form

$$\phi(r_i,s_i) = \phi(x_i,y_i).$$

For the four-node quadrilateral element, we will use a linear shape function that has the form

$$N_1(r,s) = \frac{1}{4}(1-u)(1-v),$$

$$N_2(r,s) = \frac{1}{4}(1+u)(1-v),$$

$$N_3(r,s) = \frac{1}{4}(1+u)(1+v),$$

$$N_4(r,s) = \frac{1}{4}(1-u)(1+v).$$

The shape functions we have chosen are identical to the geometric interpolation functions presented earlier. The element is referred to as an ***isoparametric element***, and we have the condition

$$N_i(r,s) = G_i(r,s).$$

11.1.4 Matrix Representation

Since we will be implementing this finite element formulation in Matlab®, let us rewrite both the geometric interpolation and the field variable interpolation in matrix form. The geometric interpolation may be written as

$$x(r,s) = \sum_{i=1}^{4} N_i(r,s)x_i = \begin{bmatrix} N_1 & N_2 & N_3 & N_4 \end{bmatrix} \begin{Bmatrix} x_1 \\ x_2 \\ x_3 \\ x_4 \end{Bmatrix},$$

$$y(r,s) = \sum_{i=1}^{4} N_i(r,s)y_i = \begin{bmatrix} N_1 & N_2 & N_3 & N_4 \end{bmatrix} \begin{Bmatrix} y_1 \\ y_2 \\ y_3 \\ y_4 \end{Bmatrix}.$$

Similarly, the interpolation of a scalar field variable may be written as

$$\phi(r,s) = \sum_{i=1}^{4} N_i(r,s)\phi_i = \begin{bmatrix} N_1 & N_2 & N_3 & N_4 \end{bmatrix} \begin{Bmatrix} \phi_1 \\ \phi_2 \\ \phi_3 \\ \phi_4 \end{Bmatrix}.$$

An even more compact form can be acheived if we define a shape function matrix, $[\boldsymbol{N}]$, and a position vector, $\{\boldsymbol{X}_n\}$, as

$$[\boldsymbol{N}] = \begin{bmatrix} N_1 & N_2 & N_3 & N_4 & 0 & 0 & 0 & 0 \\ 0 & 0 & 0 & 0 & N_1 & N_2 & N_3 & N_4 \end{bmatrix}$$

and

$$\{\boldsymbol{X}_n\} = \begin{Bmatrix} x_1 \\ x_2 \\ x_3 \\ x_4 \\ y_1 \\ y_2 \\ y_3 \\ y_4 \end{Bmatrix}.$$

We can then write the geometric interpolation as

$$\begin{Bmatrix} x(r,s) \\ y(r,s) \end{Bmatrix} = [\boldsymbol{N}]\{\boldsymbol{X}_n\}.$$

11.1.5 Integration of a Field Variable

In the two-dimensional finite element formulation, we will often integrate a function over the area of an element in the global coordinate system. Due to the distorted shape of the element in the global coordinate system, this might be computationally cumbersome. Instead, any integral in the global coordinate system can be rewritten in terms of an integral over the area of the parent element in the natural coordinate system as

$$\int_{\partial\Omega_i} \phi(x,y)\, dA = \int_{\partial\tilde{\Omega}_i} \sum_{i=1}^{4} N_i(r,s)\phi_i J\, d\tilde{A},$$

where $\phi(x,y)$ is an arbitrary function, ϕ_i is the function value at integration point i, $\partial\Omega_i$ and dA are the area and incremental area in the global coordinate system, respectively, $\partial\tilde{\Omega}_i$ and $d\tilde{A}$ are the area and an incremental area in the natural coordinate system, respectively, and J is the Jacobean. The Jacobean matrix, $\boldsymbol{J}$, is given by

$$[\boldsymbol{J}] = \begin{bmatrix} \frac{\partial x}{\partial r} & \frac{\partial y}{\partial r} \\ \frac{\partial x}{\partial s} & \frac{\partial y}{\partial s} \end{bmatrix}.$$

We will refer to the Jacobean as the determinant of the Jacobean matrix

$$J = \det \begin{bmatrix} \frac{\partial x}{\partial r} & \frac{\partial y}{\partial r} \\ \frac{\partial x}{\partial s} & \frac{\partial y}{\partial s} \end{bmatrix}.$$

The derivatives found within the Jacobean may be computed by using the geometric interpolation functions such that

$$\frac{\partial x}{\partial r} = \frac{\partial}{\partial r}\left(\sum_{i=1}^{4} N_i(r,s)x_i\right) = \sum_{i=1}^{4} \frac{\partial N_i(r,s)}{\partial r} x_i,$$

$$\frac{\partial x}{\partial s} = \frac{\partial}{\partial s}\left(\sum_{i=1}^{4} N_i(r,s)x_i\right) = \sum_{i=1}^{4} \frac{\partial N_i}{\partial s} x_i,$$

$$\frac{\partial y}{\partial r} = \frac{\partial}{\partial r}\left(\sum_{i=1}^{4} N_i(r,s)y_i\right) = \sum_{i=1}^{4} \frac{\partial N_i}{\partial r} y_i,$$

$$\frac{\partial y}{\partial s} = \frac{\partial}{\partial s}\left(\sum_{i=1}^{4} N_i(r,s)y_i\right) = \sum_{i=1}^{4} \frac{\partial N_i}{\partial s} y_i.$$

We can write the Jacobean matrix in matrix notation as

$$[\boldsymbol{J}] = \begin{bmatrix} \frac{\partial N_1}{\partial r} & \frac{\partial N_2}{\partial r} & \frac{\partial N_3}{\partial r} & \frac{\partial N_4}{\partial r} & \frac{\partial N_1}{\partial r} & \frac{\partial N_2}{\partial r} & \frac{\partial N_3}{\partial r} & \frac{\partial N_4}{\partial r} \\ \frac{\partial N_1}{\partial s} & \frac{\partial N_2}{\partial s} & \frac{\partial N_3}{\partial s} & \frac{\partial N_4}{\partial s} & \frac{\partial N_1}{\partial s} & \frac{\partial N_2}{\partial s} & \frac{\partial N_3}{\partial s} & \frac{\partial N_4}{\partial s} \end{bmatrix} \begin{bmatrix} x_1 & 0 \\ x_2 & 0 \\ x_3 & 0 \\ x_4 & 0 \\ 0 & y_1 \\ 0 & y_2 \\ 0 & y_3 \\ 0 & y_4 \end{bmatrix}.$$

The nodal values of the field variable do not change with position. Therefore, the integration of a field variable over the element area is reduced to an integration of the shape function

$$\int_{\partial\tilde{\Omega}_i} [\boldsymbol{N}]\{\boldsymbol{\phi}\} J \, d\tilde{A} = \left(\int_{\partial\tilde{\Omega}_i} [\boldsymbol{N}] J \, d\tilde{A}\right)\{\boldsymbol{\phi}\}.$$

Recalling that the coordinates of the parent element vary form -1 to 1 along both the r and s axes, we can write

$$\int_{\partial\tilde{\Omega}_i} \boldsymbol{N} J \, d\tilde{A} = \int_{-1}^{1}\int_{-1}^{1} \boldsymbol{N} J \, dr \, ds.$$

whereas the integral $\int_{\partial\tilde{\Omega}_i} \boldsymbol{N} J \, d\tilde{A}$ may be evaluated analytically, in practice it is usually numerically integrated using Gaussian quadrature.

11.1.6 Gaussian Quadrature

Gaussian quadrature is a method for the numerical integration of fields. The basic premise of numerical integration is that the integral of a function may be determined by summing a finite set of weighted functional values at discreet points. Guassian quadarature provides an integration scheme that is more efficient that Newton's method or integration using Simpson's rule for polynomial functions.

We can approximate the two-dimensional integral over an area as a sum of the functional values evaluated at specific points with specific weights such that

$$\int_A f(r,s)\, dA \cong \sum_{i=1}^{n} W_i f(r_i, s_i),$$

where n is the number of quadrature points. Selection of the number of integration points is based on the accuracy desired and the available computational resources. Increasing the number of integration points improves the accuracy but also increases the computational effort required to evaluate the integral.

Gaussian quadrature can be used to *exactly* integrate polynomials using a discrete number of functional evaluations over the limits −1 to 1. For example, if we integrate the general form of a polynomial function of order 2 in one dimension, we obtain

$$\begin{aligned}\int_{-1}^{1} (a_0 + a_1 x + a_2 x^2 + a_3 x^3)\, dx &= \left(a_0 x + \frac{a_1}{2} x^2 + \frac{a_2}{3} x^3 + \frac{a_3}{4} x^4\right)\Bigg|_{-1}^{1} \\ &= 2a_0 + \frac{2}{3} a_2.\end{aligned}$$

Similarly, evaluating the Gaussian quadrature function for two integration points gives

$$\begin{aligned}\sum_{i=1}^{n} W_i f(x_i) &= \sum_{i=1}^{2} W_i \left(a_0 + a_1 x_i + a_2 x_i^2\right) \\ &= W_1 \left(a_0 + a_1 x_1 + a_2 x_1^2\right) + W_2 \left(a_0 + a_1 x_2 + a_2 x_2^2\right) \\ &= (W_1 + W_2)\, a_0 + (W_1 x_1 + W_2 x_2)\, a_2 + \left(W_1 x_1^2 + W_2 x_2^2\right) a_2.\end{aligned}$$

We can determine the weights and positions that give exact values of the integral by comparing the two results and equating the coefficients. This gives a set of three equations for four unknowns:

$$\begin{aligned}2 &= (W_1 + W_2), \\ 0 &= (W_1 x_1 + W_2 x_2), \\ \frac{2}{3} &= (W_1 x_1^2 + W_2 x_2^2).\end{aligned}$$

Notice that we have two unknown weights, W_1 and W_2, and two unknown positions, x_1 and x_2, with only three equations. In order to find a solution, we assume symmetry in the selection of the integration points. This gives the additional constraint that $x_1 = -x_2$ allowing us to solve the system of equations.

Solving the set of equations gives us the weights $W_1 = W_2 = 1$ and integration points $x_1 = -x_2 = \sqrt{\frac{1}{3}}$. This selection of combined integration points and weights gives the exact integral for a second-order polynomial. In fact, it can be shown that n integration points are sufficient to exactly integrate a polynomial of degree $2n-1$ over the limits -1 to 1.

For a two-dimensional integral, the integral can be written as

$$\int_{-1}^{1}\int_{-1}^{1} f(r,s)\,du\,dv \cong \sum_{i=1}^{n}\sum_{j=1}^{m} W_i W_j f(r_i,s_j),$$

where n is the number of integration points along the v axis, and m is the number of integration points along the u axis. If we select $n=2$ and $m=2$, this gives a total of four integration points and four weights, W_iW_j. Enumerating these four integration points, we can write

$$\int_{-1}^{1}\int_{-1}^{1} f(r,s)\,dr\,ds = \sum_{i=1}^{4} \tilde{W}_i f(r_i,s_i),$$

where $\tilde{W}_i$ is the multiple of the two one-dimensional weights. Therefore, for the quadrilateral element with four numerical integration points, we have the following values for integration point positions and weights.

Integration point, i	r_i	s_i	$\tilde{W}_i$
1	$-\frac{1}{\sqrt{3}}$	$-\frac{1}{\sqrt{3}}$	1
2	$\frac{1}{\sqrt{3}}$	$-\frac{1}{\sqrt{3}}$	1
3	$\frac{1}{\sqrt{3}}$	$\frac{1}{\sqrt{3}}$	1
4	$-\frac{1}{\sqrt{3}}$	$\frac{1}{\sqrt{3}}$	1

11.1.7 Differentiation of a Field Variable

We will often need to find the partial derivative of an interpolated variable with respect to the global coordinates within an element. Since the functional form of the interpolated fields is known, these derivatives can be expressed as derivatives of the interpolation functions. Employing the shape functions, the derivative of a field variable can be written as

$$\frac{\partial\phi(x,y)}{\partial x} = \frac{\partial}{\partial x}\sum_{i=1}^{4} N_i(r,s)\phi_i = \sum_{i=1}^{4}\frac{\partial N_i}{\partial x}\phi_i$$

and

$$\frac{\partial\phi(x,y)}{\partial y} = \frac{\partial}{\partial y}\sum_{i=1}^{4} N_i(r,s)\phi_i = \sum_{i=1}^{4}\frac{\partial N_i}{\partial y}\phi_i.$$

While we require the derivative of the shape function with respect to the global coordinates, they are functions of the natural coordinates. Therefore, we must use the chain rule to write

$$\frac{\partial N_i}{\partial r} = \frac{\partial N_i}{\partial x}\frac{\partial x}{\partial r} + \frac{\partial N_i}{\partial y}\frac{\partial y}{\partial r},$$
$$\frac{\partial N_i}{\partial s} = \frac{\partial N_i}{\partial x}\frac{\partial x}{\partial s} + \frac{\partial N_i}{\partial y}\frac{\partial y}{\partial s}.$$

Written in matrix form, this becomes

$$\begin{Bmatrix} \frac{\partial N_i}{\partial r} \\ \frac{\partial N_i}{\partial s} \end{Bmatrix} = \begin{bmatrix} \frac{\partial x}{\partial r} & \frac{\partial y}{\partial r} \\ \frac{\partial x}{\partial s} & \frac{\partial y}{\partial s} \end{bmatrix} \begin{Bmatrix} \frac{\partial N_i}{\partial x} \\ \frac{\partial N_i}{\partial y} \end{Bmatrix} = [\boldsymbol{J}] \begin{Bmatrix} \frac{\partial N_i}{\partial x} \\ \frac{\partial N_i}{\partial y} \end{Bmatrix}.$$

Inverting the equation gives

$$\begin{Bmatrix} \frac{\partial N_i}{\partial x} \\ \frac{\partial N_i}{\partial y} \end{Bmatrix} = [\boldsymbol{J}]^{-1} \begin{Bmatrix} \frac{\partial N_i}{\partial r} \\ \frac{\partial N_i}{\partial s} \end{Bmatrix},$$

Which allows the derivative of a field variable to be written as

$$\begin{Bmatrix} \frac{\partial \phi(x,y)}{\partial x} \\ \frac{\partial \phi(x,y)}{\partial y} \end{Bmatrix} = [\boldsymbol{J}]^{-1} \begin{bmatrix} \frac{\partial N_1}{\partial r} & \frac{\partial N_2}{\partial r} & \frac{\partial N_3}{\partial r} & \frac{\partial N_4}{\partial r} \\ \frac{\partial N_1}{\partial s} & \frac{\partial N_2}{\partial s} & \frac{\partial N_3}{\partial s} & \frac{\partial N_4}{\partial s} \end{bmatrix} \begin{Bmatrix} \phi_1 \\ \phi_2 \\ \phi_3 \\ \phi_4 \end{Bmatrix}.$$

11.2 Formulation of the Governing Equations

The formulation of the finite element problem begins with the weak form of the governing equation. In the displacement-based finite element formulation, we will approximate the displacement field within an element. Therefore, we desire a governing equation which has only first order derivatives of the displacement field.

For the case of small strain elasticity, we begin with a statement of virtual work. The principle of virtual work requires that for a system in equilibrium, the virtual work must be zero. Recall that a thermo-mechanical material in equilibrium must satisfy the equation for conservation of linear momentum such that

$$\rho\dot{\boldsymbol{v}} + \nabla\cdot\boldsymbol{\sigma}^T - \boldsymbol{b} = \boldsymbol{0}.$$

Application of a virtual displacement field, $\delta\boldsymbol{u}(\boldsymbol{x})$, which is consistent with any displacement boundary conditions placed on the object, must do no virtual work

$$\delta W = \int_{\Omega(t)} \delta\boldsymbol{u}\cdot(\rho\dot{\boldsymbol{v}} + \nabla\cdot\boldsymbol{\sigma}^T - \boldsymbol{b})\, dV = 0.$$

By consistent, we mean that the virtual displacement must be zero at any point where a prescribed displacement boundary condition is applied. We may rewrite the statement of virtual work by distributing the virtual displacement such that

$$\delta W = \int_{\Omega(t)} \delta\boldsymbol{u}\cdot\rho\dot{\boldsymbol{v}}\, dV + \int_{\Omega(t)} \delta\boldsymbol{u}\cdot\nabla\cdot\boldsymbol{\sigma}^T\, dV - \int_{\Omega(t)} \delta\boldsymbol{u}\cdot\boldsymbol{b}\, dV.$$

This statement will not work for a displacement-based finite element method. The measurement of strain involves the gradient of the displacement field. We know that for any material where stress is a function of strain, the divergence of the stress will involve the divergence of the gradient of the displacement field. Therefore, we integrate by parts to eliminate the divergence term obtaining

$$\int_{\Omega(t)} \delta\boldsymbol{u}\cdot\nabla\cdot\boldsymbol{\sigma}^T\,dV = \int_{\Omega(t)} \frac{\partial\,(\delta\boldsymbol{u})}{\partial\boldsymbol{x}}:\boldsymbol{\sigma}^T\,dV - \int_{\partial\Omega(t)} \delta\boldsymbol{u}\cdot\boldsymbol{\sigma}^T\cdot\boldsymbol{n}_x\,dS.$$

Now each of the terms has only a first derivative of the displacement field. We can simplify the equation further by noting since the infinitesimal strain tensor is defined as

$$\boldsymbol{\epsilon} = \frac{1}{2}\left(\nabla\boldsymbol{u} + \boldsymbol{u}\overleftarrow{\nabla}\right),$$

then the variation in the infinitesimal strain tensor due to a variation in the displacement field is given by

$$\delta\boldsymbol{\epsilon} = \frac{1}{2}\left(\nabla\,(\delta\boldsymbol{u}) + (\delta\boldsymbol{u})\,\overleftarrow{\nabla}\right).$$

Since the Cauchy stress tensor is symmetric, we can write

$$\frac{\partial\,(\delta\boldsymbol{u})}{\partial\boldsymbol{x}}:\boldsymbol{\sigma}^T = \frac{1}{2}\left(\frac{\partial\,(\delta\boldsymbol{u})}{\partial\boldsymbol{x}} + \left(\frac{\partial\,(\delta\boldsymbol{u})}{\partial\boldsymbol{x}}\right)^T\right):\boldsymbol{\sigma} = \delta\boldsymbol{\epsilon}:\boldsymbol{\sigma}.$$

Finally, substituting the Cauchy traction vector, $\boldsymbol{t}_x = \boldsymbol{\sigma}^T\cdot\boldsymbol{n}_x$, and gathering terms in the virtual work statement gives

$$\delta W = \int_{\Omega(t)} \delta\boldsymbol{u}\cdot\rho\dot{\boldsymbol{v}}\,dV + \int_{\Omega(t)} \delta\boldsymbol{\epsilon}:\boldsymbol{\sigma}\,dV - \int_{\Omega(t)} \delta\boldsymbol{u}\cdot\boldsymbol{b}\,dV - \int_{\partial\Omega(t)} \delta\boldsymbol{u}\cdot\boldsymbol{t}_x\,dS = 0,$$

where $\delta\boldsymbol{u}$ is the virtual displacement vector and $\delta\boldsymbol{\epsilon}$ is the resulting strain increment which can be determined from the virtual displacements. If we choose to ignore inertial effects, $\dot{\boldsymbol{v}} = 0$, and we can write

$$\delta W = \int_{\Omega(t)} \delta\boldsymbol{\epsilon}:\boldsymbol{\sigma}\,dV - \int_{\Omega(t)} \delta\boldsymbol{u}\cdot\boldsymbol{b}\,dV - \int_{\partial\Omega(t)} \delta\boldsymbol{u}\cdot\boldsymbol{t}_x\,dS = 0.$$

For a linearly elastic material subject to infinitesimal deformation, we can write the constitutive equation in its general form

$$\boldsymbol{\sigma} = \mathbb{C}\cdot\boldsymbol{\epsilon},$$

where $\boldsymbol{\sigma}$ is the Cauchy stress tensor, $\boldsymbol{\epsilon}$ is the infinitesimal strain tensor, and $\mathbb{C}$ is the fourth-order stiffness tensor. Therefore, the state of virtual work becomes

$$\delta W = \int_{\Omega(t)} \delta\boldsymbol{\epsilon}:\mathbb{C}\cdot\boldsymbol{\epsilon}\,dV - \int_{\Omega(t)} \delta\boldsymbol{u}\cdot\boldsymbol{b}\,dV - \int_{\partial\Omega(t)} \delta\boldsymbol{u}\cdot\boldsymbol{t}_x\,dS = 0. \tag{11.1}$$

11.3 Plane Strain Deformation

In this section, we will develop the equations required to determine the nodal displacements for the plane strain deformation of a linear elastic solid. For a plane strain state, only the strains in the $x-y$ plane are nonzero giving

$$\epsilon_{zz} = \epsilon_{zy} = \epsilon_{zx} = 0.$$

Due to the simplicity of the strain state and the fact that all field values are constant through the thickness of the body, a two-dimensional representation of the object is sufficient for solving plane strain problems. We begin by using a Voigt representation to write the stress, σ, and strain, ϵ, tensors in vector form and the fourth-order stiffness tensor, $\mathbb{C}$, can be written as a matrix such that

$$\{\sigma\} = \begin{Bmatrix} \sigma_{xx} \\ \sigma_{yy} \\ \sigma_{xy} \end{Bmatrix}, \quad \{\epsilon\} = \begin{Bmatrix} \epsilon_{xx} \\ \epsilon_{yy} \\ \epsilon_{xy} \end{Bmatrix}, \quad \text{and} \quad [\mathbb{C}] = \begin{bmatrix} 2G+\lambda & \lambda & 0 \\ \lambda & 2G+\lambda & 0 \\ 0 & 0 & 2G \end{bmatrix},$$

where G is the shear modulus and λ is Lame's first parameter. This allows us to write the constitutive model of a linear elastic material subject to plane strain as

$$\{\sigma\} = [\mathbb{C}]\{\epsilon\}. \tag{11.2}$$

11.3.1 Statement of Virtual Work

Substitution of the constitutive model, Equation (11.2), into the statement of virtual work, Equation (11.1), gives

$$\delta W = -\int_{\Omega(t)} \{\delta\epsilon\}^T [\mathbb{C}] \{\epsilon\}\, dV + \int_{\Omega(t)} \{\delta \boldsymbol{u}\}^T \{\boldsymbol{b}\}\, dV + \int_{\partial\Omega(t)} \{\delta \boldsymbol{u}\}^T \{\boldsymbol{\tau}\}\, dS = 0.$$

11.3.2 Discretization of Space

Discretizing space by meshing the object allows us to rewrite the state of virtual work for the entire body as the sum of the individual contributions from each element as

$$\sum_{i=1}^{n_e} \left(-\int_{\Omega_e} \{\delta\epsilon\}^T [\mathbb{C}]\{\epsilon\}\, dV + \int_{\Omega_e} \{\delta \boldsymbol{u}\}^T \{\boldsymbol{b}\}\, dV + \int_{\partial\Omega_e} \{\delta \boldsymbol{u}\}^T \{\boldsymbol{\tau}\}\, dS \right) = 0, \tag{11.3}$$

where n_e is the total number of elements used.

11.3.3 Approximation of the Field Variables

Notice that the first term of the virtual work statement requires the integration of the strain field over the entire element. In order to compute this, we must have a description of the displacement field that can be differentiated to determine the strain within the element. Since we are using a displacement-based finite element formulation, the displacement is the field variable. We solve for the displacements at the nodes and then interpolate to find the displacement vectors on the interior of the element. Let us also assume that we have a four-node isoparametric quadrilateral element.

If we assume that the components of the nodal displacement vector, (u_i, v_i) are known, then the displacement at any point within an element is given by

$$u(x,y) = \sum_{i=1}^{4} N_i(r,s)\, u_i \quad \text{and} \quad v(x,y) = \sum_{i=1}^{4} N_i(r,s)\, v_i.$$

We introduce a nodal displacement vector, $\{\boldsymbol{u}_n\}$, and the nodal virtual displacement vector, $\{\delta\boldsymbol{u}_n\}$, such that

$$\{\boldsymbol{u}_n\} = \begin{Bmatrix} u_1 \\ u_2 \\ u_3 \\ u_4 \\ v_1 \\ v_2 \\ v_3 \\ v_4 \end{Bmatrix} \quad \text{and} \quad \{\delta\boldsymbol{u}_n\} = \begin{Bmatrix} \delta u_1 \\ \delta u_2 \\ \delta u_3 \\ \delta u_4 \\ \delta v_1 \\ \delta v_2 \\ \delta v_3 \\ \delta v_4 \end{Bmatrix},$$

where u_i and v_i are the displacement components in the x and y directions for node i, respectively, and δu_i and δv_i are the virtual displacement components in the x and y directions for node i, respectively. We can then write the interpolated displacements, and the interpolated virtual displacements as

$$\{\boldsymbol{u}\} = \begin{Bmatrix} u(x,y) \\ v(x,y) \end{Bmatrix} = [\boldsymbol{N}]\{\boldsymbol{u}_n\} \quad \text{and} \quad \{\delta\boldsymbol{u}\} = \begin{Bmatrix} \delta u(x,y) \\ \delta v(x,y) \end{Bmatrix} = [\boldsymbol{N}]\{\delta\boldsymbol{u}_n\}.$$

The nonzero components of the ***infinitesimal strain tensor*** can be written in terms of the displacement vectors and the interpolated field values as

$$\epsilon_{xx} = \frac{\partial u}{\partial x} = \sum_{i=1}^{4} \frac{\partial N_i}{\partial x} u_i,$$

$$\epsilon_{yy} = \frac{\partial v}{\partial y} = \sum_{i=1}^{4} \frac{\partial N_i}{\partial y} v_i,$$

$$\epsilon_{xy} = \frac{1}{2}\left(\frac{\partial u}{\partial y} + \frac{\partial v}{\partial x}\right) = \frac{1}{2}\left(\sum_{i=1}^{4} \frac{\partial N_i}{\partial y} u_i + \sum_{i=1}^{4} \frac{\partial N_i}{\partial x} v_i\right).$$

Combining these equations gives the infinitesimal strain written in terms of the nodal displacements

$$\{\boldsymbol{\epsilon}\} = [\boldsymbol{N}']\{\boldsymbol{u}_n\}, \quad \text{and} \quad \{\delta\boldsymbol{\epsilon}\} = [\boldsymbol{N}']\{\delta\boldsymbol{u}_n\}, \tag{11.4}$$

where $[\boldsymbol{N}']$ is a matrix of the shape function derivatives such that

$$[\boldsymbol{N}'] = \begin{bmatrix} \frac{\partial N_1}{\partial x} & \frac{\partial N_2}{\partial x} & \frac{\partial N_3}{\partial x} & \frac{\partial N_4}{\partial x} & 0 & 0 & 0 & 0 \\ 0 & 0 & 0 & 0 & \frac{\partial N_1}{\partial y} & \frac{\partial N_2}{\partial y} & \frac{\partial N_3}{\partial y} & \frac{\partial N_4}{\partial y} \\ \frac{1}{2}\frac{\partial N_1}{\partial y} & \frac{1}{2}\frac{\partial N_2}{\partial y} & \frac{1}{2}\frac{\partial N_3}{\partial y} & \frac{1}{2}\frac{\partial N_4}{\partial y} & \frac{1}{2}\frac{\partial N_1}{\partial x} & \frac{1}{2}\frac{\partial N_2}{\partial x} & \frac{1}{2}\frac{\partial N_3}{\partial x} & \frac{1}{2}\frac{\partial N_4}{\partial x} \end{bmatrix}.$$

11.3.4 FEM Formulation

Substitution of the field variable approximations, Equation (11.4), into the statement of virtual work, Equation (11.3), gives

$$\sum_{i=1}^{n_e}\left(-\int_{\Omega_e}([N']\{\delta u_n\})^T\,[\mathbb{C}][N']\,\{u_n\}\,dV+\int_{\Omega_e}([N]\{\delta u_n\})^T\,\{b\}\,dV\right.$$
$$\left.+\int_{\partial\Omega_e}([N]\{\delta u_n\})^T\,\{\tau\}\,dS\right)=0.$$

Noting that $([N']\{\delta u_n\})^T=\{\delta u_n\}^T[N']^T$, that the nodal virtual displacement vector, $\{\delta u_n\}$, and the nodal displacement vector, $\{u_n\}$, are constant over the entire element, and that we can combine the integral terms for each element, we can write

$$\sum_{i=1}^{n_e}\{\delta u_n\}^T\left(-\left(\int_{\Omega_e}[N']^T\,[\mathbb{C}][N']dV\right)\{u_n\}+\int_{\Omega_e}[N]^T\,\{b\}\,dV\right.$$
$$\left.+\int_{\partial\Omega_e}[N]^T\,\{\tau\}\,dS\right)=0$$

Equivalently, we may write

$$\sum_{i=1}^{n_e}\{\delta u_n\}^T\left(-[K_e]\{u_n\}+\{F_b\}+\{F_\tau\}\right)=0,$$

where $[K_e]=\int_{\Omega_e}[N']^T\,[\mathbb{C}][N']dV$ is the element stiffness matrix, $\{F_b\}=\int_{\Omega_e}[N]^T\,\{b\}\,dV$ is the body force vector, and $\{F_\tau\}=\int_{\partial\Omega_e}[N]^T\,\{\tau\}\,dS$ is the force vector due to surface tractions.

Since the statement of virtual work must be true for any virtual nodal displacements, $\{\delta u_n\}$, we have the conditions that

$$[K_e]\{u_n\}=\{F_b\}+\{F_\tau\}.$$

11.3.5 Element Stiffness Tensor

We can numerically evaluate each of these terms by first noting that all fields are constant through the thickness of the object. This allows us to factor out one of the spatial dimensions such that

$$\begin{aligned}[K_e]&=\int_{\Omega_e}[N']^T\,[\mathbb{C}][N']dV\\&=t\int_{\partial\Omega_i}[N']^T\,[\mathbb{C}]\,[N']dA\end{aligned}$$

We can then use Gaussian quadrature to numerically integrate this term. Converting the integral over global coordinates to an integral over the parent element gives

$$[K_e]=t\int_{-1}^{1}\int_{-1}^{1}[N'(r,s)]^T\,[\mathbb{C}]\,[N'(r,s)]J\,dr\,ds.$$

Two-dimensional Gaussian quadrature using four integration points gives

$$[\boldsymbol{K}_e] \cong t\sum_{j=1}^{4} \tilde{W}_j J\,[\boldsymbol{N}'(r_j,s_j)]^T\,[\mathbb{C}]\,[\boldsymbol{N}'(r_j,s_j)].$$

11.3.6 Body Force Vector

Given the assumption that all fields are constant through the thickness of the element, the body force vector can be written as an integral over the area of the element such that

$$\begin{aligned}\{\boldsymbol{F}_b\} &= \int_{\Omega_e} [\boldsymbol{N}]^T \{\boldsymbol{b}\}\,dV\\ &= t\int_{\partial\Omega_e} [\boldsymbol{N}]^T \{\boldsymbol{b}\}\,dA.\end{aligned}$$

Converting the global integral to an integral over the natural coordinate system of the parent element gives

$$\{\boldsymbol{F}_b\} = t\int_{-1}^{1}\int_{-1}^{1} [\boldsymbol{N}]^T \{\boldsymbol{b}\}\,J\,dr\,ds.$$

Two-dimensional Gaussian quadrature using four integration points gives

$$\{\boldsymbol{F}_b\} \cong t\sum_{j=1}^{4} \tilde{W}_j J\,[\boldsymbol{N}]^T \{\boldsymbol{b}\}.$$

11.3.7 Traction Force Vector

Given the traction fields are constant through the thickness of the element, the traction force vector can be written as a line integral over the contour of the element

$$\begin{aligned}\{\boldsymbol{F}_\tau\} &= \int_{\partial\Omega_e} [\boldsymbol{N}]^T \{\boldsymbol{\tau}\}\,dS\\ &= t\oint_{\partial\Omega_e} [\boldsymbol{N}]^T \{\boldsymbol{\tau}\}\,dL.\end{aligned}$$

Conversion of the global contour integral into a contour integral over the four sides of the rectangular parent element gives

$$\begin{aligned}\{\boldsymbol{F}_\tau\} = &\left(\int_{-1}^{1} [\boldsymbol{N}]^T \{\boldsymbol{\tau}_{12}\}\frac{dL}{dr}\,dr\right)_{s=-1} + \left(\int_{-1}^{1} [\boldsymbol{N}]^T \{\boldsymbol{\tau}_{23}\}\frac{dL}{ds}\,ds\right)_{r=1}\\ &+ \left(\int_{1}^{-1} [\boldsymbol{N}]^T \{\boldsymbol{\tau}_{34}\}\frac{dL}{dr}\,dr\right)_{s=1} + \left(\int_{1}^{-1} [\boldsymbol{N}]^T \{\boldsymbol{\tau}_{41}\}\frac{dL}{ds}\,ds\right)_{r=-1}.\end{aligned}$$

Note that since the parent element is rectangular, for each side either dr or ds is zero.

Also, we can compute the derivatives $\frac{dL}{ds}$ and $\frac{dL}{dr}$ by noting that in the global coordinate system the incremental contour length, dL, is given by the Pythagorean theorem such that

$$dL^2 = \left(dx^2 + dy^2\right).$$

Therefore, we have the derivatives with respect to the natural coordinates

$$\frac{dL}{ds} = \frac{1}{2}\left[\left(\frac{dx}{ds}\right)^2 + \left(\frac{dy}{ds}\right)^2\right]^{-\frac{1}{2}} \quad \text{and} \quad \frac{dL}{dr} = \frac{1}{2}\left[\left(\frac{dx}{dr}\right)^2 + \left(\frac{dy}{dr}\right)^2\right]^{-\frac{1}{2}}.$$

Each term may be integrated using a one-dimensional Gaussian quadrature with two integration points to obtain

$$\begin{aligned}\{\boldsymbol{F}_\tau\} = &\left(t\sum_{j=1}^{2} W_j \frac{dL}{dr}[\boldsymbol{N}]^T\{\boldsymbol{\tau}_{12}\}\right)_{s=-1} + \left(t\sum_{j=1}^{2} W_j \frac{dL}{ds}[\boldsymbol{N}]^T\{\boldsymbol{\tau}_{23}\}\right)_{r=1} \\ &+ \left(t\sum_{j=1}^{2} W_j \frac{dL}{dr}[\boldsymbol{N}]^T\{\boldsymbol{\tau}_{34}\}\right)_{s=1} + \left(t\sum_{j=1}^{2} W_j \frac{dL}{ds}[\boldsymbol{N}]^T\{\boldsymbol{\tau}_{41}\}\right)_{r=-1}.\end{aligned}$$

Note that each of the terms in the previous equation represents the line integral over a single side of the element.

11.3.8 Single Element Implementation

The following section describes the implementation of the numerical solution of an infinitesimal plane strain deformation problem using finite element analysis. For the sake of simplicity, a single element was constructed as shown in Figure 11.3.

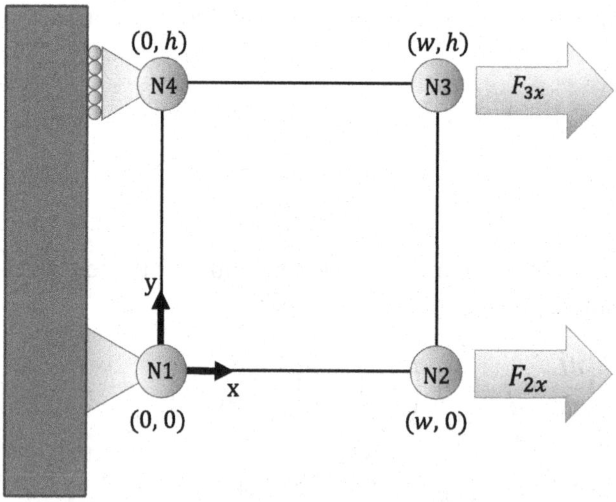

Figure 11.3. Illustration of the boundary conditions applied to a single plain strain element where $\boldsymbol{w}$ is the width of the element and $\boldsymbol{h}$ is the height.

Point loads along the x axis are applied directly to nodes 3 and 4. The horizontal displacement for nodes 1 and 4 are fixed and the vertical displacement is fixed for node 1. Note that for a plane strain problem, the displacement in the z axis is zero.

In this section, we have modeled a plate that has a width given by $w = 10$ cm, a height given by $h = 10$ cm, and a thickness of 1 cm. The modulus and Poisson's ratio were assumed to be $E = 5 \times 10^6$ Pa and $\nu = 0.4$, respectively. The graph of strain versus applied stress is shown in Figure 11.4. Note that the strain is on the vertical axis. Strain is the dependent variable in these simulations. The agreement between analytical and finite element solutions is excellent.

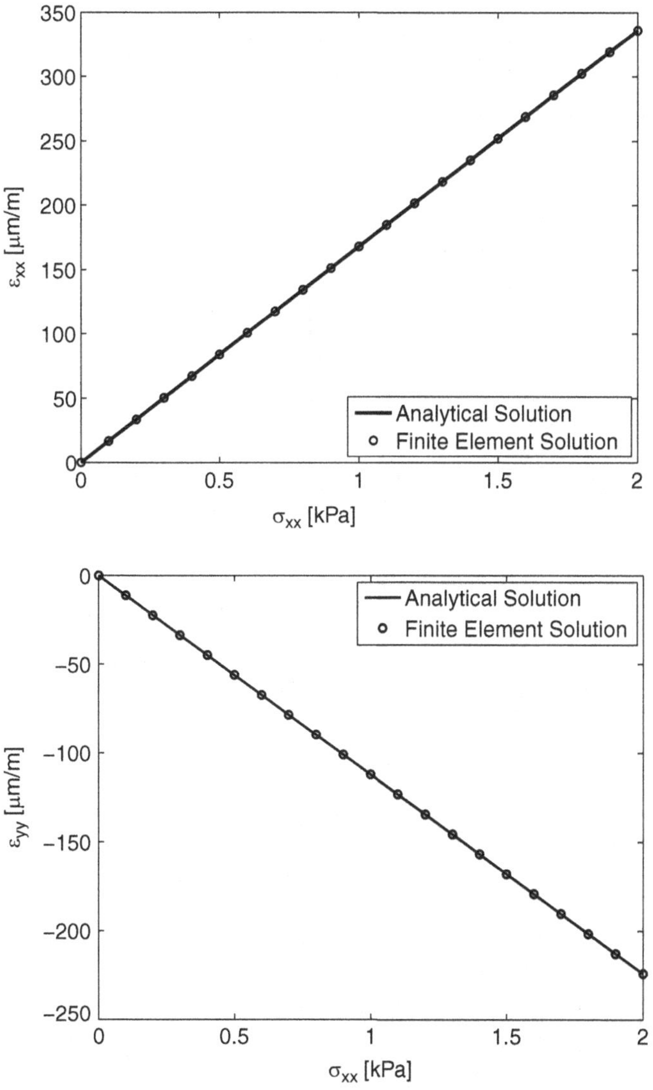

Figure 11.4. Comparison of analytical and numerical finite element solutions for plane strain deformation of a rectangular plate.

11.3.8.1 Flow Chart

1. Set up nodal positions and nodal point loads

$$[\text{nPosXY}] = \begin{bmatrix} x_1 & y_1 \\ x_2 & y_2 \\ x_3 & y_3 \\ x_4 & y_4 \end{bmatrix} \quad [\text{nForces}] = \begin{bmatrix} fx_1 & fy_1 \\ fx_2 & fy_2 \\ fx_3 & fy_3 \\ fx_4 & fy_4 \end{bmatrix}$$

2. Assemble C matrix for plane strain deformation of an isotropic material

$$[\mathbb{C}] = \begin{bmatrix} 2G+\lambda & \lambda & 0 \\ \lambda & 2G+\lambda & 0 \\ 0 & 0 & 2G \end{bmatrix}$$

3. Initialize material properties

$$E = 5 \text{ MPa}$$
$$\nu = 0.4$$

4. Set up Gaussian integration points. We use two integration points along the r axis and two along the s axis giving a total of four integration points per element.

Integration point, i	r_i	s_i	$\widetilde{W}_i$
1	$-1/\sqrt{3}$	$-1/\sqrt{3}$	1
2	$1/\sqrt{3}$	$-1/\sqrt{3}$	1
3	$1/\sqrt{3}$	$1/\sqrt{3}$	1
4	$-1/\sqrt{3}$	$1/\sqrt{3}$	1

5. Assemble the shape functions and derivatives at each integration point

$$N_1^{(J)} = \frac{1}{4}(1-r_j)(1-s_j), \quad \frac{\partial N_1^{(J)}}{\partial r} = \frac{1}{4}(-1)(1-s_j), \quad \frac{\partial N_1^{(J)}}{\partial r} = \frac{1}{4}(1-r_j)(-1)$$

$$N_2^{(J)} = \frac{1}{4}(1+r_j)(1-s_j), \quad \frac{\partial N_2^{(J)}}{\partial r} = \frac{1}{4}(+1)(1-s_j), \quad \frac{\partial N_2^{(J)}}{\partial r} = \frac{1}{4}(1+r_j)(-1)$$

$$N_3^{(J)} = \frac{1}{4}(1+r_j)(1+s_j), \quad \frac{\partial N_3^{(J)}}{\partial r} = \frac{1}{4}(+1)(1+s_j), \quad \frac{\partial N_3^{(J)}}{\partial r} = \frac{1}{4}(1+r_j)(+1)$$

$$N_4^{(J)} = \frac{1}{4}(1-r_j)(1+s_j), \quad \frac{\partial N_4^{(J)}}{\partial r} = \frac{1}{4}(-1)(1+s_j), \quad \frac{\partial N_4^{(J)}}{\partial r} = \frac{1}{4}(1-r_j)(+1)$$

6. Find the Jacobean

$$[J^{(J)}] = \begin{bmatrix} \frac{\partial N_1^{(J)}}{\partial r} & \frac{\partial N_2^{(J)}}{\partial r} & \frac{\partial N_3^{(J)}}{\partial r} & \frac{\partial N_4^{(J)}}{\partial r} \\ \frac{\partial N_1^{(J)}}{\partial s} & \frac{\partial N_2^{(J)}}{\partial s} & \frac{\partial N_3^{(J)}}{\partial s} & \frac{\partial N_4^{(J)}}{\partial s} \end{bmatrix} \begin{bmatrix} x_1 & y_1 \\ x_2 & y_2 \\ x_3 & y_3 \\ x_4 & y_4 \end{bmatrix}$$

7. Invert the Jacobean to find $\frac{\partial N}{\partial x}$ and $\frac{\partial N}{\partial y}$

$$\begin{bmatrix} \frac{\partial N_1^{(J)}}{\partial x} & \frac{\partial N_2^{(J)}}{\partial x} & \frac{\partial N_3^{(J)}}{\partial x} & \frac{\partial N_4^{(J)}}{\partial x} \\ \frac{\partial N_1^{(J)}}{\partial y} & \frac{\partial N_2^{(J)}}{\partial y} & \frac{\partial N_3^{(J)}}{\partial y} & \frac{\partial N_4^{(J)}}{\partial y} \end{bmatrix} = [J]^{-1} \begin{bmatrix} \frac{\partial N_1^{(J)}}{\partial r} & \frac{\partial N_2^{(J)}}{\partial r} & \frac{\partial N_3^{(J)}}{\partial r} & \frac{\partial N_4^{(J)}}{\partial r} \\ \frac{\partial N_1^{(J)}}{\partial s} & \frac{\partial N_2^{(J)}}{\partial s} & \frac{\partial N_3^{(J)}}{\partial s} & \frac{\partial N_4^{(J)}}{\partial s} \end{bmatrix}$$

8. Assemble $\boldsymbol{N}'$

$$[\boldsymbol{N}'^{(J)}] = \begin{bmatrix} \frac{\partial N_1^{(J)}}{\partial x} & \frac{\partial N_2^{(J)}}{\partial x} & \frac{\partial N_3^{(J)}}{\partial x} & \frac{\partial N_4^{(J)}}{\partial x} & 0 & 0 & 0 & 0 \\ 0 & 0 & 0 & 0 & \frac{\partial N_1^{(J)}}{\partial y} & \frac{\partial N_2^{(J)}}{\partial y} & \frac{\partial N_3^{(J)}}{\partial y} & \frac{\partial N_4^{(J)}}{\partial y} \\ \frac{1}{2}\frac{\partial N_1^{(J)}}{\partial y} & \frac{1}{2}\frac{\partial N_2^{(J)}}{\partial y} & \frac{1}{2}\frac{\partial N_3^{(J)}}{\partial y} & \frac{1}{2}\frac{\partial N_4^{(J)}}{\partial y} & \frac{1}{2}\frac{\partial N_1^{(J)}}{\partial x} & \frac{1}{2}\frac{\partial N_2^{(J)}}{\partial x} & \frac{1}{2}\frac{\partial N_3^{(J)}}{\partial x} & \frac{1}{2}\frac{\partial N_4^{(J)}}{\partial x} \end{bmatrix}$$

9. Compute the stiffness matrix

$$[\boldsymbol{K}_e] = [\boldsymbol{K}_e] + t\ \widetilde{W}_j \det(\boldsymbol{J}^{(J)})\ [\boldsymbol{N}'^{(J)}]^T\ [\mathbb{C}]\ [\boldsymbol{N}'^{(J)}]$$

Cycle over integration

10. Set displacement boundary conditions for degree of freedom j by making the diagonal component of the stiffness very large and the cooresponding force zero.

$$\boldsymbol{K}_e(j,j) = 1 \times 10^{12} * \boldsymbol{K}_e(j,j)$$

$$f(j) = 0$$

11. Solve the system of equations using the Matlab® factorization operator

$$\{\mathrm{u}\} = [\boldsymbol{K}_e]\backslash\{\boldsymbol{f}\}$$

11.3.8.2 Matlab® Code

```
% ---------------------------------------------------------------------------
% DESCRIPTION
% ---------------------------------------------------------------------------
% FEM simulation of a single quadralateral element with 4 numerical
% integration points. The nodes are numbered as follows:
%                        side 3
%                   4|-----------|3
%   y    side 4      |           |    side 2
%   ^                |           |
%   |               1|-----------|2
%   |                    side 1
%   |----> x
%  Node 1 has zero displacement in both the x and y direction
%  Node 4 has zero displacement in the y direction
%  A point load parallel to the x axis is appliced to nodes 2 and 3.
%
%  Variables Used in this Program
%       nnpe          - number of nodes per element
%       nPosXY(i,j)   - position of node i in the j direction
%       nForces(i,j) - point load applied to node i in the j direction
%
%       numqpt         - number of quadrature points per element
%       QPT(i,j)       - position of quadrature point i in the j direction
%
%       Jacobian       - Jacobian transformation matrix
%       invJacob       - Inverse of the Jacobian matrix
%
%       stiffness(i,j)- stiffness matrix element for degree of freedom i
%                       dof j
%       force(i)       - force vector for degree of freedom i
%
%       u(i)           - displacement for degree of freedom i
%       displace(i,j) - displacement of node i in direction j
%       strain         - strain in the element
%       stress         - stress in the element
% ---------------------------------------------------------------------------
% PREPROCESSING
% ---------------------------------------------------------------------------
clear
nnpe  = 4;                               % number of nodes per element
% ---------------------------------------------------------------------------
% Step 1. Input the x and y coordinate of each node, and components of
%         the point forces acting on each node.
nPosXY    = zeros(4, 2);  % Initialize nodal positions
nForces   = zeros(4, 2);  % Initialize external nodal forces
thickness = 0.01;         % [m] thickness of the plate
```

```
width     = 0.1;          % [m] width of the plate
height    = 0.1;          % [m] height of the plate
% ------------------------------------------------------------------------
nPosXY(1,:)=[0,      0];       % [m] node 1
nPosXY(2,:)=[width, 0];        % [m] node 1
nPosXY(3,:)=[width, height];  % [m] node 1
nPosXY(4,:)=[0,      height];  % [m] node 1
% ------------------------------------------------------------------------
nForces(1,:) = [0,  0];  % [N] node 1
nForces(2,:) = [10, 0];  % [N] node 2
nForces(3,:) = [10, 0];  % [N] node 3
nForces(4,:) = [0,  0];  % [N] node 4
% ------------------------------------------------------------------------
% Step 2. Give the material properties for the element
E      = 5.0e6;           % [Pa] elastic modulus
nu     = 0.4;             % Poisson's Ratio
% ------------------------------------------------------------------------
G      = E/(2.0*(1+nu));         % shear modulus
lambda = E*nu/((1+nu)*(1-2*nu)); % Lame constant
% ------------------------------------------------------------------------
% Step 3. Assemble the C matrix assuming plane strain deformation
C = [2.0*G+lambda, lambda,       0;
     lambda,       2.0*G+lambda, 0;
     0,            0,            G];
% ------------------------------------------------------------------------
% Step 4. Give the positions of the numerical integration points
%         in natural coordinate system (r, s).
numqpt = 4;                               % number of quadrature points
wt     = 1.00;                            % weights for quadrature points
QPT = [ -0.5773502692, -0.5773502692;     % (r,s) location for
         0.5773502692, -0.5773502692;     %     the 4 quadrature points
         0.5773502692,  0.5773502692;
        -0.5773502692,  0.5773502692;];
% ------------------------------------------------------------------------
% Step 5. Assemble the shape functions and necessary derivatives
for j = 1:numqpt;
    % r-s coordinates for the integration points
    r = QPT(j,1);  s = QPT(j,2);
    % shape functions
    sf(1,j) = 0.25*(1.0-r)*(1.0-s);
    sf(2,j) = 0.25*(1.0+r)*(1.0-s);
    sf(3,j) = 0.25*(1.0+r)*(1.0+s);
    sf(4,j) = 0.25*(1.0-r)*(1.0+s);
    % derivative of shape function wrspt to r
    dndrs(1,1,j) = 0.25*(-1.)*(1.0-s);
    dndrs(2,1,j) = 0.25*(+1.)*(1.0-s);
    dndrs(3,1,j) = 0.25*(+1.)*(1.0+s);
```

```
        dndrs(4,1,j) = 0.25*(-1.)*(1.0+s);
        % derivative of shape function wrspt to s
        dndrs(1,2,j) = 0.25*(1.0-r)*(-1.0);
        dndrs(2,2,j) = 0.25*(1.0+r)*(-1.0);
        dndrs(3,2,j) = 0.25*(1.0+r)*(+1.0);
        dndrs(4,2,j) = 0.25*(1.0-r)*(+1.0);
end
% ---------------------------------------------------------------------------
stiffness = zeros(8, 8);    % initialize element stiffness matrix
for j = 1:numqpt
    % Step 6. Find the Jacobean matrix
    Jacobian  = dndrs(:,:,j)'*nPosXY
    invJacob  = Jacobian^-1;
    % Step 7. Find the dNdX and dNdY
    dndxy     = (invJacob*dndrs(:,:,j)')';
    % Step 8. Assemble Nprime
    NP        = [dndxy(:,1)'  0, 0, 0, 0; ...
                 0, 0, 0, 0,  dndxy(:,2)'; ...
                 dndxy(:,2)', dndxy(:,1)'];
    % Step 9. Compute the stiffness matrix
    stiffness = stiffness + thickness*wt*NP'*C*NP*det(Jacobian);
end
force = [nForces(:,1); nForces(:,2)];
% ---------------------------------------------------------------------------
% Step 10. Set displacement boundary conditions
%       10a. Node 1 has zero displacement in x and y direction
stiffness(1,1) = stiffness(1,1)*1.0e12; force(1)   = 0; % zero node 1 x component
stiffness(5,5) = stiffness(5,5)*1.0e12; force(5)   = 0; % zero node 1 y component
%       10b. Node 4 has zero displacement in the x direction
stiffness(4,4) = stiffness(4,4)*1.0e12; force(4)   = 0; % zero node 4 x component
% ---------------------------------------------------------------------------
%    SOLVER
% ---------------------------------------------------------------------------
% Step 11. Solve for the nodal displacements using
%          Matlab's built in Guassian elimination
u = stiffness\force;
```

11.4 Axisymmetric Deformation

In this section, we will develop the quations required to determine the nodal displacements for the axisymmetric deformation of a linear elastic solid. For axisymmetric deformation, we can simplify the strain state such that

$$\epsilon_{z\theta} = \epsilon_{\mathrm{r}\theta} = \epsilon_{\theta\mathrm{r}} = \epsilon_{\theta z} = 0.$$

All of the fields in the problem are functions of r and z only. Therefore, the axisymmetric deformation can be represented by a mesh in the rz plane. We begin by using a Voigt representation to write the stress and strain tensors in vector form, and the fourth-order stiffness tensor can be written as a matrix such that

$$\{\sigma\}=\begin{Bmatrix}\sigma_{rr}\\ \sigma_{\theta\theta}\\ \sigma_{zz}\\ \sigma_{rz}\end{Bmatrix},\ \{\epsilon\}=\begin{Bmatrix}\epsilon_{rr}\\ \epsilon_{\theta\theta}\\ \epsilon_{zz}\\ \epsilon_{rz}\end{Bmatrix},\ \text{and}\ [\mathbb{C}]=\frac{E}{(1+\nu)(1-2\nu)}\begin{bmatrix}1-\nu & \nu & \nu & 0\\ \nu & 1-\nu & \nu & 0\\ \nu & \nu & 1-\nu & 0\\ 0 & 0 & 0 & \frac{1-2\nu}{2}\end{bmatrix}.$$

This allows us to write the constitutive model of a linear elastic material subject to plane strain as

$$\{\sigma\}=[\mathbb{C}]\{\epsilon\}. \tag{11.5}$$

11.4.1 Statement of Virtual Work

Substitution of the constitutive model, Equation (11.5), into the statement of virtual work, Equation (11.1), gives

$$\delta W=-\int_{\Omega(t)}\{\delta\epsilon\}^T[\mathbb{C}]\{\epsilon\}\,dV+\int_{\Omega(t)}\{\delta \boldsymbol{u}\}^T\{\boldsymbol{b}\}\,dV+\int_{\partial\Omega(t)}\{\delta \boldsymbol{u}\}^T\{\boldsymbol{\tau}\}\,dS=0.$$

11.4.2 Discretization of Space

Discretizing space by meshing the object allows us to rewrite the state of virtual work of the entire body as the sum of the individual contributions from each element as

$$\sum_{i=1}^{n_e}\left(-\int_{\Omega_e}\{\delta\epsilon\}^T[\mathbb{C}]\{\epsilon\}\,dV+\int_{\Omega_e}\{\delta \boldsymbol{u}\}^T\{\boldsymbol{b}\}\,dV+\int_{\partial\Omega_e}\{\delta \boldsymbol{u}\}^T\{\boldsymbol{\tau}\}\,dS\right)=0, \tag{11.6}$$

where n_e is the total number of elements used.

11.4.3 Approximation of the Field Variables

Once again, we must approximate the displacement functions using interpolation of the nodal displacements. For the axisymmetric problem, the displacements, u and v, are the displacement components in the radial and axial directions, respectively.

The displacement at any point within the solid can be written in terms of the nodal displacements as

$$u(x,y) = \sum_{i=1}^{4} N_i(r,s)\, u_i, \quad \text{and} \quad v(x,y) = \sum_{i=1}^{4} N_i(r,s)\, v_i.$$

We introduce a nodal displacement vector, $\{\boldsymbol{u}_n\}$, and the nodal virtual displacement vector, $\{\delta \boldsymbol{u}_n\}$, such that

$$\{\boldsymbol{u}_n\} = \begin{Bmatrix} u_1 \\ u_2 \\ u_3 \\ u_4 \\ v_1 \\ v_2 \\ v_3 \\ v_4 \end{Bmatrix} \quad \text{and} \quad \{\delta \boldsymbol{u}_n\} = \begin{Bmatrix} \delta u_1 \\ \delta u_2 \\ \delta u_3 \\ \delta u_4 \\ \delta v_1 \\ \delta v_2 \\ \delta v_3 \\ \delta v_4 \end{Bmatrix}.$$

As with the plane strain problem, the interpolated displacement and interpolated virtual displacement can be written as

$$\{\boldsymbol{u}\} = \begin{Bmatrix} u(x,y) \\ v(x,y) \end{Bmatrix} = [\boldsymbol{N}]\{\boldsymbol{u}_n\} \quad \text{and} \quad \{\delta \boldsymbol{u}\} = \begin{Bmatrix} \delta u(x,y) \\ \delta v(x,y) \end{Bmatrix} = [\boldsymbol{N}]\{\delta \boldsymbol{u}_n\}.$$

The nonzero components of the ***infinitesimal strain tensor*** can be written in terms of the displacement vectors and the interpolated field values as

$$\epsilon_{\mathrm{r}} = \frac{\partial u}{\partial \mathrm{r}} = \sum_{i=1}^{4} \frac{\partial N_i}{\partial \mathrm{r}} u_i,$$

$$\epsilon_z = \frac{\partial v}{\partial z} = \sum_{i=1}^{4} \frac{\partial N_i}{\partial z} v_i,$$

$$\epsilon_\theta = \frac{u}{\mathrm{r}} = \sum_{i=1}^{4} \frac{N_i}{\mathrm{r}} u_i,$$

$$\epsilon_{\mathrm{r}z} = \frac{\partial u}{\partial z} + \frac{\partial w}{\partial \mathrm{r}} = \sum_{i=1}^{4} \frac{\partial N_i}{\partial z} u_i + \sum_{i=1}^{4} \frac{\partial N_i}{\partial r} v_i,$$

$$\epsilon_{\mathrm{r}\theta} = 0,$$

$$\epsilon_{\theta z} = 0.$$

In matrix form, we can write this set of equations as

$$\{\epsilon\} = [\boldsymbol{N}']\{\boldsymbol{u}_n\}, \tag{11.7}$$

where the N′ matrix for axisymmetric deformation is given by

$$[\boldsymbol{N'}] = \begin{bmatrix} \frac{\partial N_1}{\partial \mathrm{r}} & \frac{\partial N_2}{\partial \mathrm{r}} & \frac{\partial N_3}{\partial \mathrm{r}} & \frac{\partial N_4}{\partial \mathrm{r}} & 0 & 0 & 0 & 0 \\ \frac{N_1}{\mathrm{r}} & \frac{N_2}{\mathrm{r}} & \frac{N_3}{\mathrm{r}} & \frac{N_4}{\mathrm{r}} & 0 & 0 & 0 & 0 \\ 0 & 0 & 0 & 0 & \frac{\partial N_1}{\partial z} & \frac{\partial N_2}{\partial z} & \frac{\partial N_3}{\partial z} & \frac{\partial N_4}{\partial z} \\ \frac{1}{2}\frac{\partial N_1}{\partial y} & \frac{1}{2}\frac{\partial N_2}{\partial y} & \frac{1}{2}\frac{\partial N_3}{\partial y} & \frac{1}{2}\frac{\partial N_4}{\partial y} & \frac{1}{2}\frac{\partial N_1}{\partial x} & \frac{1}{2}\frac{\partial N_2}{\partial x} & \frac{1}{2}\frac{\partial N_3}{\partial x} & \frac{1}{2}\frac{\partial N_4}{\partial x} \end{bmatrix}.$$

11.4.4 FEM Formulation

Substitution of the field variable approximations, Equation (11.7), into the statement of virtual work, Equation (11.6), gives

$$\sum_{i=1}^{n_e} \{\delta \boldsymbol{u}_n\}^T \left(-\left(\int_{\Omega_e} [\boldsymbol{N'}]^T \,[\mathbb{C}][\boldsymbol{N'}]dV \right) \{\boldsymbol{u}_n\} + \int_{\Omega_e} [\boldsymbol{N}]^T \,\{\boldsymbol{b}\}\, dV \right.$$
$$\left. + \int_{\partial\Omega_e} [\boldsymbol{N}]^T \,\{\boldsymbol{\tau}\}\, dS \right) = 0.$$

Once again, we obtain the condition

$$[\boldsymbol{K}_e]\,\{\boldsymbol{u}_n\} = \{\boldsymbol{F}_b\} + \{\boldsymbol{F}_\tau\},$$

where $[\boldsymbol{K}_e] = \int_{\Omega_e} [\boldsymbol{N'}]^T \,[\mathbb{C}][\boldsymbol{N'}]dV$ is the element stiffness matrix, $\{F_b\} = \int_{\Omega_e} [\boldsymbol{N}]^T \,\{\boldsymbol{b}\}\, dV$ is the body force vector, and $\{F_\tau\} = \int_{\partial\Omega_e} [\boldsymbol{N}]^T \,\{\boldsymbol{\tau}\}\, dS$ is the force vector due to surface tractions.

11.4.5 Element Stiffness Tensor

Since we are using cylindrical coordinates, the infinitesimal volume element, dV, is given by

$$dV = r\, dr\, dz\, d\theta = r\, dA\, d\theta.$$

The element stiffness matrix can be written as

$$[\boldsymbol{K}_e] = \int_{\Omega_e} [\boldsymbol{N'}]^T \,[\mathbb{C}][\boldsymbol{N'}]dV$$
$$= \int_0^{2\pi} \int_{\partial\Omega_i} [\boldsymbol{N'}]^T \,[\mathbb{C}]\,[\boldsymbol{N'}]\,\mathrm{r}\, dA\, d\theta.$$

There is no θ dependence in any of these quantities, we can write

$$[\boldsymbol{K}_e] = 2\pi \int_{\partial\Omega_i} [\boldsymbol{N}']^T \, [\mathbb{C}] \, [\boldsymbol{N}'] \, \mathrm{r} \, dA.$$

We can then use Gaussian quadrature to numerically integrate this term. Converting the integral over global coordinates to an integral over the parent element gives

$$[\boldsymbol{K}_e] = 2\pi \int_{-1}^{1} \int_{-1}^{1} \mathrm{r} \, [\boldsymbol{N}'(r,s)]^T \, [\boldsymbol{R}] \, [\boldsymbol{N}'(r,s)] J \, dr \, ds.$$

Two-dimensional Gaussian quadrature using four integration points gives

$$[\boldsymbol{K}_e] \cong 2\pi \sum_{j=1}^{4} \tilde{W}_j J \, \mathrm{r}_j \, [\boldsymbol{N}'\left(r_j,s_j\right)]^T \, [\boldsymbol{R}] \, [\boldsymbol{N}'\left(r_j,s_j\right)].$$

11.4.6 Body Force Vector

Again using the fact that there is no variation of the body force along the θ direction, we can write the body forces as

$$\begin{aligned} \{\boldsymbol{F}_b\} &= \int_{\Omega_e} [\boldsymbol{N}]^T \, \{\boldsymbol{b}\} \, dV \\ &= \int_0^{2\pi} \int_{\partial\Omega_e} [\boldsymbol{N}]^T \, \{\boldsymbol{b}\} \, \mathrm{r} \, dA \, d\theta \\ &= 2\pi \int_{\partial\Omega_e} [\boldsymbol{N}]^T \, \{\boldsymbol{b}\} \, \mathrm{r} \, dA. \end{aligned}$$

Converting the global integral to an integral over the parent element gives

$$\{\boldsymbol{F}_b\} = 2\pi \int_{-1}^{1} \int_{-1}^{1} [\boldsymbol{N}]^T \, \{\boldsymbol{b}\} \, \mathrm{r} J \, dr \, ds.$$

And two-dimensional Gaussian quadrature using four integration points gives

$$\{\boldsymbol{F}_b\} \cong 2\pi \sum_{j=1}^{4} \tilde{W}_j J \, \mathrm{r}_j \, [\boldsymbol{N}]^T \, \{\boldsymbol{b}\}.$$

The traction force vector becomes a line integral over the contour of the element

$$
\begin{aligned}
\{\boldsymbol{F}_\tau\} &= \int_{\partial\Omega_e} [\boldsymbol{N}]^T \{\boldsymbol{\tau}\}\, dS \\
&= \int_0^{2\pi} \oint_{\partial\Omega_e} [\boldsymbol{N}]^T \{\boldsymbol{\tau}\}\, dL\, d\theta \\
&= 2\pi \oint_{\partial\Omega_e} [\boldsymbol{N}]^T \{\boldsymbol{\tau}\}\, dL.
\end{aligned}
$$

Conversion of the global contour integral into a contour integral over the parent element gives

$$
\begin{aligned}
\{\boldsymbol{F}_\tau\} = {} & \left(\int_{-1}^{1} [\boldsymbol{N}]^T \{\boldsymbol{\tau}_{12}\}\, \mathrm{r} \frac{dL}{dr}\, dr\right)_{s=-1} + \left(\int_{-1}^{1} [\boldsymbol{N}]^T \{\boldsymbol{\tau}_{23}\}\, \mathrm{r} \frac{dL}{ds}\, ds\right)_{r=1} \\
& + \left(\int_{1}^{-1} [\boldsymbol{N}]^T \{\boldsymbol{\tau}_{34}\}\, \mathrm{r} \frac{dL}{dr}\, dr\right)_{s=1} + \left(\int_{1}^{-1} [\boldsymbol{N}]^T \{\boldsymbol{\tau}_{41}\}\, \mathrm{r} \frac{dL}{ds}\, ds\right)_{r=-1}.
\end{aligned}
$$

Since the parent element is rectangular, for each side only *dr* or *ds* is nonzero. Therefore, the integral is broken up into an integral over each side of the rectangle resulting in four integrals.

Note that since

$$
dL = \left(d\mathrm{r}^2 + dz^2\right)^2,
$$

we have

$$
\frac{dL}{ds} = \left[\left(\frac{d\mathrm{r}}{ds}\right)^2 + \left(\frac{dz}{ds}\right)^2\right]^{\frac{1}{2}}
$$

and

$$
\frac{dL}{dr} = \left[\left(\frac{d\mathrm{r}}{dr}\right)^2 + \left(\frac{dz}{dr}\right)^2\right]^{\frac{1}{2}}.
$$

Each term may be integrated using a one-dimensional Gaussian quadrature with two integration points to obtain a term similar to

$$\{\boldsymbol{F}_\tau\} = \left(2\pi\sum_{j=1}^{2} W_j \frac{dL}{dr}[\boldsymbol{N}]^T \mathrm{r}_j\{\boldsymbol{\tau}_{12}\}\right)_{s=-1} + \left(2\pi\sum_{j=1}^{2} W_j \frac{dL}{ds}[\boldsymbol{N}]^T\{\boldsymbol{\tau}_{23}\}\mathrm{r}_j\right)_{r=1}$$
$$+ \left(2\pi\sum_{j=1}^{2} W_j \frac{dL}{dr}[\boldsymbol{N}]^T\{\boldsymbol{\tau}_{34}\}\mathrm{r}_j\right)_{s=1} + \left(2\pi\sum_{j=1}^{2} W_j \frac{dL}{ds}[\boldsymbol{N}]^T\{\boldsymbol{\tau}_{41}\}\mathrm{r}_j\right)_{r=-1}.$$

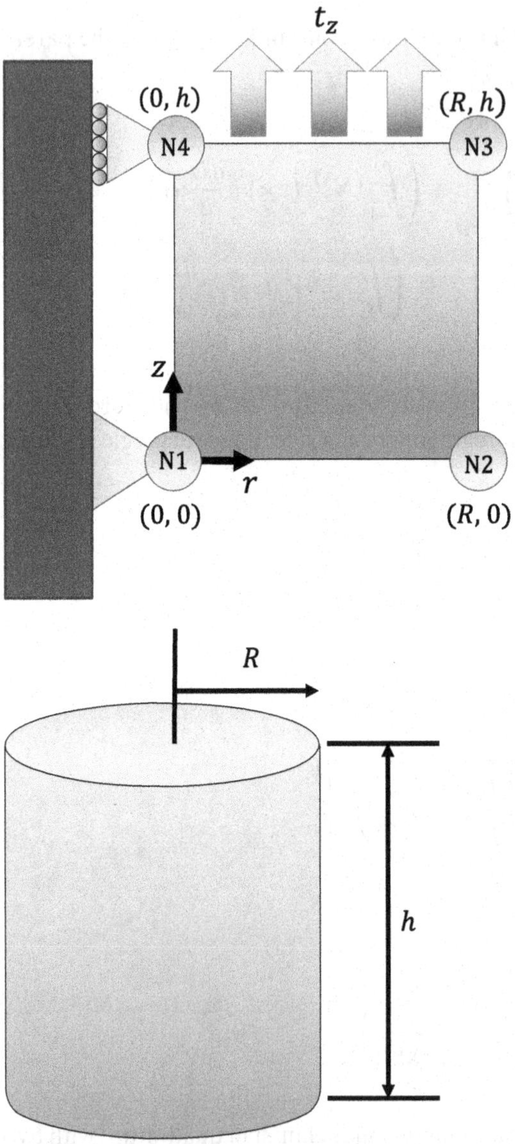

Figure 11.5. (top) Illustration of a single axisymmetric element subject to traction and displacement boundary conditions where $\boldsymbol{R}$ is the radius of the cylinder, and $\boldsymbol{h}$ is the height of the cylinder. (bottom) Solid obtained by rotating the element about the $z-z$ axis.

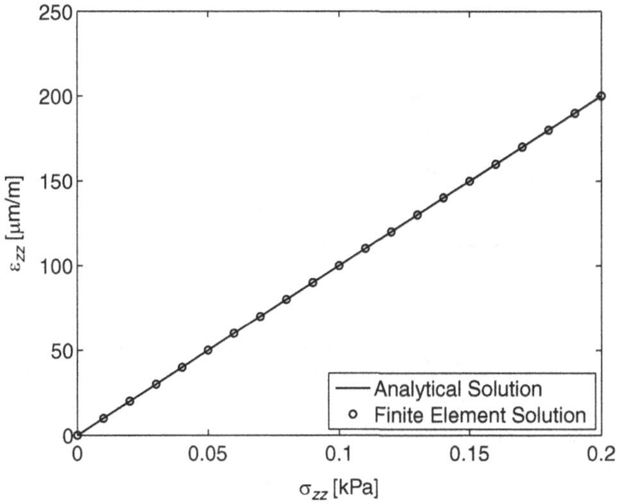

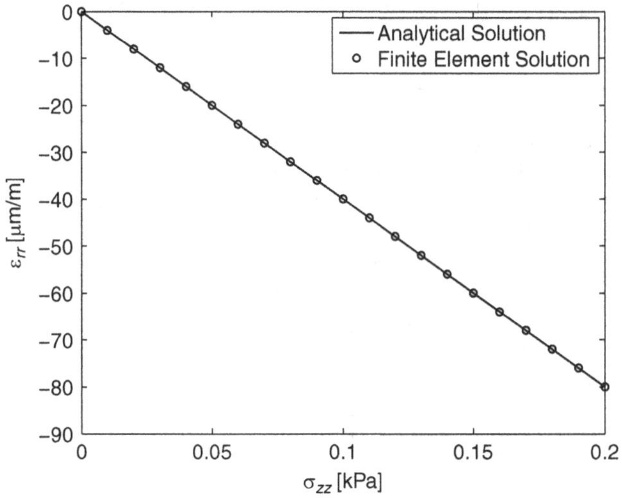

Figure 11.6. Comparison of analytical and numerical finite element solutions for axisymmetric deformation of a cylinder.

11.4.7 Single Element Implementation

The implementation of a finite element solution for infinitesimal axisymmetric deformation is presented in this section. A single element is constructed as shown in the following figure. The radial displacement of nodes 1 and 4 are constrained, and the axial displacement of node 1 is constrained. A uniform surface traction, t_z, is applied to the top surface (side 3).

In this section, we have modeled a cylinder that has a radius given by $R = 10$ cm and a height given by $h = 50$ cm. The modulus and Poisson's ratio were assumed to be $E = 5 \times 10^6$ Pa and $\nu = 0.4$, respectively. The graph of strain versus applied stress is shown in the following figure.

11.4.7.1 Flow Chart

1. Set up nodal positions and surface traction

$$[\text{nPosRZ}] = \begin{bmatrix} r_1 & z_1 \\ r_2 & z_2 \\ r_3 & z_3 \\ r_4 & z_4 \end{bmatrix} \quad \text{tractionz} = 200\ \text{Pa}$$

↓

2. Initialize material properties

$$E = 1\ \text{MPa}$$

$$\nu = 0.4$$

↓

3. Assemble C matrix for plane strain deformation of an isotropic material

$$[\mathbb{C}] = \frac{E}{(1+\nu)(1-2\nu)} \begin{bmatrix} 1-\nu & \nu & \nu & 0 \\ \nu & 1-\nu & \nu & 0 \\ \nu & \nu & 1-\nu & 0 \\ 0 & 0 & 0 & \frac{1-2\nu}{2} \end{bmatrix}$$

↓

4. Set up Gaussian integration points. We use two integration points along the r axis and two along the s axis giving a total of four integration points per element for the area integration.

Integration point, i	r_i	s_i	$\widetilde{W}_i$
1	$-1/\sqrt{3}$	$-1/\sqrt{3}$	1
2	$1/\sqrt{3}$	$-1/\sqrt{3}$	1
3	$1/\sqrt{3}$	$1/\sqrt{3}$	1
4	$-1/\sqrt{3}$	$1/\sqrt{3}$	1

Three integration points are used to for line integrals

Integration point, i	r_i	$\widehat{W}_i$
1	$-\sqrt{3/5}$	5/9
2	$\sqrt{3/5}$	8/9
3	$\sqrt{3/5}$	5/9

↓

5. Assemble the shape functions and derivatives at each integration point, $j = 1$ *to* 4

$$N_1^{(j)} = \frac{1}{4}(1-r_j)(1-s_j), \quad \frac{\partial N_1^{(j)}}{\partial r} = \frac{1}{4}(-1)(1-s_j), \quad \frac{\partial N_1^{(j)}}{\partial r} = \frac{1}{4}(1-r_j)(-1)$$

$$N_2^{(j)} = \frac{1}{4}(1+r_j)(1-s_j), \quad \frac{\partial N_2^{(j)}}{\partial r} = \frac{1}{4}(+1)(1-s_j), \quad \frac{\partial N_2^{(j)}}{\partial r} = \frac{1}{4}(1+r_j)(-1)$$

$$N_3^{(j)} = \frac{1}{4}(1+r_j)(1+s_j), \quad \frac{\partial N_3^{(j)}}{\partial r} = \frac{1}{4}(+1)(1+s_j), \quad \frac{\partial N_3^{(j)}}{\partial r} = \frac{1}{4}(1+r_j)(+1)$$

$$N_4^{(j)} = \frac{1}{4}(1-r_j)(1+s_j), \quad \frac{\partial N_4^{(j)}}{\partial r} = \frac{1}{4}(-1)(1+s_j), \quad \frac{\partial N_4^{(j)}}{\partial r} = \frac{1}{4}(1-r_j)(+1)$$

↓

6. Find the Jacobean

$$[J^{(J)}] = \begin{bmatrix} \frac{\partial N_1^{(J)}}{\partial r} & \frac{\partial N_2^{(J)}}{\partial r} & \frac{\partial N_3^{(J)}}{\partial r} & \frac{\partial N_4^{(J)}}{\partial r} \\ \frac{\partial N_1^{(J)}}{\partial s} & \frac{\partial N_2^{(J)}}{\partial s} & \frac{\partial N_3^{(J)}}{\partial s} & \frac{\partial N_4^{(J)}}{\partial s} \end{bmatrix} \begin{bmatrix} x_1 & y_1 \\ x_2 & y_2 \\ x_3 & y_3 \\ x_4 & y_4 \end{bmatrix}$$

7. Invert the Jacobean to find $\frac{\partial N}{\partial x}$ and $\frac{\partial N}{\partial y}$

$$\begin{bmatrix} \frac{\partial N_1^{(J)}}{\partial \mathrm{r}} & \frac{\partial N_2^{(J)}}{\partial \mathrm{r}} & \frac{\partial N_3^{(J)}}{\partial \mathrm{r}} & \frac{\partial N_4^{(J)}}{\partial \mathrm{r}} \\ \frac{\partial N_1^{(J)}}{\partial z} & \frac{\partial N_2^{(J)}}{\partial z} & \frac{\partial N_3^{(J)}}{\partial z} & \frac{\partial N_4^{(J)}}{\partial z} \end{bmatrix} = [J]^{-1} \begin{bmatrix} \frac{\partial N_1^{(J)}}{\partial r} & \frac{\partial N_2^{(J)}}{\partial r} & \frac{\partial N_3^{(J)}}{\partial r} & \frac{\partial N_4^{(J)}}{\partial r} \\ \frac{\partial N_1^{(J)}}{\partial s} & \frac{\partial N_2^{(J)}}{\partial s} & \frac{\partial N_3^{(J)}}{\partial s} & \frac{\partial N_4^{(J)}}{\partial s} \end{bmatrix}$$

8. Find the r coordinate of each integration point

$$r^{(J)} = \sum N_i^{(J)} r_i$$

9. Assemble $\boldsymbol{N'}$

$$[\boldsymbol{N'}^{(J)}] = \begin{bmatrix} \frac{\partial N_1^{(J)}}{\partial \mathrm{r}} & \frac{\partial N_2^{(J)}}{\partial \mathrm{r}} & \frac{\partial N_3^{(J)}}{\partial \mathrm{r}} & \frac{\partial N_4^{(J)}}{\partial \mathrm{r}} & 0 & 0 & 0 & 0 \\ \frac{N_1^{(J)}}{\mathrm{r}^{(J)}} & \frac{N_2^{(J)}}{\mathrm{r}^{(J)}} & \frac{N_3^{(J)}}{\mathrm{r}^{(J)}} & \frac{N_4^{(J)}}{\mathrm{r}^{(J)}} & 0 & 0 & 0 & 0 \\ 0 & 0 & 0 & 0 & \frac{\partial N_1^{(J)}}{\partial z} & \frac{\partial N_2^{(J)}}{\partial z} & \frac{\partial N_3^{(J)}}{\partial z} & \frac{\partial N_4^{(J)}}{\partial z} \\ \frac{1}{2}\frac{\partial N_1^{(J)}}{\partial r} & \frac{1}{2}\frac{\partial N_2^{(J)}}{\partial r} & \frac{1}{2}\frac{\partial N_3^{(J)}}{\partial r} & \frac{1}{2}\frac{\partial N_4^{(J)}}{\partial r} & \frac{1}{2}\frac{\partial N_1^{(J)}}{\partial z} & \frac{1}{2}\frac{\partial N_2^{(J)}}{\partial z} & \frac{1}{2}\frac{\partial N_3^{(J)}}{\partial z} & \frac{1}{2}\frac{\partial N_4^{(J)}}{\partial z} \end{bmatrix}$$

10. Compute the stiffness matrix

$$[\boldsymbol{K_e}] = [\boldsymbol{K_e}] + 2\pi\ \widetilde{W}_j\ \mathrm{r}^{(j)} \det(\boldsymbol{J}^{(J)})\ [\boldsymbol{N'}^{(J)}]^T\ [\mathbb{C}]\ [\boldsymbol{N'}^{(J)}]$$

Cycle over integration

11. Assemble the shape functions and derivatives for the one-dimensional numerical integration along the surface 3-4, for integration points $k = 1$ to 3

$$\hat{N}_3^{(k)} = \frac{1}{2}(1 + r_k), \qquad \frac{\partial \hat{N}_3^{(k)}}{\partial r} = \frac{1}{2}$$

$$\hat{N}_4^{(k)} = \frac{1}{2}(1 - r_k), \qquad \frac{\partial \hat{N}_4^{(k)}}{\partial r} = -\frac{1}{2}$$

12. Compute the equivalent nodal forces given the applied surface traction. Only one surface has applied forces; therefore, we may write

$$\frac{\partial \mathrm{r}}{\partial r} = \frac{\partial \hat{N}_3^{(k)}}{\partial r}\mathrm{r}_3 + \frac{\partial \hat{N}_4^{(k)}}{\partial r}\mathrm{r}_4$$

$$\frac{\partial z}{\partial s} = \frac{\partial \hat{N}_3^{(k)}}{\partial s} s_3 + \frac{\partial \hat{N}_4^{(k)}}{\partial r} s_4$$

$$\frac{dL}{dr} = \left[\left(\frac{d\mathrm{r}}{dr}\right)^2 + \left(\frac{dz}{dr}\right)^2\right]^{\frac{1}{2}}$$

$$\mathrm{r}^{(k)} = \hat{N}_3^{(k)}\mathrm{r}_3 + \hat{N}_4^{(k)}\mathrm{r}_4$$

$$\{\boldsymbol{F}_\tau\} = \left(2\pi \sum_{j=1}^{3} \hat{W}_j \frac{dL}{dr} [\boldsymbol{N}]^T \{\boldsymbol{\tau}_{34}\}\, \mathrm{r}^{(j)}\right)_{s=1}$$

$$F_{3z} = 2\pi \sum_{j=1}^{3} \hat{W}_j \frac{dL}{dr} \hat{N}_3^{(k)} \tau_{34z}\, \mathrm{r}^{(j)}$$

$$F_{4z} = 2\pi \sum_{j=1}^{3} \hat{W}_j \frac{dL}{dr} \hat{N}_4^{(k)} \tau_{34z}\, \mathrm{r}^{(j)}$$

13. Set displacement boundary conditions for degree of freedom j by making the diagonal component of the stiffness very large and the cooresponding force zero

$$\boldsymbol{K}_e(j,j) = 1 \times 10^{12} * \boldsymbol{K}_e(j,j)$$

$$f(j) = 0$$

14. Solve the system of equations using the Matlab® factorization operator

$$\{u\} = [\boldsymbol{K}_e]\backslash\{f\}$$

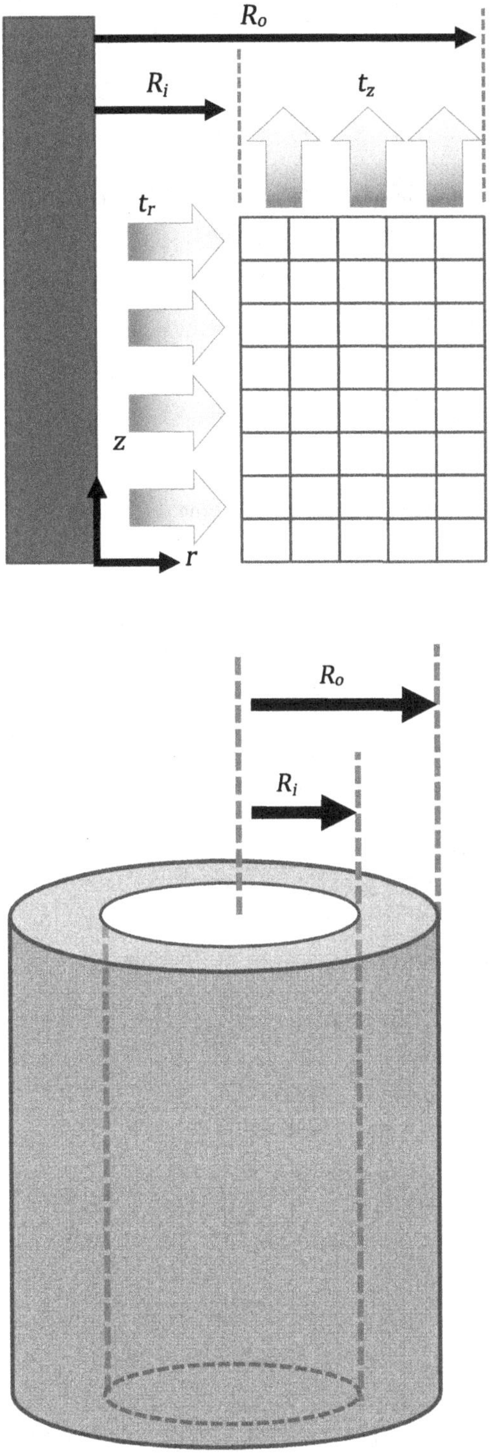

Figure 11.7. (top) Illustration of axisymmetric mesh subject to traction and displacement boundary conditions. (bottom) Solid obtained by rotating mess about the $z - z$ axis.

11.4.7.2 Matlab® Code

```
% ------------------------------------------------------------------------
% DESCRIPTION
% ------------------------------------------------------------------------
% Axisymmetric FEM simulation of a single quadralateral element with 4 numerical
% integration points. The nodes are numbered as follows:
%                         side 3
%                   4|-----------|3
%   z     side 4     |           |     side 2
%   ^                |           |
%   |               1|-----------|2
%   |                     side 1
%   |----> r
%  A point load parallel to the y axis is appliced to nodes 4 and 3.
%
%  Variables Used in this Program
%       nnpe          - number of nodes per element
%       nPosRZ(i,j)   - position of node i in the j direction
%       nForces(i,j)  - point load applied to node i in the j direction
%
%       numqpt         - number of quadrature points per element
%       QPT(i,j)       - position of quadrature point i in the j direction
%
%       Jacobian       - Jacobian transformation matrix
%       invJacob       - Inverse of the Jacobian matrix
%
%       stiffness(i,j)- stiffness matrix element for degree of freedom i
%                       dof j
%       force(i)       - force vector for degree of freedom i
%
%       u(i)           - displacement for degree of freedom i
%       displace(i,j)  - displacement of node i in direction j
%       strain         - element strain
%       stress         - element stress
% ------------------------------------------------------------------------
% PREPROCESSING
% ------------------------------------------------------------------------
clear
nnpe  = 4;       % number of nodes per element
% ------------------------------------------------------------------------
% Step 1. Input the r and z coordinate of each node, and components of
%         the surface traction acting on the element.
nPosRZ  = zeros(4, 2);            % Initialize nodal positions
nForces = zeros(4, 2);            % Initialize external nodal forces
radius    = 0.1;                  % [m] radius of the cylinder
height    = 0.5;                  % [m] height of the cylinder
```

```
% -----------------------------------------------------------------------
nPosRZ(1,:) = [0,       0];
nPosRZ(2,:) = [radius,0];
nPosRZ(3,:) = [radius,height];
nPosRZ(4,:) = [0,       height];
% -----------------------------------------------------------------------
% Specify the surface tractions
tractionz = 200;                   % [Pa] surface traction acting on side 3
                                   %        along the z axis
% -----------------------------------------------------------------------
% Step 2. Give the material properties for the element
E      = 1.0e6;                    % elastic modulus
nu     = 0.4;                      % Poisson's Ratio
% -----------------------------------------------------------------------
G      = E/(2.0*(1+nu));          % shear modulus
lambda = E*nu/((1+nu)*(1-2*nu)); % Lame constant
% -----------------------------------------------------------------------
% Step 3. Assemble the R matrix assuming plane strain deformation
C = [2.0*G+lambda, lambda,       lambda, 0;
     lambda,       2.0*G+lambda, lambda, 0;
     lambda,       lambda,       2.0*G+lambda, 0;
     0,            0,            0,      G];
% -----------------------------------------------------------------------
% Step 4. Give the positions of the numerical integration points
%         in natural (r-s) coordinate system.
% parameters for numerical integration over an area
numqpt = 4;                     % number of quadrature points
wt     = [ 1, 1, 1, 1];         % weights for quadrature points
q = sqrt(1/3);
QPT = [ -q, -q;                 % (r,s) location for
         q, -q;                 %     the 4 quadrature points
         q,  q;
        -q,  q;];
% parameters for numerical integration along a line
numsurfqpt  = 3;
q1          = sqrt(3/5);
sQPT        = [-q1, 0, q1];
wts         = [5/9, 8/9, 5/9];
% -----------------------------------------------------------------------
% Step 5. Assemble the shape functions and necessary derivatives
for j = 1:numqpt;
    % r-s coordinate for each integration point
    r = QPT(j,1);  s = QPT(j,2);
    % shape functions
    sf(1,j) = 0.25*(1.0-r)*(1.0-s);
    sf(2,j) = 0.25*(1.0+r)*(1.0-s);
    sf(3,j) = 0.25*(1.0+r)*(1.0+s);
```

```
        sf(4,j) = 0.25*(1.0-r)*(1.0+s);
        % derivative of shape function wrspt to u
        dndrs(1,1,j) = 0.25*(-1.)*(1.0-s);
        dndrs(2,1,j) = 0.25*(+1.)*(1.0-s);
        dndrs(3,1,j) = 0.25*(+1.)*(1.0+s);
        dndrs(4,1,j) = 0.25*(-1.)*(1.0+s);
        % derivative of shape function wrspt to v
        dndrs(1,2,j) = 0.25*(1.0-r)*(-1.0);
        dndrs(2,2,j) = 0.25*(1.0+r)*(-1.0);
        dndrs(3,2,j) = 0.25*(1.0+r)*(+1.0);
        dndrs(4,2,j) = 0.25*(1.0-r)*(+1.0);
end
% -------------------------------------------------------------------------
stiffness = zeros(8, 8);   % initialize element stiffness matrix
for j = 1:numqpt
    % Step 6. Find the Jacobean matrix
    Jacobian  = nPosRZ'*dndrs(:,:,j);
    invJacob  = Jacobian^-1;
    % Step 7. Find dNdR and dNdZ
    dNdRZ     = dndrs(:,:,j)*invJacob;
    % Step 8. Find the global r position for each integration point
    rpos      = nPosRZ(:,1)'*sf(:,j);
    % Step 9. Assemble Nprime
    NP        = [dNdRZ(:,1)'  0, 0, 0, 0;  ...
                 0, 0, 0, 0,  dNdRZ(:,2)'; ...
                 sf(:,j)'./rpos,  0, 0, 0, 0;  ...
                 dNdRZ(:,2)', dNdRZ(:,1)'];
    % Step 10. Compute the stiffness matrix
    stiffness = stiffness + 2*pi*rpos*wt(j)*NP'*C*NP*det(Jacobian);
end
% -------------------------------------------------------------------------
% Step 11. Define the values and derivatives of the shape functions
%          used for 1-d integration along side 3-4.
force     = zeros(8,1);
ssf3      = 0.25*(1.0+sQPT)*(2);
ssf4      = 0.25*(1.0-sQPT)*(2);
dssfdu3   = 0.25*(1) * 2;
dssfdu4   = 0.25*(-1)* 2;
% Step 12. Find the equivalent nodal forces on nodes 3 and 4
dxdxi     = dssfdu3*nPosRZ(3,1)+dssfdu4*nPosRZ(4,1);
dydyi     = dssfdu3*nPosRZ(3,2)+dssfdu4*nPosRZ(4,2);
detjs     = sqrt(dxdxi^2+dydyi^2);
r         = ssf3.*nPosRZ(3,1) + ssf4.*nPosRZ(4,1);
force(7) = sum(2*pi*wts.*ssf3.*r*detjs*tractionz);
force(8) = sum(2*pi*wts.*ssf4.*r*detjs*tractionz);
% -------------------------------------------------------------------------
% Step 13. Set displacement boundary conditions
```

```
%       13a. Node 1 has zero displacement in the x and y direction
stiffness(1,:) = 0.0; stiffness(1,1) = 1.0; force(1) = 0;
stiffness(5,:) = 0.0; stiffness(5,5) = 1.0; force(5) = 0;
%       13b. Node 4 has zero displacement in the x direction
stiffness(4,:) = 0.0; stiffness(4,4) = 1.0; force(4) = 0;
%       13c. Node 2 has zero displacement in the y direction
stiffness(6,:) = 0.0; stiffness(6,6) = 1.0; force(6) = 0;
% -------------------------------------------------------------------------
%   SOLVER
% -------------------------------------------------------------------------
% Step 14. Solve for the nodal displacements using
%           Matlab's built in Guassian elimination
u = stiffness\force;
```

11.4.8 Multiple Element Implementation

The numerical solution for infinitesimal axisymmetric deformation is presented in this section for a mesh that contains multiple elements. This is intended to illustrate the construction of the global stiffness matrix. The inner radius of the rectangular mesh is R_i and the outer diameter is R_o. A uniform surface traction, t_z, is applied to the top surface and a uniform surface traction, t_r, is applied to the inner surface. The number of nodes in the mesh is variable and can be adjusted.

In this section, we have modeled a cylinder which has a radius given by $R = 10$ cm and a height given by $h = 50$ cm using multiple elements. The modulus and Poisson's ratio were assumed to be $E = 5 \times 10^6$ Pa and $\nu = 0.4$, respectively. The graph of strain versus applied stress is shown Figure 11.8.

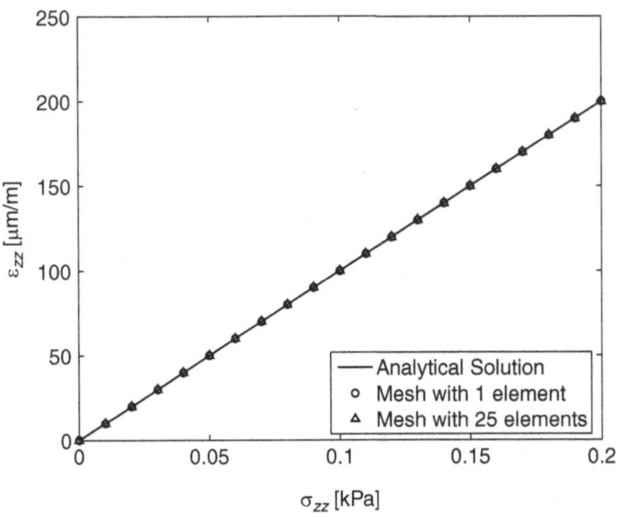

Figure 11.8. Strain versus stress curve for two axisymmetric meshes containing 1 and 25 elements.

11.4.8.1 Flow Chart

1. Input cylinder geometry, mesh parameters, material properties, and applied surface traction

↓

2. Evenly space the gridpoints across the domain

↓

3. Compute element connectivity matrix. Each row contains the identifier for the nodes making up the element; the nodes are selected so that number is counterclockwise

↓

4. Assign surface traction to the segments along the top of the mesh and to the inner surface of the mesh

↓

5. Assign fixed boundary conditions to nodes

↓

6. Assemble C matrix for plane strain deformation of an isotropic material

$$[\mathbb{C}] = \frac{E}{(1+\nu)(1-2\nu)}\begin{bmatrix} 1-\nu & \nu & \nu & 0 \\ \nu & 1-\nu & \nu & 0 \\ \nu & \nu & 1-\nu & 0 \\ 000 & & & \frac{1-2\nu}{2} \end{bmatrix}$$

↓

7. Setup Gaussian integration points. We use two integration points along the r axis and two along the s axis giving a total of four integration points per element for the area integration

Integration point, i	r_i	s_i	$\bar{W}_i$
1	$-1/\sqrt{3}$	$-1/\sqrt{3}$	1
2	$1/\sqrt{3}$	$-1/\sqrt{3}$	1
3	$1/\sqrt{3}$	$1/\sqrt{3}$	1
4	$-1/\sqrt{3}$	$1/\sqrt{3}$	1

Three integration points are used to for line integrals

Integration point, i	r_i	$\hat{W}_i$
1	$-\sqrt{3/5}$	5/9
2	$\sqrt{3/5}$	8/9
3	$\sqrt{3/5}$	5/9

↓

8. Assemble the shape functions and derivatives at each integration point, $j = 1\ to\ 4$

$$N_1^{(J)} = \frac{1}{4}(1-r_j)(1-s_j), \quad \frac{\partial N_1^{(J)}}{\partial r} = \frac{1}{4}(-1)(1-s_j), \quad \frac{\partial N_1^{(J)}}{\partial r} = \frac{1}{4}(1-r_j)(-1)$$

$$N_2^{(J)} = \frac{1}{4}(1+r_j)(1-s_j), \quad \frac{\partial N_2^{(J)}}{\partial r} = \frac{1}{4}(+1)(1-s_j), \quad \frac{\partial N_2^{(J)}}{\partial r} = \frac{1}{4}(1+r_j)(-1)$$

$$N_3^{(J)} = \frac{1}{4}(1+r_j)(1+s_j), \quad \frac{\partial N_3^{(J)}}{\partial r} = \frac{1}{4}(+1)(1+s_j), \quad \frac{\partial N_3^{(J)}}{\partial r} = \frac{1}{4}(1+r_j)(+1)$$

$$N_4^{(J)} = \frac{1}{4}(1-r_j)(1+s_j), \quad \frac{\partial N_4^{(J)}}{\partial r} = \frac{1}{4}(-1)(1+s_j), \quad \frac{\partial N_4^{(J)}}{\partial r} = \frac{1}{4}(1-r_j)(+1)$$

↓

9. Find the Jacobean

$$[J^{(j)}] = \begin{bmatrix} \frac{\partial N_1^{(j)}}{\partial r} & \frac{\partial N_2^{(j)}}{\partial r} & \frac{\partial N_3^{(j)}}{\partial r} & \frac{\partial N_4^{(j)}}{\partial r} \\ \frac{\partial N_1^{(j)}}{\partial s} & \frac{\partial N_2^{(j)}}{\partial s} & \frac{\partial N_3^{(j)}}{\partial s} & \frac{\partial N_4^{(j)}}{\partial s} \end{bmatrix} \begin{bmatrix} x_1 & y_1 \\ x_2 & y_2 \\ x_3 & y_3 \\ x_4 & y_4 \end{bmatrix}$$

10. Invert the Jacobean to find $\frac{\partial N}{\partial x}$ and $\frac{\partial N}{\partial y}$

$$\begin{bmatrix} \frac{\partial N_1^{(j)}}{\partial \mathrm{r}} & \frac{\partial N_2^{(j)}}{\partial \mathrm{r}} & \frac{\partial N_3^{(j)}}{\partial \mathrm{r}} & \frac{\partial N_4^{(j)}}{\partial \mathrm{r}} \\ \frac{\partial N_1^{(j)}}{\partial z} & \frac{\partial N_2^{(j)}}{\partial z} & \frac{\partial N_3^{(j)}}{\partial z} & \frac{\partial N_4^{(j)}}{\partial z} \end{bmatrix} = [J]^{-1} \begin{bmatrix} \frac{\partial N_1^{(j)}}{\partial r} & \frac{\partial N_2^{(j)}}{\partial r} & \frac{\partial N_3^{(j)}}{\partial r} & \frac{\partial N_4^{(j)}}{\partial r} \\ \frac{\partial N_1^{(j)}}{\partial s} & \frac{\partial N_2^{(j)}}{\partial s} & \frac{\partial N_3^{(j)}}{\partial s} & \frac{\partial N_4^{(j)}}{\partial s} \end{bmatrix}$$

11. Find the r coordinate of each integration point

$$\mathrm{r}^{(j)} = \sum N_i^{(j)}\, \mathrm{r}_i$$

12. Assemble $\boldsymbol{N'}$

$$[\boldsymbol{N'}^{(j)}] = \begin{bmatrix} \frac{\partial N_1^{(j)}}{\partial \mathrm{r}} & \frac{\partial N_2^{(j)}}{\partial \mathrm{r}} & \frac{\partial N_3^{(j)}}{\partial \mathrm{r}} & \frac{\partial N_4^{(j)}}{\partial \mathrm{r}} & 0 & 0 & 0 & 0 \\ \frac{N_1^{(j)}}{\mathrm{r}^{(j)}} & \frac{N_2^{(j)}}{\mathrm{r}^{(j)}} & \frac{N_3^{(j)}}{\mathrm{r}^{(j)}} & \frac{N_4^{(j)}}{\mathrm{r}^{(j)}} & 0 & 0 & 0 & 0 \\ 0 & 0 & 0 & 0 & \frac{\partial N_1^{(j)}}{\partial z} & \frac{\partial N_2^{(j)}}{\partial z} & \frac{\partial N_3^{(j)}}{\partial z} & \frac{\partial N_4^{(j)}}{\partial z} \\ \frac{1}{2}\frac{\partial N_1^{(j)}}{\partial \mathrm{r}} & \frac{1}{2}\frac{\partial N_2^{(j)}}{\partial \mathrm{r}} & \frac{1}{2}\frac{\partial N_3^{(j)}}{\partial \mathrm{r}} & \frac{1}{2}\frac{\partial N_4^{(j)}}{\partial \mathrm{r}} & \frac{1}{2}\frac{\partial N_1^{(j)}}{\partial z} & \frac{1}{2}\frac{\partial N_2^{(j)}}{\partial z} & \frac{1}{2}\frac{\partial N_3^{(j)}}{\partial z} & \frac{1}{2}\frac{\partial N_4^{(j)}}{\partial z} \end{bmatrix}$$

13. Compute the stiffness matrix

$$[\boldsymbol{K_e}] = [\boldsymbol{K_e}] + 2\pi\ \widetilde{W}_j\ \mathrm{r}^{(j)} \det(\boldsymbol{J}^{(j)})\ [\boldsymbol{N'}^{(j)}]^T\ [\mathbb{C}]\ [\boldsymbol{N'}^{(j)}]$$

Cycle over

14. Add element stiffness components to the global stiffness matrix

Cycle over

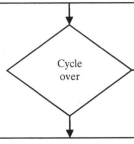

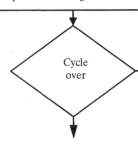

15a. Find the equivalent nodal forces on the top surface of the cylinder.

The shape functions are given by

$$\hat{N}_3^{(k)} = \frac{1}{2}(1 + r_k), \qquad \frac{\partial \hat{N}_3^{(k)}}{\partial r} = \frac{1}{2}$$

$$\hat{N}_4^{(k)} = \frac{1}{2}(1 - r_k), \qquad \frac{\partial \hat{N}_4^{(k)}}{\partial r} = -\frac{1}{2}$$

$$\frac{\partial \mathrm{r}}{\partial r} = \frac{\partial \hat{N}_3^{(k)}}{\partial r}\mathrm{r}_3 + \frac{\partial \hat{N}_4^{(k)}}{\partial r}\mathrm{r}_4 \quad \text{,and} \quad \frac{\partial z}{\partial s} = \frac{\partial \hat{N}_3^{(k)}}{\partial s}s_3 + \frac{\partial \hat{N}_4^{(k)}}{\partial r}s_4$$

$$\frac{dL}{dr} = \left[\left(\frac{d\mathrm{r}}{dr}\right)^2 + \left(\frac{dz}{dr}\right)^2\right]^{\frac{1}{2}}$$

$$\mathrm{r}^{(k)} = \hat{N}_3^{(k)}\mathrm{r}_3 + \hat{N}_4^{(k)}\mathrm{r}_4$$

The equivalent nodal forces are

$$F_{3z} = 2\pi \sum_{k=1}^{3} \hat{W}_j \frac{dL}{dr}\hat{N}_3^{(k)}\tau_{34z}\mathrm{r}^{(k)} \quad \text{and} \quad F_{4z} = 2\pi \sum_{k=1}^{3} \hat{W}_j \frac{dL}{dr}\hat{N}_4^{(k)}\tau_{34z}\mathrm{r}^{(k)}$$

15b. Find the equivalent nodal forces on the inner surface of the cylinder. The shape functions are given by

$$\hat{N}_1^{(k)} = \frac{1}{2}(1 + s_k), \qquad \frac{\partial \hat{N}_1^{(k)}}{\partial r} = -\frac{1}{2}$$

$$\hat{N}_4^{(k)} = \frac{1}{2}(1 + s_k), \qquad \frac{\partial \hat{N}_4^{(k)}}{\partial r} = \frac{1}{2}$$

$$\frac{\partial \mathrm{r}}{\partial r} = \frac{\partial \hat{N}_1^{(k)}}{\partial s}\mathrm{r}_3 + \frac{\partial \hat{N}_4^{(k)}}{\partial s}\mathrm{r}_4 \quad \text{and} \quad \frac{\partial z}{\partial s} = \frac{\partial \hat{N}_1^{(k)}}{\partial s}s_3 + \frac{\partial \hat{N}_4^{(k)}}{\partial s}s_4$$

$$\frac{dL}{ds} = \left[\left(\frac{d\mathrm{r}}{ds}\right)^2 + \left(\frac{dz}{ds}\right)^2\right]^{\frac{1}{2}}$$

$$\mathrm{r}^{(k)} = \hat{N}_3^{(k)}\mathrm{r}_3 + \hat{N}_4^{(k)}\mathrm{r}_4$$

The equivalent nodal forces are

$$F_{1r} = 2\pi \sum_{k=1}^{3} \hat{W}_j \frac{dL}{dr}\hat{N}_1^{(k)}\tau_{14r}\mathrm{r}^{(k)} \quad \text{and} \quad F_{4r} = 2\pi \sum_{k=1}^{3} \hat{W}_j \frac{dL}{dr}\hat{N}_4^{(k)}\tau_{14r}\mathrm{r}^{(k)}$$

14. Set displacement boundary conditions for degree of freedom j by making the diagonal component of the stiffness very large and the corresponding force zero

$$\boldsymbol{K}_e(j,j) = 1 \times 10^{12} * \boldsymbol{K}_e(j,j)$$

$$f(j) = 0$$

15. Solve the system of equations using the Matlab® factorization operator

$$\{\boldsymbol{u}\} = [\boldsymbol{K}_e]\backslash\{\boldsymbol{f}\}$$

11.4.8.2 Matlab® Code

```
% ----------------------------------------------------------------------
% DESCRIPTION
% ----------------------------------------------------------------------
% Axisymmetric FEM simulation of a single quadralateral element with 4 numerical
% integration points. The nodes are numbered as follows:
%             (nrY)---(nrY+1)--- ...      nrY*nrX
%                 .
%                 .
%             (nrX+1)--(nrX+2)--- ...      2*nrX
%                1 ------ 2 ----- 3 - 4 ... nrX
%
%   z
%   ^
%   |
%   |
%   |----> r
%  A point load parallel to the y axis is appliced to nodes 4 and 3.
%
%  Variables Used in this Program
%       nnpe          - number of nodes per element
%       nPosRZ(i,j)   - position of node i in the j direction
%       nForces(i,j) - point load applied to node i in the j direction
%
%       numqpt        - number of quadrature points per element
%       QPT(i,j)      - position of quadrature point i in the j direction
%
%       Jacobian      - Jacobian transformation matrix
%       invJacob      - Inverse of the Jacobian matrix
%
%       stiffness(i,j)- stiffness matrix element for degree of freedom i
%                       dof j
%       force(i)      - force vector for degree of freedom i
%
%       u(i)          - displacement for degree of freedom i
%       displace(i,j) - displacement of node i in direction j
%       strain        - element strain
%       stress        - element stress
% ----------------------------------------------------------------------
% PREPROCESSING
% ----------------------------------------------------------------------
clear
% ----------------------------------------------------------------------
% Step 1. Input the r and z coordinate of each node, and components of
%         the surface traction acting on the element.
% ----------------------------------------------------------------------
% Mesh parameters
nrR   = 5; % number of nodes in the X direction
nrZ   = 5; % number of nodes in the Y direction
nnpe  = 4; % number of nodes per element (4 - node quads used)
```

```
% ----------------------------------------------------------------------------
% Cylinder geometry
innerRadius = 0.0;      % [m] mesh innerRadius
outerRadius = 0.02;     % [m] mesh outerRadius
height      = 0.1;      % [m] mesh height
% ----------------------------------------------------------------------------
% Material properties for the element
E       = 1.0e9;        % [Pa] Elastic modulus
nu      = 0.35;         % Poisson's Ratio
% ----------------------------------------------------------------------------
% Traction Components
tractionZTop   = 1e6;   % [Pa] traction applied to top surface
tractionRInner =  0;    % [Pa] traction applied to inner surface
% ----------------------------------------------------------------------------
% Step 2. Compute nodal positions based on geometry and mesh parameters
nPosRZ  = zeros(nrR*nrZ, 2);       % Initialize nodal positions
for j = 1:nrZ
    for i = 1:nrR
        nodeNum(i,j) = i+(j-1)*nrR;
        nPosRZ(nodeNum(i,j),1) = innerRadius+(outerRadius-innerRadius)*1.0/(nrR-1)*(i-1);
        nPosRZ(nodeNum(i,j),2) = height*1.0/(nrZ-1)*(j-1);
    end
end
% ----------------------------------------------------------------------------
% Step 3. Compute the element connectivity matrix
count = 1;
for j = 1:nrZ-1
    for i = 1:nrR-1
        elem(count, 1) = nodeNum(i,   j);     % bottom left node
        elem(count, 2) = nodeNum(i+1, j);     % bottom right node
        elem(count, 3) = nodeNum(i+1, j+1);   % top    left node
        elem(count, 4) = nodeNum(i,   j+1);   % top    right node
        count = count+1;
    end
end
nel   = length(elem(:,1)); % number of elements in the system
% ----------------------------------------------------------------------------
% Step 4. Assemble the surface traction matrix
%       Top Surface
trac = zeros(nrR-1,4);
for i = 1:nrR-1
    %          node1          node2            tractionR tractionZ [Pa]
    trac(i,:) = [nodeNum(i,nrZ) nodeNum(i+1,nrZ) 0         tractionZTop];
end
%       Inner Surface
trac2 = zeros(nrZ-1,4);
for j = 1:nrZ-1
    %          node1         node2            tractionR      tractionZ [Pa]
    trac2(j,:) = [nodeNum(1,j) nodeNum(1,j+1) tractionRInner        0];
end
```

```
% ---------------------------------------------------------------------
% Step 5. Input the node numbers which have zero displacement boundary
%         conditions
fixZ = [nodeNum(:,1)];  % these nodes have fixed Z coordinate
fixR = [nodeNum(1,:)];  % these nodes have fixed R coordinate
% ---------------------------------------------------------------------
% Step 6. Assemble the C matrix for an isotropic linear elastic material
G      = E/(2.0*(1+nu));          % shear modulus
lambda = E*nu/((1+nu)*(1-2*nu)); % Lame constant
% ---------------------------------------------------------------------
C = [2.0*G+lambda, lambda,       lambda, 0;
     lambda,       2.0*G+lambda, lambda, 0;
     lambda,       lambda,       2.0*G+lambda, 0;
     0,            0,            0,      G];
% ---------------------------------------------------------------------
% Step 7. Give the positions of the numerical integration points
%         in r-s coordinate system.
% parameters for numerical integration of an area
numqpt = 4;                  % number of quadrature points
wt     = [ 1, 1, 1, 1];      % weights for quadrature points
q = sqrt(1/3);
QPT = [ -q, -q;              % (r,s) location of
         q, -q;              %     the 4 quadrature points
         q,  q;
        -q,  q;];
% parameters for numerical integration along a line
numsurfqpt  = 3;
q1          = sqrt(3/5);
sQPT        = [-q1, 0, q1];
wts         = [5/9, 8/9, 5/9];
% ---------------------------------------------------------------------
% Step 8. Assemble the shape functions and necessary derivatives
for j = 1:numqpt;
    r = QPT(j,1);  s = QPT(j,2);
    % shape functions
    sf(1,j) = 0.25*(1.0-r)*(1.0-s);
    sf(2,j) = 0.25*(1.0+r)*(1.0-s);
    sf(3,j) = 0.25*(1.0+r)*(1.0+s);
    sf(4,j) = 0.25*(1.0-r)*(1.0+s);
    % derivative of shape function wrspt to r
    dndrs(1,1,j) = 0.25*(-1.)*(1.0-s);
    dndrs(2,1,j) = 0.25*(+1.)*(1.0-s);
    dndrs(3,1,j) = 0.25*(+1.)*(1.0+s);
    dndrs(4,1,j) = 0.25*(-1.)*(1.0+s);
    % derivative of shape function wrspt to s
    dndrs(1,2,j) = 0.25*(1.0-r)*(-1.0);
    dndrs(2,2,j) = 0.25*(1.0+r)*(-1.0);
    dndrs(3,2,j) = 0.25*(1.0+r)*(+1.0);
    dndrs(4,2,j) = 0.25*(1.0-r)*(+1.0);
end
```

```
% -----------------------------------------------------------------------
% Step 9. Assemble the stiffness tensor and the force Marix
nnodes      = length(nPosRZ(:,1)); % number of nodes in the system
glStiffness = zeros(2*nnodes);     % initialize the global stiffness matrix
% cycle through each element
for k = 1:nel
    stiffness   = zeros(8, 8);     % initialize the element stiffness matrix
    % cycle over each integration point within the element
    for j = 1:numqpt
        % Step 9. Compute the Jacobian matrix
        Jacobian  = nPosRZ(elem(k,:),:)'*dndrs(:,:,j);
        invJacob  = Jacobian^-1;
        % Step 10. Compute dndr and dndZ
        dndrz     = dndrs(:,:,j)*invJacob;
        % Step 11. Find r distance for each integration point
        r         = nPosRZ(elem(k,:),1)'*sf(:,j);
        % Step 12. Assemble NP
        NP        = [dndrz(:,1)'  0, 0, 0, 0;  ...
                     0, 0, 0, 0,  dndrz(:,2)'; ...
                     sf(:,j)'./r,  0, 0, 0, 0;  ...
                     dndrz(:,2)', dndrz(:,1)'];
        % Step 13. Assemble the element stiffness tensor
        stiffness = stiffness + 2*pi*r*wt(j)*NP'*C*NP*det(Jacobian);
    end
% -----------------------------------------------------------------------
% Step 14. Add element stiffness tensor to the global stiffness matrix
    for l = 1:nnpe
        l1 = elem(k,l);
        l2 = elem(k,l) + nnodes;
        for m = 1:nnpe
            m1 = elem(k,m);
            m2 = elem(k,m) + nnodes;
            glStiffness(l1,m1) = glStiffness(l1,m1) +  stiffness(l, m);
            glStiffness(l1,m2) = glStiffness(l1,m2) +  stiffness(l, m+nnpe);
            glStiffness(l2,m1) = glStiffness(l2,m1) +  stiffness(l+nnpe, m);
            glStiffness(l2,m2) = glStiffness(l2,m2) +  stiffness(l+nnpe, m+nnpe);
        end
    end
end
% -----------------------------------------------------------------------
% Step 15. Find the equivalent nodal forces
% 15a. Traction on top surface (r, s=1)
force     = zeros(2*nnodes,1);
ssf1      = 0;
ssf2      = 0;
ssf3      = 0.25*(1.0+sQPT)*(2);
ssf4      = 0.25*(1.0-sQPT)*(2);
dssfdu3   = 0.5;
dssfdu4   = -0.5;
for i = 1:length(trac(:,1))
```

```
    % compute dLdr
    drdr     = dssfdu3*nPosRZ(trac(i,1),1)+dssfdu4*nPosRZ(trac(i,2),1);
    dzdr     = dssfdu3*nPosRZ(trac(i,1),2)+dssfdu4*nPosRZ(trac(i,2),2);
    dLdr     = sqrt(drdr^2+dzdr^2)
    % find the r coordinate for each integration point
    r         = ssf3.*nPosRZ(trac(i,1),1) + ssf4.*nPosRZ(trac(i,2),1);
    % select the correct degree of freedom and assign the equivalent
    % nodal forces
    dof1r = trac(i,1);
    dof2r = trac(i,2);
    force(dof1r) = force(dof1r)+sum(2*pi*wts.*ssf3.*r*dLdr*trac(i,3));
    force(dof2r) = force(dof2r)+sum(2*pi*wts.*ssf4.*r*dLdr*trac(i,3));
    dof1z = trac(i,1)+nnodes;
    dof2z = trac(i,2)+nnodes;
    force(dof1z) = force(dof1z)+sum(2*pi*wts.*ssf3.*r*dLdr*trac(i,4));
    force(dof2z) = force(dof2z)+sum(2*pi*wts.*ssf4.*r*dLdr*trac(i,4));
end
% -------------------------------------------------------------------
% 15b. Traction on inner surface (r=-1, s)
ssf1      = 0.25*2*(1.0-sQPT);
ssf2      = 0;
ssf3      = 0;
ssf4      = 0.25*2*(1.0+sQPT);
dssfdu1  = -0.5;
dssfdu4  =  0.5;
for i = 1:length(trac2(:,1))
    % compute dLds
    drds     = dssfdu1*nPosRZ(trac2(i,1),1)+dssfdu4*nPosRZ(trac2(i,2),1);
    dzds     = dssfdu1*nPosRZ(trac2(i,1),2)+dssfdu4*nPosRZ(trac2(i,2),2);
    dLds     = sqrt(drds^2+dzds^2);
    % find the r coordinate for each integration point
    r        = ssf1.*nPosRZ(trac2(i,1),1) + ssf4.*nPosRZ(trac2(i,2),1);
    % select the correct degree of freedom and assign the equivalent
    % nodal forces
    dof1r = trac2(i,1);
    dof2r = trac2(i,2);
    force(dof1r) = force(dof1r)+sum(2*pi*wts.*ssf1.*r*dLds*trac2(i,3));
    force(dof2r) = force(dof2r)+sum(2*pi*wts.*ssf4.*r*dLds*trac2(i,3));
    dof1z = trac2(i,1)+nnodes;
    dof2z = trac2(i,2)+nnodes;
    force(dof1z) = force(dof1z)+sum(2*pi*wts.*ssf1.*r*dLds*trac2(i,4));
    force(dof2z) = force(dof2z)+sum(2*pi*wts.*ssf4.*r*dLds*trac2(i,4));
end
% -------------------------------------------------------------------
% Step 16. Set displacement boundary conditions
for i = 1: length(fixR)
    dof = fixR(i);
    glStiffness(dof,:)   = 0.0;
    glStiffness(dof,dof) = 1.0;
    force(dof)           = 0;
```

```
end
for i = 1:length(fixZ)
    dof = fixZ(i) + nnodes;
    glStiffness(dof,:)   = 0.0;
    glStiffness(dof,dof) = 1.0;
    force(dof)           = 0;
end
% ------------------------------------------------------------------------
%   SOLVER
% ------------------------------------------------------------------------
% Step 17. Solve for the nodal displacements using
%          Matlab's built in Guassian elimination
u = glStiffness\force;
```

11.5 Infinitesimal Plane Strain FEM with Material Nonlinearity

In the most general case, nonlinearity in the finite element approximation arises due to both geometric and material nonlinearity. In the previous sections of this chapter, we have discussed the formulation of the finite element approximation to the governing equations for linear elastic materials subject to infinitesimal deformations. In this section, we introduce material nonlinearity into the finite element formulation while restricting ourselves to infinitesimal deformation. Therefore, we will be discussing problems that involve material nonlinearity but not geometric nonlinearity. For clarity, we will restate some of the results previously obtained in Section 11.3.

11.5.1 Statement of Virtual Work

Let us assume that we have a nonlinear elastic material for which the constitutive model for the Cauchy stress, $\boldsymbol{\sigma}$, may be written in the form

$$\boldsymbol{\sigma} = \boldsymbol{\sigma}(\boldsymbol{C}),$$

where $\boldsymbol{C}$ is the right Cauchy deformation tensor. In other words, the stress is a general function of the deformation. Substitution of the constitutive model into the statement of virtual work, Equation (11.1), gives

$$\delta W = -\int_{\Omega(t)} \{\delta\boldsymbol{\epsilon}\}^T \{\boldsymbol{\sigma}(\boldsymbol{C})\}\, dV + \int_{\Omega(t)} \{\delta\boldsymbol{u}\}^T \{\boldsymbol{b}\}\, dV + \int_{\partial\Omega(t)} \{\delta\boldsymbol{u}\}^T \{\boldsymbol{\tau}\}\, dS = 0,$$

where we have chosen to explicitly state the functional dependence of the stress tensor on the deformation right Cauchy deformation tensor.

11.5.2 Discretization of Space

Discretizing space by meshing the object allows us to rewrite the state of virtual work of the entire body as the sum of the individual contributions from each element as

$$\sum_{i=1}^{n_e} \left(-\int_{\Omega_e} \{\delta\boldsymbol{\epsilon}\}^T \{\boldsymbol{\sigma}(\boldsymbol{C})\}\, dV + \int_{\Omega_e} \{\delta\boldsymbol{u}\}^T \{\boldsymbol{b}\}\, dV + \int_{\partial\Omega_e} \{\delta\boldsymbol{u}\}^T \{\boldsymbol{\tau}\}\, dS \right) = 0. \quad (11.8)$$

11.5.3 Approximation of the Field Variables

Once again, we assume that we have meshed the spatial dimension with four-node isoparametric quadrilateral elements. The displacement field may be interpolated within the element using the equations

$$u(x,y) = \sum_{i=1}^{4} N_i(r,s)\, u_i \quad \text{and} \quad v(x,y) = \sum_{i=1}^{4} N_i(r,s)\, v_i,$$

where the nodal displacements are given by u_i and v_i, and the shape functions are given by $N_i(r,s)$. We introduce a nodal displacement vector, $\{\boldsymbol{u}_n\}$, and the nodal virtual displacement vector, $\{\delta \boldsymbol{u}_n\}$, such that

$$\{\boldsymbol{u}_n\} = \begin{Bmatrix} u_1 \\ u_2 \\ u_3 \\ u_4 \\ v_1 \\ v_2 \\ v_3 \\ v_4 \end{Bmatrix} \quad \text{and} \quad \{\delta \boldsymbol{u}_n\} = \begin{Bmatrix} \delta u_1 \\ \delta u_2 \\ \delta u_3 \\ \delta u_4 \\ \delta v_1 \\ \delta v_2 \\ \delta v_3 \\ \delta v_4 \end{Bmatrix},$$

where u_i and v_i are the displacement components in the x and y directions for node i, respectively, and δu_i and δv_i are the virtual displacement components in the x and y directions for node i, respectively. We can then write the interpolated displacements, and the interpolated virtual displacements as

$$\{\boldsymbol{u}\} = \begin{Bmatrix} u(x,y) \\ v(x,y) \end{Bmatrix} = [\boldsymbol{N}]\{\boldsymbol{u}_n\} \quad \text{and} \quad \{\delta \boldsymbol{u}\} = \begin{Bmatrix} \delta u(x,y) \\ \delta v(x,y) \end{Bmatrix} = [\boldsymbol{N}]\{\delta \boldsymbol{u}_n\}.$$

The nonzero components of the ***infinitesimal strain tensor*** can be written in terms of the displacement vectors and the interpolated field values as

$$\epsilon_{xx} = \frac{\partial u}{\partial x} = \sum_{i=1}^{4} \frac{\partial N_i}{\partial x} u_i,$$

$$\epsilon_{yy} = \frac{\partial v}{\partial y} = \sum_{i=1}^{4} \frac{\partial N_i}{\partial y} v_i,$$

$$\epsilon_{xy} = \frac{1}{2}\left(\frac{\partial u}{\partial y} + \frac{\partial v}{\partial x}\right) = \frac{1}{2}\left(\sum_{i=1}^{4} \frac{\partial N_i}{\partial y} u_i + \sum_{i=1}^{4} \frac{\partial N_i}{\partial x} v_i\right).$$

Combining these equations gives the infinitesimal strain written in terms of the nodal displacements

$$\{\boldsymbol{\epsilon}\} = [\boldsymbol{N}']\{\boldsymbol{u}_n\} \quad \text{and} \quad \{\delta \boldsymbol{\epsilon}\} = [\boldsymbol{N}']\{\delta \boldsymbol{u}_n\}, \tag{11.9}$$

where $[\boldsymbol{N}']$ is a matrix of the shape function derivatives such that

$$[\boldsymbol{N}'] = \begin{bmatrix} \frac{\partial N_1}{\partial x} & \frac{\partial N_2}{\partial x} & \frac{\partial N_3}{\partial x} & \frac{\partial N_4}{\partial x} & 0 & 0 & 0 & 0 \\ 0 & 0 & 0 & 0 & \frac{\partial N_1}{\partial y} & \frac{\partial N_2}{\partial y} & \frac{\partial N_3}{\partial y} & \frac{\partial N_4}{\partial y} \\ \frac{1}{2}\frac{\partial N_1}{\partial y} & \frac{1}{2}\frac{\partial N_2}{\partial y} & \frac{1}{2}\frac{\partial N_3}{\partial y} & \frac{1}{2}\frac{\partial N_4}{\partial y} & \frac{1}{2}\frac{\partial N_1}{\partial x} & \frac{1}{2}\frac{\partial N_2}{\partial x} & \frac{1}{2}\frac{\partial N_3}{\partial x} & \frac{1}{2}\frac{\partial N_4}{\partial x} \end{bmatrix}.$$

11.5.4 FEM Formulation

Substitution of the field variable approximations, Equation (11.9), into the statement of virtual work, Equation (11.8), gives

$$\sum_{i=1}^{n_e}\left(-\int_{\Omega_e}([\boldsymbol{N}']\{\delta\boldsymbol{u}_n\})^T\{\boldsymbol{\sigma}(\boldsymbol{C})\}\,dV+\int_{\Omega_e}([\boldsymbol{N}]\{\delta\boldsymbol{u}_n\})^T\{\boldsymbol{b}\}\,dV\right.$$
$$\left.+\int_{\partial\Omega_e}([\boldsymbol{N}]\{\delta\boldsymbol{u}_n\})^T\{\boldsymbol{\tau}\}\,dS\right)=0.$$

Noting that $([\boldsymbol{N}']\{\delta\boldsymbol{u}_n\})^T=\{\delta\boldsymbol{u}_n\}^T[\boldsymbol{N}']^T$, that the nodal virtual displacement vector, $\{\delta\boldsymbol{u}_n\}$, and the nodal displacement vector, $\{\boldsymbol{u}_n\}$, are constant over the entire element, and that we can combine the integral terms for each element, we can write

$$\sum_{i=1}^{n_e}\{\delta\boldsymbol{u}_n\}^T\left(-\left(\int_{\Omega_e}[\boldsymbol{N}']^T\{\boldsymbol{\sigma}(\boldsymbol{C})\}\,dV\right)+\int_{\Omega_e}[\boldsymbol{N}]^T\{\boldsymbol{b}\}\,dV+\int_{\partial\Omega_e}[\boldsymbol{N}]^T\{\boldsymbol{\tau}\}\,dS\right)=0.$$

Equivalently, we may write

$$\sum_{i=1}^{n_e}\{\delta\boldsymbol{u}_n\}^T\left(-[\boldsymbol{K}_{nl}]+\{\boldsymbol{F}_b\}+\{\boldsymbol{F}_\tau\}\right)=0,$$

where $[\boldsymbol{K}_{nl}]=\int_{\Omega_e}[\boldsymbol{N}']^T\{\boldsymbol{\sigma}(\boldsymbol{C})\}dV$ is the element stiffness matrix, $\{F_b\}=\int_{\Omega_e}[\boldsymbol{N}]^T\{\boldsymbol{b}\}\,dV$ is the body force vector, and $\{F_\tau\}=\int_{\partial\Omega_e}[\boldsymbol{N}]^T\{\boldsymbol{\tau}\}\,dS$ is the force vector due to surface tractions.

Since the statement of virtual work must be true for any virtual nodal displacements, $\{\delta\boldsymbol{u}_n\}$, we have the conditions that

$$[\boldsymbol{K}_{nl}]=\{\boldsymbol{F}_b\}+\{\boldsymbol{F}_\tau\}.$$

Unfortunately, for most complex models, this equation may not be solved directly but must instead be solved iteratively. This equation may be solved using the direct substitution method or the Newton-Raphson method, both of which are outlined next.

11.5.4.1 Direct Substitution Method

If we assume that we have an initial guess for the nodal displacements, $\boldsymbol{u}_n^i$, then, we can determine a better approximation, $\boldsymbol{u}_n^{i+1}$, by solving the following equation:

$$\left[\boldsymbol{K}_e^i\left(\boldsymbol{u}_n^i\right)\right]\left\{\boldsymbol{u}_n^{i+1}\right\}=\{\boldsymbol{F}_b\}+\{\boldsymbol{F}_\tau\}.$$

11.5.4.2 Newton-Raphson Method

The Newton-Raphson method is an iterative method that produces improved solutions based on previous iterations. Let us assume that we have an initial guess for the nodal displacements, u^i. Then one measure of the error associated with this initial guess is the residual error, $\{\boldsymbol{R}^i\}$, of iteration, i, given by

$$\left\{\boldsymbol{R}^i\right\}=\left[\boldsymbol{K}_e^i\right]\left\{\boldsymbol{u}^i\right\}-\left\{\boldsymbol{F}_b^i\right\}-\left\{\boldsymbol{F}_\tau^i\right\}.$$

Note that the exact solution would have a residual error of zero. However, we will obtain some approximation of the true solution, which has some nonzero residual. We will improve the solution by iterating until the residual becomes acceptably small.

If we perform a Taylor series expansion of the residual about an initial solution, $\boldsymbol{u}^i$, we obtain

$$\{\boldsymbol{R}\} = \left\{\boldsymbol{R}^i\right\} + \left[\frac{\partial \boldsymbol{R}}{\partial \boldsymbol{u}^i}\right]\left\{\Delta \boldsymbol{u}^{i+1}\right\} + O\left(\left\{\Delta \boldsymbol{u}^{i+1}\right\}^2\right),$$

where the difference between the solution in iteration $i+1$ and iteration i is $\Delta \boldsymbol{u}^{i+1}$ which is given by

$$\left\{\Delta \boldsymbol{u}^{i+1}\right\} \equiv \left(\left\{\boldsymbol{u}^{i+1}\right\} - \left\{\boldsymbol{u}^i\right\}\right).$$

Neglecting higher order terms, we obtain

$$\{\boldsymbol{R}\} = \left\{\boldsymbol{R}^i\right\} + \left[\frac{\partial \boldsymbol{R}}{\partial \boldsymbol{u}^i}\right]\left\{\Delta \boldsymbol{u}^{i+1}\right\}.$$

We are searching for an improved solution so we set the residual to zero and solve for the new displacements such that

$$\{\boldsymbol{R}\} = 0 \cong \left\{\boldsymbol{R}^i\right\} + \left[\frac{\partial \boldsymbol{R}}{\partial \boldsymbol{u}^i}\right]\left\{\Delta \boldsymbol{u}^{i+1}\right\}.$$

Rearranging the equation gives

$$\left\{\Delta \boldsymbol{u}^{i+1}\right\} = -\left[\frac{\partial \boldsymbol{R}}{\partial \boldsymbol{u}^i}\right]^{-1}\left\{\boldsymbol{R}^i\right\},$$

or

$$\left\{\boldsymbol{u}^{i+1}\right\} = \left\{\boldsymbol{u}^i\right\} - \left[\frac{\partial \boldsymbol{R}}{\partial \boldsymbol{u}^i}\right]^{-1}\left\{\boldsymbol{R}^i\right\}.$$

The tangent stiffness matrix, $[\boldsymbol{K}_t]$, is defined as

$$\left[\boldsymbol{K}_T^{i+1}\right] = \left[\frac{\partial \boldsymbol{R}}{\partial \boldsymbol{u}^i}\right].$$

Therefore, we obtain an equation for the updated solution

$$\left\{\boldsymbol{u}^{i+1}\right\} = \left\{\boldsymbol{u}^i\right\} - \left[\boldsymbol{K}_T^{i+1}\right]^{-1}\left\{\boldsymbol{R}^i\right\}.$$

11.5.4.3 Convergence

The convergence of any nonlinear scheme may be gauged by determining the quantity

$$\frac{\left\|\Delta \boldsymbol{u}^{i+1}\right\|}{\left\|\boldsymbol{u}^i\right\|} \leq \epsilon.$$

As the change in displacement values decreases below some threshold value, ϵ, the algorithm is terminated, and an approximation of the solution is obtained.

11.5.4.4 Tangent Stiffness Matrix for a Hyperelastic Material

The tangent stiffness matrix is defined as

$$\left[\boldsymbol{K}_T^{i+1}\right] = \left[\frac{\partial \boldsymbol{R}}{\partial \boldsymbol{u}^i}\right] = \int_{\Omega_e} [\boldsymbol{N}']^T \left[\mathbb{C}_T^{(i)}\right] [\boldsymbol{N}'] \, dV.$$

The tangent modulus is defined as

$$\mathbb{C}_T^{(i)} = \frac{\partial \sigma \left(\epsilon^{(i)}\right)}{\partial \epsilon}.$$

Recall that for a hyperelastic material, we may write the stress in terms of the strain energy density such that the second Piola-Kirchhoff stress is given by

$$\boldsymbol{P}(\epsilon) = \frac{\partial W}{\partial \epsilon}.$$

To first approximation, we have

$$\mathbb{C}_T^{(i)} = \frac{\partial^2 W}{\partial \epsilon \, \partial \epsilon} = 4 \frac{\partial^2 W}{\partial \boldsymbol{C} \, \partial \boldsymbol{C}}.$$

For an isotropic hyperelastic material, we may write the strain energy density in terms of the invariants of the right Cauchy deformation gradient such that

$$W = W\left(I_C, II_C, III_C\right).$$

Therefore, for this special case, we may write

$$\mathbb{C}_T^{(i)} = 4 \frac{\partial^2 \boldsymbol{W}}{\partial \boldsymbol{C} \, \partial \boldsymbol{C}} = 2 \frac{\partial \boldsymbol{\sigma}}{\partial \boldsymbol{C}} = 2 \frac{\partial \boldsymbol{\sigma}}{\partial \boldsymbol{C}} = 2 \frac{\partial}{\partial \boldsymbol{C}} \left(2 \left(\frac{\partial W}{\partial I_C} \frac{\partial I_C}{\partial \boldsymbol{C}} + \frac{\partial W}{\partial II_C} \frac{\partial II_C}{\partial \boldsymbol{C}} + \frac{\partial W}{\partial III_C} \frac{\partial III_C}{\partial \boldsymbol{C}} \right) \right).$$

Expanding using the chain rule gives

$$\begin{aligned}
\mathbb{C}_T^{(i)} = 4 \Bigg[& \left(\frac{\partial^2 W}{\partial I_C^2} \frac{\partial I_C}{\partial \boldsymbol{C}} + \frac{\partial^2 W}{\partial I_C \partial II_C} \frac{\partial II_C}{\partial \boldsymbol{C}} + \frac{\partial^2 W}{\partial I_C \partial III_C} \frac{\partial III_C}{\partial \boldsymbol{C}} \right) \frac{\partial I_C}{\partial \boldsymbol{C}} \\
& + \left(\frac{\partial^2 W}{\partial II_C \partial I_C} \frac{\partial I_C}{\partial \boldsymbol{C}} + \frac{\partial^2 W}{\partial II_C^2} \frac{\partial II_C}{\partial \boldsymbol{C}} + \frac{\partial^2 W}{\partial II_C \partial III_C} \frac{\partial III_C}{\partial \boldsymbol{C}} \right) \frac{\partial II_C}{\partial \boldsymbol{C}} \\
& + \left(\frac{\partial^2 W}{\partial III_C \partial I_C} \frac{\partial I_C}{\partial \boldsymbol{C}} + \frac{\partial^2 W}{\partial III_C \partial II_C} \frac{\partial II_C}{\partial \boldsymbol{C}} + \frac{\partial^2 W}{\partial III_C^2} \frac{\partial III_C}{\partial \boldsymbol{C}} \right) \frac{\partial III_C}{\partial \boldsymbol{C}} \\
& + \frac{\partial W}{\partial I_C} \frac{\partial^2 I_C}{\partial \boldsymbol{C}^2} + \frac{\partial W}{\partial II_C} \frac{\partial^2 II_C}{\partial \boldsymbol{C}^2} + \frac{\partial W}{\partial III_C} \frac{\partial^2 III_C}{\partial \boldsymbol{C}^2} \Bigg].
\end{aligned}$$

The second derivatives of the invariants are given by

$$\frac{\partial^2 I_C}{\partial C_{nm} \partial C_{pq}} = \frac{\partial \left(\delta_{pq}\right)}{\partial C_{nm}} = 0$$

and

$$\begin{aligned}
\frac{\partial^2 II_C}{\partial C_{nm} \partial C_{pq}} &= \frac{\partial \left(C_{ii}\delta_{pq} - C_{qp}\right)}{\partial C_{nm}} \\
&= \delta_{in}\delta_{im}\delta_{pq} - \delta_{qn}\delta_{pm} \\
&= \delta_{nm}\delta_{pq} - \delta_{qn}\delta_{pm}
\end{aligned}$$

and

$$\frac{\partial^2 III_C}{\partial C_{nm}\partial C_{pq}} = \frac{\partial\left(C_{qi}C_{ip} - I_C C_{qp} + II_C\,\delta_{pq}\right)}{\partial C_{nm}}$$
$$= \delta_{qn}\delta_{im}C_{ip} + C_{qi}\delta_{in}\delta_{pm} - I_C\delta_{qn}\delta_{pm} - \delta_{nm}C_{qp} + \left(C_{ii}\delta_{nm} - C_{mn}\right)\delta_{pq}$$
$$= \delta_{qn}C_{mp} + C_{qn}\delta_{pm} - I_C\delta_{qn}\delta_{pm} - \delta_{nm}C_{qp} + C_{ii}\delta_{nm}\delta_{pq} - C_{mn}\delta_{pq}.$$

Substitution into the equation for the tangent modulus gives

$$\left[\mathbb{C}_T^{(i)}\right]_{ijkl} = 4\begin{bmatrix}\frac{\partial I_C}{\partial C_{ij}} & \frac{\partial II_C}{\partial C_{ij}} & \frac{\partial III_C}{\partial C_{ij}}\end{bmatrix}\begin{bmatrix}\frac{\partial^2 W}{\partial I_C^2} & \frac{\partial^2 W}{\partial I_C\partial II_C} & \frac{\partial^2 W}{\partial I_C\partial III_C}\\ \frac{\partial^2 W}{\partial I_C\partial II_C} & \frac{\partial^2 W}{\partial II_C^2} & \frac{\partial^2 W}{\partial II_C\partial III_C}\\ \frac{\partial^2 W}{\partial I_C\partial III_C} & \frac{\partial^2 W}{\partial II_C\partial III_C} & \frac{\partial^2 W}{\partial III_C^2}\end{bmatrix}\begin{bmatrix}\frac{\partial I_C}{\partial C_{kl}}\\ \frac{\partial II_C}{\partial C_{kl}}\\ \frac{\partial III_C}{\partial C_{kl}}\end{bmatrix}$$
$$+4\begin{bmatrix}\frac{\partial W}{\partial I_C} & \frac{\partial W}{\partial II_C} & \frac{\partial W}{\partial III_C}\end{bmatrix}\begin{bmatrix}\frac{\partial^2 I_C}{\partial C_{ij}\partial C_{kl}}\\ \frac{\partial^2 II_C}{\partial C_{ij}\partial C_{kl}}\\ \frac{\partial^2 III_C}{\partial C_{ij}\partial C_{kl}}\end{bmatrix}.$$

11.6 Plane Strain Finite Deformation

In this section, we will derive the finite element formulation for a problem that has both material nonlinearity and geometric nonlinearity. We begin by reformulating the virtual work statement in the reference configuration. Recall that the Green strain tensor, $\boldsymbol{E}$, is given by

$$\boldsymbol{E} = \frac{1}{2}\left(\boldsymbol{C} - \boldsymbol{I}\right).$$

The statement of virtual work written in terms of the reference configuration is

$$\delta W = \int_{\Omega_0} \delta\boldsymbol{u}\cdot\left(\rho_o\,\dot{\boldsymbol{v}}_o - \nabla_X\cdot\boldsymbol{P} - \rho_o\boldsymbol{b_o}\right)dV_o.$$

Integrating this expression by parts and relating the first Piola-Kirchhoff stress tensor to the second Piola-Kirchhoff stress tensor, $\boldsymbol{P} = \boldsymbol{F}\cdot\boldsymbol{S}$, gives

$$\delta W = \int_{\Omega(t)} \delta\boldsymbol{u}\cdot\rho_o\dot{\boldsymbol{v}}_{\boldsymbol{o}}\,dV + \int_{\Omega_0}\frac{\partial\delta\boldsymbol{u}}{\partial\boldsymbol{X}} : \left(\boldsymbol{F}\cdot\boldsymbol{S}\right)dV_o - \int_{\partial\Omega(t)}\delta\boldsymbol{u}\cdot\boldsymbol{t}_X\,dS_o - \int_{\Omega_0}\delta\boldsymbol{u}\cdot\rho_o\boldsymbol{b}_o\,dV_o.$$

Note that the derivative of the virtual displacements with respect to the reference coordinates is equal to the increment in the deformation gradient such that

$$\frac{\partial\delta\boldsymbol{u}}{\partial\boldsymbol{X}} = \delta\boldsymbol{F}.$$

In addition, since the Green strain tensor can be written as $\boldsymbol{E} = \frac{1}{2}\left(\boldsymbol{F}^T\cdot F - \boldsymbol{I}\right)$, the variation in the Green strain tensor can be written as

$$\delta\boldsymbol{E} = \frac{1}{2}\left(\delta\boldsymbol{F}^T\cdot\boldsymbol{F} + \boldsymbol{F}^T\cdot\delta\boldsymbol{F}\right).$$

Therefore, we can write

$$\frac{\partial \delta \boldsymbol{u}}{\partial \boldsymbol{X}} : (\boldsymbol{F} \cdot \boldsymbol{S}) = \boldsymbol{F}^T \cdot \delta \boldsymbol{F} : \boldsymbol{S}.$$

Due to symmetry of $\boldsymbol{S}$, we can write

$$\boldsymbol{F}^T \cdot \delta \boldsymbol{F} : \boldsymbol{S} = \frac{1}{2}\left(\boldsymbol{F}^T \cdot \delta \boldsymbol{F} + \delta \boldsymbol{F}^T \cdot \boldsymbol{F}\right) : \boldsymbol{S} = \delta \boldsymbol{E} : \boldsymbol{S}.$$

The virtual work statement becomes

$$\delta W = \int_{\Omega(t)} \delta \boldsymbol{u} \cdot \rho_o \dot{\boldsymbol{v}}_o \, dV + \int_{\Omega_o} \delta \boldsymbol{E} : \boldsymbol{S} \, dV_o - \int_{\partial\Omega(t)} \delta \boldsymbol{u} \cdot t_X \, dS_o - \int_{\Omega_o} \delta \boldsymbol{u} \cdot \rho_o \boldsymbol{b}_o \, dV_o.$$

11.6.1 Total Lagrangian Method

The equations for the total Lagrangian method are formulated in terms of the reference configuration. The reference positions are interpolated within the element such that

$$\begin{Bmatrix} X(r,s) \\ Y(r,s) \end{Bmatrix} = \begin{bmatrix} N_1 & N_2 & N_3 & N_4 & 0 & 0 & 0 & 0 \\ 0 & 0 & 0 & 0 & N_1 & N_2 & N_3 & N_4 \end{bmatrix} \begin{Bmatrix} X_1 \\ X_2 \\ X_3 \\ X_4 \\ Y_1 \\ Y_2 \\ Y_3 \\ Y_4 \end{Bmatrix}.$$

The displacements, $u(r,s)$ and $v(r,s)$, are given by

$$\begin{Bmatrix} u(r,s) \\ v(r,s) \end{Bmatrix} = \begin{bmatrix} N_1 & N_2 & N_3 & N_4 & 0 & 0 & 0 & 0 \\ 0 & 0 & 0 & 0 & N_1 & N_2 & N_3 & N_4 \end{bmatrix} \begin{Bmatrix} u_1 \\ u_2 \\ u_3 \\ u_4 \\ v_1 \\ v_2 \\ v_3 \\ v_4 \end{Bmatrix}.$$

The incremental displacement is given by

$$\{\delta \boldsymbol{u}\} = [\boldsymbol{N}]\{\delta \boldsymbol{u}_n\}.$$

Taking the gradient of the displacements, we obtain

$$\left\{\frac{\partial \boldsymbol{u}}{\partial \boldsymbol{X}}\right\} = \left[\frac{\partial \boldsymbol{N}}{\partial \boldsymbol{X}}\right]\{\boldsymbol{u}_n\}.$$

The incremental strains can then be written as

$$\delta \boldsymbol{E} = \frac{1}{2}\left(\left(\frac{\partial \delta \boldsymbol{u}}{\partial \boldsymbol{X}}\right)^T \cdot \boldsymbol{F} + \boldsymbol{F}^T \cdot \left(\frac{\partial \delta \boldsymbol{u}}{\partial \boldsymbol{X}}\right)\right).$$

Expanding this equation gives

$$\{\delta \boldsymbol{E}\} = \begin{bmatrix} \alpha_{111}^{(1)} & \alpha_{112}^{(1)} & \alpha_{113}^{(1)} & \alpha_{114}^{(1)} & \alpha_{211}^{(1)} \\ \alpha_{211}^{(2)} & \alpha_{212}^{(2)} & \alpha_{213}^{(2)} & \alpha_{214}^{(2)} & \alpha_{221}^{(2)} \\ \left(\alpha_{121}^{(1)} + \alpha_{111}^{(2)}\right) & \left(\alpha_{122}^{(1)} + \alpha_{112}^{(2)}\right) & \left(\alpha_{123}^{(1)} + \alpha_{113}^{(2)}\right) & \left(\alpha_{124}^{(1)} + \alpha_{114}^{(2)}\right) & \left(\alpha_{221}^{(1)} + \alpha_{211}^{(2)}\right) \end{bmatrix.$$

$$\times \begin{matrix} \alpha_{212}^{(1)} & \alpha_{213}^{(1)} & \alpha_{214}^{(1)} \\ \alpha_{222}^{(2)} & \alpha_{223}^{(2)} & \alpha_{224}^{(2)} \\ \left(\alpha_{222}^{(1)} + \alpha_{212}^{(2)}\right) & \left(\alpha_{223}^{(1)} + \alpha_{213}^{(2)}\right) & \left(\alpha_{224}^{(1)} + \alpha_{214}^{(2)}\right) \end{bmatrix} \begin{Bmatrix} \delta u_1 \\ \delta u_2 \\ \delta u_3 \\ \delta u_4 \\ \delta v_1 \\ \delta v_2 \\ \delta v_3 \\ \delta v_4 \end{Bmatrix},$$

where

$$\alpha_{ijk}^{(1)} = F_{ij} \frac{\partial N_k}{\partial X} \quad \text{and} \quad \alpha_{ijk}^{(2)} = F_{ij} \frac{\partial N_k}{\partial Y}.$$

Equivalently, we may write this as

$$\{\delta \boldsymbol{E}\} = [\boldsymbol{N}'] \{\delta \boldsymbol{u}_n\}$$

Where

$$[\boldsymbol{N}'] = \begin{bmatrix} \alpha_{111}^{(1)} & \alpha_{112}^{(1)} & \alpha_{113}^{(1)} & \alpha_{114}^{(1)} & \alpha_{211}^{(1)} \\ \alpha_{211}^{(2)} & \alpha_{212}^{(2)} & \alpha_{213}^{(2)} & \alpha_{214}^{(2)} & \alpha_{221}^{(2)} \\ \left(\alpha_{121}^{(1)} + \alpha_{111}^{(2)}\right) & \left(\alpha_{122}^{(1)} + \alpha_{112}^{(2)}\right) & \left(\alpha_{123}^{(1)} + \alpha_{113}^{(2)}\right) & \left(\alpha_{124}^{(1)} + \alpha_{114}^{(2)}\right) & \left(\alpha_{221}^{(1)} + \alpha_{211}^{(2)}\right) \end{matrix}$$

$$\times \begin{matrix} \alpha_{212}^{(1)} & \alpha_{213}^{(1)} & \alpha_{214}^{(1)} \\ \alpha_{222}^{(2)} & \alpha_{223}^{(2)} & \alpha_{224}^{(2)} \\ \left(\alpha_{222}^{(1)} + \alpha_{212}^{(2)}\right) & \left(\alpha_{223}^{(1)} + \alpha_{213}^{(2)}\right) & \left(\alpha_{224}^{(1)} + \alpha_{214}^{(2)}\right) \end{bmatrix}.$$

The virtual work statement becomes

$$\sum_{n_e} \{\delta \boldsymbol{u}_n\}^T \left(\int_{\Omega(t)} \rho_o [\boldsymbol{N}]^T \{\dot{\boldsymbol{v}}_o\}\, dV + \int_{\Omega_o} [\boldsymbol{N}']^T [\boldsymbol{S}]\, dV_o - \int_{\partial\Omega(t)} [\boldsymbol{N}]^T \{\boldsymbol{t}_X\}\, dS_o \right.$$

$$\left. - \int_{\Omega_o} \rho_o [\boldsymbol{N}]^T \{\boldsymbol{b}_o\}\, dV_o \right) = 0.$$

Assuming static equilibrium, we may write

$$\sum_{n_e} \{\delta \boldsymbol{u}_n\}^T \left(\int_{\Omega_o} [\boldsymbol{N}']^T [\boldsymbol{S}]\, dV_o - \int_{\partial\Omega(t)} [\boldsymbol{N}]^T \{\boldsymbol{t}_X\}\, dS_o - \int_{\Omega_o} \rho_o [\boldsymbol{N}]^T \{\boldsymbol{b}_o\}\, dV_o \right) = 0.$$

Using a Newton-Raphson method to solve these equations, we can write the residual error in iteration $\boldsymbol{R}^i$ as

$$\left\{\boldsymbol{R}^i\right\} = \int_{\Omega_o} [\boldsymbol{N}']^T [\boldsymbol{S}]\, dV_o - \int_{\partial\Omega(t)} [\boldsymbol{N}]^T \{\boldsymbol{t}_X\}\, dS_o - \int_{\Omega_o} \rho_o [\boldsymbol{N}]^T \{\boldsymbol{b}_o\}\, dV_o.$$

If we define the difference between two solutions as

$$\left\{\Delta u^{i+1}\right\} \equiv \left(\left\{u^{i+1}\right\} - \left\{u^{i}\right\}\right).$$

If we perform a Taylor series expansion about the solution, we obtain

$$\{R\} = \left\{R^{i}\right\} + \left[\frac{\partial R}{\partial u^{i}}\right]\left\{\Delta u^{i+1}\right\} + O\left(\left\{\Delta u^{i+1}\right\}^{2}\right).$$

If we neglect higher order terms and set the residual to zero, we have

$$\{R\} = 0 \cong \left\{R^{i}\right\} + \left[\frac{\partial R}{\partial u^{i}}\right]\left\{\Delta u^{i+1}\right\}.$$

The tangent stiffness matrix, $\left[\tilde{K}_t\right]$, is defined as

$$\begin{aligned}
\left[\tilde{K}_T^{i+1}\right] &= \left[\frac{\partial R}{\partial u^{i}}\right] \\
&= \frac{\partial}{\partial u^{i}}\left(\int_{\Omega_0} [N']^T\,[S]\,dV_o\right) \\
&= \int_{\Omega_0} \left[\frac{\partial N'}{\partial u^{i}}\right]^T [S]\,dV_o + \int_{\Omega_0} [N']^T \left[\frac{\partial S}{\partial u^{i}}\right] dV_o \\
&= K_G^{i} + K_T^{i},
\end{aligned}$$

where K_G^i is the geometric stiffness matrix, and K_T^i is the tangent stiffness matrix.

Since $\left[\frac{\partial S}{\partial u^i}\right]\{du\} = \left[\frac{\partial S}{\partial E^i}\right] dE = 2\left[\frac{\partial S}{\partial C^i}\right][N']\{du\}$, we can write the tangent stiffness matrix as

$$\begin{aligned}
K_T^{i} &= \int_{\Omega_0} [N']^T \left[\frac{\partial S}{\partial u^{i}}\right] dV_o \\
&= \int_{\Omega_0} [N']^T \left[\frac{\partial S}{\partial E^{i}}\right][N']\,dV_o \\
&= \int_{\Omega_0} 2[N']^T \left[\frac{\partial S}{\partial C^{i}}\right][N']\,dV_o.
\end{aligned}$$

The geometric part, K_G^i, is given by

$$K_G^{i} = \int_{\Omega_0} \left[\frac{\partial N'}{\partial u^{i}}\right]^T [S]\,dV_o.$$

Therefore, we can determine the improved solution by using the tangent stiffness matrix such that

$$\left\{\Delta u^{i+1}\right\} = -\left[\tilde{K}_T^{i+1}\right]^{-1}\left\{R^{i}\right\}.$$

Since $\{\Delta u^{i+1}\} = \{u^{i+1}\} - \{u^i\}$, we may write

$$\left\{u^{i+1}\right\} = \left\{u^{i}\right\} - \left[\tilde{K}_T^{i+1}\right]^{-1}\left\{R^{i}\right\}.$$

11.6.2 Updated Lagrangian Method

In contrast to the total Lagrangian method, the equations in the updated Lagrangian method are formulated in terms of the current configuration. Recall that the current nodal coordinates can be found in terms of the reference positions, $\{X_n\}$, and displacements, $\{u_n\}$, such that

$$\{x_n\} = \{X_n\} + \{u_n\}.$$

For the updated Lagrangian method, we will use shape functions to interpolate current coordinates such that

$$x = \left[\bar{N}\right]\{x_n\}.$$

Transforming the increment in the deformation gradient to current coordinates gives

$$\delta\boldsymbol{F} = \frac{\partial \delta \boldsymbol{u}}{\partial \boldsymbol{X}} = \frac{\partial \delta \boldsymbol{u}}{\partial \boldsymbol{x}} \cdot \frac{\partial \boldsymbol{x}}{\partial \boldsymbol{X}} = \frac{\partial \delta \boldsymbol{u}}{\partial \boldsymbol{x}} \cdot \boldsymbol{F}.$$

Therefore, we have

$$\begin{aligned}
\delta\boldsymbol{E} &= \delta\left(\frac{1}{2}\left(\boldsymbol{F}^T \cdot \boldsymbol{F} - \boldsymbol{I}\right)\right) \\
&= \left(\frac{1}{2}\left(\delta\boldsymbol{F}^T \cdot \boldsymbol{F} + \boldsymbol{F}^T \cdot \delta\boldsymbol{F}\right)\right) \\
&= \frac{1}{2}\left(\boldsymbol{F}^T \cdot \left(\frac{\partial \delta \boldsymbol{u}}{\partial \boldsymbol{x}}\right)^{\boldsymbol{T}} \cdot \boldsymbol{F} + \boldsymbol{F}^T \cdot \left(\frac{\partial \delta \boldsymbol{u}}{\partial \boldsymbol{x}}\right) \cdot \boldsymbol{F}\right) \\
&= \boldsymbol{F}^T \cdot \delta\epsilon \cdot \boldsymbol{F}.
\end{aligned}$$

Writing the work term in terms of the Cauchy stress gives

$$\delta\boldsymbol{E} : \boldsymbol{S} = \boldsymbol{F}^T \cdot \delta\epsilon \cdot \boldsymbol{F} : J\,\boldsymbol{F}^{-1} \cdot \boldsymbol{\sigma} \cdot \boldsymbol{F}^{-T} = J\,\delta\epsilon : \boldsymbol{\sigma}.$$

Also, since the tangent modulus may be written as $\mathbb{C}_T = \frac{\partial \boldsymbol{S}}{\partial \boldsymbol{E}} = \frac{\partial^2 W}{\partial \boldsymbol{E}^2}$ we may write

$$\delta\boldsymbol{E} : \mathbb{C}_T : d\boldsymbol{E} = \boldsymbol{F}^T \cdot \delta\epsilon \cdot \boldsymbol{F} : \mathbb{C}_T : \boldsymbol{F}^T \cdot \delta\epsilon \cdot \boldsymbol{F} = \delta\epsilon : \bar{\mathbb{C}_T} : \delta\epsilon,$$

where

$$F_{ji}F_{kl}\left(\mathbb{C}_T\right)_{ilmn}F_{om}F_{pn} = \left(\bar{\mathbb{C}}_T\right)_{jkop}.$$

Therefore, the tangent stiffness matrix may be written as

$$\boldsymbol{K}_T^i = \int_{\Omega_0} [\boldsymbol{N}']^T\,[\mathbb{C}_T][\boldsymbol{N}']\,dV_o = \int_{\Omega} \left[\bar{N}'\right]^T \left[\bar{\mathbb{C}_T}\right]\left[\bar{N}'\right] dV.$$

Similarly, the geometric part may be written as

$$\boldsymbol{K}_G^i = \int_{\Omega_0} \left[\frac{\partial \boldsymbol{N}'}{\partial \boldsymbol{u}^i}\right]^T [\boldsymbol{S}]\,dV_o = \int_{\Omega} \left[\frac{\partial \bar{N}'}{\partial \boldsymbol{u}^i}\right]^T [\boldsymbol{S}]\,dV.$$

11.6.3 Updated Lagrangian Method Single Element Implementation

11.6.3.1 Flow Chart

1. Set up nodal positions and nodal point loads

$$[\text{nPosXY}] = \begin{bmatrix} x_1 & y_1 \\ x_2 & y_2 \\ x_3 & y_3 \\ x_4 & y_4 \end{bmatrix} \quad [\text{nForces}] = \begin{bmatrix} fx_1 & fy_1 \\ fx_2 & fy_2 \\ fx_3 & fy_3 \\ fx_4 & fy_4 \end{bmatrix}$$

↓

2. Initialize material properties

$$\alpha_1 = 10^6$$
$$\alpha_2 = 20^6$$

↓

3. Set up Gaussian integration points: We use two integration points along the r axis and two along the s axis giving a total of four integration points per element

Integration point, i	r_i	s_i	$\tilde{W}_i$
1	$-1/\sqrt{3}$	$-1/\sqrt{3}$	1
2	$1/\sqrt{3}$	$-1/\sqrt{3}$	1
3	$1/\sqrt{3}$	$1/\sqrt{3}$	1
4	$-1/\sqrt{3}$	$1/\sqrt{3}$	1

↓

4. Assemble the shape functions and derivatives at each integration point

$$N_1^{(J)} = \frac{1}{4}(1-r_j)(1-s_j), \quad \frac{\partial N_1^{(J)}}{\partial r} = \frac{1}{4}(-1)(1-s_j), \quad \frac{\partial N_1^{(J)}}{\partial r} = \frac{1}{4}(1-r_j)(-1)$$

$$N_2^{(J)} = \frac{1}{4}(1+r_j)(1-s_j), \quad \frac{\partial N_2^{(J)}}{\partial r} = \frac{1}{4}(+1)(1-s_j), \quad \frac{\partial N_2^{(J)}}{\partial r} = \frac{1}{4}(1+r_j)(-1)$$

$$N_3^{(J)} = \frac{1}{4}(1+r_j)(1+s_j), \quad \frac{\partial N_3^{(J)}}{\partial r} = \frac{1}{4}(+1)(1+s_j), \quad \frac{\partial N_3^{(J)}}{\partial r} = \frac{1}{4}(1+r_j)(+1)$$

$$N_4^{(J)} = \frac{1}{4}(1-r_j)(1+s_j), \quad \frac{\partial N_4^{(J)}}{\partial r} = \frac{1}{4}(-1)(1+s_j), \quad \frac{\partial N_4^{(J)}}{\partial r} = \frac{1}{4}(1-r_j)(+1)$$

↓

5. Assemble the nodal force vector. Apply load in increments.

$$[force] = \frac{loadIncrement}{nrIncrements} * [nForces]$$

↓

6. Find the Jacobean

$$[J^{(J)}] = \begin{bmatrix} \frac{\partial N_1^{(J)}}{\partial r} & \frac{\partial N_2^{(J)}}{\partial r} & \frac{\partial N_3^{(J)}}{\partial r} & \frac{\partial N_4^{(J)}}{\partial r} \\ \frac{\partial N_1^{(J)}}{\partial s} & \frac{\partial N_2^{(J)}}{\partial s} & \frac{\partial N_3^{(J)}}{\partial s} & \frac{\partial N_4^{(J)}}{\partial s} \end{bmatrix} \begin{bmatrix} x_1 & y_1 \\ x_2 & y_2 \\ x_3 & y_3 \\ x_4 & y_4 \end{bmatrix}$$

7. And its inverse

$$[\boldsymbol{invJacob}] = [J]^{-1}$$

8. Invert the Jacobean to find $\frac{\partial N}{\partial x}$ and $\frac{\partial N}{\partial y}$

$$\begin{bmatrix} \frac{\partial N_1^{(J)}}{\partial x} & \frac{\partial N_2^{(J)}}{\partial x} & \frac{\partial N_3^{(J)}}{\partial x} & \frac{\partial N_4^{(J)}}{\partial x} \\ \frac{\partial N_1^{(J)}}{\partial y} & \frac{\partial N_2^{(J)}}{\partial y} & \frac{\partial N_3^{(J)}}{\partial y} & \frac{\partial N_4^{(J)}}{\partial y} \end{bmatrix} = [J]^{-1} \begin{bmatrix} \frac{\partial N_1^{(J)}}{\partial r} & \frac{\partial N_2^{(J)}}{\partial r} & \frac{\partial N_3^{(J)}}{\partial r} & \frac{\partial N_4^{(J)}}{\partial r} \\ \frac{\partial N_1^{(J)}}{\partial s} & \frac{\partial N_2^{(J)}}{\partial s} & \frac{\partial N_3^{(J)}}{\partial s} & \frac{\partial N_4^{(J)}}{\partial s} \end{bmatrix}$$

↓

9. Assemble $\boldsymbol{N'}$

$$[\boldsymbol{N}'^{(J)}] = \begin{bmatrix} \frac{\partial N_1^{(J)}}{\partial x} & \frac{\partial N_2^{(J)}}{\partial x} & \frac{\partial N_3^{(J)}}{\partial x} & \frac{\partial N_4^{(J)}}{\partial x} & 0 & 0 & 0 & 0 \\ 0 & 0 & 0 & 0 & \frac{\partial N_1^{(J)}}{\partial y} & \frac{\partial N_2^{(J)}}{\partial y} & \frac{\partial N_3^{(J)}}{\partial y} & \frac{\partial N_4^{(J)}}{\partial y} \\ \frac{1}{2}\frac{\partial N_1^{(J)}}{\partial y} & \frac{1}{2}\frac{\partial N_2^{(J)}}{\partial y} & \frac{1}{2}\frac{\partial N_3^{(J)}}{\partial y} & \frac{1}{2}\frac{\partial N_4^{(J)}}{\partial y} & \frac{1}{2}\frac{\partial N_1^{(J)}}{\partial x} & \frac{1}{2}\frac{\partial N_2^{(J)}}{\partial x} & \frac{1}{2}\frac{\partial N_3^{(J)}}{\partial x} & \frac{1}{2}\frac{\partial N_4^{(J)}}{\partial x} \end{bmatrix}$$

↓

10. Assemble the nonlinear $\boldsymbol{N'}$

$$[\boldsymbol{N}'^{(J)}] = \begin{bmatrix} \frac{\partial N_1^{(J)}}{\partial x} & \frac{\partial N_2^{(J)}}{\partial x} & \frac{\partial N_3^{(J)}}{\partial x} & \frac{\partial N_4^{(J)}}{\partial x} & 0 & 0 & 0 & 0 \\ \frac{\partial N_1^{(J)}}{\partial y} & \frac{\partial N_2^{(J)}}{\partial y} & \frac{\partial N_3^{(J)}}{\partial y} & \frac{\partial N_4^{(J)}}{\partial y} & 0 & 0 & 0 & 0 \\ 0 & 0 & 0 & 0 & \frac{\partial N_1^{(J)}}{\partial x} & \frac{\partial N_2^{(J)}}{\partial x} & \frac{\partial N_3^{(J)}}{\partial x} & \frac{\partial N_4^{(J)}}{\partial x} \\ 0 & 0 & 0 & 0 & \frac{\partial N_1^{(J)}}{\partial y} & \frac{\partial N_2^{(J)}}{\partial y} & \frac{\partial N_3^{(J)}}{\partial y} & \frac{\partial N_4^{(J)}}{\partial y} \end{bmatrix}$$

↓

11. Compute the deformation gradient

$$[\boldsymbol{F}] = \begin{bmatrix} 1 + \sum_{i=1}^{4} \frac{\partial N_i^{(J)}}{\partial x} u_i & \sum_{i=1}^{4} \frac{\partial N_i^{(J)}}{\partial y} u_i & 0 \\ \sum_{i=1}^{4} \frac{\partial N_i^{(J)}}{\partial x} v_i & 1 + \sum_{i=1}^{4} \frac{\partial N_i^{(J)}}{\partial y} v_i & 0 \\ 0 & 0 & 1 \end{bmatrix} \quad \text{and} \quad \boldsymbol{C} = \boldsymbol{F}^T \cdot \boldsymbol{F}$$

↓

12. Find the invariants of C and their derivatives

$$I_C = tr\boldsymbol{C}, \qquad II_C = \frac{1}{2}\left(tr(\boldsymbol{C})^2 - tr(\boldsymbol{C}^2)\right), \qquad \text{and } III_C = \det \boldsymbol{C}$$

$$\frac{\partial I_C}{\partial \boldsymbol{C}} = \boldsymbol{I}, \qquad \frac{\partial II_C}{\partial \boldsymbol{C}} = \frac{\partial \left\{\frac{1}{2}\left(tr(\boldsymbol{C})^2 - tr(\boldsymbol{C}^2)\right)\right\}}{\partial \boldsymbol{C}}, \quad \text{and } \frac{\partial III_C}{\partial \boldsymbol{C}} = III_C\, \boldsymbol{C}^{-1}$$

$$\frac{\partial^2 II_C}{\partial C_{nm}\, \partial C_{pq}} = \delta_{nm}\delta_{pq} - \delta_{qn}\delta_{pm} \qquad \text{and} \quad \frac{\partial^2 III_C}{\partial C_{nm} \partial C_{pq}} = \delta_{qn} C_{mp} + C_{qn}\delta_{pm} - I_C \delta_{qn}\delta_{pm} - \delta_{nm} C_{qp} + C_{ii}\delta_{nm}\delta_{pq} - C_{mn}\delta_{pq}$$

↓

13. Compute the strain energy function and its derivatives

$$W = \alpha_1(I_C - 3 - \log(III_C)) + \alpha_2(III_C - 1)^2$$

14. Assemble the tangent stiffness

$$[\mathbb{C}_T]_{ijkl} = 4\begin{bmatrix}\frac{\partial I_C}{\partial C_{ij}} & \frac{\partial II_C}{\partial C_{ij}} & \frac{\partial III_C}{\partial C_{ij}}\end{bmatrix}\begin{bmatrix}\frac{\partial^2 W}{\partial I_C^2} & \frac{\partial^2 W}{\partial I_C \partial II_C} & \frac{\partial^2 W}{\partial I_C \partial III_C} \\ \frac{\partial^2 W}{\partial I_C \partial II_C} & \frac{\partial^2 W}{\partial II_C^2} & \frac{\partial^2 W}{\partial II_C \partial III_C} \\ \frac{\partial^2 W}{\partial I_C \partial III_C} & \frac{\partial^2 W}{\partial II_C \partial III_C} & \frac{\partial^2 W}{\partial III_C^2}\end{bmatrix}\begin{bmatrix}\frac{\partial I_C}{\partial C_{kl}} \\ \frac{\partial II_C}{\partial C_{kl}} \\ \frac{\partial III_C}{\partial C_{kl}}\end{bmatrix} + 4\begin{bmatrix}\frac{\partial W}{\partial I_C} & \frac{\partial W}{\partial II_C} & \frac{\partial W}{\partial III_C}\end{bmatrix}\begin{bmatrix}\frac{\partial^2 I_C}{\partial C_{ij}\partial C_{kl}} \\ \frac{\partial^2 II_C}{\partial C_{ij}\partial C_{kl}} \\ \frac{\partial^2 III_C}{\partial C_{ij}\partial C_{kl}}\end{bmatrix}.$$

15. Push forward the tangent stiffness

$$\bar{\mathbb{C}}_{jkop} = F_{ji}F_{kl}\mathbb{C}_{ilmn}F_{om}F_{pn}$$

16. Convert to Voigt notation

$$[\boldsymbol{Dt}] = \begin{bmatrix}\bar{\mathbb{C}}_{1111} & \bar{\mathbb{C}}_{1122} & 0 \\ \bar{\mathbb{C}}_{1122} & \bar{\mathbb{C}}_{2222} & 0 \\ 0 & 0 & \bar{\mathbb{C}}_{1212}\end{bmatrix}$$

17. Compute the Cauchy stress

$$\boldsymbol{\sigma} = 2J^{-1}\left(\left(\frac{\partial W}{\partial I_C} + I_C\frac{\partial W}{\partial II_C}\right)\boldsymbol{B} - \frac{\partial W}{\partial II_C}\boldsymbol{B}^2 + III_C\frac{\partial W}{\partial III_C}\mathbf{I}\right)$$

18. Compute the tangent stiffness and the geometric stiffness

$$\boldsymbol{K}_T^i = \int_\Omega [\bar{\boldsymbol{N}}']^T\,[\boldsymbol{Dt}][\bar{\boldsymbol{N}}']\;dV$$

$$\boldsymbol{K}_G^i = \int_\Omega \left[\frac{\partial \bar{\boldsymbol{N}}'}{\partial \boldsymbol{u}^i}\right]^T\,[\boldsymbol{S}]\;dV_o$$

19. Compute the effective load vector

$$[Peff] = [force] - [K_e]$$

20. Set displacement boundary conditions for degree of freedom j by making the diagonal component of the stiffness matrix equal one and all others set to zero

$$\boldsymbol{K}_{tot}(j,j) = 1$$

$$Peff(j) = 0$$

21. Solve for new incremental displacements

$$\Delta u = K_{tot}\backslash Peff$$

22. Update displacements

$$u = u + du$$

11.6.3.2 Matlab® Code

```
% ---------------------------------------------------------------------
% DESCRIPTION
% ---------------------------------------------------------------------
% FEM simulation of a single quadralateral element with 4 numerical
% integration points. The nodes are numbered as follows:
%                        side 3
%                  4|-----------|3
%   y    side 4     |           |    side 2
%   ^               |           |
%   |              1|-----------|2
%   |                    side 1
%   |----> x
%  Node 1 has zero displacement in both the x and y direction
%  Node 4 has zero displacement in the y direction
%  A point load parallel to the x axis is appliced to nodes 2 and 3.
%
%  Variables Used in this Program
%       nnpe          - number of nodes per element
%       nPosXY(i,j)   - position of node i in the j direction
%       nForces(i,j) - point load applied to node i in the j direction
%
%       numqpt         - number of quadrature points per element
%       QPT(i,j)       - position of quadrature point i in the j direction
%
%       Jacobian       - Jacobian transformation matrix
%       invJacob       - Inverse of the Jacobian matrix
%
%       stiffness(i,j)- stiffness matrix element for degree of freedom i
%                       dof j
%       Ke             - element stiffness matrix
%       Kg             - geometric stiffness matrix
%       Kt             - tangent modulus matrix
%       force(i)       - force vector for degree of freedom i
%
%       u(i)           - displacement for degree of freedom i
%       displace(i,j) - displacement of node i in direction j
%       strain         - strain in the element
%       stress         - stress in the element
%
%       F              - deformation gradient
%       C              - right Cauchy deformation tensor
% ---------------------------------------------------------------------
% PREPROCESSING
% ---------------------------------------------------------------------
clear
nnpe = 4;                    % number of nodes per element
I    = eye(3,3);             % identity matrix
% ---------------------------------------------------------------------
% Step 1. Input the x and y coordinate of each node, and components of
```

```
%         the point forces acting on each node.
nPosXY    = zeros(4, 2);  % Initialize nodal positions
nForces   = zeros(4, 2);  % Initialize external nodal forces
thickness = 0.01;         % [m] thickness of the plate
% -------------------------------------------------------------------------
nPosXY(1,:)=[0, 0];         % [m] node 1
nPosXY(2,:)=[1, 0];         % [m] node 1
nPosXY(3,:)=[1, 1];         % [m] node 1
nPosXY(4,:)=[0, 1];         % [m] node 1
% -------------------------------------------------------------------------
nForces(1,:) = [0,     0];   % [N] node 1
nForces(2,:) = [10000, 0];   % [N] node 2
nForces(3,:) = [10000, 0];   % [N] node 3
nForces(4,:) = [0,     0];   % [N] node 4
nrIncrements = 50;          % input the desired number of load increments
% -------------------------------------------------------------------------
% Step 2. Give the material parameters for the Neo-Hookean model
alpha_1 = 10^6;
alpha_2 = 20^6;
% -------------------------------------------------------------------------
% Step 3. Give the positions of the numerical integration points
%         in natural coordinate system (r, s).
numqpt = 4;                                 % number of quadrature points
wt     = [1, 1, 1, 1];                      % weights for quadrature points
QPT = [ -0.5773502692, -0.5773502692;   % (r, s) location of
         0.5773502692, -0.5773502692;   %     the 4 quadrature points
         0.5773502692,  0.5773502692;
        -0.5773502692,  0.5773502692;];
% -------------------------------------------------------------------------
% Step 4. Assemble the shape functions and necessary derivatives
for j = 1:numqpt;
    r = QPT(j,1);  s = QPT(j,2);
    % shape functions
    sf(1,j) = 0.25*(1.0-r)*(1.0-s);
    sf(2,j) = 0.25*(1.0+r)*(1.0-s);
    sf(3,j) = 0.25*(1.0+r)*(1.0+s);
    sf(4,j) = 0.25*(1.0-r)*(1.0+s);
    % derivative of shape function wrspt to r
    dndrs(1,1,j) = 0.25*(-1.)*(1.0-s);
    dndrs(2,1,j) = 0.25*(+1.)*(1.0-s);
    dndrs(3,1,j) = 0.25*(+1.)*(1.0+s);
    dndrs(4,1,j) = 0.25*(-1.)*(1.0+s);
    % derivative of shape function wrspt to s
    dndrs(1,2,j) = 0.25*(1.0-r)*(-1.0);
    dndrs(2,2,j) = 0.25*(1.0+r)*(-1.0);
    dndrs(3,2,j) = 0.25*(1.0+r)*(+1.0);
    dndrs(4,2,j) = 0.25*(1.0-r)*(+1.0);
end
% -------------------------------------------------------------------------
%   SOLVER
```

```
% ----------------------------------------------------------------------------
u  = zeros(8,1); % initialize the displacement vector
% ----------------------------------------------------------------------------
for loadIncrement = 1:nrIncrements
  iteration = 0;                % initialize the iteration counter
  error     = 1;                % initialize the error
  while (error > 1e-6)
    iteration = iteration + 1; % increment the iteration counter
    % ------------------------------------------------------------------------
    % Step 5. Assemble the nodal force vector
    force = loadIncrement/nrIncrements*[nForces(:,1); nForces(:,2)];
    % ------------------------------------------------------------------------
    Ke      = zeros(8, 1);   % initialize element stiffness matrix
    Kg      = zeros(8, 8);   % initialize geometric stiffness matrix
    Kt      = zeros(8, 8);   % initialize tangent modulus matrix

    for j = 1:numqpt
      % Step 6. Find the Jacobean
      Jacobian  = dndrs(:,:,j)'*(nPosXY+[u(1:4) u(5:8)]);
      % Step 7. Invert the Jacobean
      invJacob  = Jacobian^-1;
      % Step 8. Find the dNdX and dNdY
      dndxy     = (invJacob*dndrs(:,:,j)')';
      % Step 9. Assemble Nprime
      NP        = [dndxy(:,1)'  0, 0, 0, 0;  ...
                   0, 0, 0, 0,  dndxy(:,2)'; ...
                   dndxy(:,2)', dndxy(:,1)'];
      % Step 10. Assemble the Nprime nonlinear
      NPNL      = [dndxy(:,1)', 0, 0, 0, 0;  ...
                   dndxy(:,2)', 0, 0, 0, 0;  ...
                   0, 0, 0, 0,  dndxy(:,1)'; ...
                   0, 0, 0, 0,  dndxy(:,2)'];
      % Step 11. Deformation gradient
      F = [1+[dndxy(:,1)',0,0,0,0]*u,   [dndxy(:,2)',0,0,0,0]*u, 0; ...
             [0,0,0,0,dndxy(:,1)']*u, 1+[0,0,0,0,dndxy(:,2)']*u, 0; ...
             0,                          0,                          1];
      C         = F'*F;   % Right Cauchy deformation tensor
      % Step 12. Find the invariants and the derivatives of the
      %          invariants of C
      IC     = trace(C);
      IIC    = trace(C)^2 - trace(C*C');
      IIIC   = det(C);
      dIdC   = I;
      dIIdC  = IC * I - C';
      dIIIdC = IIIC * C^-1;
      % Step 13. Input the strain energy function and its derivatives
      W         = alpha_1*(IC-3-log(IIIC)) + alpha_2*(IIIC-1)^2;
      dWdI      = alpha_1;
      dWdII     = 0;
      dWdIII    = -alpha_1/IIIC + alpha_2*2*(IIIC-1);
```

```
d2WdI2      = 0;
d2WdII2     = 0;
d2WdIII2    = alpha_1/IIIC^2 + alpha_2*2;
d2WdIdII    = 0;
d2WdIdIII   = 0;
d2WdIIdIII = 0;
for m = 1:2
    for n = 1:2
        for p = 1:2
            for q = 1:2
                d2IdC2(m,n,p,q)    = 0;
                d2IIdC2(m,n,p,q)   = I(n,m)*I(p,q)-I(q,n)*I(p,m);
                d2IIIdC2(m,n,p,q) = I(q,n)*C(m,p)+C(q,n)*I(p,m) ...
                              - IC*I(q,n)*I(p,m) - I(n,m)*C(q,p) ...
                              + IC*I(n,m)*I(p,q) - C(m,n)*I(p,q);
            end
        end
    end
end
% Step 14. Assemble the tangent stiffness
M  = [d2WdI2    d2WdIdII   d2WdIdIII; ...
      d2WdIdII  d2WdII2    d2WdIIdIII; ...
      d2WdIdIII d2WdIIdIII d2WdIII2;];
v3 = [dWdI      dWdII      dWdIII];

for m = 1:2
 for n = 1:2
   for p = 1:2
     for q = 1:2
       v1 = [dIdC(m,n)         dIIdC(m,n)         dIIIdC(m,n)];
       v2 = [dIdC(p,q);        dIIdC(p,q);        dIIIdC(p,q)];
       v4 = [d2IdC2(m,n,p,q); d2IIdC2(m,n,p,q); d2IIIdC2(m,n,p,q)];
       Ct(m,n,p,q) = v1*M*v2+v3*v4;
     end
   end
 end
end
% Step 15. Push forward the tangent stiffness
for m = 1:2
 for n = 1:2
  for p = 1:2
   for q = 1:2
    ct(m,n,p,q) = 0;
    for r = 1:2
     for s = 1:2
      for t = 1:2
       for v = 1:2
         ct(m,n,p,q) = ct(m,n,p,q)+ ...
                        F(m,r)*F(n,s)*Ct(r,s,t,v)*F(p,t)*F(q,v);
       end
```

```
            end
          end
        end
      end
    end
  end
end
% Step 16. Convert to Voigt notation
Dt            =     4*[ct(1,1,1,1)  ct(1,1,2,2) 0;
                       ct(1,1,2,2)  ct(2,2,2,2) 0;
                       0            0           2*ct(1,2,1,2)];
% Step 17. Compute the Cauchy stress
B            = F*F';
Cauchy       = 2/det(F)*((dWdI+IC*dWdII)*B-dWdII*B*B+IIIC*dWdIII*I);
stressMatrix = [Cauchy(1,1), Cauchy(2,1), 0, 0;
                Cauchy(2,1), Cauchy(2,2), 0, 0;
                0,           0,           Cauchy(1,1), Cauchy(2,1);
                0,           0,           Cauchy(2,1), Cauchy(2,2)];
stressVector = [Cauchy(1,1); Cauchy(2,2); Cauchy(2,1)];
% Step 18. Compute the tangent stiffness, the geometric
%          stiffness
Kt  = Kt + thickness*wt(j)*NP'*Dt*NP*det(Jacobian);
Kg  = Kg + thickness*wt(j)*NPNL'*stressMatrix*NPNL*det(Jacobian);
Ke  = Ke + thickness*wt(j)*NP'*stressVector*det(Jacobian);
end
% -----------------------------------------------------------------
% Step 19. Compute the effective load vector
Peff   = force - Ke;
% -----------------------------------------------------------------
% Step 20. Set displacement boundary conditions
%      20.1. Node 1 has zero displacement in the x and y direction
Ktot      = Kt + Kg;
Ktot(1,:) = 0.0; Ktot(1,1) = 10^4; Peff(1) = 0;
Ktot(5,:) = 0.0; Ktot(5,5) = 10^4; Peff(5) = 0;
%      20.2. Node 4 has zero displacement in the x direction
Ktot(4,:) = 0.0; Ktot(4,4) = 10^4; Peff(4) = 0;
%      20.3. Node 2 has zero displacement in the y direction
Ktot(6,:) = 0.0; Ktot(6,6) = 10^4 ; Peff(6) = 0;
% -----------------------------------------------------------------
% Step 21. Solve for new incremental displacements
du = Ktot\Peff;
% -----------------------------------------------------------------
% Step 22. Update displacements
u = u + du;
% -----------------------------------------------------------------
% step 23. Store quatities for plotting convergence criteria
error                   = norm(Peff);
normResidual(iteration) = norm(Peff);
% -----------------------------------------------------------------
fprintf(1,'Load Step %d, Iteration %d, Residual %f. \n', ...
```

```
                        loadIncrement, iteration, error);
  end
  % -------------------------------------------------------------------------
  % Save the geometry for each load increment
  Increment(loadIncrement).Displacement = u;
  Increment(loadIncrement).force        = force;
  Increment(loadIncrement).stress       = Cauchy;
  Increment(loadIncrement).error        = error;
end
```

12 Appendix

12.1 Introduction to Matlab®

The Matlab® software package allows for the manipulation and visualization of tensor equations, partial differential equations, and experimental data. In the text, you will find code that can be run in Matlab®. It is assumed that you are already familiar with the Matlab® program; however, a basic primer is included here. If you require additional information, you will find many helpful documents and tutorials online. Most importantly, the Matlab® program includes an extensive help document that you can access through the menu, or by typing "help" followed by the command name.

Commands may be entered directly at the command prompt found within the command window. The command prompt is denoted by the double arrows, >>. Variables may be defined and assigned new values using the command prompt. For example, defining a scalar variable can be achieved by entering the following text at the command prompt.

```
>> alpha = 2.0
```

We have now defined a variable named "alpha" which has a value of 2.0. You may then use this variable in another equation.

```
>> beta = alpha^2
```

Vector quantities can be defined as column vectors using the following syntax.

```
>> a = [ 1.5; 2.3; 6.2]
```

We have defined a column vector, $\boldsymbol{a}$, which has three components $a_1 = 1.5$, $a_2 = 2.3$, and $a_3 = 6.2$.

We will often need to generate a sequence of numbers in vector form. The following simple command generates a row vector with values from 0 to 10 in increments of 1.

```
>> a = [ 0: 1 : 10]
```

This is equivalent to typing

```
>> a = [ 0 1 2 3 4 5 6 7 8 9 10]
```

Notice that the separator between start, increment, and end is a colon. Matlab® variables are case sensitive so be careful selecting variable names.

12.2 Reference Tables

Table 12.1. *Symbol definitions*

Symbol		Name
$\{\boldsymbol{e}_1, \boldsymbol{e}_2, \boldsymbol{e}_3\}$		Orthonormal basis set
$\{\boldsymbol{E}_1, \boldsymbol{E}_2, \boldsymbol{E}_3\}$		Referential orthonormal basis set
$\mathcal{K}$		Kinetic energy
$\mathfrak{U}$		Potential energy
μ_i		Chemical potential of species i
n_i		Number of atoms of species i
$\boldsymbol{X}$		Reference position of a material point
$\boldsymbol{x}$		Current position of a material point
$\boldsymbol{v}$	$\dfrac{\partial \boldsymbol{x}(\boldsymbol{X},t)}{\partial t}$	Velocity
$\boldsymbol{u}$	$\boldsymbol{x}(\boldsymbol{X},t) - \boldsymbol{X}$	Displacement
$\boldsymbol{L}$	$\dfrac{\partial \boldsymbol{v}(\boldsymbol{x},t)}{\partial \boldsymbol{x}}$	Velocity Gradient
$\boldsymbol{D}$	$\frac{1}{2}\left(\boldsymbol{L}+\boldsymbol{L}^T\right)$	Rate of deformation tensor
$\boldsymbol{W}$	$\frac{1}{2}\left(\boldsymbol{L}-\boldsymbol{L}^T\right)$	Vorticity tensor
$\boldsymbol{F}$	$\dfrac{\partial \boldsymbol{x}(\boldsymbol{X},t)}{\partial \boldsymbol{X}}$	Deformation gradient
$\boldsymbol{U}$	$\sqrt{\boldsymbol{F}^T \cdot \boldsymbol{F}}$	Right stretch tensor
$\boldsymbol{V}$	$\sqrt{\boldsymbol{F} \cdot \boldsymbol{F}^T}$	Left stretch tensor
$\boldsymbol{R}$	$\boldsymbol{F} \cdot \boldsymbol{U}^{-1}$	Orthogonal tensor
$\boldsymbol{C}$	$\boldsymbol{F}^T \cdot \boldsymbol{F}$	Right Cauchy deformation tensor
$\boldsymbol{B}$	$\boldsymbol{F} \cdot \boldsymbol{F}^T$	Left Cauchy deformation tensor
η	$\frac{1}{2}\left(\mathbf{I}-\boldsymbol{B}^{-1}\right)$	Almansi strain tensor
$\boldsymbol{E}$	$\frac{1}{2}(\boldsymbol{C}-\boldsymbol{I})$	Green strain tensor
$\boldsymbol{\varepsilon}$		Infinitesimal strain tensor
σ		Cauchy stress tensor
$\boldsymbol{P}$	$J\,\sigma \cdot \boldsymbol{F}^{-T}$	First Piola-Kirchhoff stress tensor
$\boldsymbol{S}$	$J\boldsymbol{F}^{-1} \cdot \sigma \cdot \boldsymbol{F}^{-T}$	Second Piola-Kirchhoff stress tensor
$\boldsymbol{t}_{\boldsymbol{x}}^{\boldsymbol{n}}$	$\sigma^T \cdot \boldsymbol{n}_x$	Cauchy traction vector
$\boldsymbol{t}_{\boldsymbol{X}}^{\boldsymbol{n}}$	$\boldsymbol{P} \cdot \boldsymbol{n}_X$	First Piola-Kirchhoff traction vector
$\boldsymbol{t}_{\boldsymbol{X}}^{*}$	$\boldsymbol{S} \cdot \boldsymbol{n}_X$	Second Piola-Kirchhoff traction vector
η		Entropy per unit mass
$\mathfrak{D}$		Internal dissipation
$\rho_o(\boldsymbol{X})$		Referential density
$\rho(\boldsymbol{x},t)$		Spatial density
m		Mass
$\boldsymbol{b}_x(\boldsymbol{x},t)$		Spatial body force per unit mass
θ		Temperature
e		Internal energy per unit mass
ψ		Helmholtz free energy
$\mathbf{n}^{(i)}, \hat{\boldsymbol{n}}^{(i)}$		Right eigenvector, and normalized right eigenvector
$\mathbf{m}^{(i)}, \hat{\boldsymbol{m}}^{(i)}$		Left eigenvector, and normalized left eigenvector
λ_i		Eigenvalue and stretch
ε_{ijk}		Permutation symbol
δ_{ij}		Kronecker delta
c_v		Specific heat at constant volume

Table 12.2. *Table of physical constants*

Symbol	Value	Name
k_B	$1.38 \times 10^{-23}\,\mathrm{J/K^{-1}}$	Boltzmann's constant
N_A	6.022×10^{23} atoms / mol	Avagadro's constant
R_g	$8.31\,\mathrm{J/K\,mol}$	Universal gas constant
R_s	$R_s = \dfrac{R_g}{molar\,mass} = \dfrac{k_b}{molecular\,mass}$	Specific gas constant

Table 12.3. *Vector and tensor Identities*

$$
\begin{aligned}
&(\mathbf{A}\cdot\mathbf{B})\cdot\mathbf{C} = \mathbf{A}\cdot(\mathbf{B}\cdot\mathbf{C}) = \mathbf{A}\cdot\mathbf{B}\cdot\mathbf{C}\\
&\mathbf{A}^2 = \mathbf{A}\cdot\mathbf{A}\\
&(\mathbf{A}+\mathbf{B})\cdot\mathbf{C} = \mathbf{A}\cdot\mathbf{C}+\mathbf{B}\cdot\mathbf{C}\\
&\mathbf{A}\cdot\mathbf{B}\neq\mathbf{B}\cdot\mathbf{A}\\
&(\mathbf{A}\cdot\mathbf{B})^T = \mathbf{B}^T\cdot\mathbf{A}^T\\
&(\alpha\mathbf{A}+\beta\mathbf{B})^T = \alpha\mathbf{A}^T+\beta\mathbf{B}^T\\
&(\mathbf{A}\cdot\mathbf{B})^{-1} = \mathbf{B}^{-1}\cdot\mathbf{A}^{-1}\\
&\left(\mathbf{A}^{-1}\right)^T = \left(\mathbf{A}^T\right)^{-1}\\
&\\
&\mathbf{a}^2 = \mathbf{a}\cdot\mathbf{a}\\
&\mathbf{a}\cdot\mathbf{b} = \mathbf{b}\cdot\mathbf{a}\\
&\mathbf{a}\cdot\mathbf{0} = 0\\
&\mathbf{a}\cdot(\alpha\mathbf{b}+\beta\mathbf{c}) = \alpha(\mathbf{a}\cdot\mathbf{b})+\beta(\mathbf{a}\cdot\mathbf{c})\\
&\mathbf{a}\cdot(\mathbf{A}\cdot\mathbf{b}) = \mathbf{b}\cdot\left(\mathbf{A}^T\cdot\mathbf{a}\right)\\
&\mathbf{A}:\mathbf{B} = tr\left(\mathbf{A}^T\mathbf{B}\right)\\
&\mathbf{a}\times\mathbf{b} = -\mathbf{b}\times\mathbf{a}\\
&\alpha\mathbf{a}\times\mathbf{b} = \mathbf{a}\times\alpha\mathbf{b} = \alpha(\mathbf{a}\times\mathbf{b}) = -\alpha(\mathbf{b}\times\mathbf{a})\\
&\mathbf{a}\cdot(\mathbf{b}\times\mathbf{c}) = \mathbf{b}\cdot(\mathbf{c}\times\mathbf{a})\\
&\\
&\mathbf{a}\otimes\mathbf{b} = \mathbf{a}\mathbf{b}^T\\
&(\mathbf{a}\otimes\mathbf{b})^T = \mathbf{b}\otimes\mathbf{a}\\
&\mathbf{c}\cdot(\mathbf{a}\otimes\mathbf{b}) = \mathbf{b}(\mathbf{a}\cdot\mathbf{c})\\
&(\mathbf{a}\otimes\mathbf{b})\cdot\mathbf{c} = \mathbf{a}(\mathbf{b}\cdot\mathbf{c})\\
&(\mathbf{a}\otimes\mathbf{b}):(\mathbf{c}\otimes\mathbf{d}) = (\mathbf{a}\cdot\mathbf{c})(\mathbf{b}\cdot\mathbf{d})\\
&(\mathbf{c}\otimes\mathbf{d}):\mathbf{A} = \mathbf{c}\cdot\mathbf{A}\cdot\mathbf{d}\\
&tr(\mathbf{A}) = \mathbf{A}:\mathbf{I}\\
&\\
&tr(\mathbf{A}\cdot\mathbf{B}) = tr(\mathbf{B}\cdot\mathbf{A})\\
&tr(\mathbf{A}+\mathbf{B}) = tr(\mathbf{A})+tr(\mathbf{B})\\
&\\
&\det(\mathbf{A}\cdot\mathbf{B}) = \det\mathbf{A}\,\det\mathbf{B}\\
&\det(\mathbf{A}^T) = \det\mathbf{A}\\
&\det(\mathbf{A}^{-1}) = 1/\det\mathbf{A}\\
&\det(\mathbf{A}^n) = (\det\mathbf{A})^n\\
&\det(\alpha\mathbf{A}) = \alpha^3\det\mathbf{A}\\
&\epsilon_{ijk}A_{ir}A_{js}A_{kt} = \epsilon_{rst}\det A\\
&\\
&\frac{\partial\mathbf{A}^{-1}}{\partial\alpha} = -\boldsymbol{A}^{-1}\frac{\partial\mathbf{A}}{\partial\alpha}\boldsymbol{A}^{-1}\\
&\frac{\partial}{\partial\alpha}(\det\boldsymbol{A}) = \det\mathbf{A}\,tr\left(\mathbf{A}^{-1}\frac{\partial\mathbf{A}}{\partial\alpha}\right)\\
&\frac{\partial}{\partial\alpha}(\mathbf{A}\cdot\boldsymbol{B}) = \mathbf{A}\cdot\frac{\partial\mathbf{B}}{\partial\alpha}+\frac{\partial\mathbf{A}}{\partial\alpha}\cdot\mathbf{B}\\
&\frac{\partial}{\partial\alpha}(\mathbf{a}\otimes\mathbf{b}) = \frac{\partial\mathbf{a}}{\partial\alpha}\otimes\mathbf{b}+\mathbf{a}\otimes\frac{\partial\mathbf{b}}{\partial\alpha}
\end{aligned}
\tag{12.1}
$$

Table 12.4. *Tensor, matrix, index, and Matlab® notation*

Tensor notation	Matrix notation	Index notation	Matlab® notation	
$\mathbf{D}=\mathbf{A}\cdot\mathbf{B}$	$\mathbf{D}=\mathbf{AB}$	$D_{ij}=A_{ik}B_{kj}$	$\mathbf{D}=\mathbf{A}{*}\mathbf{B}$	12.2
$\mathbf{D}=\mathbf{A}+\mathbf{B}$	$\mathbf{D}=\mathbf{A}+\mathbf{B}$	$D_{ij}=A_{ij}+B_{ij}$	$\mathbf{D}=\mathbf{A}+\mathbf{B}$	12.3
$\mathbf{D}=\mathbf{A}\cdot\mathbf{B}\cdot\mathbf{C}$	$\mathbf{D}=\mathbf{ABC}$	$D_{il}=A_{ij}B_{jk}C_{kl}$	$\mathbf{D}=\mathbf{A}{*}\mathbf{B}*\mathbf{C}$	12.4
$\mathbf{D}=\mathbf{A}^2$	$\mathbf{D}=\mathbf{A}^2$	$D_{ik}=A_{ij}A_{jk}$	$\mathbf{D}=\mathbf{A}$^2	12.5
$\mathbf{D}=\mathbf{A}\cdot\mathbf{C}+\mathbf{B}\cdot\mathbf{C}$	$\mathbf{D}=\mathbf{AC}+\mathbf{BC}$	$D_{ik}=A_{ij}C_{jk}+B_{ij}C_{jk}$	$\mathbf{D}=\mathbf{A}{*}\mathbf{C}+\mathbf{B}*\mathbf{C}$	12.6
$\mathbf{D}=\mathbf{A}^T\cdot\mathbf{B}$	$\mathbf{D}=\mathbf{A}^T\mathbf{B}$	$D_{ij}=A_{ki}B_{kj}$	$\mathbf{D}=\mathbf{A}'{*}\mathbf{B}$	12.7
$\mathbf{D}=\alpha\mathbf{A}^T+\beta\mathbf{B}^T$	$\mathbf{D}=\alpha\mathbf{A}^T+\beta\mathbf{B}^T$	$D_{ij}=\alpha A_{ji}+\beta B_{ji}$	$\mathbf{D}=\alpha*\mathbf{A}'+\beta{*}\mathbf{B}'$	12.8
$\mathbf{D}=(\mathbf{A}\cdot\mathbf{B})^{-1}$	$\mathbf{D}=\mathbf{B}^{-1}\mathbf{A}^{-1}$	$D_{ij}=B^{-1}_{ik}A^{-1}_{kj}$	$\mathbf{D}=(\mathbf{A}{*}\mathbf{B})$^ -1	12.9
$\mathbf{D}=\mathbf{A}^{-1}$	$\mathbf{D}=\mathbf{A}^{-1}$	$D_{ij}=A^{-1}_{ij}$	$\mathbf{D}=\mathbf{A}$^ -1	12.10
$\mathbf{D}=\mathbf{A}^T$	$\mathbf{D}=\mathbf{A}^T$	$D_{ij}=A_{ji}$	$\mathbf{D}=\mathbf{A}'$	12.11
$\mathbf{d}=\mathbf{A}\cdot\mathbf{a}$	$\mathbf{d}=\mathbf{Aa}$	$d_i=A_{ij}a_j$	$\mathbf{d}=\mathbf{A}{*}\mathbf{a}$	12.12
$\mathbf{d}=\mathbf{a}\cdot\mathbf{A}$	$\mathbf{d}=\mathbf{a}^T\mathbf{A}$	$d_i=A_{ij}a_i$	$\mathbf{d}=\mathbf{a}'*\mathbf{A}$	12.13
$\mathbf{d}=\mathbf{a}\cdot(\mathbf{A}\cdot\mathbf{b})$	$\mathbf{d}=\mathbf{a}^T\mathbf{Ab}$	$d_i=A_{ij}a_ib_j$	$\mathbf{d}=\mathbf{a}'*\mathbf{A}*\mathbf{b}$	12.14
$\alpha=\mathbf{A}{:}\mathbf{B}$	$\alpha=\mathbf{A}{:}\mathbf{B}$	$\alpha=A_{ij}B_{ij}$	$\alpha=sum\,(sum\,(\mathbf{A}.*\mathbf{B}))$	12.15
$\alpha=\mathbf{a}\cdot\mathbf{b}$	$\alpha=\mathbf{a}\cdot\mathbf{b}$	$\alpha=a_ib_i$	$\alpha=dot\,(\mathbf{a},\mathbf{b})$	12.16
$\mathbf{d}=\mathbf{a}\times\mathbf{b}$	$\mathbf{d}=\mathbf{a}\times\mathbf{b}$	$d_i=\epsilon_{ijk}a_jb_k$	$\mathbf{d}=cross\,(\mathbf{a},\mathbf{b})$	12.17
$\mathbf{D}=\mathbf{a}\otimes\mathbf{b}$	$\mathbf{D}=\mathbf{ab}^\mathbf{T}$	$D_{ij}=a_ib_j$	$\mathbf{D}=\mathbf{a}*\mathbf{b}'$	12.18
$\alpha=tr\,(\mathbf{A})$	$\alpha=tr\,(\mathbf{A})$	$\alpha=A_{ii}$	$\alpha=trace\,(\mathbf{A})$	12.19
$\alpha=\det(\mathbf{A})$	$\alpha=\det(\mathbf{A})$	$\alpha=\epsilon_{ijk}A_{i1}A_{j2}A_{k3}$	$\alpha=\det(\mathbf{A})$	12.20

Table 12.5. *Grad, div, curl in Cartesian coordinates*

∇		$\boldsymbol{e}_i\dfrac{\partial}{\partial x_i}$
$\nabla\Phi$	*grad* Φ	$\Phi_{,i}\boldsymbol{e}_i$
$\boldsymbol{a}\overleftarrow{\nabla}$	*grad* $\boldsymbol{a}$	$a_{i,j}\boldsymbol{e}_i\otimes\boldsymbol{e}_j$
$\nabla\cdot\boldsymbol{a}$	*div* $\boldsymbol{a}$	$a_{i,i}$
$\boldsymbol{A}\overleftarrow{\nabla}$	*grad* $\boldsymbol{A}$	$A_{ij,k}\boldsymbol{e}_i\otimes\boldsymbol{e}_j\otimes\boldsymbol{e}_k$
$\nabla\cdot\boldsymbol{A}$	*div* $\boldsymbol{A}$	$A_{ij,i}\boldsymbol{e}_j$
$\nabla\times\boldsymbol{a}$	*curl* $\boldsymbol{a}$	$\varepsilon_{ijk}a_{j,i}\boldsymbol{e}_k$
$\nabla^2\phi$	$\Delta\phi$	$\phi_{,ii}$
$\nabla^2\boldsymbol{v}$	$\Delta\boldsymbol{v}$	$v_{j,ii}\boldsymbol{e}_j$

The matrix components of a tensor can be defined using the following syntax.

```
>> A = [ 1, 2, 3; 2, 1, 3; 3, 3, 1]
```

Here we have defined a tensor, **A,** which has the components $A_{11}=1$, $A_{12}=A_{21}=2$, $A_{13}=A_{31}=3$, $A_{23}=A_{32}=3$, $A_{22}=1$, and $A_{33}=1$. You may have noticed that once you press enter, Matlab® will display the current variable assignment as a 3×3 matrix. This can be helpful when checking your assignments. If you do not want to display the results of a calculation or assignment, you may include a semicolon at the end of the command. For example, the following command will not display the assignment.

```
>> A = [ 1, 2, 3; 2, 1, 3; 3, 3, 1];
```

Table 12.6. *Grad, div, curl in cylindrical coordinates*

∇		$e_r \frac{\partial}{\partial x_r} + e_\theta \frac{1}{r}\frac{\partial}{\partial x_\theta} + e_z \frac{\partial}{\partial x_z}$
$\nabla\phi$	*grad* Φ	$e_r \frac{\partial \phi}{\partial x_r} + e_\theta \frac{1}{r}\frac{\partial \phi}{\partial x_\theta} + e_z \frac{\partial \phi}{\partial x_z}$
$\boldsymbol{a}\overleftarrow{\nabla}$	*grad* $\boldsymbol{a}$	$\begin{bmatrix} \frac{\partial a_r}{\partial x_r} & \left[\frac{1}{r}\frac{\partial a_r}{\partial x_\theta} - \frac{a_\theta}{r}\right] & \frac{\partial a_r}{\partial x_z} \\ \frac{\partial a_\theta}{\partial x_r} & \left[\frac{1}{r}\frac{\partial a_\theta}{\partial x_\theta} + \frac{a_r}{r}\right] & \frac{\partial a_\theta}{\partial x_z} \\ \frac{\partial a_z}{\partial x_r} & \left[\frac{1}{r}\frac{\partial a_z}{\partial x_\theta}\right] & \frac{\partial a_z}{\partial x_z} \end{bmatrix}$
$\nabla\cdot\boldsymbol{a}$	*div* $\boldsymbol{a}$	$\left(\frac{\partial (r a_r)}{\partial x_r} + \frac{1}{r}\frac{\partial a_\theta}{\partial x_\theta} + \frac{\partial a_z}{\partial x_z}\right)$
$\nabla\cdot\boldsymbol{A}$	*div* $\boldsymbol{A}$	$\left(\frac{1}{r}\frac{\partial}{\partial x_r}(rA_{rr}) + \frac{1}{r}\frac{\partial A_{\theta r}}{\partial x_\theta} + \frac{\partial A_{zr}}{\partial x_z} - \frac{1}{r}A_{\theta\theta}\right) e_r + \left(\frac{1}{r}\frac{\partial A_{\theta\theta}}{\partial x_\theta} + \frac{\partial A_{z\theta}}{\partial x_z} + \frac{1}{r}\frac{\partial}{\partial x_r}(rA_{r\theta}) + \frac{1}{r}A_{\theta r}\right) e_\theta + \left(\frac{\partial A_{zz}}{\partial x_z} + \frac{1}{r}\frac{\partial}{\partial x_r}(rA_{rz}) + \frac{1}{r}\frac{\partial A_{\theta z}}{\partial x_r}\right) e_z$
$\nabla\times\boldsymbol{a}$	*curl* $\boldsymbol{a}$	$\left(\frac{1}{r}\frac{\partial a_z}{\partial x_\theta} - \frac{\partial a_\theta}{\partial x_z}\right) e_r + \left(\frac{\partial a_r}{\partial x_z} - \frac{\partial a_z}{\partial x_r}\right) e_\theta + \frac{1}{r}\left(\frac{\partial (r a_\theta}{\partial x_r} - \frac{\partial a_r}{\partial x_\theta}\right) e_z$
$\nabla^2\phi$		$\frac{1}{r}\frac{\partial}{\partial x_r}\left(r\frac{\partial \phi}{\partial x_r}\right) + \frac{1}{r^2}\frac{\partial^2 \phi}{\partial x_\theta^2} + \frac{\partial \phi}{\partial x_z^2}$

Table 12.7. *Grad, div, curl in spherical coordinates*

∇	$e_r \frac{\partial}{\partial x_r} + e_\theta \frac{1}{r}\frac{\partial}{\partial x_\theta} + e_\phi \frac{1}{r\sin\theta}\frac{\partial}{\partial x_\phi}$

Vector and tensor math can be written in compact form in the Matlab® program. For example, the tensor equation, $\boldsymbol{x} = \boldsymbol{A}\cdot\boldsymbol{a}$**,** can be solved using the simple syntax found that follows:

```
>> x = A * a
```

Upon pressing enter; Matlab® will display the resulting value for vector $\boldsymbol{x}$.

```
x =
24.7000
23.9000
17.6000
```

Table 12.4 contains the translation of vector and tensor products into their corresponding Matlab® commands. The commands for the trace, determinant, dot product, cross product, and more can be found within this table.

We will often use the plotting functions within Matlab® to visualize the results of our calculations. Several important plotting functions include the ***plot***, ***contour***, and ***quiver*** commands. The *plot* function will plot x-y data. For example, if we wish to plot the function $y = 2.2\sin(2\pi t)$ over the interval $0 \le t \le 10$, we would use the following syntax:

```
>> t = [0: 0.1: 10];
>> y = 2.2*sin (2*pi*t);
>> plot(t,y)
```

If we wish to plot a multidimensional function such as $f(x,y) = \sin(x) + \cos(y)$ within the interval $0 \le x \le 1$ and $0 \le y \le 1$, we can use the contour function.

```
>> [x, y] = meshgrid([0:0.1:1], [0:0.1:1]);
>> f = sin(x) + cos(y);
>> contour(x,y,f)
```

Notice that the grid of x, y points on which the function f is evaluated is generated using the meshgrid command.

Finally, we may plot two-dimensional vector fields using the quiver function. For example, we may plot the vector function $\boldsymbol{v}(x_1,x_2) = x_1x_2\boldsymbol{e}_1 + x_2\boldsymbol{e}_2$ over the interval $0 \le x_1 \le 1$ and $0 \le x_2 \le 1$ using the quiver function.

```
>> [x1, x2] = meshgrid([0:0.1:1], [0:0.1:1]);
>> v1 = x1.*x2;
>> v2 = x2;
>> quiver(x1, x2, v1, v2)
```

Instead of entering a series of commands at the command prompt, we can create and save Matlab® scripts. To do this, we first click on FILE→NEW→BLANK M-FILE to open the Matlab® editor. Type in the desired Matlab® commands. Finally, click on FILE→SAVE to save the resulting Matlab® file. Pay attention to the location in which you place the file and the name you give it since you may wish to reuse this file at some later point. You can run the script by clicking on DEBUG→SAVE FILE AND RUN, or by pressing the F5 key. Doing this will execute the commands contained within your script in the order in which they are written. You should pay attention to the inclusion or exclusion of the semicolon at the end of the line. You may hide or show the intermediate results of any calculation by selectively using the semicolon. Also, the values of variables will be stored until they are either overwritten or cleared using the ***clear*** command. Similarly, the ***clf*** command may be used to clear the contents of the figure window.

Many of the scripts included in this text make use of ***if*** statements and ***for*** loops. The ***if*** statement is used to selectively execute a command if a specified condition is met. An example of the if statement would be as follows:

```
>> if (x < 0)
>> x = 0;
>> end
```

In this case, if the variable x has a value less than zero, the variable is set to zero.

The ***for*** loop is used to repeat a loop a specified number of times. For example, if we wish to compute $y_i = x_i^2$ for $i = 0$ to 3, we could write

```
>> for (i=0:1:3)
>> y(i) = x(i)^2
>> end
```

We could achieve the same result with the compact notation

```
>> y=x.^2
```

The .^operator is used to raise each component of the vector or matrix to the specified power. Similarly, the operator .* represents component wise multiplication, and the operator ./ represents component wise division.

Index

abbreviated summation convention, 2
acceleration, 38
Almansi strain tensor, 64
alternating symbol, 4
Amontons, Guillaume, 124
Archimedes, 147
Avogadro, Amadeo, 124

basis vector, 2
Bernoulli equation, 160
Bernoulli, Daniel, 129, 148
body forces, 74
Boyle, Robert, 123

Cauchy deformation tensor, 63
Cauchy stress tensor, 77
Cauchy traction vector, 76
Cauchy, Augustin Louis, 184
Cayley-Hamilton theorem, 24
chain rule, 26
characteristic equation, 12
Clapeyron, Beniot, 123
Clausius-Duhem inequality, 96
conduction, 120
conjugate properties, 99
conservation equations
  angular momentum, 90
  energy, 93
  linear momentum, 88
  mass, 87
constitutive model, 105
  principles of, 106
convection, 120
convective derivative, 55
coordinate transformation, 18
Couette viscometer, 162
cross product, 4
current configuration, 36

da Vinci, Leonardo, 183
deformation gradient, 45, 58, 61
density, 74
determinant, 10
directional derivative, 26
divergence theorem, 30
dot product, 3
dummy index, 3
dyadic product, 8

eigenvalue, 12
eigenvector, 12
  left, 13
  right, 13
energy equation, 93
enthalpy, 120
entropy equation, 96
equations of motion, 36
Euler strain tensor, 64
evolution equation, 119

finite deformation, 183
finite difference method
  backward difference approximation, 136
  forward difference approximation, 136
finite element method, 269
first Piola-Kirchhoff stress tensor, 80
first Piola-Kirchhoff traction vector, 76
fluids
  Newtonian, 146
  non-Newtonian, 146
Fourier's inequality, 108
free index, 3
freely jointed chain, 197

Galilei, Galileo, 183
Gibbs free energy, 120
gradient, 26
gradient theorem, 31
Green strain tensor, 64

heat conduction, 120
heat flux vector, 86, 105, 120
Helmholtz free energy, 97, 120
homogenous deformation, 54
Hooke, Robert, 184

hyperelastic material model, 183
hyperelasticity, 188

ideal gas, 122
incompressibility, 54, 195
incompressible fluid, 159, 161
index notation, 2
infinitesimal strain tensor, 65
internal dissipation, 96
internal energy, 120
internal variables, 119
invariants, 24
inviscid fluid, 148
isochoric deformation, 54
isothermal finite elasticity, 183
isotropic functions, 115
isotropic solid, 190
isotropic tensor, 20

Jacobian, 51
Joule, James Prescott, 122
jump conditions, 100

kinetic energy, 93, 100
Kronecker delta, 3

Lagrange multiplier, 82
Lagrangian strain tensor, 64
Landau potential, 120
left Cauchy deformation tensor, 63
left stretch tensor, 59
Levi-Civita symbol, 4
linear thermoelastic materials, 183

material derivative, 55
material description, 36
material frame indifference, 109
material models
  Aruda Boyce 8 chain, 197
  Bingham model, 162
  Cross model, 162
  generalized Mooney-Rivlin, 197
  generalized neo-Hookean, 197
  generalized Yeoh, 197
  incompressible Gent, 197
  incompressible neo-Hookean, 197, 211
  incompressible Yeoh, 197
  power-law, 161
material nonlinearity, 183
material point, 35
material symmetry, 114

Nanson's Formula, 52
Navier, Claude-Louis, 148
Navier-Stokes equation, 158
Newtonian fluid, 148, 155
non-Newtonian fluid, 160
normal component, 81
normal stress, 81
normalized eigenvector, 13

objectivity
  principle of, 109
  scalar function, 110
  tensor function, 110
  vector function, 110
orthonormal, 2

particle trajectory, 38, 55
permutation symbol, 4
Poisson, Simeon Denis, 148
polar decomposition theorem, 58
polymer, 184, 196
potential energy, 93
principal axes, 82
principal values, 12

radiative heat transfer, 120
rate of deformation tensor, 66
reference configuration, 35
Reynolds' transport theorem, 68
right Cauchy deformation tensor, 63
right stretch tensor, 59
rigid body motion, 55
rotation, 59

scalar
  field, 25
  product, 3
Searle viscometer, 162
second law of thermodynamics, 96
second-order tensor, 8
second Piola-Kirchhoff stress tensor, 80
second Piola-Kirchhoff traction vector, 76
shape function, 273
shear stress, 81
skew-symmetric tensor, 9
spatial derivative, 56
spatial description, 37
spectral decomposition, 15
spectral representation, 15
spin tensor, 67
Stefan-Boltzmann law, 121
strain
  Almansi, 64
  Euler, 64
  Green, 64
  infinitesimal, 65
  Lagrangian, 64
stress power, 99
stress tensor
  Cauchy, 77
  deviatoric part, 83
  first Piola-Kirchhoff, 80
  isotropic part, 83
  second Piola-Kirchhoff, 80
stress vector, 75
stretch ratio, 63
stretching tensor, 66

substantial derivative, 55
summation convention, 2
surface forces, 74
surface traction, 75
symmetric tensor, 9

tensor
  contraction, 9
  decomposition, 59
  determinant, 10
  deviatoric part, 82
  dyadic product, 8
  field, 25
  first order, 1
  gradient, 27
  inverse, 25
  isotropic part, 82
  norm, 9
  product, 8
  second order, 8
  skew-symmetric, 9
  spectral decomposition, 15
  symmetric, 9
  trace, 9
  transformation rule, 20
  transpose, 9
  two-point, 46
traction vector, 75
  Cauchy, 76
  first Piola-Kirchhoff, 76
  nominal, 76
  second Piola-Kirchhoff, 76
  true, 76
trajectory, 38
transformation matrix, 19
transpose, 9
transverse isotropy, 114, 193

unit vector, 2

vector, 1
  field, 25
  gradient, 27
  transformation rules, 20
velocity, 38
velocity gradient tensor, 65
viscometer, 162
vorticity tensor, 67

Young, Thomas, 184

Printed in the United States
By Bookmasters